NANOSTRUCTURED MATERIALS

Processing, Properties and Potential Applications

NANOSTRUCTURED MATERIALS

Processing, Properties and Potential Applications

Edited by

Carl C. Koch

North Carolina State University
Raleigh, North Carolina

NOYES PUBLICATIONS

WILLIAM ANDREW PUBLISHING
Norwich, New York, U.S.A.

Library of Congress Catalog Card Number: 2001096788
ISBN: 0-8155-1451-4
Printed in the United States

Published in the United States of America by
Noyes Publications / William Andrew Publishing
13 Eaton Avenue
Norwich, NY 13815
1-800-932-7045
www.williamandrew.com
www.knovel.com

10 9 8 7 6 5 4 3 2 1

NOTICE

MATERIALS SCIENCE AND PROCESS TECHNOLOGY SERIES

Series Editors

Electronic Materials and Process Technology

CHARACTERIZATION OF SEMICONDUCTOR MATERIALS, Volume 1: edited by Gary E. McGuire

CHEMICAL VAPOR DEPOSITION FOR MICROELECTRONICS: by Arthur Sherman

CHEMICAL VAPOR DEPOSITION OF TUNGSTEN AND TUNGSTEN SILICIDES: by John E. J. Schmitz

CHEMISTRY OF SUPERCONDUCTOR MATERIALS: edited by Terrell A. Vanderah

CONTACTS TO SEMICONDUCTORS: edited by Leonard J. Brillson

DIAMOND CHEMICAL VAPOR DEPOSITION: by Huimin Liu and David S. Dandy

DIAMOND FILMS AND COATINGS: edited by Robert F. Davis

DIFFUSION PHENOMENA IN THIN FILMS AND MICROELECTRONIC MATERIALS: edited by Devendra Gupta and Paul S. Ho

ELECTROCHEMISTRY OF SEMICONDUCTORS AND ELECTRONICS: edited by John McHardy and Frank Ludwig

ELECTRODEPOSITION: by Jack W. Dini

HANDBOOK OF CARBON, GRAPHITE, DIAMONDS AND FULLERENES: by Hugh O. Pierson

HANDBOOK OF CHEMICAL VAPOR DEPOSITION, Second Edition: by Hugh O. Pierson

HANDBOOK OF COMPOUND SEMICONDUCTORS: edited by Paul H. Holloway and Gary E. McGuire

HANDBOOK OF CONTAMINATION CONTROL IN MICROELECTRONICS: edited by Donald L. Tolliver

HANDBOOK OF DEPOSITION TECHNOLOGIES FOR FILMS AND COATINGS, *Second Edition:* edited by Rointan F. Bunshah

HANDBOOK OF HARD COATINGS: edited by Rointan F. Bunshah

HANDBOOK OF ION BEAM PROCESSING TECHNOLOGY: edited by Jerome J. Cuomo, Stephen M. Rossnagel, and Harold R. Kaufman

HANDBOOK OF MAGNETO-OPTICAL DATA RECORDING: edited by Terry McDaniel and Randall H. Victora

HANDBOOK OF MULTILEVEL METALLIZATION FOR INTEGRATED CIRCUITS: edited by Syd R. Wilson, Clarence J. Tracy, and John L. Freeman, Jr.

HANDBOOK OF PLASMA PROCESSING TECHNOLOGY: edited by Stephen M. Rossnagel, Jerome J. Cuomo, and William D. Westwood

HANDBOOK OF POLYMER COATINGS FOR ELECTRONICS, *Second Edition:*by James Licari and Laura A. Hughes

HANDBOOK OF REFRACTORY CARBIDES AND NITRIDES: by Hugh O. Pierson

HANDBOOK OF SEMICONDUCTOR SILICON TECHNOLOGY: edited by William C. O'Mara, Robert B. Herring, and Lee P. Hunt

HANDBOOK OF SEMICONDUCTOR WAFER CLEANING TECHNOLOGY: edited by Werner Kern

HANDBOOK OF SPUTTER DEPOSITION TECHNOLOGY: by Kiyotaka Wasa and Shigeru Hayakawa

HANDBOOK OF THIN FILM DEPOSITION PROCESSES AND TECHNIQUES, *Second Edition:* edited by Krishna Seshan

HANDBOOK OF VACUUM ARC SCIENCE AND TECHNOLOGY: edited by Raymond L. Boxman, Philip J. Martin, and David M. Sanders

HANDBOOK OF VLSI MICROLITHOGRAPHY, *Second Edition:* edited by John N. Helbert

HIGH DENSITY PLASMA SOURCES: edited by Oleg A. Popov

HYBRID MICROCIRCUIT TECHNOLOGY HANDBOOK, *Second Edition:* by James J. Licari and Leonard R. Enlow

IONIZED-CLUSTER BEAM DEPOSITION AND EPITAXY: by Toshinori Takagi

MOLECULAR BEAM EPITAXY: edited by Robin F. C. Farrow

NANOSTRUCTURED MATERIALS: edited by Carl. C. Koch

SEMICONDUCTOR MATERIALS AND PROCESS TECHNOLOGY HANDBOOK: edited by Gary E. McGuire

ULTRA-FINE PARTICLES: edited by Chikara Hayashi, R. Ueda and A. Tasaki

WIDE BANDGAP SEMICONDUCTORS: edited by Stephen J. Pearton

Related Titles

ADVANCED CERAMIC PROCESSING AND TECHNOLOGY, Volume 1: edited by Jon G. P. Binner

CEMENTED TUNGSTEN CARBIDES: by Gopal S. Upadhyaya

CERAMIC CUTTING TOOLS: edited by E. Dow Whitney

CERAMIC FILMS AND COATINGS: edited by John B. Wachtman and Richard A. Haber

CORROSION OF GLASS, CERAMICS AND CERAMIC SUPERCONDUCTORS: edited by David E. Clark and Bruce K. Zoitos

FIBER REINFORCED CERAMIC COMPOSITES: edited by K. S. Mazdiyasni

FRICTION AND WEAR TRANSITIONS OF MATERIALS: by Peter J. Blau

HANDBOOK OF CERAMIC GRINDING AND POLISHING: edited by Ioan D. Marinescu, Hans K. Tonshoff, and Ichiro Inasaki

HANDBOOK OF HYDROTHERMAL TECHNOLOGY: edited by K. Byrappa and Masahiro Yoshimura

HANDBOOK OF INDUSTRIAL REFRACTORIES TECHNOLOGY: by Stephen C. Carniglia and Gordon L. Barna

MECHANICAL ALLOYING FOR FABRICATION OF ADVANCED ENGINEERING MATERIALS: by M. Sherif El-Eskandarany

SHOCK WAVES FOR INDUSTRIAL APPLICATIONS: edited by Lawrence E. Murr

SOL-GEL TECHNOLOGY FOR THIN FILMS, FIBERS, PREFORMS, ELECTRONICS AND SPECIALTY SHAPES: edited by Lisa C. Klein

SOL-GEL SILICA: by Larry L. Hench

SPECIAL MELTING AND PROCESSING TECHNOLOGIES: edited by G. K. Bhat

SUPERCRITICAL FLUID CLEANING: edited by John McHardy and Samuel P. Sawan

Contributors

Karl T. Aust
Department of Metallurgy and
Materials Science
University of Toronto
Toronto, Ontario
Canada

Ulrich Brossmann
Institut für Technische Physik
Technische Universität Graz
A-8010 Graz, Austria

Gan-Moog Chow
Department of Materials Science
National University of Singapore
Kent Ridge, Singapore

Philip Clapp
Institute of Materials Science
University of Connecticut
Storrs, CT

Uwe Erb
Department of Metallurgy and
Materials Science
University of Toronto
Toronto, Ontario
Canada

Jürgen Eckert
IFW Dresden
Institute of Metallic Materials
Dresden, Germany

Hans J. Fecht
Center for Energy Technology
Universitat Ulm
Ulm, Germany

Joanna Groza
Department of Chemical Engineering
and Materials Science
University of California
Davis, CA

Akihisa Inoue
Institute for Materials Research
Tohoku University
Sendai, Japan

Carl C. Koch
Materials Science and Engineering Department
North Carolina State University
Raleigh, NC

Lynn K. Kurihara
Naval Research Laboratory
Washington, D.C.

Maggy L. Lau
Department of Chemical and Biochemical Engineering and Materials Science
University of California
Irvine, CA

Enrique J. Lavernia
Department of Chemical and Biochemical Engineering and Materials Science
University of California
Irvine, CA

Akihiro Makino
Central Research Laboratory
Alps Electric Co. Ltd.
Nagaoka, Japan

Gino Palumbo
Integran Technologies, Inc.
Toronto, Ontario
Canada

Hans-Eckhardt Schaefer
Institut für Theoretische und Angewandte Physik
Universität Stuttgart
D-70569 Stuttgart, Germany

Michel Trudeau
Hydro-Quebec Research Institute
Varennes, Quebec
Canada

Raphael Tsu
Department of Electrical and Computer Engineering
University of North Carolina
Charlotte, NC

Julia R. Weertman
Department of Materials Science and Engineering
Northwestern University
Evanston, IL

Roland Würschum
Institut für Technische Physik
Technische Universität Graz
A-8010 Graz, Austria

Qi Zhang
Department of Electrical and Computer Engineering
University of North Carolina
Charlotte, NC

Preface

INTRODUCTION

Nanostructure science and technology has become an identifiable, if very broad and multidisciplinary, field of research and emerging applications in recent years. It is one of the most visible and growing research areas in materials science in its broadest sense. Nanostructured materials include atomic clusters, layered (lamellar) films, filamentary structures, and bulk nanostructured materials. The common thread to these various material forms is the nanoscale dimensionality, i.e., at least one dimension less than 100 nm, more typically less than 50 nm. In some cases, the physics of such nanoscale materials can be very different from the macroscale properties of the same substance. The different, often superior, properties that can then occur are the driving force behind the explosion in research interest in these materials. While the use of nanoscale dimensions to optimize properties is not new, as will be outlined below, the present high visibility and definition of the field is mainly attributable to the pioneering work of Gleiter and coworkers in the early 1980s.[1] They synthesized nanoscale grain size materials by the in situ consolidation of atomic clusters. The studies of clusters preceded the work by researchers such as Uyeda.[2] The International Technology Research Institute, World Technology Division (WTEC), supported a panel study of research and development status and trends in nanoparticles, nanostructured materials, and nanodevices during 1996–1998. The main results of this study have been published.[3] This report attempted to cover the very broad field of nanostructure science and technology and included assessments of the areas of synthesis and assembly, dispersions and coatings, high surface area materials, functional nanoscale devices, bulk nanostructured materials, and biologically related aspects of nanoparticles, nanostructured materials, and nanodevices. A conclusion of the report is that while many aspects of the field existed well before it was identified as a field in the last decade, three related scientific/technological advances have made it a coherent area of research. These are:

1. New and improved synthesis methods that allow control of the size and manipulation of the nanoscale "building blocks."
2. New and improved characterization tools for study at the nanoscale (e.g., spatial resolution, chemical sensitivity).
3. Better understanding of the relationships between nanostructure and properties and how these can be engineered.

With the recent intense interest in the broad field of nanostructure science and technology, a number of books, articles, and conference proceedings have been published. A partial listing of these publications is given in the bibliography, starting with the review of Gleiter in 1989. The justification for yet another book in this expanding field is two-fold. Since many areas of the field are moving rapidly with increased understanding from both experiment and simulation studies, it would appear useful to record another "snapshot" of the field. It will be assumed that by the time of publication certain information may become obsolete, but at least most of the background will still be useful to researchers and students. Second, since the field is so broad, spanning the study of atomic clusters to bulk, and materials from biological to metallic structures, the book has been designed to focus mainly on those areas of synthesis, characterization, and properties relevant to applications that require bulk, and mainly inorganic materials. An exception is the article by Tsu on electronic and optoelectronic materials. Before a brief description of the chapters and organization of the book is presented, a historical perspective will be given to suggest how the field has developed and what new information has been provided by reaching the limit of the nanoscale.

HISTORICAL PERSPECTIVE

Nanoscale microstructural features are not new, either in the natural world or in materials engineering. There are examples of nanoscale ferromagnetic particles found in microorganisms, e.g., 50 nm Fe_3O_4 in the organism *A. magnetotactum.*[4] A number of examples exist of improvement in mechanical properties of structural materials when a fine microstructure was developed. Early in the last century, when "microstructures" were revealed primarily with the optical microscope, it was recognized that refined microstructures, for example, small grain sizes, often provide attractive properties such as increased strength and toughness in structural materials. A classic example of property enhancement due to a refined microstructure—with features too small to resolve with the optical microscope—was age-hardening of aluminum alloys. The phenomenon, discovered by Alfred Wilm in 1906, was essentially explained by Merica, Waltenberg, and Scott in 1919,[5] and the microstructural features responsible were first inferred by the x-ray studies of Guinier and Preston in 1938. With the advent of transmission electron microscopy (TEM) and sophisticated x-ray diffraction methods, it is now known that the fine precipitates responsible for age-hardening, in Al-4%Cu alloys, for example, are clusters of Cu atoms—Guinier-Preston (GP) Zones—and the metastable partially coherent θ' precipitate.[6][7] Maximum hardness is observed with a mixture of GPII (or θ'', coarsened GP zones) and θ', with the dimensions of the θ' plates, typically about 10 nm in thickness by 100 nm in diameter. Therefore, the important microstructural feature of age-hardened aluminum alloys is nanoscale.

Critical length scales often determine optimum properties which are structure sensitive. Mechanical properties such as strength and hardness are typical and as above, microstructural features such as precipitates or dispersoids are most effective when their dimensions are nanoscale. In ferromagnetic materials, the coercive force has been found to be a maximum if spherical particles (e.g., Fe_3C in Fe) which act as domain wall pinners have a diameter about equal to the domain wall thickness, i.e., about 50 nm.[8] Similarly, in type II superconductors, it has been found that fluxoid pinning, which determines the magnitude of the critical current density, is most effective when the pinning centers typically have dimensions of the order of the superconducting coherence length for a given material. For the high field superconductors, the coherence length is usually about 10–20 nm and indeed the commercial superconductors have pinning centers that approximate these dimensions. In Nb_3Sn, the grain boundaries are the major pinning sites and optimum critical current densities are obtained when the grain sizes are about 50 nm.[9] Many other examples could be given of the long term use of nanoscale materials in fields such as catalysis.

ORGANIZATION

The scientific/technological advances that have focused the field into a broad but coherent field were given above. In this book, the new or improved synthesis methods that are one of the cornerstones of the field will be reviewed in Part I. In Part II, selected properties of nanostructured materials will be covered. Potential applications of nanostructured materials will be described as appropriate throughout the book.

In Ch. 1, Chow and Kurihara present an overview of the chemical synthesis and processing of nanostructured particles, films, and coatings. This includes particles from all materials classes, that is metals, ceramics, organic materials, etc. The chemical methods described include aqueous, non-aqueous, sonochemical, precursor, organometallic, hydrolysis, hydrothermal, and sol-gel methods. Other methods discussed are host-derived hybrid materials, surfactant membrane mediated synthesis, and a variety of films and coatings.

Lau and Lavernia describe the thermal spray processing of nanostructured materials. This method has the potential for early commercialization of coatings with nanocrystalline microstructures and superior properties. The chapter provides an overview of thermal sprayed coatings produced from nanocrystalline feedstock powders. The various routes for producing the nanocrystalline feedstock powders are discussed. The structure and properties of the nanocrystalline coatings are considered in the light of retention of a nanoscale microstructure during processing. A review of theoretical models to predict and optimize the thermal spraying parameters for optimized coatings is presented.

Fecht considers in his chapter the preparation of nanostructured materials and composites by solid-state processing methods which involve plastic mechanical deformation. The use of ball-milling of powders has become a popular method of producing nanocrystalline materials because of the simplicity of the equipment and the possibility to scale-up from laboratory to tonnage quantities of material. Fecht describes the use of mechanical attrition for production of nanocrystalline materials in a wide variety of materials including metals, ceramics, polymer blends, and composites. The stability at elevated temperatures is discussed for the nanocrystalline microstructures made by these methods. The production of nanocrystalline materials by severe plastic deformation induced by methods other than milling are also discussed. Such methods are bulk plastic deformation, rolling, and friction.

A major problem with nanocrystalline materials made in particulate form is the requirement for consolidation into bulk for most applications. The consolidation must provide theoretical density and strong particulate bonding while not unduly coarsening the nanocrystalline microstructure. Groza reviews powder consolidation methods in her chapter. She reviews the thermodynamics and kinetics of nanopowder densification. This includes the driving force for densification, surface energy, sintering mechanisms, activation energies, and scaling laws. The role of surface contamination with impurities during sintering is emphasized. The processes of cold compaction, pressureless and pressure-assisted sintering, and full densification methods are described with the goal of maintaining the nanoscale microstructure.

While Chs. 1 and 3 describe processing methods for nanocrystalline materials that result in particulates that require subsequent compaction, i.e., "two-step processing," there are one-step processing methods available that eliminate the need for compaction with its attendant problems. A notable and commercially attractive one-step method is electrodeposition. Erb, Aust, and Palumbo describe the process, structure, and properties of nanostructured materials made by electrodeposition. While electrodeposition is an old industrial process, it has only been in the last decade or so that it has been systematically applied to the synthesis of nanocrystalline materials. This chapter describes the processing methods as well as the structure and properties of the electrodeposited nanostructured materials. Comparisons are presented for the structure and properties with those of nanostructured materials made by other methods. Examples of industrial applications of electrodeposited nanostructured materials are given.

Clapp reviews the growing area of computer simulation of nanomaterials. This comprises "virtual processing," so is placed in Part I. Because of the difficulties involved with preparation of artifact-free nanocrystalline materials of the smallest grain sizes (<25 nm), simulation studies can be of great benefit and complement experimental work. This review begins with studies of the stability of individual isolated nanoparticles, then moves to a variety of subjects related to interface properties in nanomaterials, and finally covers simulations of three-dimensional nanograin materials.

Part II of the book deals with selected properties of nanocrystalline materials. Diffusion in nanocrystalline materials is reviewed by Wurschum, Brossmann, and Schaefer. The results of diffusion studies of nanocrystalline metals are presented and discussed. Correlations between diffusion and grain growth, that is nanostructure stability, are described. Examples for specific systems include the Fe-based soft magnetic "Finemet" alloys, hydrogen diffusion in nanocrystalline metals, and ^{18}O diffusion in nanocrystalline ZrO_2. The latter is compared with diffusion studies in other nanocrystalline ceramics.

Trudeau presents a review of recent developments for nanocrystalline materials in gas reactive applications. Three research areas are discussed. They are:

1. Catalysis and electrocatalysis
2. Semiconductor gas sensors
3. Hydrogen storage materials

The advantages of nanocrystalline materials are described in each case.

The magnetic properties of nanocrystalline materials provide the possibility of near-term application with, in particular, outstanding soft-magnetic properties. Inoue and Makino review examples of the important nanocrystalline soft magnetic materials. These have been developed by either partial crystallization of rapidly solidified (melt spun) amorphous precursors, or by sputtering of nanocrystalline/amorphous films. The systems described include Fe-M-B (M = Zr, Hf, or Nb) and Fe-M-O (M = Zr, Hf, or rare earth element). The improvement of the soft magnetic properties by alloying additions is discussed. A number of potential applications of these superior soft magnetic materials are presented.

Mechanical properties of nanocrystalline materials have been studied extensively since early work suggested revolutionary improvements in both strength and ductility. The state-of-the-art in this area is reviewed by Weertman. This review describes the various models that have been proposed for mechanical behavior of nanocrystalline materials. The important subject of the microstructural characterization of nanocrystalline materials is covered. Then the experimentally determined deformation behavior of nanostructured materials is presented.

While the results of mechanical property studies have been disappointing so far on single phase nanocrystalline materials, there appears to be promise for multiphase nanocrystalline materials. The structure, formation, and mechanical behavior of two-phase nanostructured materials are reviewed by Eckert. The methods and process variables used to produce bulk two-phase nanostructured materials are described. The mechanical behavior of such materials is then discussed for both room and elevated temperature testing. In some cases, it appears that multiphase nanocrystalline materials can offer combinations of both high strength and ductility compared to single phase materials in which ductility is very limited for grain sizes of about 25 nm or less.

The subject of "functional" nanostructured materials for electronic and optoelectronic materials is a large and important area. While this field is not stressed in this book, it was felt that a chapter outlining some of the important features of this area should be included. Tsu and Zhang give examples of functional nanocrystalline materials, that is, typically thin films or quantum dots for electronic or optoelectronic applications. An in-depth treatment of several topics related to silicon semiconductors is given. This includes the physics of nanostructured materials which covers the dielectric constant, the capacitance, doping and exiton binding energies of a nanoparticle. In addition, possible devices requiring nanoscale features are described. Such devices are light emitting diodes (LEDs) and quantum field effect transistors (QD-FETs).

REFERENCES

1. Gleiter, H., *Progress in Materials Science*, 33:223–315 (1989)
2. Uyeda, R., *Progress in Materials Science*, 35:1–96 (1991)
3. Siegel, R. W., Hu, E., and Roco, M. C., (eds.), *Nanostructure Science and Technology*, Kluwer Academic Publishers, Dordrecht, Netherlands (1999)
4. Kirschvink, J. L., Koyayashi-Kirschvink, A., and Woodford, B. J., *Proc. Nat'l. Acad. Sci.*, USA, 89:7683–7687 (1992)
5. Mehl, R. F., and Cahn, R. W., *Historical Development, Physical Metallurgy,* 3rd ed., pp. 1–35, North Holland (1983)
6. Silcock, J. M., Heal, T. J., and Hardy, H. K., *J. Institute of Metals*, 82:239 (1953–54)
7. Cohen, J. B., *Metall. Trans. A.*, 23A:2685 (1992)
8. Swisher, J. H., English, A. T., and Stoffers, R. C., *Trans. ASM,* 62:257 (1969)
9. Scanlan, R. M., Fietz, W. A., and Koch, E. F., *J. Appl. Phys.*, 46:2244 (1975)

BIBLIOGRAPHY

Gleiter, H., *Nanocrystalline Materials, Progress in Materials Science*, 33:223–315 (1989)

Siegel, R. W., Nanostructured Materials—Mind Over Matter, *Nanostructured Materials*, 3:1 (1993)

Hadjipanayis, G. C., and Siegel, R. W., *Nanophase Materials: Synthesis-Properties-Appplications*, Kluwer Press, Dordrecht, Netherlands (1994)

Siegel, R. W., *Nanophase Materials*, in: *Encyclopedia of Applied Physics*, (G. L. Trigg, ed.), 11:1–27, VCH, Weinheim (1994)

Gleiter, H., Nanostructured Materials: State of the Art and Perspectives, *NanoStructured Materials,* 6:3 (1995)

Edelstein, A. S., and Cammarata, R. C., (eds.), *Nanomaterials*: *Synthesis, Properties, and Appplications*, Institute of Physics, Bristol (1996)

Carl Koch
Raleigh, North Carolina

October, 2001

Contents

Part I
Processing

Part II Properties

Part I

Processing

1

Chemical Synthesis and Processing of Nanostructured Powders and Films

Gan-Moog Chow and Lynn K. Kurihara***

1.0 INTRODUCTION

The performance of materials depends on their properties. The properties in turn depend on the atomic structure, composition, microstructure, defects, and interfaces, which are controlled by thermodynamics and kinetics of the synthesis. A current paradigm of synthesizing and processing of advanced materials emphasizes the tailored assembly of atoms and particles, from the atomic or molecular scale to the macroscopic scale. Nanostructured materials, often characterized by a physical dimension of 1–100 nm (such as grain size) and a significant amount of surfaces and interfaces, have been attracting much interest because of their demonstrated or anticipated unique properties compared to conventional materials. Nanostructured materials can be made by attrition of parent coarse-grained materials using the top-down approach from the macroscale to the nanoscale, or conversely, by assembly of atoms or particles using the bottom-up approach. The control of arrangement of atoms from the nanoscale to the macroscale is indeed the strength of materials chemistry. Therefore,

*Formerly at Naval Research Laboratory, Washington, DC, USA.
**Also at Potomac Research International, Fairfax, VA, USA.

it is not surprising that increasing attention has been paid to the chemical synthesis and processing of nanostructured materials.[1]–[8]

Chemical reactions for material synthesis can be carried out in the solid, liquid, or gaseous state.[9] The more conventional solid-state synthetic approach is to bring the solid precursors (such as metal oxides or carbonates) into close contact by grinding and mixing, and to subsequently heat treat this mixture at high temperatures to facilitate diffusion of atoms or ions in the chemical reaction. The diffusion of atoms depends on the temperature of the reaction and grain boundary contacts. The transport across grain boundaries is also affected by impurities and defects located there. The mixing and grinding steps are usually repeated throughout the heat cycle, and generally involve a great deal of effort to mix materials at the nanoscale and also to prepare fresh surfaces for further reactions. For systems that do not contain means to inhibit grain growth (such as grain growth inhibitors and immiscible composites), grain growth at elevated temperature reactions leads to solids with large grain size.

Compared to solid-state synthesis, diffusion of matter in the liquid or gas phase is typically and advantageously many orders of magnitude larger than in the solid phase, thus the synthesis of nanostructured materials can be achieved at lower temperatures. Lower reaction temperatures also discourage detrimental grain growth. Many materials can be synthesized in aqueous or nonaqueous solutions. For example, water is one of the best known and most common solvents. There are three general classes of aqueous reactions: acid/base reaction, precipitation, and reduction/oxidation (redox). In basic chemistry terms, starting materials of a chemical reaction are called the *reactants* and the material to which the reactants are converted the *products*. The reactants can be solids, liquids, or gases in any combination, in the form of single elements or multi-component compounds. A multi-element compound is usually called a *precursor*. In a precursor, the components of the final product are in a "mixture" with atomic scale mixing. Many precursors can be prepared by precipitation reactions. In precipitation reactions, solutions of two or more electrolytes are mixed and an insoluble precipitate or a gelatinous precipitate forms.

In chemical synthesis of materials, one should always use caution when handling reactants and precursors, reaction by-products and post-reaction wastes, particularly when complex and hazardous chemicals are involved. Special procedures may be required to remove any entrapped impurities from the products and to avoid post-synthesis contamination. Although many laboratory-scale reactions can be scaled up to economically produce large quantities of materials, the laboratory-scale reaction parameters may not be linearly related to that of large-scale reaction. The synthesis parameters such as temperature, pH, reactant concentration, and time should be ideally correlated with factors such as supersaturation, nucleation and growth rates, surface energy, and diffusion coefficients, in order to ensure the reproducibility of reactions.

Chemistry is based on the manipulations of atoms and molecules, and indeed has a very long history in the synthesis of materials comprising nanostructures. The fields of colloids and catalysts are such examples. The recent popularity of "nanoscience" not only revitalized the use of many "old" chemical methods, but also motivated many "new" and "modified" ones to be continually developed for the synthesis of nanostructured materials. The scope of chemical synthesis and processing of nanostructured materials is very wide, spanning structural, optical, electronic, magnetic, biological, catalytic, and biomedical materials. In this chapter, a comprehensive review of every aspect of this field is not possible. An overview of chemical synthesis and processing of nanostructured particles, films, and coatings is given, with selected examples of metals, ceramics, and hybrid materials. The chapter is organized according to the classes of materials and types of synthetic approaches. However, due to the fact that many advanced materials are hybrid and are prepared using multidisciplinary techniques, clear distinction is not always possible. Interested readers are encouraged to consult the cited references,[1]–[8] archival journals,[10] and conference proceedings[11] for further details.

2.0 PARTICLES

2.1 Nucleation and Growth

The synthesis of particles in a solution occurs by chemical reactions that result in the formation of stable nuclei and subsequent particle growth. The term *precipitation* is often used to describe this series of events. The reactants are introduced frequently as solids or liquids, and sometimes as gases, in aqueous or nonaqueous solvents that can have a wide range of dielectric constants. The phenomenon of precipitation of solids in solution has been well studied.[12][13] Elemental or multicomponent particles can be precipitated. When a multicomponent material is desired, special attention is required to control co-precipitation conditions in order to achieve chemical homogeneity of the final product. This is because different ions often precipitate under different conditions of pH and temperatures, and have different solubility product constants.

Upon addition of reagents such as reducing or oxidizing agents to the solution containing the reactants, chemical reactions occur and the solution becomes supersaturated with the product. The supersaturation drives the chemical system to a far departure from the minimum free energy configuration. The thermodynamics equilibrium state of the system is restored by condensation of nuclei of the reaction product. Two types of nucleation can occur. *Homogeneous nucleation* does not involve any foreign species as nucleating aids. *Heterogeneous nucleation*, on the other hand, allows the formation of nuclei on foreign species.

Kinetic factors compete with the thermodynamics of the system in a growth process.[14] Kinetic factors such as reaction rates, transport rates of reactants, accommodation, removal, and redistribution of matter compete with the influence of thermodynamics in particle growth. The reaction and transport rates are affected by concentration of reactants, temperature, pH, the order in which the reagents are added to the solution, and mixing. The structure and crystallinity of the particle can be influenced by reaction rates and impurities. Particle morphology is influenced by factors such as supersaturation, nucleation and growth rates, colloidal stability, recrystallization, and the aging process. Generally, supersaturation has a predominant influence on the morphology of precipitates. At low supersaturation, the particles are small, compact, and well-formed, and the shape depends on crystal structure and surface energies. At high supersaturation levels, large and dendritic particles form. At even larger supersaturation, smaller but compacted, agglomerated particles form.[13] The growth in solution is interface-controlled when the particle is small; after reaching a critical size, it becomes diffusion-controlled.[15]

2.2 Stable Dispersion and Agglomeration

In a supersaturated solution when the nuclei form at nearly the same time, subsequent growth of these nuclei results in formation of particles with a very narrow size distribution, provided that secondary nucleation does not occur later.[16] Homogeneous nucleation as a single event requires the use of proper concentrations of reagents. Foreign nuclei should be removed before reaction to prevent heterogeneous nucleation that may otherwise result in a wide size distribution of particles. This narrow size distribution can be maintained as long as agglomeration and Ostwald ripening of particles in solution does not occur. The formation of stable colloids and dispersion of agglomerated particles have been extensively investigated.[17] The terms *colloids* and *sols* refer to the dispersion of particles (with particle sizes less than 100 nm) within a continuous fluid matrix. The ultrafine particles approach and then separate from each other by Brownian motion, and as a result, settling of particles out of solution does not occur. It should be noted that random agglomeration between particles may still occur by Brownian motion. Agglomerates or particles larger than 100 nm tend to settle out of solution.

In aqueous solvents, particles that possess a surface oxide layer or a hydrated surface may become charged under appropriate conditions. Electrostatic repulsion, with a force proportional to the inverse of second power of separation distance, occurs between two particles carrying the same charge. The attractive van der Waals force is proportional to the inverse of the distance with an exponent of 3–6. The net attractive or repulsive force between the particles in such a suspension is the sum of the electrostatic repulsion and the attractive van der Waals forces. The DLVO theory (Derjaguin, Landau, Verwey, and Overbeek) describes the effects of attraction and repulsion of particles as a function of separation distance.[18] On the DLVO plot of potential energy vs the separation distance of particles, there exists a positive potential energy peak, which separates the negative potential energy of primary minimum and secondary minimum. The height of the potential energy peak must be $\geq$ 25 millivolts (corresponding to the thermal energy of Brownian motion at 20°C) at ambient conditions, in order for a dispersion of particles to be stable. In an appropriate solvent, an electric double layer is formed surrounding the particle. The stable distance of particle separation depends not only on the charges of the particles, but also the concentration of other ions in the diffuse region of the double layer. When there is a sufficient number of such ions or ions with multiple charges in the diffuse layer, the charge repulsion will be neutralized. The collapse of the double layer leads to particle contacts and agglomeration.[18]

Nanostructured particles possess large surface areas and often form agglomerates as a result of attractive van der Waals forces and the tendency of the system to minimize the total surface or interfacial energy. *Coagulation* refers to the formation of strong, compact aggregates (corresponding to the primary minimum on the DLVO plot of potential energy vs particle separation), and *flocculation* refers to the formation of a loose network of particles (corresponding to the secondary minimum on the

DLVO plot). Agglomeration of particles can occur during any of the following stages: synthesis, drying, handling, and processing. In many applications and processing where dispersed particles or stabilized dispersions are required, undesirable agglomeration in each synthesis and processing step must be prevented. To produce unagglomerated particles, surfactants can be used to control the dispersion during chemical synthesis, or disperse as-synthesized agglomerated fine particles.

A *surfactant* is any substance that lowers the surface or interfacial tension of the medium in which it is dissolved. A surfactant is a surface-active agent that needs not be completely soluble and may decrease surface or interfacial tension by spreading over the surface. It has an amphipathic structure in that solvent, i.e., a lyophobic (solvent repulsive) and lyophilic group (solvent attractive). Surfactants are classified as anionic, cationic, zwitterionic (bearing both positive and negative charges), or non-ionic (bearing no charges). Surfactant effectiveness is measured by the maximum reduction in surface or interfacial tension by the surfactant, whereas, surfactant efficiency refers to the surfactant concentration that is needed to reduce the surface or interfacial tension by a certain amount from that of the pure solvents. For example, water and oil can be dispersed in each other if a suitable surfactant is used to stabilize the microemulsion. The surfactant establishes itself at and defines the boundary between the two liquids. The relative quantity of a surfactant determines the amount of surface that can be covered and, therefore, the extent to which the size and number of droplets of one liquid is dispersed in the other. When the major component is apolar (oil), the dispersion is one in which the water (polar) phase forms the droplets or reverse micelles. The polar head group of the surfactant is pointing inward toward the water phase while the hydrocarbon tail is pointing outward into the oil phase. The radius of the water droplet is related to the amount of water and surfactant. Figure 1 shows some examples of surfactant membrane structures.

Repulsive interparticle forces are needed to prevent the particles from agglomeration during synthesis. A common method is to disperse the particles by electrostatic repulsion resulting from the interactions between the electric double layers surrounding the particles. This can be achieved by adjusting the pH of the solution or adsorbing charged surfactant molecules on the particle surfaces. Such stabilization is generally effective in dilute systems of aqueous or polar organic media, and is very sensitive to the pH and effects of other electrolytes in the solution. At the *isoelectric point*, the pH where the particles have no net surface charges, agglomeration may occur. The isoelectric point varies for different materials.

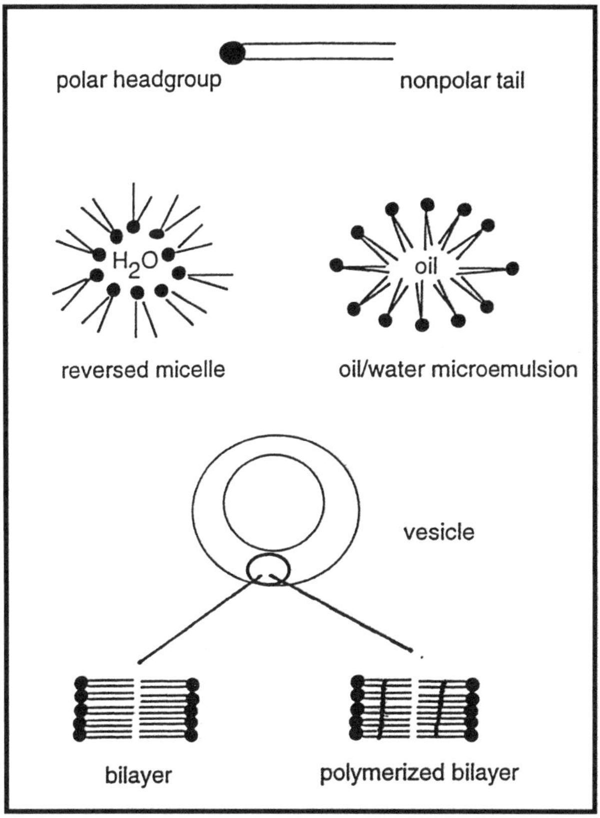

Figure 1. Examples of self-assembled surfactant membrane microstructures. The range of diameter: 5–100 nm for micelles and microemulsions, 100–800 nm for multilamellar vesicles, and 30–60 nm for single bilayer vesicles. Vesicle bilayers can be polymerized.

In most nonaqueous solvents without significant ionization, electrostatic repulsion has a lesser contribution to stabilization of particles. Another stabilization approach involves the steric forces produced by adsorbed surfactant on particle surfaces. The lyophilic, non-polar chains of surfactant molecules extend into the solvent and interact with each other. The interactions of non-polar chains have much less van der Waals attraction and provide a steric hindrance to interparticle approach. For optimized steric stabilization, the size of surfactant molecules must be large enough to be a barrier without entangling each other. When the particles approach one another, the stretched-out lyophilic chains of the adsorbed surfactant are forced into a smaller spatial confinement. This interaction leads to a thermodynamically unfavorable decrease of the entropy of the system, thus, the particles will be prevented from approaching each other by this entropic repulsion. Entropic stabilization becomes even more stable when the temperature of the dispersion is increased. Steric stabilization can occur in the absence of electric barriers and is effective in both aqueous and nonaqueous media. It is also less sensitive to impurities or trace additives than electrostatic stabilization and is particularly effective in dispersing high concentrations of particles.

Dry, high-surface-area powders agglomerate by van der Waals forces and hydrogen bonds. When these agglomerates need to be used in a dispersed form during subsequent processing, deagglomeration can be achieved by breaking the agglomerates using methods such as milling or ultrasonication in an appropriate solvent containing a suitable surfactant for dispersion.[19] The deagglomerated powders may then be carried out in a liquid for further processing such as injection molding and polymer-based casting.

2.3 Metals, Intermetallics, Alloys, and Composites

Fine metal particles are used in electronic and magnetic materials, explosives, catalysts, pharmaceuticals, and in powder metallurgy. In chemical synthesis of metal powders, many reducing agents such as sodium formate, formic acid, borohydride, sodium hydrosulfite, and hydrazines can be used in aqueous and nonaqueous media. Because of high reactivity of nanostructured metals due to the large surface area, special care must be taken during washing, filtering, and drying of nanostructured powders to avoid hydrolysis or oxidation.

Aqueous Methods. Water has a high permittivity which makes it a good solvent for polar or ionic compounds. Therefore, many chemical reactions take place in aqueous media. Precious, elemental metal powders for electronic applications can be prepared by adding liquid reducing agents to aqueous solutions of respective salts at adjusted pH.[20] Nanostructured amorphous alloys, or crystalline alloys, and composites can also be prepared using aqueous chemistry. For example, aqueous potassium borohydride reduction was used to make ultrafine amorphous Fe-Co-B alloy powders for applications in ferrofluids and magnetic memory systems.[21] The amorphous phase was formed when the reaction was carried out below the glass transition temperature and stabilized by a high concentration of boron atoms. The reaction medium often dictates the kind of product(s) formed. When sodium borohydride was used to prepare Co-B alloy, the solvent had an important role in determining the final product. In the aqueous case, nanoscale Co_2B particles were the primary product,[22] whereas, in the nonaqueous reduction of Co ions in diglyme, nanostructured Co particles were formed.[23] Aqueous and nonaqueous borohydride reduction were also used to synthesize nanoparticles of Fe, FeB and Fe_2B.[24]

Metastable phases can be formed by fast kinetics in a chemical reaction. For example, iron and copper are immiscible in the equilibrium state. Metastable alloys of Fe-Cu can be synthesized using far-from-equilibrium processing techniques such as vapor phase quenching or mechanical alloying. Nanocrystalline Fe_xCu_{100-x} (x is at%) alloys and composite powders were synthesized by reducing aqueous solutions of ferrous chloride and cupric chloride (in various molar ratio) by sodium borohydride solution.[25] The reaction was carried out at room temperature with stirring for only 5 min. The ratio between Fe and Cu atoms was very close to that in the original aqueous solutions. When x = 40, only fcc metastable alloys were formed. At higher Fe concentration such as x about 70, phase separation of fcc-Cu and bcc-Fe took place. The formation of Cu_2O was observed and its concentration scaled with that of Cu. The powders were agglomerates of nanocrystallites. The crystallites were between 30 to 45 nm for alloys, and between 10 to 15 for Fe and 30 to 40 nm for Cu in the composites. As-synthesized alloys were magnetically soft with coercivities ranging from 10 to 40 Oe, due to the lack of nearest neighbors interaction in solid solutions. As-synthesized composite powders had coercivity as high as 400 Oe. The formation of crystalline phases was controlled by decreasing the boron concentration in the powders when a higher molarity of borohydride was used in the reaction.

Another example is the Co-Cu systems that form terminal solid solutions. Nanostructured Co_xCu_{100-x} powders were synthesized by sodium borohydride reduction of aqueous cobalt and cupric chloride solutions.[26] As-synthesized powders were a mixture of fcc and amorphous phases. The concentration of amorphous phase increased with the ratio of Co/Cu. The annealed powders phase separated to fcc-Co and fcc-Cu at about 500°C. Annealing led to significant surface sintering and some grain growth (grain size about 30 to 40 nm), and boron impurity was found to segregate at surfaces of sintered powders. The coercivity behavior of the powders, similar to that of nanostructured Co-Cu films prepared by annealing sputtered alloy films, increased with annealing temperature to a maximum of 620 Oe.

Although the aqueous approach to making metal powders is not new, its use in the synthesis of nanoscale metal powders requires special attention to avoid undesirable contaminated products. Impurities such as salts and other reaction by-products may not be completely removed, even by repeated washing procedures, if they are entrapped inside the particles or agglomerates during a fast and ill-controlled reaction. Washing to remove soluble salts may result in hydrolysis and oxidation of metal particles. Subsequent drying often requires vacuum-assisted procedures to avoid oxidation.

Nonaqueous Methods. Many reactants and reducing agents used in aqueous synthesis of nanoscale metal particles can also be used in nonaqueous solvents for the same purpose. For example, sodium borohydride can be used to reduce copper chloride in tetrahydrofuran (THF) to prepare nanostructured Cu particles. Similar to the aqueous approach, residual salts need to be removed from the product. A nonaqueous synthesis method known as the *polyol process* has been used to make finely dispersed single elemental metallic particles such as Cu, Ni, and Co in the micron and submicron size range.[27] In this method, precursor compounds such as oxides, nitrates, and acetates are either dissolved or suspended in ethylene glycol or diethylene glycol. The mixture is heated to reflux between 180°C and 194°C. During the reaction, the precursors are reduced and metal particles precipitate out of solution. Submicron size particles can be synthesized by increasing the reaction temperature or inducing heterogeneous nucleation via adding foreign nuclei or forming foreign nuclei in-situ. The reaction temperature affects the nucleation and growth in the production of submicron gold particles.[28] A higher temperature favored the nucleation step and this, in turn, favored the monodispersity of particles when more nuclei were formed. Nanocrystalline powders such as Fe, Co, Ni, Cu, Ru, Rh, Pd, Ag, Sn, Re, W, Pt, Au, Fe-Cu, Co-Cu, and Ni-Cu were also synthesized using this method[29]–[36] with different salt precursors. In many cases, the use of nucleating aids to assist the formation of nanoparticles was not required.[29]–[33] Figure 2 shows the relationship of average particle size and refluxing temperature of various powders and films synthesized by the polyol method.

For example, nanostructured powders of Co_xCu_{100-x} ($4 \leq x \leq 49$ at%)[31][32] were synthesized by reacting cobalt acetate tetrahydrate and copper acetate hydrate in various proportions in ethylene glycol. The mixtures were refluxed at 180–190°C for 2 h. The powders precipitated out of solution, and were subsequently collected and dried. The reaction rate was slower and the reaction time was much longer than that of the aqueous borohydride reduction for Co-Cu synthesis discussed above.[26] In this case, slower kinetics did not favor formation of a metastable solid solution. Since Cu was more reducible than Co, nucleation of Cu occurred first, and Co subsequently nucleated on Cu crystallites. X-ray diffraction, often conventionally used to study alloy formation, showed some evidence to suggest that metastable alloys could have formed. Diffraction peaks due to fcc Cu were detected in all samples with different copper concentrations, but Co diffraction peaks were not detected until x = 19 at%. To confirm the structure of powders, studies of the local atomic environment were performed using extended x-ray absorption fine structure (EXAFS) spectroscopy and solid-state nuclear magnetic resonance (NMR). The results from these investigations and vibrating sample magnetometry (VSM) ruled out the formation of metastable alloys, but confirmed the synthesis of nanocomposites of Co-Cu. The powders were agglomerated (Fig. 3), as in the case of powders prepared using aqueous borohydride.

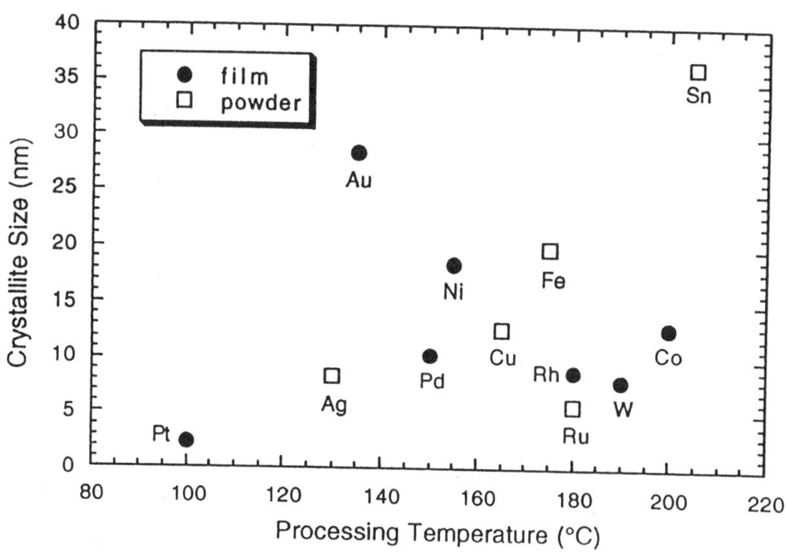

Figure 2. Average particle size vs processing temperature for some metals prepared using the polyol method.

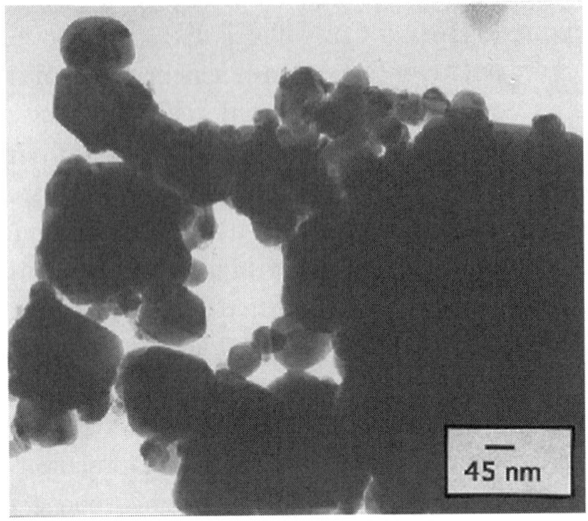

Figure 3. A TEM (transmission electron microscopy) micrograph of CoCu powders synthesized by the polyol method.

The polyol method has also been shown as a useful preparative technique for the synthesis of nanocrystalline alloys and bimetallic clusters.[37]–[39] Nanocrystalline $Ni_{25}Cu_{75}$ powders with grain size of 8 nm was prepared by reducing acetates of nickel and copper in ethylene glycol without nucleating agents.[29][33] The diffraction peaks of $Ni_{25}Cu_{75}$ followed Vegard's law of solid solution. Nickel clusters were prepared using Pt or Pd as nucleation agents.[34] The nucleating agent was added 10 minutes after the nickel-hydroxide-PVP-ethylene glycol solution began refluxing. The Ni particle size was reduced from about 140 nm to 30 nm when a nucleating agent was used. Reduction of particle size was also obtained by decreasing the nickel hydroxide concentration and by the use of PVP. Nickel prepared without nucleating agents had an oxidation temperature of 370°C. Smaller nickel particles synthesized with nucleating aids oxidized at a lower temperature of 260°C, as expected from the higher surface area of finer particles. Desorption studies showed the adsorbed surface species were CO moieties and H_2O, and nitrogen–containing species were not observed. This indicated that ethylene glycol, not the polymer, was adsorbed on the surface of particles. The ethylene glycol had only a half monolayer coverage. When this protective glycol was completely removed from the surface, oxidation occurred. The Ni-Pd and Ni-Pt particles had a 7–9 nm Pd nucleus and a 6–8 nm Pt nucleus, respectively. Oxidation studies showed that some alloying of Ni with Pt occurred. Cobalt nickel alloys of 210 to 260nm particle sizes were also prepared using either silver or iron as nucleating agents.[35][36] The CoNi alloy particles had densities and saturation magnetizations close to the bulk values, and showed a shift to higher resonance frequencies as the Co/Ni increased. This was also observed in the Fe-Co-Ni particles[36] that were 50–150 nm in size.

Polymer protected bimetallic clusters were also formed using a modified polyol process.[37]–[39] The modification included the addition of other solvents and sodium hydroxide. In the synthesis of Cu/Pt or Cu/Pd which had average diameters between 1–2 nm, copper sulfate, PVP, and ethylene glycol were mixed with either palladium acetate in dioxane or chloroplatinic acid in water. Sodium hydroxide was also added to the glycol solution. The glycol and organic solvents were removed from solution by acetone or filtration. It was found that PVP did not protect the Cu particles from agglomeration and clusters were not formed. The Cu particles had sizes ranging from 3–250 nm. In the case of platinum group metals, monodispersed cluster formation was reported. Monodispersed clusters of Cu/Pt or Cu/Pd were formed, and Pt or Pd was found on the surface of clusters. The bimetallic clusters had a single catalytic activity and selectivity over monometallic species for the same catalytic reaction. For example, when Cu/Pd clusters were used as catalysts in hydration of acrylonitrile, the only product detected was acylamide, and cyanohydrin was not formed. This showed a 100% selectivity for amide formation using this bimetallic catalyst, whereas Pd clusters showed no catalytic activity for the same reaction.

Compared to aqueous methods, the polyol approach resulted in the synthesis of metallic nanoparticles protected by surface adsorbed glycol, thus minimizing the oxidation problem. The use of a nonaqueous solvent such as the polyol also further reduced the problem of hydrolysis of fine metal particles that often occurred in the aqueous case.

Sonochemical Methods. Ultrasound has been used in chemical synthesis of nanostructured materials. High energy sonochemical reactions, without any molecular coupling of the ultrasound with the chemical species, are driven by the formation, growth and collapse of bubbles in a liquid. This acoustic cavitation involves a localized hot spot of temperature of about 5000 K, a pressure of ~1800 atm and a subsequent cooling rate of about 10^9 K/sec, due to implosive collapse of a bubble in the liquid.[40] Generally, volatile precursors in low vapor pressure solvents are used to optimize the yield. Ultrasonic irradiation is carried out with an ultrasound probe, such as a titanium horn operating at 20 kHz.

Nanostructured particles for catalytic applications were sonochemically synthesized using volatile organometallic precursors.[41] These powders had a surface area which was over a hundred times greater than powders commercially available. The sonication was typically carried out for 3 h at different temperatures in inert argon atmosphere. For example, 3–8 nm amorphous iron particles on silica support were synthesized at 20°C from iron pentacarbonyl [$Fe(CO)_5$], decane and silica gel. Porous aggregates of 10–20 nm particles of Fe-Co alloys were prepared at 0°C from $Fe(CO)_5$, $Co(CO_3)(NO)$ and decane. Porous aggregates of 2 nm Mo_2C particles were obtained by heat treating the precursor powders sonochemically synthesized from molybdenum hexacarbonyl and hexadecane at 90°C. These high-surface area nanostructured materials were active heterogeneous catalysts for hydrocarbon reforming and CO hydrogenation. The use of volatile carbonyl compounds requires the use of equipment that handles air-sensitive chemicals.

A conventional coarse-grained intermetallic such as $MoSi_2$ is limited by its brittleness at normal environmental temperatures, despite its attractive properties such as low density, high temperature strength and oxidation resistance. Nanostructured intermetallics offer promising potentials such as improved ductility, strength and fracture toughness. Co-reduction of $MoCl_5$, $SiCl_4$, and NaK alloys in hexane was carried out sonochemically to synthesize precursor powders of $MoSi_2$.[42] Enhanced mixing and reactions of the reactants of heterogeneous mixtures were achieved by ultrasound irradiation. Instead of a few days that are conventionally taken for such reaction, the ultrasound-driven synthesis can be completed in a few hours. The precursor powders were then annealed in a low vacuum at 900°C to sublime impurities of NaCl and KCl, and to form nanocrystalline $MoSi_2$ particles. Both low and high temperature phases of $MoSi_2$ were present. The crystallite size was in the range of 16–31 nm, depending on the time of heat treatment. These powders were consolidated using hot pressing and hot isostatic pressing under various conditions. Although nanoscale grain size was retained without significant grain growth during consolidation, a significant amount of porosity (about 20%) persisted. The microhardness and compression strength of consolidated samples were higher than coarse-grained materials. However, perhaps due to high porosity, improvement of low temperature ductility was not observed. It was noted that scaling up the powder production from a 5 g-batch to a larger quantity resulted in incorporation of Si-deficient impurity phases of silicides. It was suggested that impurities were formed because of incomplete mixing in the larger batch solution. This example showed that, in chemical synthesis of materials, parameters for production of a large quantity may not be linearly scaled with that used for a smaller quantity.

Precursor Methods. As previously described, the conventional approach to making alloys and composites is to first grind and mix the solid precursors using some mechanical means, and then carry out appropriate chemical reactions to obtain final products. The communition and mixing in solid state are generally limited to submicron level. Consequently, material diffusion in chemical synthesis is limited to this spatial scale, which has a direct influence on the time and temperature of reactions and the final chemical homogeneity of the product. With great efforts such as high energy milling, solid state mixing of constituents at the atomic scale is possible. If the precursors can be mixed at the atomic level, the synthesis reactions can be carried out at shorter times and reduced temperatures due to the shorter distance for material diffusion. Intimate contact of constituents at the atomic scale also provides a better means to control the stoichiometry and homogeneity of the final product. These advantages are the motivation for synthesizing precursor materials which have the constituents as atomic neighbors (for example, as in a compound). These precursors are subsequently subjected to thermo-chemical reactions to synthesize alloys and composites with improved properties compared to the same materials obtained by traditional solid state reactions.

Organometallic Methods. An organometallic compound is one which has a direct metal to carbon bond. Advantages of using organometallic compounds are that precursors can be made that have the constituents in molecular proximity to each other and that can be decomposed at relatively low temperatures to form the final product desired. These reactions can be used in fine chemical synthesis. The biggest disadvantage of this approach is that most of the reactions involve air sensitive reactants as well as the final precursor, therefore, the glove box or schlenck line techniques must be used. Because of the air-sensitive nature of some of the reactants, greater care must also be taken in preparation of solvents and the choice of atmospheres. Generally, organometallic routes produce only small amounts of material.

There has been a renewed interest in the synthesis of magnetic nanoclusters with unique properties such as enhanced magnetization.[43] Cobalt nanoclusters has been prepared by reduction of organometallic compounds of $Co(n3\text{-}C_8H_{13})(n4\text{-}C_8H_{12})$ in hydrogen and in the presence of PVP.[44] Dried colloids were non-interacting superparamagnetic particles with enhanced magnetization relative to that of bulk value. By varying the decomposition temperature, the size of particles could be controlled.

Organometallic precursors were also used to prepare other colloidal metals with a variation of size and structure. For example, cuboctahedral and icosahedral platinum nanoparticles of 1.2–1.5 nm were stabilized by CO, CO and THF, or CO, PPh3 and THF.[45] Using CO alone or with THF stabilized the cuboctahedral structure. When the phosphine ligand was also incorporated into the synthesis, the icosahedral structure resulted.

Multicomponent materials can be prepared using precursors synthesized from solution chemistry. For example, conventional M50 steel (with a typical chemical composition of 4.5% Mo, 4.0% Cr, 1.0% V, 0.8% C, and balance of Fe, in wt%) is extensively used as main-shaft bearings in aircraft gas turbine engines because of its good resistance to wear and tempering, and rolling contact fatigue. The large carbide particles in these materials serve as the fatigue crack initiation sites. Nanostructured M50 steel materials with smaller carbides and grain size are expected to have enhanced properties because of the reduction of flaw size of carbides and matrix grain size. Nanoscale precursor powders of M50 type steel for bearing applications have been chemically synthesized by either thermal decomposition of organometallic precursors or co-reduction of metal halides.[46][47] Thermal decomposition of $Fe(CO)_5$, $Cr(Et_xC_6H_{6-x})_2$, $Mo(Et_xC_6H_{6-x})_2$, and $V(CO)_6$ was carried out in decalin. In co-reduction, sodium borohydride or lithium triethyl borohydride was used to reduce $FeCl_3$, $MoCl_3$, $CrCl_3$, and VCl_3 in THF. Impurity by-products such as NaCl or LiCl were removed by sublimation at about 950°C and 700°C, respectively. The thermal decomposition of organometallic precursors was easier and more cost efficient to scale up for the production of large quantities of these precursor powders, and it did not produce residual impurities that would require removal at higher temperatures prior to powder consolidation. The structural and microstructural developments of the powders were controlled by subsequent consolidation such as hot pressing or hot-isostatic pressing. During consolidation, amorphous precursor powders transformed to nanocrystalline M50 type structure with precipitation of carbides, and, simultaneously, pressure–assisted sintering and densification of powders occurred.

The consolidated bulk samples, obtained by hot pressing the amorphous precursor powders between 700°C and 850°C at 275 MPa for 0.5 to 2 h, comprised a matrix of α-Fe with a grain size ranging from 5 to 70 nm and clusters of 10 nm Mo_2C precipitates. Large precipitates of about 100 nm were located at the triple point grain boundaries of smaller matrix grains. Smaller carbide precipitates within large matrix grains had little effect on preventing grain growth. The consolidation results, using hot pressing and hot isostatic pressing, indicated that increasing pressure and pressing temperature did not have significant effect on reducing the density of porosity (about 5%) in all samples. Both normal and abnormal grain growth of matrices and precipitates occurred with increasing temperature. The retention of nanostructures in powder consolidation of multicomponent engineering materials with full density at a practical pressure range remains challenging. Preliminary mechanical properties of nanostructured M50 compacts showed improved hardness and yield strength.[48]

Nanostructured powders of Ni/Cr alloys for wear and corrosion resistance applications were prepared using chemical precursors. Reductive decomposition of an organic solution of metal chloride by sodium triethylborohydride led to coprecipitation of nanoscale mixed powders of Ni and Cr. These amorphous powders were subsequently annealed at about 300°C to form nanocrystalline Ni/Cr alloy powders with grain size between 10 to 18 nm. When the precursor powders were washed in deoxygenated water to remove by-product of NaCl, the formation of Ni-Cr_2O_3 powders was observed after heat treatment. The oxidation of Cr was believed to be caused by reactions involving chemisorbed O-H groups of water on powder surfaces. When the precursor powder surface was passivated by deoxygenated mineral oil before washing, oxidation of Cr did not take place during heat treatment. It was suggested that a protective surface carbide due to chemisorbed C-H groups from the mineral oil was formed. Carburization of precursor powders in methane at about 880°C resulted in formation of nanostructured Ni-Cr_3C_2 cermet powders. The grain size of Ni and Cr_3C_2 was 18 and 30 nm, respectively.[49]

Refractory carbide composites for cutting and drilling tools and wear parts such as WC-Co were prepared using a precursor approach. The precursor Co(ethylenediamine)$_3$$WO_4$ was synthesized by aqueous precipitation of $CoCl_2$ and H_2WO_4 in ethylenediamine. Nanoporous and nanophase W-Co powders were obtained by reductive decomposition. Carburization using CO_2-Co gas converted W-Co to WC-Co powders.[50] Homogeneous molecular precursor powders containing W and Co mixed on the atomic scale were also synthesized by aqueous coprecipitation of sodium tungstate $[(NH_4)_2WO_4]$ or ammonium metatungstate $[(NH_4)_6(H_2W_{12}O_{40})]$ with ammonium salts of di-cobalt anion $[(H_2Co_2W_{11}O_{40})^{8-}]$.[51] The synthesized precursor salt $[(NH_4)_8(H_2Co_2W_{11}O_{40})]$ was subsequently reduced in H_2 and carburized in a mixture of H_2/CH_4 to obtain WC-Co cermet powders, with particle sizes of 70–300 nm. After the powders were sintered, grain growth led to the final microstructures of 0.5 μm.

2.4 Ceramics

Chemical methods such as precipitation[2][52] and sol-gel processing[4][53]–[55] can be used to synthesize nanostructured ceramic powders. As-synthesized powders, depending on the synthesis technique used, may require subsequent heat treatment for dehydration, removal of organics, and controlled crystallization to form oxides with desirable structure and crystallite size.

Hydrolysis. Generally, precipitation from solution involves the formation of an insoluble hydroxide which can then be converted to its oxide by heat-assisted dehydration. Chemical techniques for production of monodispersed colloidal particles (with particle size larger than submicron) has been extensively investigated.[56] For example, metal (hydrous) oxide particles were synthesized by forced hydrolysis involving controlled deprotonation of hydrated cations. Coated colloidal particles can also be prepared using forced hydrolysis. By controlling hydrolysis of aluminum salts where solute complexes were formed instead of precipitation, particles such as hematite, chromia, and titania were coated with a layer of aluminum (hydrous) oxide.[57] The deposition of coating occurred by surface precipitation, which was controlled by available surface area of particle substrate and concentrations of complexes forming the coating. The thickness of coating was varied by mass parameters and reaction time of the process.

Hydrothermal. In hydrothermal techniques, the reaction mixture is heated above the boiling point of water in an autoclave or other closed system and the sample is exposed to steam at high pressures. The precipitation of zirconia under hydrothermal conditions yields powders that have a narrow size distribution and a controllable composition and morphology.[58] Hydrothermally treated $Zr(SO_4)_2$ at 250°C in the presence of either $MgSO_4$ or $(NH_4)_2SO_4$ led to the formation of acicular monoclinic zirconia. The zirconia particles have particle dimensions of 0.3 to 1.3 microns in length and 0.1 to 0.2 microns in width.

Sol-Gel Methods. Sol-gel processing is not a new technique. As early as the mid 1800s, it was reported that silicon tetrachloride, when left standing in an open container, hydrolyzed and turned into a gel.[59][60] After this time, biologists did much of the work with gels and colloids. In the early 1930s, aerogels were discovered.[61] Since the 1950s, sol-gel techniques have been used for phase equilibrium studies which opened up the field of ceramics.[62]–[65]

Sol-gel processes can be used to prepare a variety of materials, including: glass, powders, films, fibers, and monoliths. Traditionally, sol-gel process involves hydrolysis and condensation of metal alkoxides. Metal alkoxides have the general formula $M(OR)_x$ and an alkoxide ion is the conjugate base of an alcohol. The general synthesis of metal alkoxides involves the reaction of metal species (a metal, metal hydroxide, metal oxide, or metal halide) with an alcohol. Metal alkoxides are good precursors because they readily undergo hydrolysis; that is, the hydrolysis step replaces an alkoxide with a hydroxide group from water and a free alcohol is formed.

Eq. (1) $$M(OR)_x + H_2O \rightarrow M(OR)_{x-1}(OH) + ROH$$

Once hydrolysis has occurred the sol can react further and condensation (polymerization) occurs.

Eq. (2) $$2M(OR)_{x-1}(OH) \rightarrow (RO)_{x-1}M\text{-}O\text{-}M(OR)_{x-2}OH + ROH$$

Eq. (3) $$2M(OR)_{x-1}(OH) \rightarrow (RO)_{x-1}M\text{-}O\text{-}M(OR)_{x-1} + H_2O$$

It is these condensation reactions that lead to gel formation. In condensation two hydrolyzed fragments join together and either an alcohol or water is released (Eqs. 2 and 3). Condensation occurs by either nucleophilic substitution or nucleophilic addition.

Factors that need to be considered in a sol-gel process are solvent, temperature, precursors, catalysts, pH, additives, and mechanical agitation. These factors can influence the kinetics, growth reactions, hydrolysis, and condensation reactions.[55] The solvent influences the kinetics and conformation of the precursors, and the pH affects the hydrolysis and condensation reactions. Acidic conditions favor hydrolysis, which means that fully or nearly fully hydrolyzed species are formed before condensation begins. Under acidic conditions there is a low crosslink density which yields a denser final product when the gel collapses. Basic conditions favor condensation reactions, therefore, condensation begins before hydrolysis is complete. The pH also affects the isoelectric point and the stability of the sol. These, in turn, affect the aggregation and particle size. By varying the factors that influence the reaction rates of hydrolysis and condensation, the structure and properties of the gel can be tailored.

Sol-gel routes can be used to prepare pure, stoichiometric, dense, equiaxed, and monodispersed particles. For example, amorphous TiO_2

particles with average particle size of 0.07 to 0.3 μm were prepared by controlled hydrolysis of titanium tetraisopropoxide.[66] Monodispersed amorphous powders were synthesized using titanium tetraethoxide, where only homogeneous single nucleation occurred without flocculation. Depending on the reaction kinetics and subsequent post-synthesis aging of particles derived by sol-gel, nanoscale oxide particles can be amorphous or crystalline. These nanostructured oxide particles can be carburized or nitrided to form non-oxide powders at reduced temperatures and shorter reaction time due to the significant amount of highly active surfaces.

For example, nanostructured AlN powders for thermal management in electronics applications were synthesized by nitridation of oxide precursor powders.[67][68] The oxide powders were prepared by hydrolysis of aluminum tri-sec-butoxide at room temperature to favor precipitate formation instead of gel formation. The dry precursor (AlOOH) powders were calcined and subsequently nitrided in ammonia at temperatures up to 1100°C for 10 h. Since small particles favored the diffusion-controlled nitridation kinetics, nanostructured AlN powders were synthesized at temperatures of 400 to 600°C lower than that used in conventional carbothermal nitridation or direct nitridation of large grained oxide or aluminum powders. Nitride formation was favored when the oxide precursor powders were either amorphous or highly disordered. Commercially available nanocrystalline alumina powders could not be fully nitrided under the same nitridation conditions used for the sol-gel derived amorphous precursors. As-synthesized AlN powders were consolidated by pressureless sintering. Significant densification was observed for these nanostructured AlN powders when compared to the sintering of commercial coarse-grained AlN powder under the same conditions. This result is expected from nanosized powders of high-surface area (specific surface area is inversely proportional to the particle size). However, fully dense nanostructured AlN bulk solids could not be obtained by pressureless sintering due to adverse agglomeration of powders with a mixed pore size. The agglomerated powders contained pores with a random size distribution, and the largest of these pores hindered sintering to higher density.

It should be noted that this sol-gel process is particularly attractive for the synthesis of multicomponent particles with binary or ternary compositions using double alkoxides (two metals in one molecule), or mixed alkoxides (with mixed metaloxane bonds between two metals). Atomic homogeneity is not easily achieved by coprecipitating colloidal hydroxides from a mixture of salt solutions since it is difficult to construct double metaloxane bonds from metal salts.[54] Hybrid materials such as metal-oxide, organics-oxide can be prepared using the sol gel approach. For example, controlled nanoheterogeneity can be achieved in metal/ceramic nanocomposites.[69] A solution of metal salt, $(RO)_3Si(CH_2)_3A$ (A is functional organic group) and $Si(OR)_4$ was used in sol-gel processing. The ethylene diamine derivative such as $(RO)_3Si(CH_2)_3NHCH_2CH_2NH_2$(DIAMO) was used to form stable complexes with most transition metals. The complexation of metal ions and anchoring of the resulting metal complexes to the oxide matrix took place. The oxidation of the metal complex containing gels was then carried out at high temperatures and resulted in formation of composites of nanoscale metal oxide particles in an oxide matrix. Reduction of metal oxide particles in hydrogen provided the metal-ceramic nanocomposite powders such as Cu (< 5 nm) in silica. The metal particles, a few nanometers in size with a very narrow size distribution even for high metal loading, were statistically distributed in the oxide matrix without any agglomeration, as a result of anchoring the metal complexes to the oxide matrix. The narrow particle size distribution could not be achieved if the sol-gel processing was performed without complexation of metal ions.

Hybrid organic-inorganic materials can also be prepared by sol gel approach. These materials can be divided into three types:

1. Organics used to impregnate a porous oxide gel
2. Organics incorporated into the liquid sol-gel solution and subsequently trapped in the oxide matrix after gelation
3. Organics which react with the sol-gel precursors in solution and subsequently form chemical bond to the oxide after gelation[70]

Other nanostructured hybrid materials such as nanoconfined proteins and enzymes were prepared by encapsulation of biomolecules in sol-gel processing.[71] It was found that the dopant biomolecules served as templates for the formation of self-specific pores around them. The stable biomolecules were accessible to the external environment due to the porous nature of the oxide matrix, and biorecognition of the hybrid materials was feasible.

Multicomponent gels can be thermochemically converted to form nanocomposite nitride powders. For example, a precomposite gel, prepared by reacting iron chloride hexahydrate, urea, boric acid in water under strongly basic conditions at 150°C, was reacted in ammonia at 500°C to obtain Fe_xN/BN (x = 3 or 4) nanocomposite powder.[72] Similarly, AlN_xBN_{100-x} ($0 \le x \le 100$) nanocomposite powders were synthesized by pyrolysis of the precomposite gels (with aluminum chloride hexahydrate as Al source) in ammonia.[73] The compositional and thermal effects on these AlN-BN nanocomposite powders were studied.[74] It was found that BN was amorphous when its concentration was below 35 mol%. For higher BN concentrations, mixtures of amorphous and turbostratic-BN were observed. Annealing the powders led to a more ordered hexagonal BN phase. Independent of compositions, grain growth of AlN (both normal and abnormal) and BN occurred. Both crystallites retained their nanometer size up to 1600°C. The normal grain growth of AlN phase was impeded by the second phase of immobile inclusion particles of BN. When normal grain growth of AlN was inhibited by inclusions, only large AlN grains with boundary curvatures larger than average could move past the inclusions. As a result, larger grains experienced exaggerated grain growth. These results indicated that initial crystallization of AlN in the composite powders produced grains with a mixed size distribution.

Non-Hydrolytic Sol-Gel. In non-hydrolytic sol-gel methods, the traditional hydrolysis and condensation reactions are replaced by direct condensation reactions or transesterification reactions. The reaction may be direct condensation of metal halide and alkoxides[75][76] or condensation of a metal alkoxide and a metal carboxylate.[77] The transesterification reaction involves reacting a trialkylsilyl acetate and a metal alkoxide. Non-hydrolytic reactions do not involve water or polar solvents. In the traditional sol-gel methods, the condensation reaction may be reversible and special efforts are required to completely remove water or polar solvents.

2.5 Host-Derived Hybrid Materials

Porous or layered ceramics can be used as host materials in which nanoparticles are synthesized. The resulting hybrids may have novel properties due to the microstructure of the host. For example, minerals such as silicate and aluminosilicate can form unique structures. Continued linking of these structures forms a complex framework that contain mesoscopic pores and channels known as zeolites. Zeolites, in natural or synthetic forms, are aluminosilicates with a three dimensional structure (Fig. 4). The basic units of the zeolite framework are tetrahedra of SiO_4 and

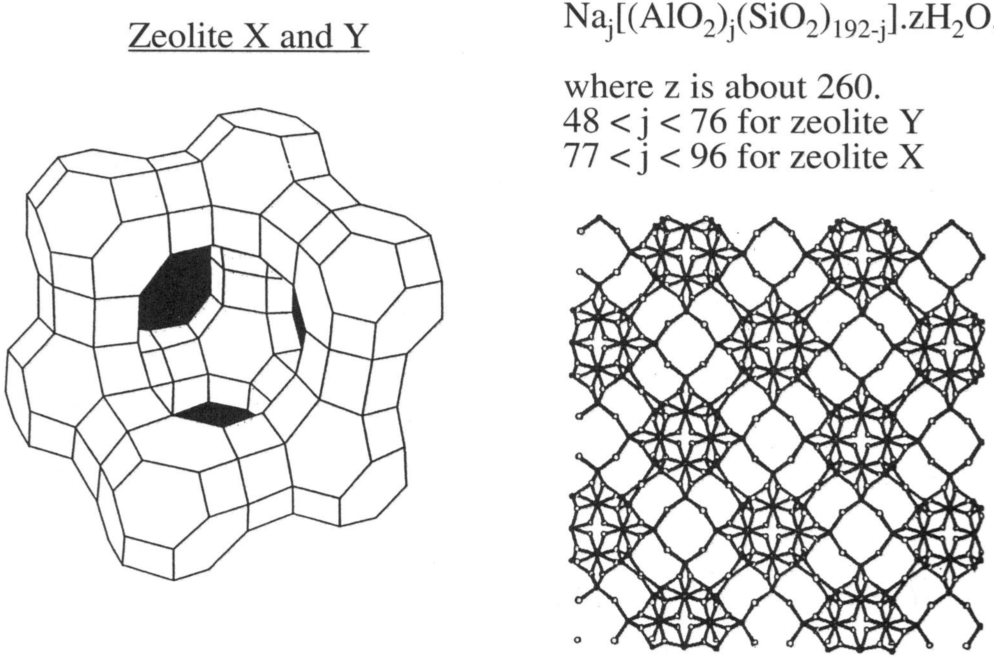

Figure 4. A general zeolite structure. *(Courtesy of Dr. Y. W. Lee, Defense Science Organization National Laboratories, Republic of Singapore.)*

AlO_4. These tetrahedra are linked through corner oxygen atoms and create the three dimensional structure. These pores and channels can be used as molecular sieves, ion exchange medium, and catalysts.

Transition metal clusters can be chemically deposited in supports of zeolites and other layered minerals for catalytic applications. For example, manganese oxide octahedral molecular sieves (OMS)[78][79] with MnO_6 octahedra as structural units were synthesized. The size of tunnel structures, ranging from 0.4-1 nm, was varied by using cations as templates. One such compound, based on the hollandite structure, consists of two sheets of MnO_6 octahedra joined at edges to form a 2×2 tunnel structure with Ba^{2+} as the countercation in the tunnel. OMS-1 is a 3×3 tunnel structure similar to todorokite with a 0.69 nm tunnel. OMS-2 is similar to cryptomelane, a 2×2 tunnel structure with K^+ as the countercation. These structures showed a unique catalytic activity. Besides tunnel structures, layered materials (OL) can also be formed by the MnO_6 sheets. The synthesis conditions[80] were adjusted to form different tunnel sizes. For example, thermally treating the Ba-OL material led to hollandite (2×2 tunnel) but hydrothermal treatment resulted in formation of a 2×3 tunnel structure.

By changing the tunnel size, it is possible to change the electronic, catalytic, and structural properties of the molecular sieves. Thermal properties and the average oxidation state of the Mn depended on synthetic conditions.[79] For example, oxygen evolution of hollandite prepared by refluxing occurred at 300°C , whereas, in the calcined sample this evolution took place at 900°C. Calcining resulted in a structure with a lower average Mn oxidation state (3.68), compared to that obtained by refluxing (3.80) or autoclaving (3.94). The lower oxidation state indicated there existed more Mn^{3+} or Mn^{2+}, which had a higher cation exchange ability. If the countercation was a divalent cation like copper or iron, ESR results show that there was a higher concentration of Mn^{2+}. The OMS and OL structures with divalent cations as the countercations had different catalytic properties. For CO oxidation, Cu-OMS-1 was more active than Cu-OL-1. This is due to the fact that molecular sieve had a larger surface area in the tunnel structure with available surface oxygen species. Besides the molecular sieve materials and the layered materials, MnO_6 also formed nanofibers.[81][82] The diameter of the fibers ranged from a few nanometers to 25 nm with lengths of tens of nanometers to a micron. The fibers had a cryptomelane type structure with the general formula KMn_8O_{16}. The fibers formed in a bird's nest morphology with improved porosity (Fig. 5).

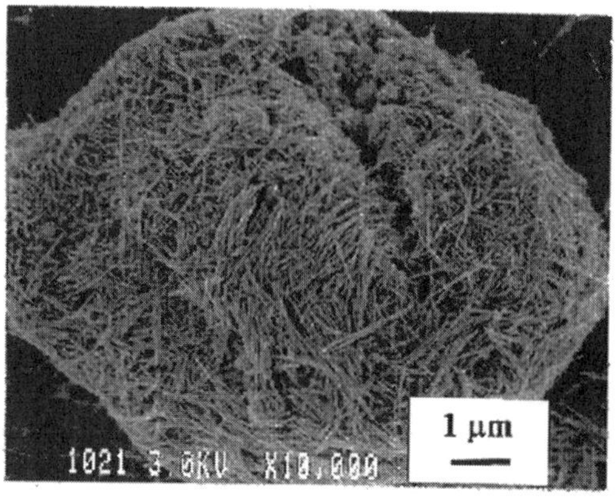

Figure 5. A SEM (scanning electron microscopy) micrograph of the hollandite nanophase MnO_2 "bird's nest" structure. *(Courtesy of Dr. T.D. Xiao, Inframat Corporation, USA.)*

Organic materials can also be used as templates to prepare host materials. Mesoporous transition metal oxides were synthesized using liquid crystal templating.[83] These hexagonal-packed transition-metal oxide molecular sieves (M-TMS-1) were based on the aluminosilcate family MCM-41. By using transition metals instead of AlO_4 and SiO_4, the catalytic, electronic and magnetic properties could be varied because of variable oxidation states of the metals. Two methods were used to prepare these materials. The first was a sol-gel technique where titanium alkoxide was modified with a acetylactonate to change the hydrolysis rates before templating with the surfactant to form the TMS. The second method involved formation of a discreet bond between the alkoxide and the organic templating material before hydrolysis. Using the second method, materials with ultrahigh surface areas were obtained. The phase of the material depended on the chain length of surfactant molecules. Longer chain length favored the formation of the cubic phase. When Nb was used as the transition metal, the resulting compound remained stable when the surfactant was removed.[84] In this case, niobium alkoxide reacted with the surfactant, and the Nb-N linkage controlled condensation reactions. The resulting Nb-TMS-1 was a hexagonally packed structural analog to MCM-41 with 0.27 nm pore size that retained its porosity and crystallinity to 400°C.

Soft chemistry (also known as *chemie douce*), can be used to modify an existing compound under relatively mild conditions to produce a closely related material. Chemie douce was first introduced in the

1970s.[85][86] Some of the chemical reactions that can be considered soft chemistry include: cation exchange, dehydroxylation, redox, and intercalation. Metastable states can be prepared under "soft" conditions. For example, a guest molecule can be inserted into the host without altering the host's structure by intercalation reactions (Fig. 6). Layered materials such as clays, graphite and MS_2 (M is transition metal) were used as hosts.[87] Layered ammonium zinc molybdate materials were prepared at room temperature.[88] The molybdate material was intercalated into a calcined layered double hydroxide (LDH). The structure of the LDH was based on edge sharing octahedra of Mg^{2+} forming sheets. The LDH in the carbonate form was thermally decomposed to a mixed oxide and then the anion was intercalated into the mixed oxide.

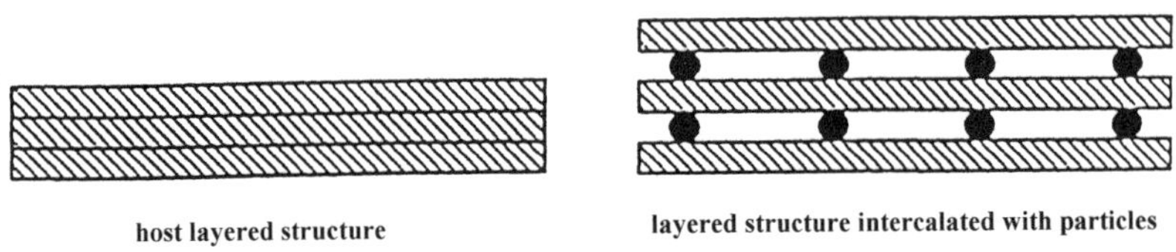

Figure 6. An example of an intercalated structure.

Another example, montmorillonite, a crystalline and expandable hydrated layered aluminosilicate, allows cation exchange reactions to take place in the interlayers due to the cationic substitution in the octahedral sheets. Catalytic active sites were chemically intercalated in the interlayers of montmorillonite.[89] Copper acetate hydroxide hydrate in excess of the ion-exchange concentration was incorporated in aluminosilicate interlayers using sodium hydroxide. The exchange of copper acetate hydroxide hydrate interlayers propped apart the silicate layers. Copper metal clusters of about 0.5 nm were then intercalated in the interlayers by in-situ reduction of copper acetate hydroxide in ethylene glycol at refluxing temperature of about 195°C for 2–6 h. The cluster size was deduced by the difference between the thickness of silicate layer (0.96 nm) and the basal spacing of the reduced sample (1.44 nm). Intercalation of metal clusters by hydrogen reduction was unsuccessful because of the collapse of the layers and expulsion of metal complexes from the interlayers to the external surface.

Nickel and silver were also intercalated into montmorillonite using the polyol method.[90] When nickel was intercalated, the (100) spacing of the clay was increased to 1.69 nm. After removing the glycol from the samples by washing, this (100) spacing decreased to 1.4 nm. Since nickel was detected both before and after removal of glycol from samples, the change in lattice spacing of clay was attributed to the intercalation of glycol. The TEM results showed 10 nm Ni particles, which were much larger than that calculated from interspacing separation. These particles increased to 30 nm with increasing Ni concentration, and many of them were found on the surface of the clay but not in the layers. When silver was used to intercalate the clay, 0.5 nm clusters were estimated from the interlayer separation, but 25–33 nm particles were seen in the TEM. The polyol approach was also used to reduce the metal ions encapsulated in a ceramic host to the metal particles.[30][32] A solution of potassium hexachloroplatinate (IV) was doped into an alumina gel formed by the Yolda's method.[91] The Pt doped gel was allowed to age and then heat treated at 700°C for 2 h to form a γ-alumina matrix. This γ-alumina/Pt precursor composite was suspended and reduced in ethylene glycol at 180°C for 2 h. The nanocomposite of 5 nm Pt particles in alumina was formed.

2.6 Stabilized Dispersions

In a solution with colloidal particles, solvent molecules, ions, or electric charges may help stabilize the particles against agglomeration, particularly for ceramic materials. In the case of metal colloids, these factors only provide limited protection. In practice, for making stable ceramics and metal colloids, dispersant are commonly used in the synthesis to provide more reliable stabilization. Dispersants are often surfactants. They can adsorb on the surface or form an envelope around the particle to provide either electrostatic or steric repulsion. For example, stable dispersions of 5–15 nm Fe colloidal particles were synthesized by thermolysis of iron pentacarbonyl in dilute solutions of polymers.[92] The functional polymer was catalytic for the decomposition of the carbonyl and induced particle nucleation in its domain. Monodispersed Pd-Cu particles of 3–5 nm were stabilized by poly(vinylpyrrolidone) (PVP) in refluxing the mixtures of metal acetate precursors and PVP in 2-ethoxyethanol.[93] Monodispersed 14 nm Ni powders were prepared by reducing nickel hydroxide in the polyol with PVP, where PVP was used to prevent particle sintering. When both PVP and nucleating agents such as Pd or Pt (introduced in the reaction as metal salts) were used, smaller particles of 30 nm were made by heterogeneous nucleation.[94] Gold colloids as giant clusters[95] with an average size of 20 nm (± 10%) were synthesized by reducing $AuCl_3$ with sodium citrate. Stabilization of gold colloids was achieved by using a water soluble derivative of triphenylphosphine (used in cluster synthesis). These negatively charged molecules formed an envelope around the particles and provided electrostatic repulsion.

If dispersants are not used during synthesis and as-synthesized particles are agglomerated, surfactants can be used in post-synthesis processing to disperse the agglomerated particles. Deagglomeration is accomplished by breaking the agglomerates by milling or ultrasonication in a suitable solvent and surface-active dispersant.[96] For example, agglomerated nanoscale AlN particles were dispersed using N-methylpyrrolidone (NMP) and stabilized in a polyimide matrix.[97] The proposed deagglomeration mechanism[98] involved initial breaking of large agglomerates into smaller ones and subsequent stabilization of these smaller agglomerates by interactions with the NMP molecules using ultrasonication and high speed mechanical agitation. Poly(amic acid), a polyelectrolyte and a precursor to polyimide, was then added and adsorbed onto the surface of AlN. A part of the polymer chain eventually diffused into the pores of the smaller agglomerates and further deagglomeration occurred by exfoliation. It was suggested that mechanisms such as depletion stabilization, steric stabilization, and electrosteric stabilization were involved. The polyimide-AlN nanocomposites showed a decrease of thermal expansion and increase of hardness and Young's modulus, with an increase of the AlN concentration in the composite.

2.7 Surfactant Membrane Mediated Synthesis

Self-assembled surfactant structures can be used as a reactor to synthesize nanoparticles.[99] In an appropriate solution, for example, an aqueous medium, surfactant molecules orient themselves so that contact of the non-polar tails of the molecules with the solvent is minimized. The polar headgroups of the molecules are attracted to water by electrostatic and hydrogen bond interactions. These interactions allow the molecules to self-assemble into membrane structures with minimum energy configuration. Self-assembly can be used to form monolayer films, Langmuir-Blodgett films, micelles, reversed micelles, vesicles, and tubules.

For example, glycerol monooleate, a lipid with a single fatty acid tail and a glycerol headgroup, forms bicontinuous cubic phases with size-controlled aqueous channels that are connected in a three dimensionally periodic network. Size-controlled 4 nm Pd nanoparticles were synthesized by a polyol type reaction of a solution of tetrachloropalladate and glycerol monooleate.[100] The hydroxyl groups in the glycerol headgroup reduced the metal salt to metal at room temperature. The narrow aqueous channels of the bicontinuous phase constrained the particle size of Pd.

Membrane structures such as reverse micelles and vesicles can be used as nanoreactors to synthesize metallic and ceramic nanoparticles. Precipitation can be carried out inside the hollow compartments of membrane structures. Compared to bulk precipitation, better chemical homogeneity is achieved since the reactions occur in a better controlled environment. The size of the membrane structures can be manipulated to control the particle size and size distribution. The particles are encapsulated inside the membrane and are prevented from agglomeration with other particles by the membrane which acts as a barrier. Since it is possible to incorporate functional groups on vesicle surfaces, nanoparticles may be carried inside functionalized membranes for targeted applications.

Reverse micelles (sometimes known as water-in-oil microemulsions) are a single layer of surfactant molecules entrapping solubilized water pools in a hydrocarbon solvent. The size of the water pool depends on the amount of entrapped water at a given surfactant concentration. Nanoscale metal colloids such as Cu,[101] Au,[102] and semiconductors such as CdSe[103] were made using the reverse micelle method. Ceramic particles can be prepared by sol-gel reactions in microemulsions.[104][105] By using w/o techniques with the Stober synthesis of silica,[106] monodispersed spherical silica particles were synthesized. It was found that the hydrolysis rate was the rate-limiting step when a microemulsion was employed. By changing the ratio of concentration of water to surfactant and thus the size of water droplet nanoreactor, the particle size of silica was varied. Ternary and quaternary nanostructured mixed oxides such as $BaTiO_3$, $BaZrO_3$, $SrTiO_3$, and $Ba_xSr_{1-x}TiO_3$ were also synthesized in microemulsions.[107]

Vesicles are closed bilayer membrane assemblies. Multilamellar vesicles have diameters in the range of 100–800 nm. Single bilayer vesicles are 30–60 nm in diameter. Sonication or extrusion can be used to control the vesicle size. Extrusion, however, is limited by preparation of only a small quantity of vesicles. Vesicles are more stable than micelles. Nanoscale oxide particles can be synthesized using the vesicle-mediated approach. Metal ions and lipid mixtures form vesicles upon sonication. After removal of exogenous ions, anions are added and allowed to diffuse through the membrane layers and intravesicular precipitation occurs. Due to preferential anion diffusion across the membrane, generally only oxides such as silver oxide,[108] iron oxide,[109] and aluminum oxide[110] have been prepared.

Dispersed nanocrystalline metal particles can be prepared using polymerized phospholipid vesicles.[111][112] The non-cross-linked polymerization of vesicles resulted in many individual polymer chains in the membrane structure. This enhanced the structural integrity of vesicles and provided breaks in the polymer network through which both anions and cations could diffuse across the polymerized membranes (thus removing the restriction to only oxide synthesis as in the case of non-polymerized vesicles). In this approach, vesicles made from mixed phospholipids (negatively charged and zwitterionic) were UV polymerized. Positively charged Pd ions were selectively and chemically attached to the negatively charged headgroup molecule on the surface of the interior compartment of membrane. After adding Au salt and a reducing agent to the solution, Au ions diffused across the membrane. The interior membrane-bound Pd species acted as catalysts for initiation of autocatalytic electroless metallization of Au, which led to formation of nanocrystalline Au particles inside the polymerized vesicles.

When unpolymerized vesicles were used to make nanoscale Au particles using this electroless method,[113] Pd catalysts were bound both to the internal and external membrane surfaces and electroless metallization occurred on external surfaces of vesicles. The osmotic pressure (built up inside unpolymerized vesicles due to the difference of salt concentrations across the membranes) and the external electroless reaction weakened the structural integrity of these vesicles, which eventually ruptured. In this case, the lipid molecules served as a dispersant during the synthesis of Au particles. The dispersion property of the nanocrystalline Au particles depended on the thermal disordering of phospholipid mixture, which became more effective as a dispersant with increasing temperature. Many multiply-twinned particles (MTP) were formed in reactions carried out at 25°C and 40°C for 3 h (Fig. 7). The formation of MTP by rapid quenching in vapor deposition is well known. Vapor deposition is generally a "far-from-equilibrium" process. The condensation of supersaturated vapor into a solid has a maximum departure from equilibrium (ΔG) of approximately 160 kJ/N_a, where N_a is Avogardro's number. The precipitation of solid from a supersaturated liquid has a ΔG of 8 kJ/N_a.[114] Although the deposition kinetics in solution chemistry differs significantly from that of vapor phase, MTPs of Au can be synthesized in both cases.

Semiconductor crystallites of 2–10 nm have bulk-like internal lattices but their optical spectra show resolved discrete features. In bulk species, the electronic energy is continuous but, as the particle size gets below 6 nm, the electronic levels are discrete due to quantum confinement of the carriers. This leads to the third order non-linear optical effects. Agglomeration of the particles must be avoided in preparation of these materials by dispersing the particles in polymers,[115] inside a multi-layer,[116] pores,[117] and vesicles.[118] For example, cubic nanocrystals of CdS were prepared in phosphatidylcholine (PC) vesicles.[118] The vesicle's internal diameter controlled the particle size of CdS. The absorption spectrum of CdS using PC had exciton energy shifts to shorter wavelengths.

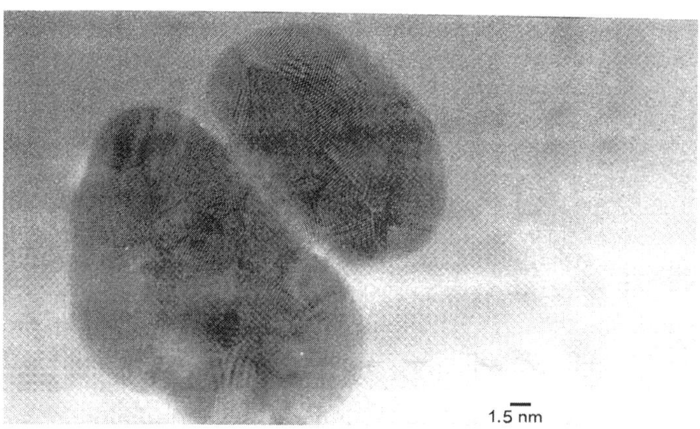

Figure 7. A HRTEM (high resolution transmission electron microscopy) micrograph of gold colloids synthesized using unpolymerized vesicles.

3.0 FILMS AND COATINGS

3.1 Metals

Nanostructured metal coatings can be deposited by aqueous and nonaqueous methods. Both electrolytic and electroless plating have been used.

Aqueous Electroplating. In electrodeposition, the deposition of a pure metal or alloy from a electrolyte solution occurs on the cathode when an external current is applied to the plating system. An electrically conductive substrate is required. Nanostructured metal coatings of pure metals, alloys and composites were deposited using electrodeposition.[119]–[122] It can produce porosity-free coatings which do not require subsequent densification. Nanostructured grains were deposited when the plating variables such as bath composition, pH, temperature, and current density were controlled so that nucleation of new grains was favored over grain growth.

Aqueous Electroless Deposition. In the electroless approach,[123] electrons are generated by chemical reactions without the supply of external current as in the case of electroplating. Unlike electrodeposition, electrical conductors are not required as substrates. Electroless deposition can occur by the following mechanisms:

1. Deposition by ion or charge exchange
2. Deposition by contacting the metal to be coated
3. Autocatalytic deposition on catalytically active surfaces from solutions containing reducing agents

In an autocatalytic electroless process, a non-catalytic surface on which electroless metal is to be deposited is initially coated with catalyst particles. The catalyst particles are typically colloids such as Pd (with diameter $\geq$ 2 nm) encapsulated in a tin oxide shell. The catalyst is first adsorbed on the substrate surface, followed by chemical stripping of the protective surface oxide to expose the catalytic Pd core. Reduction of a soluble metal ion or complex by a soluble reducing agent present in the plating solution leads to deposition of metal atoms at the surface. Metal ions are reduced by electrons provided by reducing agents. Subsequently, each deposited metal layer becomes the catalyst for deposition of the next layer, hence the name *autocatalytic metallization*.

For example, electroless metallization was used to deposit nanostructured Ni-P or Ni-B coatings (phosphorous or boron was in the composition of the plating solution).[123] Depending on factors such as post-deposition heat treatment and compositions, amorphous or crystalline structure and a range of grain sizes (2 to 100 nm) could be obtained. The control of the size of the bound catalysts is the principle determining factor in controlling the particle size of nanostructured electroless deposit.[124] Chemical modification of the substrate surface to allow for binding smaller catalysts led to a three- to fourfold reduction in the particle size of the electroless deposit.

Using the electroless approach, nanostructured metal coatings were also deposited on self-assembled biomolecular structures. Rhapidosomes, protein tubules which are about 17 nm in diameter and 400 nm in length, were electroless metallized using molecular catalysts instead of conventional colloidal catalysts. The tubule surfaces were initially catalyzed by treating with a $(PdCl_4)^{2-}$ or $Pd(Ac)_2$ solution. Several amino acids residues such as cysteine, histidine, and tyrosine could reduce the catalysts molecules to Pd crystallites (3 nm) on the rhapidosome surface. Randomly oriented nanocrystals of Ni (about 10 nm) were deposited on the surfaces by this electroless plating method.[125] Self-assembled phospholipid hollow tubules, with an average diameter of 0.5 μm and 50 to 80 μm long, are interesting materials due to the large shape anisotropy. The tubules were metallized by electroless plating with nanoscale Ni (Fig. 8). The Ni coated tubules were magnetically aligned and cast into a polymer matrix to form a composite. The magnetic anisotropy of electroless metallized Ni or permalloy tubules (with metal grain size about 2–4 nm, depending on metallization conditions) and the alignment of metallized tubules in the magnetic tubule-polymer composite were investigated.[126] The metallized tubules were used to fabricate an ungated vacuum field emission cathode structures for generation of a macroscopic electron beam current. The local electric field enhancement was achieved by exploiting the large aspect ratio of the metallized tubules, the radius of curvature, and the thickness of metal coatings at the edge of the metallized tubules.[127][128] Selective removal of this polymer matrix resulted in a composite base template of oriented exposed metallized tubules, and subsequent surface electrical contact was achieved by a thin sputtered gold film. The resultant microstructures demonstrated vacuum field emission of current $I > 10$ μA at relatively low applied macroscopic electric fields (~60–150 kV/cm).

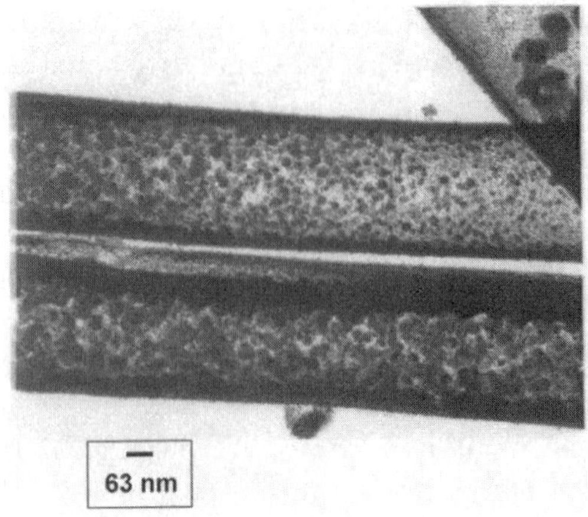

Figure 8. A TEM micrograph of Ni metallized tubules.

Nonaqueous Electroless Deposition. Aqueous electrodeposition and electroless deposition are not attractive approaches to metallization of substrates which can be detrimentally hydrolyzed or oxidized in aqueous solution. A nonaqueous coating process capable of producing fine-grained deposits has been recently developed using the polyol method. For example, Cu coatings were deposited on AlN substrates,[129] and Co was deposited on WC substrates.[130] Unlike traditional aqueous electroless metallization, this process does not require the adsorption of catalysts on electrically insulative substrates to initiate metallization. A surface study of grazing incidence asymmetric Bragg (GIAB) scattering and small angle x-ray scattering at glancing incidence revealed that the surface of Cu metal film consisted of 4 nm particles.[129]

3.2 Ceramics

Nanostructured ceramic oxide films and coatings can be deposited using sol-gel type methods.[4][131][132] Sol-gel processing is especially adaptable for film formation. Films and coatings represent the earliest commercial use of sol-gel processing. Sol-gel techniques offer the following advantages in coatings: control of microstructure, pore size, and surface area. By controlling these parameters, the film properties can be tailored (Fig. 9). Thin films use very little raw materials and can be processed quickly, and very large and irregularly shaped surfaces can also be coated. Dense pinhole-free layers can be prepared at low temperatures using sol-gel processing. Porous films can be prepared by changing the reaction conditions. This approach is particularly useful in obtaining homogeneous multicomponent coatings. To avoid cracking caused by large capillary stress during evaporation of solvent in the drying process, either slow evaporation (slow process) or supercritical drying (fast process) is used. As-deposited oxide coatings are typically amorphous. Thermal and thermochemical post-synthesis treatment can be carried out to obtain nanostructured oxide, carbide, or nitride coatings. Hybrid coatings can be fabricated by doping the sol with material of a different phase, followed by gelation and densification. For example, nanocomposite thin films for optical applications such as Ag/$BaTiO_3$ and Ag/PZT were prepared by spin coating the precursor solutions. The precursor solution was doped with silver ions, stabilized by adding organic chelating agents. Silver particles of 1–20 nm were formed in the ceramic matrix after heat treatment. Red shift of the optical absorption band of the metal particles was observed.[133]

Sol-gel films can be deposited by spraying, dip coating, and spin coating. Viscosity of the sol can be increased by aging, which can affect the deposition. The amount of porosity in a sol-gel derived film can be controlled by the pH. A higher porosity is achieved when base-catalyzed sols are used. For example, nanoporous thin films of silica were prepared by dip-coating a silica sol onto a substrate and drying at or near room temperature.[134] The film thickness followed predicted theory as a function of the dip speed. Other factors that influenced the film thickness were aging, R (H_2O/precursor), and pH. By drying the films near room temperature, the porosity of the gel was preserved. As the ratio of water to TEOS was increased under acidic conditions, there was an increase in film thickness. This was attributed to fast hydrolysis under acidic conditions. As the ratio, R, was varied between 2 and 6 the film thickness increased from 300 to 550 nm. Caution must be taken to not increase the R value too high, which will dilute the solution. At the lower R value, continued hydrolysis occurred in the film due to atmospheric moisture. Langmuir-Blodgett techniques can be used to prepare either closed packed or nanoporous TiO_2 thin films of a known thickness and porosity. Using successive compression and expansion cycles at different pressures, monolayers that had different average particles per area could be prepared. The porosity was controlled by controlling the average spacing between the crystallites in the monolayer.[135]

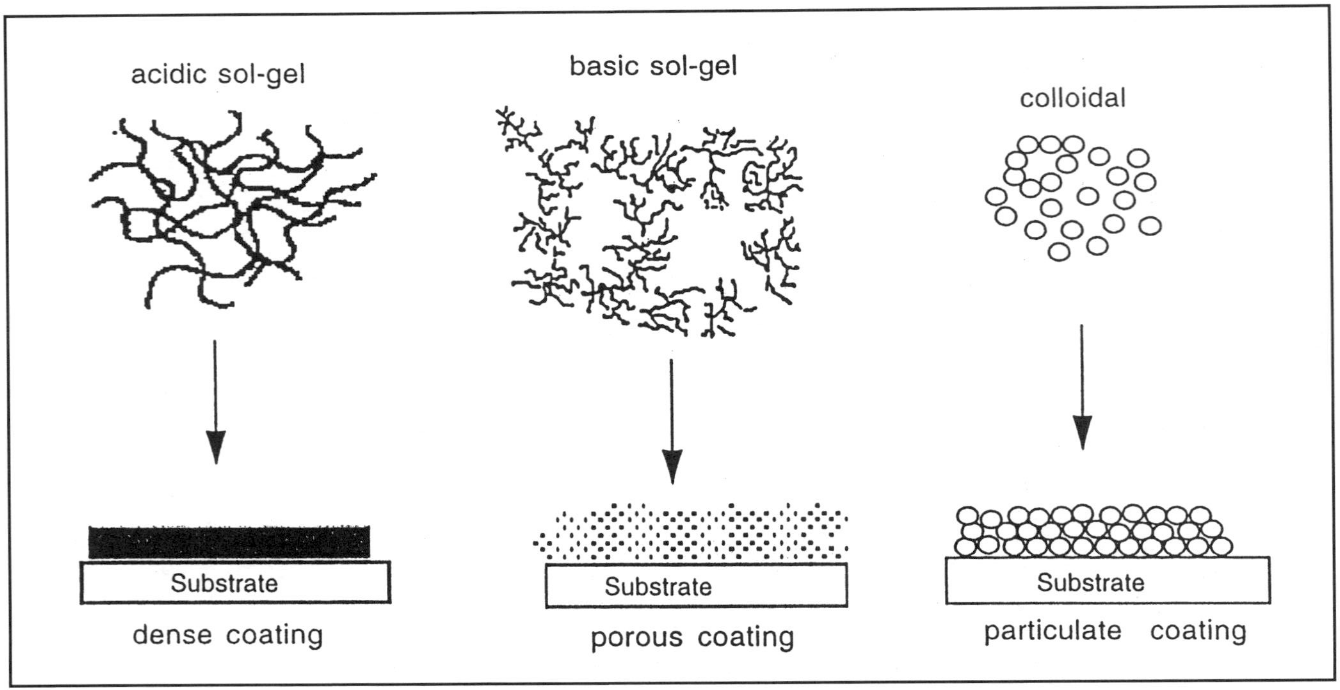

Figure 9. An example of sol-gel processing conditions on film formation.

Because the large capillary stress during solvent evaporation can cause cracking, the sol-gel process has been traditionally used for preparing thin films and coatings. To prepare thick coatings, the problems of shrinkage and cracking and the limitation of coating thickness can be mitigated by increasing particle loading in the sol-gel process.[136] This approach involved dispersing large ceramic powders in sol-gel solution, and applying the mixture onto the substrate by various techniques such as dipping and spraying. Coatings with thicknesses up to 200 μm were fabricated. The sol-gel film formed strong bonds to both oxide powders and substrates by interaction with functionalized surface hydroxyl groups on the oxide powders and the oxide layer of substrates. The strong bond reduced cracking. The shrinkage problem associated with the conventional sol-gel approach was minimized due to the high loading of ceramic powders. A two dimensional sol-gel process was also used to fabricate thick films of titania.[137] In this process, the traditional sol-gel hydrolysis and condensation reactions took place at an air-water interface. The gel films formed could then be deposited onto substrates using Langmuir Blodgett techniques.

4.0 SUMMARY

Nanostructured particles, films, and coatings can be synthesized using solution chemistry. The ability to manipulate atoms and molecules in the liquid phase provides a powerful arsenal for synthesis of tailor-designed nanomaterials using a bottom-up approach. Indeed there has been a great deal of progress in chemical synthesis and processing, and active multidisciplinary efforts are continually pursued. More work needs to be done in the area of processing of nanoparticles and coatings. Fundamental understanding of interfacial interactions of these high-surface materials, particularly the interfacial stability of the hybrid materials, is essential in order to design and control the properties of the materials.

REFERENCES

1. Chow, G. M., *Chemical Synthesis and Processing of Nanostructured Powders and Films*, NATO Advanced Study Institute on Nanostructured Materials Science and Technology, St. Petersburg, Russia (Aug. 1997), NATO ASI Series 3. High Technology-Vol. 50, Kluwer Publications, Netherlands (1998)
2. Chow, G. M., and Gonsalves, K. E., *Particle Synthesis by Chemical Routes*, (A. S. Edelstein, and R. C. Cammarata, eds.), *Nanomaterials: Synthesis, Properties, and Applications*, Institute of Physics Publishing, Bristol and Philadelphia, pp. 55–71 (1996)
3. Herron, N., and Wang, Y., *Synthesis of Semiconductor Nanoclusters*, (A. S. Edelstein, and R. C. Cammarata, eds.), *Nanomaterials: Synthesis, Properties, and Applications*, Institute of Physics Publishing, Bristol and Philadelphia, pp. 73–88 (1996)
4. Klein, L. C., *Processing of Nanostructured Sol-Gel Materials*, (A. S. Edelstein, and R. C. Cammarata, eds.), *Nanomaterials: Synthesis, Properties, and Applications*, Institute of Physics Publishing, Bristol and Philadelphia, pp. 147–164 (1996)
5. Rolison, D. R., *Chemical Properties*, (A. S. Edelstein, and R. C. Cammarata, eds.), *Nanomaterials: Synthesis, Properties, and Applications*, Institute of Physics Publishing, Bristol and Philadelphia, pp. 305–321 (1996)
6. Chow, G. M., and Gonsalves, K. E., (eds.), *Nanotechnology: Molecularly Designed Materials*, American Chemical Society Symposium Series 622, Washington, DC (1996)
7. Special issue: Nanostructured Materials, *Chemistry of Materials,* 8, Washington, DC (1996)
8. Gonsalves, K. E., Chow, G. M., Xiao, T. D., and Cammarata, R. C., (eds.), *Molecularly Designed Ultrafine/Nanostructured Materials*, Materials Research Society Symposium Proceedings 351, Pittsburgh, PA (1994)
9. Ellis, A. B., Geselbracht, M. J., Johnson, B. J., Lisensky, G. C., and Robinson, W. R., *Teaching General Chemistry: A Materials Science Companion*, American Chemical Society, Washington, DC (1993)
10. For example, see *J. Mater. Res., Material Science and Engineering A, Nanostructured Materials, Chemistry of Materials, J. Am. Cer. Soc., J. Am. Chem. Soc., Langmuir, J. Colloids and Interface Sci., J. Electrochem. Soc., Appl. Phys. Lett., Science, Nature*, etc.
11. For example, see relevant *Materials Research Society Symposium Proceedings, Proceedings of the American Society for Metals, Minerals, and Materials*, etc.

12. Nielsen, A. E., *Kinetics of Precipitation*, Pergamon Press, London, NY (1964)
13. Walton, A. G., *The Formation and Properties of Precipitates*, Robert Krieger Publishing Company, Huntington, NY (reprint edition) (1979)
14. Lagally, M. G., An Atomic-Level View of Kinetic and Thermodynamic Influences in The Growth of Thin Films: A Review, *Jpn. J. Appl. Phys.*, 32:1493–1501 (1993)
15. Turnbull, D., The Kinetics of Precipitation of Barium Sulfate From Aqueous Solutions, *Acta Metall.*, 1:684–691 (1953)
16. LaMer, V. K., and Dinegar , R. H., Theory, Production, and Mechanism of Formation of Monodispersed Hydrosols, *J. Am. Chem. Soc.*, 72:4847–4854 (1950)
17. For example, see Shanefield, D. J., *Organic Additives and Ceramic Processing, with Applications in Powder Metallurgy, Ink, and Paint,* Kluwer Academic Publishers, Boston, Dordrecht, London (1995)
18. Derjaguin, B. V., and Landau, J. D., *Acta Physicochim.*, URSS 14:633 (1941); Verwey, E. J. W., and Overbeek, J. T. G., *Theory of The Stability of Lyophobic Colloids,* Elsevier, Amsterdam (1948)
19. For example, see Shanefield, D. J., *Organic Additives and Ceramic Processing,* Kluwer Academic Publishers, Boston, London (1995); *Metals Handbook Ninth Edition*, Vol. 7, Amer. Soc. for Met, Metals Park, Ohio (1984)
20. Yang, K. C., and Rowan, B. D., *Production of Gold, Platinum, and Palladium Powders*, in: *Metals Handbook Ninth Edition*, Amer. Soc. for Met., 7:148–151 Metals Park, OH, (1984)
21. van Wonterghem, J., Morup, S., Koch, C. J. W., Charles, S. W., and Wells, S., Formation of Ultra-Fine Amorphous Alloy Particles by Reduction in Aqueous Solution, *Nature*, 322:622–623 (1986)
22. Glavee, G. N., Klabunde, K. J., Sorensen, C. M., and Hadjipanayis, G. C., Borohydride Reduction of Cobalt Ions in Water. Chemistry Leading to Nanoscale Metal, Boride, or Borate Particles, *Langmuir*, 9:162–169 (1993)
23. Glavee, G. N., Klabunde, K. J., Sorensen, C. M., and Hadjipanayis, G. C., Sodium Borohydride Reduction of Cobalt Ions in Nonaqueous Media: Formation of Ultrafine Particle (Nanoscale) of Cobalt Metal, *Inorganic Chemistry*, 32:474-477 (1993)
24. Glavee, G. N., Klabunde, K. J., Sorenson, C. M., and Hadjipanayis, G. C., Chemistry of Borohydride Reduction of Iron (II) and Iron (III) Ions in Aqueous and Nonaqueous Media, Formation on Nanoscale Fe, FeB, Fe_2B Powders, *Inorganic Chemistry*, 34:28–35 (1995)

25. Chow, G. M., Ambrose, T., Xiao, J. Q., Twigg, M. E., Baral, S., Ervin, A. M., Qadri, S. B., and Feng, C. R., Chemical Precipitation and Properties of Nanocrystalline Fe-Cu Alloy and Composite Powders, *Nanostructured Materials*, 1:361–368 (1992)

26. Chow, G. M., Ambrose, T., Xiao, J., Kaatz, F., and Ervin, A., Nanostructured Co-Cu Powders Via a Chemical Route, *Nanostructured Materials*, 2:131–138 (1993)

27. Fievet, F., Lagier, J. P., and Figlarz, M., *Preparing Monodisperse Metal Powders in Micrometer and Submicrometer Size by The Polyol Process*, Materials Research Society Bulletin, 14:29–34 (1989)

28. Silvert, P. Y., and Tekaia-Elhsissen, K., Synthesis of Monodispersed Submicronic Gold Particles by The Polyol Process, *Solid State Ionics*, 82:53–60 (1995)

29. Kurihara, L. K., Chow, G. M., and Schoen, P. E., Nanocrystalline Metallic Powders and Films Produced by The Polyol Method, *Nanostructured Materials*, 5:607–613 (1995)

30. Chow, G. M., Schoen, P. E., and Kurihara, L. K., *Nanostructured Metallic Powders and Films Via an Alcoholic Solvent Process*, Navy Case No. 76,572, US Patent No. 5,759,230 (June 2, 1998)

31. Chow, G. M., Kurihara, L. K., Kemner, K. M., Schoen, P. E., Elam, W. T., Ervin, A., Keller, S., Zhang, Y. D., Budnick, J., and Ambrose, T., Structural, Morphological, and Magnetic Study of Nanocrystalline Cobalt-Copper Powders Synthesized by The Polyol Process, *J. Mater. Res.*, 10:1546–1554 (1995)

32. Chow, G. M., Kurihara, L. K., and Schoen, P. E., *Synthesis of Nanostructured Composite Particles Using A Polyol Process*, Navy Case No. 77,467, US Patent application pending (1996)

33. Kurihara, L. K., Chow, G. M., Lawrence, S. H., and Schoen, P. E., Polyol-derived Synthesis of Nanocrystalline Metals, *Symposium Proceedings of Processing and Properties of Nanocrystalline Materials,* p. 49, The Minerals, Metals, and Materials Society, Warrendale, PA (1996)

34. Hedge, M. S., Larcher, D., Dupont, L., Beaudoin, B., Tekaia-Elhsissen, K., and Tarascon, J. M., Synthesis and Chemical Reactivity of Polyol Prepared Monodispersed Nickel Powders, *Solid State Ionics*, 93:33–50 (1997)

35. Viau, G., Fievet-Vincent, F., Fievet, F., Toneguzzo, P., Ravel, F., and Acher, O., Size Dependence of Microwave Permeability of Spherical Ferromagnetic Particles, *J. Appl. Phys.*, 81:2749–2754 (1997)

36. Toneguzzo, Ph., Acher, O., Viau, G., Fievet-Vincent, F., and Fievet, F., Observations of Exchange Resonance Modes on Submicrometer Sized Ferromagnetic Particles, *J. Appl. Phys.*, 81:5546–5548 (1997)

37. Toshima, N., and Lu, P., Synthesis and Catalysis of Colloidal Dispersions of Pd/Ni Bimetallic Clusters, *Chemistry Letters,* pp. 729–730 (1996)

38. Toshima, N., and Wang, Y., Preparation and Catalysis of Novel Colloidal Dispersions of Cu/noble Metal Bimetallic Clusters, *Langmuir*, 10:4574–4580 (1994)

39. Toshima, N., and Wang, Y., Novel Preparation, Characterization, and Catalytic Properties of Polymer Protected Cu/Pd Bimetallic Colloids, *Chemistry Letters,* pp. 1611–1614 (1993)

40. Suslick, K. S., Sonochemistry, *Science*, 247:1439–1445 (1990)

41. Suslick, K. S., Hyeon, T., Fang, M., and Cichowlas, A. A., Sonochemical Synthesis of Nanostructured Catalysts, *Materials Science and Engineering*, A204:186–204 (1995)

42. Trentler, T. J., Suryanarayanan, R., Sastry, S. M. L., and Buhro, W. E., Sonochemical Synthesis of Nanocrystalline Molybdenum Disilicide, *Materials Science and Engineering*, A204:193–196 (1995)

43. Chen, J. P., Sorensen, C. M., Klabunde, K. J., and Hadjipanayis, G. C., Enhanced Magnetization of Nanoscale Colloidal Co Particles, *Phys. Rev. B*, 51:11529–11532 (1995)

44. Osuna, J., de Cara, D., Amiens, C., Chaudret, B., Snoeck, E., Respaud, M., Broto, J. M., and Fert, A., Synthesis and Characterization, and Magnetic Properties of Cobalt Nanoparticles from an Organometallic Precursor, *J. Phys. Chem.*, 180:14571–14574 (1996)

45. Rodriquez, A., Amiens, C., Chaudert, B., Casanove, M. J., Lecante, P., and Bradley, J. S., Synthesis and Isolation of Cubooctohedral and Icosohedral Platinum Nanoparticles, Ligand-Dependant Structures, *Chemistry of Materials*, 8:1978–1986 (1996)

46. Gonsalves, K. E., Xiao, T. D., Chow, G. M., and Law, C. C., Synthesis and Processing of Nanostructured M50 Type Steel, *Nanostructured Materials*, 4:139–147 (1994)

47. Feng, C. R., Chow, G. M., Rangarajan, S. P., Chen, X, Gonsalves, K. E., and Law, C., TEM and HRTEM Characterization of Nanostructured M50 Type Steel, *Nanostructured Materials*, 8:45–54 (1997)

48. Gonsalves, K. E, Rangarajan, S. P., Law, C. C., Garcia-Ruiz, A., and Chow, G. M., "M50 Nanostructured Steel—The Chemical Synthesis and Characterization of Powders and Compacts, *Nanostructured Materials*, 9:169–172 (1997)

49. Xiao, T. D., Torban, S., Strutt, P. R., and Kear, B. H., Synthesis of Nanostructured Ni/Cr and Ni-Cr_3C_2 Powders by an Organic Solution Reaction Method, *Nanostructured Materials*, 7:857–871 (1996)

50. McCandlish, L. E., Kear, B. H., and Kim, B. K., Chemical Processing of Nanophase WC-Co Composite Powders, *Materials Science and Technology*, 6:953–957 (1990)

51. Wahlberg, S., Grenthe, I., and Muhammed, M., Nanostructured Hard Material Composites by Molecular Engineering: 1. Synthesis Form Soluble Tungstate Salts, *Nanostructured Materials*, 9:105–108 (1997)

52. Gallagher, P. K., *Chemical Synthesis,* in: *Engineered Materials Handbook,* Vol. 4, *Ceramics and Glasses,* ASM International, USA, pp. 52–64 (1991)

53. Shoup, R. D., Sol-Gel Processes, in: *Engineered Materials Handbook,* Vol. 4, *Ceramics and Glasses,* ASM International, USA, pp. 445–452 (1991)

54. Pierre, A.C., Sol-Gel Processing of Ceramic Powders, *Ceramic Bulletin,* 70:1281–1288 (1991)

55. Brinker, C. J., and Sherrer, G. W., *Sol-Gel Science*, Academic Press, NY (1990)

56. Matijevic, E., Production of Monodispersed Colloidal Particles, *Ann. Rev. Mater. Sci.*, 15:483–516 (1985)

57. Kratohvil, S., and Matjevic, E., Preparation and Properties of Coated, Uniform, Inorganic Colloidal Particles: 1. Aluminum (Hydrous) Oxide on Hematite, Chromia, and Titania, *Advanced Ceramic Materials*, 2:798–803 (1987)

58. Li, C., Yamai, I., Murase, Y., and Kato, E., Formation of Acicular Monoclinic Zirconia Particles Under Hydrothermal Conditions, *J. Am. Ceram. Soc.*, 72:1479–1482 (1989)

59. Ebelmen, J., *J. Ann. Chim. Phys. Series* 3, 57, and 311 (1846)

60. Graham, T., On the Properties of Silicic Acid and Other Analogous Colloidal Substances, *J. Chem. Soc. London*, 17:318–327 (1864)

61. Kistler, S. S., Coherent Expanded Aerogels and Jellies, *Nature*, 127:741(1931)

62. Roy, R., and Osborn, E. F., The System Al_2O_3-SiO_2-H_2O, *American Mineralogist*, 39:853–885 (1954)

63. DeVries, R. C., Roy, R., and Osborn, E. F., Phase Equilibria in The System CaO-TiO_2 -SiO_2., *J. Am. Ceram. Soc.*, 38:158–171 (1955)

64. Roy, R., Aids in Hydrothermal Experimentation: II, Methods of Making Mixtures for Both "Dry" and "Wet" Phase Equilibrium Studies, *J. Am. Ceram. Soc.*, 39:145–146 (1956)

65. Roy, R., Gel Route to Homogeneous Glass Preparation, *J. Am. Ceram. Soc.*, 52:344 (1969)

66. Barringer, E. A., and Bowen, H. K., *Formation, Packing, and Sintering of Monodisperse TiO_2 Powders*, Communications of the American Ceramic Society, C-199–201 (1982)

67. Kurihara, L. K., Chow, G. M., Choi, L. S., and Schoen, P. E., *Chemical Synthesis and Processing of Nanostructured Aluminum Nitride*, (T. P. Battle and H. Henein, eds.), *Processing and Handling of Powders and Dusts*, The Minerals, Metals, and Materials Society, pp. 3–12, Warrendale, PA (1997)

68. Kurihara, L. K., Chow, G. M., and Schoen, P. E., Nanostructured Ceramic Nitride Powders and A Method of Making The Same, Navy Case No. 77,219, US Patent application pending (1997)

69. Schubert, U., Schwertfeger, F., and Gorsmann, C., *Sol-gel Materials with Controlled Nanoheterogeneity*, in: *Nanotechnology: Molecularly Designed Materials*, (G. M. Chow, and K. E. Gonsalves, eds.), Ch. 25, pp. 366–381, American Chemical Society Symposium Series 622, Washington, DC (1996)

70. Mackenzie, J. D., *Hybrid Organic-Inorganic Materials*, in: *Hybrid Organic-Inorganic Composites*, (J. E. Mark, C. Y. C. Lee, and P. A. Bianconi, eds.), Ch. 17, pp. 226–236, American Chemical Society Symposium Series 585, Washington, DC (1995)

71. Dave, B. C., Dunn, B., Valentine, J. S., and Zink, J.I., Nanoconfned Proteins and Enzymes: Sol-Gel Based Biomolecular Materials, in: *Nanotechnology: Molecularly Designed Materials* (G. M. Chow, and K. E. Gonsalves, eds.), Ch. 24, pp. 351–365, American Chemical Society Symposium Series 622, Washington, DC (1996)

72. Gonsalves, K. E., Chow, G. M., Zhang, Y., Budnick, J. I., and Xiao, T. D., Iron Nitride/Boron Nitride Magnetic Nanocomposite Powders, *Advanced Materials*, 6:291–292 (1994)

73. Xiao, T. D., Gonsalves, K. E., and Strutt, P. R., Synthesis of Aluminum Nitride/Boron Nitride Materials, *J. Am. Ceram. Soc.*, 76:987–992 (1993)

74. Chow, G. M., Xiao, T. D., Chen, X., and Gonsalves, K. E., Compositional and Thermal Effects on Chemically Processed AlN-BN Nanocomposite Powders, *J. Mater. Res.*, 9:168–175 (1994)

75. Corriu, R., Leclerq, D., Lefevre, P., Mutin, P. H., and Vioux, A., Preparation of Monolithic Binary Oxide Gels By A Nonhydrolytic Sol-Gel Process, *Chemistry of Materials*, 4:961–963 (1992)

76. Corriu, R., Leclerq, D., Lefevre, P., Mutin, P. H., and Vioux, A., Preparation of Monolithic Gels From Silicon Halides By A Nonhydrolytic Sol-Gel Process, *J. Non-Cryst. Solids*, 146:301–303 (1992)

77. Jansen, M., and Guenther, E., Oxide Gels and Ceramics Prepared By A Nonhydrolytic Sol-Gel Process, *Chemistry of Materials*, 7:2110–2114 (1995)

78. Yin, Y. G., Xu, W. Q., DeGuzman, R., Suib, S. L., and O'Young, C. L., Studies of Stability and Reactivity of Synthetic Cryptomelane-like Manganese Oxide Octahedral Molecular Sieves, *Inorganic Chemistry*, 33:4384–4389 (1994)

79. DeGuzman, R., Shen, Y. F., Neth, E. J., Suib, S. L., O'Young, C. L., Levine, S., and Newsam, J. M., Synthesis and Characterization of Octahedral Molecular Sieves (OMS-2) Having The Hollandite Structure, *Chemistry of Materials*, 6:815–821(1994)

80. Shen, Y. F., Suib, S. L., and O'Young, C. L., Cu Containing Octahedral Molecular Sieves and Octahedral Layered Materials, *J. Catalysis*, 161m, pp. 115–122 (1996)

81. Xiao, T. D., Bokhimi, Benaissa, M., Perez, R., Strutt, P. R., and Yacaman, M. J., Microstructural Characteristics of Chemically Processed Manganese Oxide Nanofibres, *Acta Mater.*, 45:1685–1693 (1997)

82. Benaissa, M., Yacaman, M. J., Xiao, T. D., and Strutt, P. R., Microstructural Study of Hollandite-type MnO_2 Nano-Fibers, *Appl. Phys. Lett.*, 70:2120–2123 (1997)

83. Antonelli, D. M., Nakahira, A., and Ying, J. Y., Ligand-assisted Liquid Crystal Templating In Mesoporous Niobium Oxide Molecular Sieves, *Inorganic Chemistry*, 35:3126–3136 (1996)

84. Antonelli, D. M., and Ying, J. Y., Synthesis of a Stable Hexagonally Packed Mesoporous Niobium Oxide Molecular Sieve Through a Novel Ligand-assisted Templating Mechanism., *Angew Chem. Int. Ed. Engl.*, 35:426–429 (1996)

85. Rouxel, J., Chemical Reactivity of Low Dimensional Solids, *Chemica Scripta*, 28:33–40 (1988)

86. Livage, J., Sol-gel Processing of Metal Oxides, *Chemica Scripta*, 28:9–13 (1988)

87. Ellis, A. B., Geselbracht, M. J., Johnson, B. J., Lisensky, G. C., and Robinson, W. R., *Teaching General Chemistry: A Materials Science Companion*, pp. 345–347, American Chemical Society, Washington, DC (1993)

88. Levin, D., Soled, S. L., and Ying, J. Y., Chemie Douce Synthesis of a Layered Ammonium Zinc Molybdate, *Chemistry of Materials*, 8:836–843 (1996)

89. Malla, P. B., Ravindranathan, P., Komarneni, S., and Roy, R., Reduction of Copper Acetate Hydroxide Hydrate Interlayers in Montmorillonite by a Polyol Process: A New Approach in the Preparation of Metal-Supported Catalysts, *J. Mater. Chem.*, 2:559–565 (1992)

90. Ayyappan, S., Subbannna, G. N., Srinivasa Gopalan, R., and Rao, C. N. R., Nanoparticles of Nickel and Silver Produced by the Polyol Reduction of the Metal Salts Intercalated in Montmorillonite, *Solid State Ionics*, 84:271–281 (1996)
91. Yoldas, B. E., Alumina Sol Preparation from Alkoxides, *J. Am. Ceram. Soc.*, 54:389–390 (1975)
92. Smith, T. W., and Wychick, D., Colloidal Iron Dispersions Prepared via the Polymer-Catalyzed Decomposition of Iron Pentacarbonyl, *J. Phys. Chem.*, 84:1621–1629 (1980)
93. Bradley, J. S., Hill, E. W., Klein, C., Chaudret, B., and Duteil, A., Synthesis of Monodispersed Bimetallic Palladium-Copper Nanoscale Colloids, *Chemistry of Materials*, 5:254–256 (1993)
94. Hedge, M. S., Larcher, D., Dupont, L., Beaudoin, B., Tekaia-Elhsissen, K., and Tarascon, J. M., Synthesis and Chemical Reactivity of Polyol Prepared Monodisperse Nickel Powders, *Solid State Ionics*, 93:33–50 (1997)
95. Schmid, G., Clusters and Colloids: Bridges Between Molecular and Condensed Material, *Endeavour*, 14:172–178 (1990)
96. Shanefield, D.J., *Organic Additives and Ceramic Processing, with Applications in Powder Metallurgy, Ink, and Paint*, Kluwer Academic Publishers, Boston, Dordrecht, London (1995)
97. Chen, X, Gonsalves, K. E., Chow, G. M., and Xiao, T. D., Homogeneous Dispersion of Nanostructured Aluminum Nitride in a Polyimide Matrix, *Advanced Materials*, 6:481–484 (1994)
98. Gonsalves, K. E., Chen, X, and Baraton, M. I., Mechanistic Investigation of the Preparation of Polymer/Ceramic Nanocomposites, *Nanostructured Materials*, 9:181–184 (1997)
99. Fendler, J. H., Atomic and Molecular Clusters in Membrane Mimetic Chemistry, *Chemical Reviews*, 87:877–899 (1987)
100. Puvvada, S., Baral, S, Chow, G. M., Qadri, S. B., and Ratna, B. R., Synthesis of Palladium Metal Nanoparticles in the Bicontinuous Cubic Phase of Glycerol Monooleate, *J. Am. Chem. Soc.*, 116:2135–2136 (1994)
101. Lisiecki, I., and Pileni, M. P., *Synthesis of Copper Metallic Clusters Using Reverse Micelles as Microreactors*, 115:3887–3896 (1993)
102. Wilcoxon, J. P., Williamson, R. L., and Baughman, R., Optical Properties of Gold Colloids Formed in Inverse Micelles, *J. Chem. Phys.*, 98: 9933–9950 (1993)
103. Koran, A. R., Hull, R., Opila, R. L., Bawendi, M. G., Steigerwald, M. L., Carroll, P. J., and Brus, L. E., Nucleation and Growth of CdSe on ZnS Quantum Crystallite Seeds, and Vice Versa, In Inverse Micelle Media, *J. Am. Chem. Soc.*, 112:1327–1332 (1990)

104. Osseo-Asare, K., and Arriagada, F. J., Preparation of SiO_2 Nanoparticles in a Non-ionic Reverse Micellar System, *Colloids and Surfaces*, 50:321–339 (1990)

105. Chang, C. L., and Fogler, H. S., Kinetics of Silica Particle Formation in Nonionic W/O Microemulsions From TEOS, *AICHE Journal*, 42:3153–3163 (1996)

106. Stober, W., Fink, A., and Bohn, A., Controlled Growth of Monodispersed Spheres in the Micron Size Range, *J. Colloid and Interfacial Sci.*, 26:62–69 (1968)

107. Herrig, H. and Hempelmann, R., Microemulsion Mediated Synthesis of Ternary and Quaternary Nanoscale Mixed Oxide Ceramic Powders, *Nanostructured Materials*, 9:241–244 (1997)

108. Mann, S., and Williams, R. J. P., Precipitation Within Unilamellar Vesicles, Part 1, Studies of Silver Oxide Formation, *J. Chem. Soc.*, Dalton Transactions, pp. 311–316 (1983)

109. Mann, S., and Hannington, J. P., Formation of Iron Oxides in Unilamellar Vesicles, *J. Colloid and Interface Sci.*, 122:326–335 (1988)

110. Bhandarkar, S., and Bose, A., Synthesis of Submicrometer Crystals of Aluminum Oxide by Aqueous Intravesicular Precipitation, *J. Colloid and Interface Sci.*, 135:531–538 (1990)

111. Chow, G. M., Markowitz, M. A., and Singh, A., Synthesizing Submicrometer and Nanoscale Particles via Self-assembled Molecular Membranes, *J. Miner., Met., and Mater. Soc.*, 45:62–65 (1993)

112. Markowitz, M. A., Chow, G. M., and Singh, A., Polymerized Phospholipid Membrane Mediated Synthesis of Metal Nanoparticles, *Langmuir*, 10:4905–4102 (1994)

113. Chow, G. M., Markowitz, M. A., Rayne, R., Dunn, D. N., and Singh, A., Phospholipid Mediated Synthesis and Characterization of Gold Nanoparticles, *J. Colloid and Interface Sci.*, 183:135–142 (1996)

114. Froes, F. H., Suryanarayana, C., Russell, K. C., and Ward-Close, C. M., *Far From Equilibrium Processing of Light Metals*, in: *Novel Techniques in Synthesis and Processing of Advanced Materials*, (J. Singh, and S. M.Copley, eds.), The Minerals, Metals, and Materials Society, pp. 1–21 (1995)

115. Yang, Y., Huang, J., Liu, S., and Shen, J., Preparation, Characterization, and Electroluminescence of ZnS Nanocrystals in a Polymer Matrix, *J. Mater. Chem.*, 7:131–133 (1997)

116. Ichinose, I, Kimizaka, N., and Kunitake, T., Formation of a Novel CdS Cluster in An Organic Multilayer Template: A Case of An Organic/Inorganic Superlattice, *J. Phys. Chem.*, 99:3736–3742 (1995)

117. Zelenski, C. M., Hornyak, G. L., and Dorhout, P. K., Synthesis and Characterization of CdS Particles Within a Nanoporous Aluminum Oxide Template, *Nanostructured Materials*, 9:173–176 (1997)

118. Korgel, B. A., and Monboriquette, H. G., Synthesis of Size-Monodisperse CdS Nanocrystals Using Phoshatidylcholine Vesicles as True Reaction Compartments, *J. Phys. Chem.*, 100:346–351 (1996)

119. Ross, C. A., Electrodeposited Multilayer Thin Films, *Annual Rev. Mater. Sci.*, 24:159–188 (1994)

120. Erb, U., Electrodeposited Nanocrystals: Synthesis, Structure, Properties and Future Applications, *Canadian Metallurgical Quarterly*, 34:275–280 (1995)

121. Cheung, C., Wood, D., and Erb, U., *Applications of Electrodeposited Nanocrystals*, in: *Processing and Properties of Nanocrystalline Materials*, (C. Suryanarayana, J. Singh, and F. H. Froes, eds.), The Minerals, Metals, and Materials Society, pp. 479–489 (1996)

122. Palumbo, G., Gonzalez, F., Brennenstuhl, A. M., Erb, U. Shmayda, W., and Lichtenberger, P. C., In-situ Nuclear Steam Generator Repair Using Electrodeposited Nanocrystalline Nickel, *Nanostructured Materials*, 9:737–746 (1997)

123. Riedel, W., *Electroless Nickel Plating*, Finishing Publications Ltd., Stevenage, Hertfordshire, England (1991)

124. Brandow, S. L., Dressick, W. J., Marrian, C. R. K., Chow, G. M., and Calvert, J. M., The Morphology of Electroless Ni Deposition on a Colloidal Pd (II) Catalyst, *J. Electrochem. Soc.*, 142:2233–2243 (1995)

125. Chow, G. M., Pazirandeh, M., Baral, S., and Campbell, J. R., TEM & HRTEM Characterization of Metallized Nanotubules Derived From Bacteria, *Nanostructured Materials*, 2:495–503 (1993)

126. Krebs, J. J., Rubenstein, M., Lubitz, P., Harford, M. Z., Baral, S., Shashidhar, S., Ho, Y. S., Chow, G. M., and Qadri, S., Magnetic Properties of Permalloy-Coated Organic Tubules, *J. Appl. Phys.*, 70:6404–6406 (1991)

127. Chow, G. M., Stockton, W. B., Price, R., Baral, S., Ting, A. C., Ratna, B. R., Schoen, P. E., Schnur, J. M., Bergeron, G. L., Czarnaski, M. A., Hickman, J. J., and Kirkpatrick, D. A., Fabrication of Biologically Based Microstructure Composites for Vacuum Field Emission, *Mater. Sci. and Eng.*, A158:1–6 (1992)

128. Kirkpatrick, D. A., Bergeron, G. L., Czarnaski, M. A., Hickman, J. J., Chow, G. M., Price, R., Ratna, B. R., Schoen, P. E., Stockton, W. B., Baral, S., Ting, A. C., and Schnur, J. M., Demonstration of Vacuum Field Emission from a Self-assembling Bimolecular Microstructure Composite, *Appl. Phys. Lett.*, 60:1556–1558 (1992)

129. Chow, G. M., Kurihara, L. K., Feng, C. R., Schoen, P. E., and Martinez-Miranda, L. J., Alternative Approach to Electroless Cu Metallization of AlN By a Nonaqueous Polyol Process, *Appl. Phys. Lett.*, 70:2315–2317 (1997)

130. Eriksson, G., Siegbahn, H., Andersson, S., Turkki, T., and Muhammed, M., The Reduction of $Co2^{+}$ by Polyalcohols in the Presence of WC Surfaces Studied by XPS, *Mater. Res. Bull.*, 32:491–499 (1997)

131. Klein, L. C. (ed.), *Sol Gel Technology for Thin Films, Fibers, Preforms, Electronics, and Specialty Shapes*, Noyes Publications, NJ (1988)

132. Brinker, C. J., Hurd, A. J., Schunk, P. R., Frye, G. C., and Ashley, C. S., Review of Sol-Gel Film Formation, *J. Non-Cryst. Solids*, 147–148:424–436 (1992)

133. Zhou, J., Li, L., Gui, Z., and Zhang, X., Ferroelectric Thin Films Embedding Nanoscale Metal Particles: A Novel Process of Functional Composites, *Ferroelectrics*, 196:85–88 (1997)

134. McDonogh, C., Sheridan, F., Butler, T. M., and MacCraith, B. D., Characterization of Sol-Gel Derived Silica Films, *J. NonCryst. Solids*, 194:72–77 (1996)

135. Doherty, S., and Fitmaurice, D., Preparation and Characterization of Transparent Nanocrystalline TiO_2 Films Possessing Well-Defined Morphologies, *J. Phys. Chem.*, 100:10732–19738 (1996)

136. Barrow, D. A., Petroff, T. E., and Sayer, M., Thick Ceramic Coatings Using a Sol-Gel Based Ceramic-Ceramic 0-3 Composite, *Surface and Coatings Technology*, 76–77:113–118 (1995)

137. Moriguchi, I., Maeda, H., Teraoke, Y., and Kagwa, S., Preparation of a TiO_2 Nanoparticulate Film Using a Two Dimensional Sol-Gel Process, *Chemistry of Materials*, 9:1050–1057 (1997)

2

Thermal Spray Processing of Nanocrystalline Materials

Maggy L. Lau and Enrique J. Lavernia

1.0 INTRODUCTION

Significant interest has been generated recently in the field of nanoscale (also described as nanocrystalline or nanophase) materials, in which the grain size is usually in the range of 1–100 nm. The sudden burst of enthusiasm stems, not only from the outstanding properties that can be obtained in such materials, but also from the realization that early skepticism about the ability to produce high-quality, unagglomerated nanoscale powder was unfounded. There are literally dozens of methods utilized by over sixty companies involved in nanocrystalline materials in the United States alone, some of which are commercialized.[1] Accordingly, the focus is shifting from synthesis to processing, and the manufacture of useful coatings and structures from these powders. The potential applications span the entire spectrum of technology, from thermal barrier coatings for turbine blades to wear-resistant rotating parts. The application of nanocrystalline materials used as powder feedstock for thermal spraying has been facilitated by the wide range of powder sources available, including: vapor condensation, solution precipitation, combustion synthesis, sol-gel processing, thermochemical synthesis, and mechanical alloying/milling.[2]–[4] Among the various synthesis techniques, vapor condensation and solution precipitation methods have been successfully scaled up to produce tonnage quantities of nanoscale ceramic, metal, and composite powders for commercial usage.[5] The resultant thermally sprayed coatings have been shown to exhibit unique and often enhanced physical and mechanical properties in comparison to the coatings produced by current technology.

Inspection of the available literature shows that improvements in physical performance have been documented for several metallic and cermet based nanostructured coatings. For instance, microhardness values of thermally sprayed nanocrystalline Ni, Inconel 718, and 316-stainless steel coatings have been shown to be approximately 20%, 60%, and 36% higher than the conventional (i.e., coatings sprayed with conventional micron-sized powders).[4]–[6] The potential impact is significant; for example, there are approximately 1500 weld overlays in a single ship.[2]–[4] The anticipated service life cycle of these weld parts could be extended if a nanocrystalline coating, with the associated improvements in hardness and wear characteristics, could be used. Furthermore, it has been estimated that a significant proportion of the valve stem failures in ships is due to steam erosion. The improved wear properties of nanocrystalline coatings are ideally suited for this particular application. This and other examples suggest that the applications of thermally sprayed nanocrystalline coatings can extend to a broad range of industries, and cover a wide range of materials that could be used to fabricate diverse components.[2]–[4] According to a recently published technical market study, the projected U.S. market for nanostructured coatings is estimated to reach $6,000,000 with an average annual growth rate of 43.1% during 1996–2001.[7] Thereby, nanostructured coatings are contributing an ever-increasing role in many industrial applications as this century begins.

Published studies documenting the improvements of physical properties in nanocrystalline coatings also reveal a sobering fact. That is, the behavior of a nanostructured material during thermal spraying is rendered with complex factors such as the ones listed below.

- *Morphology of feedstock powders:* Often nanostructured materials are generated as agglomerates, hollow spheres or flakes, and their behavior during powder injection as well as during flight is not completely understood.

- *Thermal stability of nanostructured powders:* Nanostructured characteristics of the starting powder should be retained in the final coatings in order to maximize performance. Available results suggest that under certain conditions it is possible to generate nano-crystalline grains by rapid nucleation during flight and/impingement.

- *Thermal and momentum behavior of nanostructured powder:* Although preliminary results obtained for nanostructured materials using parameters that are optimized for conventional powders (typically spherical) are encouraging, optimization of thermal spraying for nanostructured materials has not been accomplished. Optimization of chemistry, morphology, and coating thickness, for example, should lead to the attainment of physical performance heretofore unattainable with conventional coatings.

In view of the above, the objective of the present paper is to provide an overview of recent advancements in the field of high-performance nanostructured coatings, paying particular attention to underlying fundamental issues.

2.0 SYNTHESIS OF NANOCRYSTALLINE POWDER FOR THERMAL SPRAYING

Nanostructure materials are defined as solids with a characteristic length in the nanometer regime (10^{-9} m) in at least one dimension.[8] Various methods such as inert gas condensation, sputtering, plasma spray synthesis, vapor deposition, electrodeposition, rapid solidification, and spark erosion can be used to produce nanocrystalline materials.[5] Among these various techniques, attritor milling and co-precipitation methods are used currently to manufacture sufficiently large quantities (in kilogram/batch) of micrometer-sized particles with nanostructure grains.[9] Gas condensation, mechanical alloying/milling, crystallization of amorphous alloys, chemical precipitation, and spray conversion processing are used to synthesize three-dimensional equiaxed nanocrystallites. Mechanical alloying/milling techniques have been used to produce large quantities for possible commercial use.[5]

Mechanical alloying is a high-energy ball-milling process in which elemental blended powders are continuously welded and fractured to achieve alloying at the atomic level.[10] Existing factors influencing the mechanical alloying/milling processes include milling time, charge ratio, milling environment, and the internal mechanics specific to each mill.[10] A recent study of the microstructural evolution of nanocrystalline Ni produced by mechanical milling indicates that milling environment and temperature strongly influence the deformation behavior of the powders. Secondary electron images (Fig. 1a–b) attained from the SEM analysis show the morphological evolution of Ni milled in methanol (Fig. 1a) and in liquid nitrogen (process known as cryomilling) for 10 hrs (Fig. 1b).[6] The BET surface area of as-received Ni powders, methanol milled, and cryomilled (10 hrs) powders determined by Particle Surface Analyzer (Coulter Corp., Miami, FL) were 0.043, 0.106, and 0.285 m^2/g respectively. The formation of irregular flake-shaped agglomerates is attributed to the continuous welding and fracturing of the powder particles during the mechanical milling process.[11] The rate of structural refinement is dependent on the mechanical energy input and the work hardening of the material.[12] Cryomilling reduces oxygen contamination from the atmosphere and minimizes the heat generated during milling; therefore, fracturing is favored over welding, especially in milling of ductile materials.[13] The grain size of Ni has been sufficiently refined when milling is conducted in liquid nitrogen instead of methanol, as listed in Table 1. Other characteristics of the Ni powder properties that resulted from milling in room temperature methanol and liquid nitrogen are listed also in Table 1.[6]

(a) (b)

Figure 1. Ni powders milled in *(a)* methanol and *(b)* liquid nitrogen for 10 hrs.

Table 1. Characteristic Powders Properties of Mechanical Milled Ni in Different Milling Conditions

Milling environment	Milling time (hr)	Agglomerate size (μm)			Grain size (nm)	Aspect ratio
		d_{10}*	d_{50}*	d_{90}*		
methanol	5	27	80	144	90 ± 65	1.67
methanol	10	15	35	77	82 ± 25	2.77
liquid nitrogen	10	20	43	84	19 ± 11	1.47

*d_{10}, d_{50}, d_{90}—Particle size corresponding to 10, 50, and 90 wt% of the cumulative particle size distribution.

Although various techniques are available for producing nanostructured powders, the powders exhibit different particle sizes and morphologies. The challenge lies in producing powders that are suitable for thermal spraying. This requirement implies a narrow size distribution, as well as spheroidal morphology. As an example, nanocrystalline WC-Co powders produced by solution precipitation exhibit major differences in powder morphology as well as particle sizes in comparison to those synthesized by the mechanical milling method. Nanocrystalline WC-Co powders produced by solution precipitation followed by spray drying and high temperature carburization processes (Nanodyne Inc., New Brunswick, NJ) consist of hollow spherical morphology with an average size of 50 μm.[14] The WC grains, with a grain size range of 20–40 nm, are distributed around the walls of the particles. On the other hand, nanocrystalline WC-Co formed by mechanical milling yields irregular-shaped nanograins.[15] The average grain size of WC-Co powders decreases from 169 nm after 2 hrs milling to 14 nm after 20 hrs of room temperature milling in hexane (C_6H_{14}).[15] The formation of the nanocrystalline carbides is attributed to the continuous fracturing process during mechanical milling. The carbide particles were brittle-fractured into fragments with sharp facets which then incorporated into the Co binder phase.[15] The repeated fracturing process eventually leads to the formation of a composite structure. The milling time influences the final size because the grain structure is further refined as the milling process continues. Similar microstructural changes were observed in the mechanical milling of Cr_3C_2-NiCr cermet in hexane, shown in Fig. 2.[15]

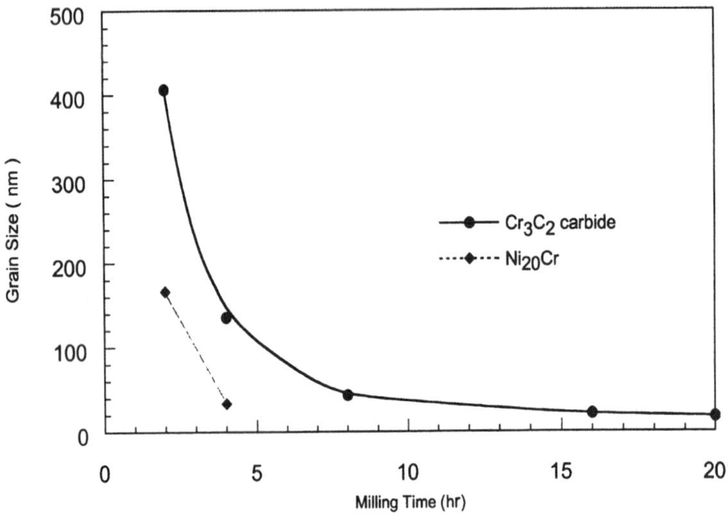

Figure 2. Microstructural evolution of mechanically milled Cr_3C_2-NiCr cermet in hexane.[15]

Recently, Karthikeyan, et al.,[16] successfully utilized plasma spraying to synthesize various nanocrystalline ceramic powders, which included Al_2O_3, ZrO_2, and Y_2O_3-ZrO_2. This process utilized an open atmospheric plasma spray to inject atomized liquid precursor chemicals into the plasma jet.[16] Nanoclusters form as a result of the rapid solidification and are collected as powders after passing through an electrostatic precipitator.[16] The powders form micron-sized aggregates with irregularly shaped blocks, which differ from the ceramic powders with a hollow-shell morphology produced from the co-precipitation processes.[16] The grain size of the ceramic powders produced by plasma spraying pyrolysis ranges between 1 and 50 nm.[16] The distribution of the grain size depends on the chemistry of the liquid precursors; aqueous solutions were observed to produce a wider distribution of large grains as compared to those from organometallic solutions.[16] Therefore, although this technique offers potential benefits in producing large quantities of nanoceramic powders, further study is underway to optimize the spraying parameters of the process.

The particle size of nanocrystalline powders produced using the above described synthesis routes is often too small or irregular to be used as the feedstock for thermal spraying. Therefore, further agglomeration is sometimes required to provide powders with a controllable and flowable particle size range for thermal spraying. Although a variety of agglomeration techniques are available, finding the appropriate binding agent to agglomerate the nanocrystalline powders is essential to ensure that the crystal structure and the morphology are preserved. Various organic binding agents, such as methyl cellulose, polyvinyl alcohol, carboxy-methyl cellulose, and polyethylene glycol have been reported to agglomerate powders effectively.[17] The amount of binder used depends upon the size of the starting powders and the percentage of binder in a binder-powders mixture.[17] In particular, sodium carboxy-methyl cellulose has been reported to maintain the powder particles in suspension, and the resultant slurry can subsequently be processed in a dryer to produce loosely agglomerated powders.[17] Recently, nanocrystalline cermet WC-Co and Cr_3C_2-NiCr powders produced by mechanical milling in hexane have been successfully agglomerated by mixing with a 2 wt% methyl cellulose solution in water to form a slurry.[15]–[18] The particle size of the mechanically milled (20 hrs) powders was less than 10 μm, which is too small for thermal spraying. The slurry was then annealed for 24 to 48 hrs at a temperature of 373 K. The water was evaporated and the slurry formed solid agglomerates. The agglomerates were subsequently sieved through a sub-50 μm mesh for thermal spraying. The size of the WC-Co agglomerates was approximately 40 μm which is suitable for HVOF (high-velocity oxy-fuel) spraying.[15]

3.0 THERMAL SPRAYING

Thermal spraying is a coating process used to produce metallic, non-metallic and ceramic coatings in which a spray of molten or semi-molten solid particles generated from a thermal source are deposited onto a substrate by mechanical bonding.[17]–[19] The microstructure of the coating results from rapid solidification of the particulates.[17] In principle, powders, rods, and wires which do not sublimate or decompose at temperatures close to their melting points can be used as spraying materials.[19] Metals and alloys in the form of rods or wires are commonly used in arc spraying (AS) and flame spraying (FS).[17] Powders of metals, alloys, ceramic oxides, cermets, and carbides are often used in thermal spraying to produce a homogeneous microstructure in the resulting coating. In most cases, the sprayed surface should be degreased, masked, and roughened prior to spraying to maximize the bonding strength between the coating and the sprayed material. Today, flame spraying (FS), atmospheric plasma spraying (APS), arc spraying (AS), detonation gun (D-gun) spraying, high-velocity oxy-fuel spraying (HVOF), vacuum plasma spraying (VPS), and controlled atmosphere plasma spraying (CAPS) are widely used to produce various coatings for various industrial applications. In general, the heat source for thermal spraying processes may be generated by an electrical or a chemical (combustion) source. Table 2 shows the various types of thermal spraying processes.[19]

Table 2. Various Types of Thermal Spraying Processes

Heat source: electrical	Heat source: chemical (combustion)
Plasma spray	Flame spray
Wire-arc spray	High velocity oxygen fuel spray
	Detonation gun spray

3.1 Coating Characteristics

The objective of all thermal spraying processes is to generate surface coatings for surface protection and to extend the service life of the protected parts. The coating characteristics determine the quality of the coating which is often characterized by its microstructure (volume fraction of porosity, unmelted particles, and the presence of oxide phases and other extraneous impurities), macrohardness (Rockwell B or C), and microhardness (Diamond Pyramid Hardness, Vickers, or Knoop).[19] The quality of a coating can also be determined by its bond strength (adhesive, cohesive, sliding), corrosion and wear resistance, thermal shock resistance, thermal conductivity, and dielectric strength.[19] The distribution of these microstructural features influences the physical properties of the coating. The microstructure of a coating results from the rapid impingement of molten or semi-molten droplets propelled onto the substrate surface during the thermal spray process. The temperature fields of the solidifying particle and the environment surrounding the substrate ultimately affect the final microstructure of the coating.[17]

Thermally sprayed coatings produced using nanocrystalline powders as feedstock yield distinctive characteristics compared to conventional coatings. Some of the microstructural features are evident by optical microscopy. For instance, in a recent study, nanocrystalline 316-stainless steel coatings were produced by HVOF processing of nanocrystalline 316-stainless powders prepared by mechanical milling in liquid nitrogen.[20][21] Backscattered electron images revealed higher porosity in the nanocrystalline coatings when compared to those of conventional coatings processed using identical spraying parameters, as shown in Fig. 3a–d. The coating porosity strongly influenced the ultimate physical and mechanical properties. Several mechanisms are thought to be responsible for the observed porosity. For example, porosity depends on the pressure created on the surface of the substrate during droplet impingement.[22] A number of studies have reported that porosity decreases with increasing particle velocity and temperature.[22] A high droplet temperature before impingement, created by a short spraying distance, for example, decreases the amount of porosity due to the increased fluidity of molten particles.[22] Table 3 lists the characteristics of a nanostructured 316-stainless steel coating which include porosity and microhardness as determined on cross sectional areas.[20][21] An increase in microhardness is observed in the various nanocrystalline coatings when compared to those of conventional counterparts. Transmission electron microscopy was used to analyze the cross sections of the stainless steel coatings sprayed using cryomilled powders, yielding some interesting observations. For instance, lamellae with random crystallographic orientation were observed in the cross-sectional view, as shown in Fig. 4. The thickness of the lamellae ranged from 40 to 400 nm. Selected area diffraction analysis indicated the presence of various oxide phases (Cr_2O_3, FeO, Fe_2O_3, and γ-Fe_2O_3) besides cubic stainless steel type 304 in the cryomilled coating. The presence of the lamellar structure indicates that some of the nanocrystalline particles melted during thermal spraying. Furthermore, the columnar grains observed in each lamellar layer suggest that the nucleation is heterogeneous, with the solidification front in the direction towards the top of the lamella. The formation of columnar grains is also attributed to the increased temperature of the residual liquid due to a recalescence effect.[17] In addition, large grains with a grain size of 115 ± 40 nm and an aspect ratio of 1.45 were also observed, as illustrated in Fig. 4.

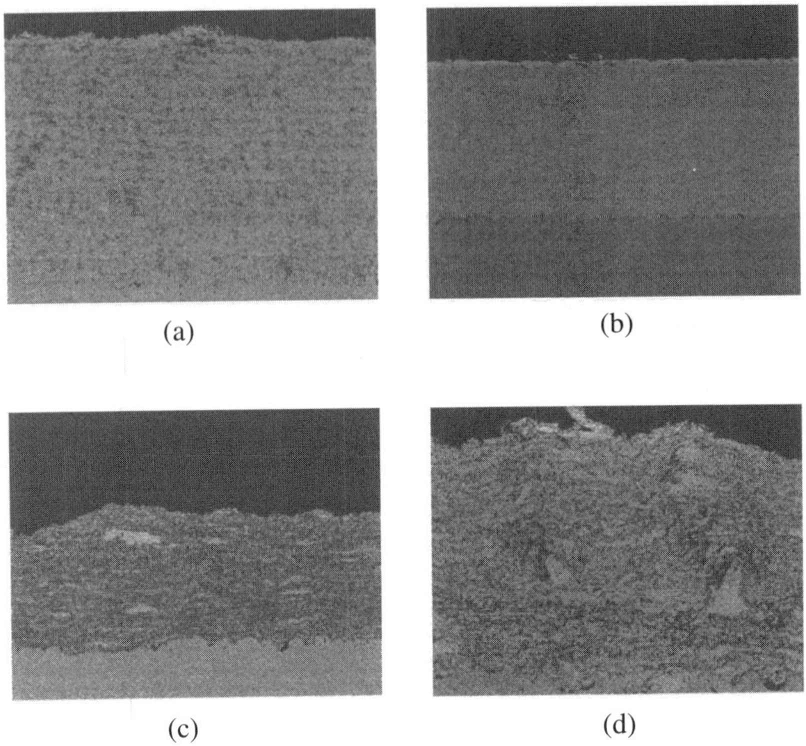

Figure 3. Backscattered electron images of (*a*) conventional 316-stainless steel coating sprayed in air; (*b*) conventional coating sprayed in nitrogen; (*c*) cryomilled (10 hrs) coating sprayed in air; and (*d*) cryomilled (10 hrs) coating sprayed in nitrogen.

Table 3. Physical Properties of Nanocrystalline 316-Stainless Steel Coatings

Milling time (hrs)	Milling environ-ment	Spraying environ-ment	Porosity (%)	Micro-hardness 300g load (DPH)	Increase in hardness (%)
10	methanol	Air	14	650	48
10	methanol	Nitrogen	10	497	31
10	liquid nitrogen	Air	16	573	31
10	liquid nitrogen	Nitrogen	15	440	16

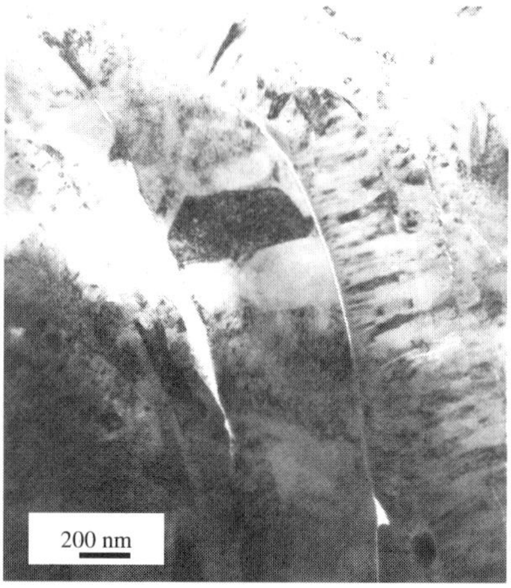

Figure 4. TEM bright field image of cross-section of cryomilled 316-stainless steel coating.

The origin of improved hardness in nanocrystalline materials has been the topic of intense study, and continuous modifications to the Hall-Petch theory, based on different dislocation-grain boundary interaction mechanisms, have been developed to explain the observed results.[5] The enhanced microhardness observed in various nanocrystalline coatings may be caused by multiple factors. First, strengthening by grain size refinement, or the Hall-Petch model, may be assumed on the basis that a dislocation network density within the nanoscale grains transfers slip through the grain boundaries.[23] Second, since the agglomerates possess larger surface areas when compared to those of conventional spherical powders, it is speculated that more chemical reactions will occur during thermal spraying of nanocrystalline powders relative to that experienced by conventional powders. To that effect, it is evident that spraying the powders in air generates coatings with higher microhardness than those obtained by spraying in N_2 gas. Accordingly, it is speculated that the formation of oxides and nitrides, as second phase particles, is responsible for the observed increase in the microhardness values in various nanocrystalline coatings. Moreover, comparison between the heat of formation values of various oxides and nitrides indicates that oxides are more likely to form than nitrides. Therefore, the formation of second phase particles is more favorable when oxygen is used instead of nitrogen during thermal spraying, leading to the observed differences in microhardness in the various nanocrystalline coatings.

Recently, Stewart, et al.,[24][25] examined the wear performance of conventional and nanostructured WC-Co cermets coatings produced by high-velocity oxy-fuel (HVOF) thermal spraying. Both the conventional WC-17 wt% Co (sintered and crushed with a carbide grain size of 2 μm; Praxair Surface Technology, Indianapolis) and nanocomposite WC-Co (carbide grain size of approximately 75 nm; Nanocarb, Nanodyne, New Brunswick, NJ) powder feedstocks were thermally sprayed by HVOF using identical spraying conditions. Microstructural characterization of the coatings by various techniques showed that WC and by-products such as tungsten hemicarbide (W_2C) and W were present in both coatings. In addition, an amorphous Co-rich binder phase containing W and C was observed. The nanocomposite coating contained a smaller fraction of unreacted WC than that of the conventional coating, mainly attributed to differences in the morphology of the powder feedstock and the initial WC grain size.[25] Three body abrasive wear tests performed on the coating material indicated that the wear of both coatings is caused by an indentation-induced sub-surface mechanism. The nanocomposite coating exhibited inferior wear resistance in comparison to that of the conventional coating under all the conditions examined.[25] The low wear resistance in the nanocomposite coating was mainly caused by the loss of ductility in the amorphous Co-rich binder phase. Furthermore, the differences in the wear behavior of the coatings could be explained in terms of differences in powder characteristics, the extent of reaction and decarburization during spraying, and the resultant microstructure in the coating during rapid solidification of the particles at high cooling rates.[25] Similar findings were observed by He, et al.,[15] in the microstructural evolution of HVOF sprayed WC-Co coatings using nanocrystalline WC-Co powders prepared by mechanical milling. Regions of amorphous Co were observed during TEM analysis. It is apparent that the physical properties of the coatings produced from the nanocrystalline powders differ noticeably from those of the conventional powders. The associated grain size refinement processes not only not reduce the grain structure of the powders, but also increase the surface area of the powder particles. Furthermore, decarburization and the presence of dissolved carbides in the Co matrix were observed to be more prominent in the thermally sprayed coating using nanocrystalline WC-Co powders than those in the conventional cermet coating.[15] Although W_2C is harder than WC, the sub-carbide is prone to cracking and hence is undesirable in thermally sprayed coatings. The inherent increase in the amount of W_2C is most likely attributed to the increased surface areas of the nanocrystalline powders resulting from the mechanical milling process.

The underlying mechanism governing the formation of the amorphous/nanocrystalline structure in coatings sprayed using nanocrystalline powders requires further study. A recent study by Verdon, et al.,[26] on the microstructural evolution of thermally sprayed WC-Co coatings by HVOF, indicated that the extent of carbide transformation during thermal spraying depends strongly on the morphology of the powders and the spraying conditions. The W_2C carbides observed in the TEM analysis were free of defects, unlike the WC that were surrounded with dislocations and stacking faults.[26] The W_2C morphology suggested that the decarburization process occurs at the surface of WC. As carbon diffuses outward of the WC grains, carbon oxidation can also take place at the surface which leads to further carbon loss.[26] The extent of carbide dissolution is also related to spraying conditions, such as the type of fuel gas used. The decarburization was more pronounced when H_2 instead of C_3H_8 was used as the fuel gas for spraying.[26] The dissociation temperature of H_2 gas is lower than that of propylene (C_3H_8) gas, which results in a higher thermal conduction at the same temperature, in comparison to that of propylene. As a result, the heat transfer between the flame gas and the powder particles increases, which promotes the decarburization process to occur rapidly.[26]

In the case of the HVOF coatings produced using nanocomposite Cr_3C_2-NiCr powders, optical microscopy showed a smooth coating surface in comparison to the coatings produced using the conventional Cr_3C_2-NiCr powders.[18] In addition, TEM studies performed on the nanocomposite coating indicated an average particle size of approximately 24 nm, identified as Cr_3C_2 by SAD patterns. The nanostructured coating exhibited an average microhardness of 1020 DPH under a 300 gram load, which corresponds to more than a 20% increase in microhardness, in comparison to the conventional coating produced by current technology. The enhancement in hardness in the nanostructured coating was thought to be related to the presence of oxide phases formed during the spraying process. In addition, microcracks were observed around the indentation during microhardness testing of a conventional coating under an applied load of 500 grams. No apparent microcracks were observed in the nanocomposite coatings, which indicates that the toughness of the nanocomposite Cr_3C_2-NiCr coating is higher than that of the conventional coating.[18]

Available studies of thermal spraying with nanostructured ceramic feedstock powders are limited due to the poor flowability of available powders. Recently, Cetegen and Yu[27] attempted to spray nanocrystalline 7 wt%-Y_2O_3 stabilized ZrO_2 (YSZ) with DC arc plasma spraying and measured the in-situ temperature, velocity, and particle size profile. The nanocrystalline YSZ powder was composed of loose clusters (grain sizes ranging between 10 to 50 nm) with a diameter of less than 30 μm.[2][27] In-situ particle size measurements, conducted using a particle doppler phase analyzer (PDPA) during plasma spraying yielded a high degree of uncertainty due to the inherent surface roughness associated with the nanocrystalline YSZ agglomerates. Furthermore, the majority of the powders propelled outside the plasma core region due to the hollow shell morphology of the nanopowders.[27] The thermal contact between the plasma arc and the nanocrystalline YSZ particles was thereby inefficiently reduced in comparison to that between the arc and the conventional YSZ, as confirmed by measured temperature profile obtained by a two-color pyrometer.[27] Although nanocrystalline ceramic powders offer the advantage of lower sintering temperature as compared to conventional ceramic powders, further studies are needed to optimize the morphology of the powders in order to elucidate the potential benefits of using nanoceramic powders for thermal spraying.

4.0 MODELING

The quality of coatings produced by thermal spraying techniques depends on the optimization of the spraying parameters such as: temperature, velocity, and degree of solidification; gas type, flow rate, and pressure; substrate material and temperature.[28] Due to the complexity of the various factors involved, the spray parameters are generally optimized by trial and error. The process is often laborious and expensive requiring years of accumulation and adjustment of data and coating properties. Accordingly, modeling represents an attractive approach to minimize experimentation and optimize performance.

The thermal, phase change, and chemical reaction history of the nanocrystalline agglomerates during HVOF spraying is a necessary input for the prediction of the evolution of a nanocrystalline structure. As mentioned above, the gas dynamics and thermal behavior aspects of HVOF have been the topic of many excellent investigations in the recent past.[29]–[34] Consequently, the analysis proposed herein focuses on particle behavior. Although most agglomerates have an irregular morphology, much can be learned from a preliminary, spherically symmetric model provided that an appropriate equivalent diameter is used and that the correlation used to estimate relevant exchange coefficients (e.g., drag or heat transfer coefficient) are suitable for irregularly-shaped particles.

4.1 Particle Dynamics

The prediction of the particle trajectory requires determination of the relative velocity of the particle/gas. To this end, a simplified equation of motion is integrated.

Eq. (1) $$\frac{dV_d}{dt} = g\left(1 - \frac{\varrho_g}{\varrho_d}\right) + \frac{3}{4D}\frac{\varrho_g}{\varrho_d}\left|V_d - V_g\right|\left(V_d - V_g\right)C_{drag}$$

In this equation the added mass and Basset history terms are neglected. A second integration is performed to determine the particle position. The gas velocity field is obtained from correlated experimental measurements (or CFD simulations). The correlation used to estimate the particle drag coefficient includes non-sphericity effects.[35]

4.2 In-Flight Heat Transfer

The thermal energy transfer in the agglomerate is described using an unsteady diffusion equation:

Eq. (2) $$\frac{\partial T}{\partial t} = a\frac{1}{r^2}\frac{\partial}{\partial r}\left(r^2\frac{\partial T}{\partial r}\right)$$

Most agglomerates will undergo at least partial melting, in-flight chemical reaction and, possibly, re-solidification, so this equation has to be solved over three regions within the particle. The position is determined by heat balances and since these interfaces are moving, it is useful to transform Eq. (2) in order to immobilize these interfaces mathematically. To this end, a mapping transformation is used that results in the appearance of a pseudo-convective term in the transformed equation:

Eq. (3) $$\frac{\partial T}{\partial t} + A(\xi,t)\frac{\partial T}{\partial \xi} = B(\xi,t)\frac{\partial}{\partial \xi}\left[C(\xi,t)\frac{\partial T}{\partial \xi}\right]$$

where A, B, and C are functions of time, the mapped coordinate, ξ, and the time-dependent interface positions, as well as their velocities. This equation can then be solved numerically, together with the boundary conditions at the particle surface, heat balances at the interfaces, and symmetry condition at the particle center. The convective heat transfer coefficient at the particle surface may be estimated as a function of the particle's Reynolds and Prandtl Numbers, using a correlation such as the Ranz-Marshall correlation.[36] Here, melting is an equilibrium process, and solidification, if it occurs, is also an equilibrium process. However, if the cooling rates were high enough to induce rapid solidification, this would be included in the model.

The thermal energy from the combustion process of the fuel and oxygen rapidly heats the powder particles to a molten or semi-molten state.[37] The powder particles propelled into the flame undergo acceleration and a significant amount of heating before contacting the substrate. Therefore, the microstructural evolution of the sprayed coating, and the resulting properties of the coatings, are influenced by both momentum and thermal transport between the flame gas and the powder particles during flight.[17] Recently, a mathematical model has been developed to study the thermal behavior of cryomilled stainless steel particles during HVOF thermal spraying.[38] The dynamic processes are further complicated in the case of thermal spraying of nanocrystalline 316-stainless steel powders as the particle morphology deviates from the conventional spherical powders normally used for thermal spraying.[38] In the case of cryomilled stainless steel powders, for example, the Biot Numbers are much smaller than 0.01, which indicates particle heating without significant thermal gradients. Therefore, the distribution of thermal energy in the particle is assumed to be uniform. The calculated thermal behavior of cryomilled stainless steel particles during HVOF thermal spraying is shown in Fig. 5.[38] Three different particle sizes corresponding to d_{10}, d_{50}, and d_{90} are selected. The particles first experience rapid heating within the distance corresponding to the barrel length of approximately 70 mm. The particles then proceed to slow heating and eventually to cooling due to mixing with the entrained air environment. The heat transfer due to convection is two orders of magnitude higher than that from radiation. According to the simulation, the particle temperature increases with decreasing particle size. In particular, the particles with a diameter of 28 μm have the highest heat transfer efficiency, reaching a temperature of about 2100 K. This is mainly attributed to the morphology associated with the cryomilled powders. For instance, the thickness of the particles corresponding to d_{10} (d = 28 μm) is 1.2 μm, which is less than half the size of the thickness of the particles corresponding to d_{50} (d = 50 μm). Therefore, the heat transfer rate is expected to be higher for the d_{10} particles than for the d_{50} particles.[38]

The cryomilled 316-stainless steel powders with a particle size of less than 50 μm will most likely melt according to the aforementioned simulation results since the melting temperature range for type 316-stainless steel is between 1648 and 1673 K.[39] With an estimated cooling rate of about 10^5–10^6 K/s, depending on the particle size, the molten particles will start to solidify before particle impingement. The rapid solidification rate often leads to the formation of a coating with lamellar structure, with columnar grains growing in the direction perpendicular to the heat flow, as observed in the cryomilled coating by TEM analysis.[38]

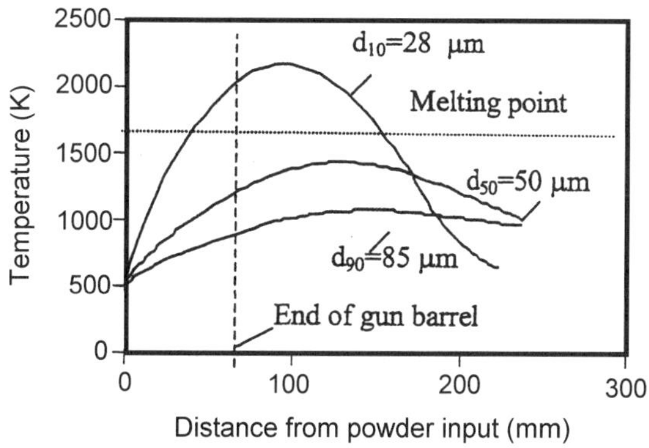

Figure 5. Variation in particle temperature along the gun barrel distance for cryomilled stainless-steel type 316 particles during HVOF thermal spraying.[38]

4.3 Oxidation Behavior

Oxide phases formed in the thermally sprayed coatings can influence the performance of the coating.[40] Oxidation typically arises from the presence of oxygen in air and/or excess oxygen in the combustion products during HVOF spraying.[41] Oxide formation in alloys typically involves ternary semiconducting layers, in which the growth rate is dependent on the defect concentration and the dissolution of solute metal ions into the oxide layer of the solvent element.[42][43] Various oxidation mechanisms, such as the linear, parabolic, cubic, or logarithmic rate laws, have been formulated to explain the kinetics of oxide growth in various alloy systems. However, the oxidation rates are complex and often involve more than one factor.[42][43] Oxidation studies were recently conducted on the thermal spraying of cryomilled 316-stainless steel particles with varying particle sizes.[38] According to the results from the mathematical modeling, the cryomilled particles with a particle size of less than 50 μm will melt and proceed to solidify prior to particle impingement. Therefore, oxides forming during in-flight of the gas-solid phase system will be insignificant in comparison to the oxides forming during gas-liquid phase oxidation. The approximate time for the particles to reach the melting temperature is 2.5×10^{-5} sec, almost six times shorter than the time a liquid particle travels to reach the substrate.[38] Due to the high temperature associated with the flame gas, the mass transfer rate strongly depends on the chemical rate constant of the oxygen at the surface and the thermodynamic equilibrium concentration of oxygen at the droplet temperature.[44][45] The oxide layer formed around the droplet has been estimated to be approximately 1.8 nm.[38] The oxidation behavior following particle impact can be estimated from Wagner's equation for oxidation.[40][42][45] In the case of cryomilled powders with a particle size between 28 and 50 μm, the approximate thickness of the oxide scale developed ranges from 50 to 110 nm.[38] Hence, oxidation is dominant during splat formation after particle impact on the substrate.[38] Cryomilled powders with a particle size larger than 50 μm will not melt during thermal spraying, and the oxidation mechanism will be related to a diffusion controlled oxidation mechanism.

5.0 CONCLUSIONS

The present review serves as an overview of the current understanding of thermally sprayed coatings produced by the use of nanocrystalline feedstock powders produced by various synthesis techniques. The preliminary results described above highlight the importance of powder synthesis during the optimization of process parameters in order to produce powders with a high chemical homogeneity and nanocrystalline grain structure with enhanced thermal stability. The development of a cost-effective route to produce nanocrystalline feedstock powders suitable for thermal spraying is essential to fully utilize this technology for commercial applications in the future.

Nanocrystalline coatings have been observed to exhibit different physical and mechanical attributes than those of conventional coatings of comparable composition. The development of a theoretical formulation is needed, accounting for the morphology of the nanocrystalline powders, capable of predicting microstructural evolution during thermal spraying. The formulation and application of robust models to optimize the experimental parameters are also necessary for the thermal spraying of nanocrystalline systems. The application of current available diagnostic tools will ensure the reproducibility of results.

ACKNOWLEDGMENTS

The authors would like to acknowledge the financial support by the Office of Naval Research under grants No. N00014-94-1-0017, N00014-97-1-0844, and N00014-98-1-0569.

REFERENCES

1. Cheung, C., Wood, D., and Erb, U., in: *Processing and Properties of Nanocrystalline Materials*, (C. Suryanarayana, J. Singh, and F. H. Froes, eds.), p. 479, The Minerals, Metals and Materials Society, Warrendale, PA (1996)
2. Kear, B. H., and Strutt, P. R., Nanostructures: The Next Generation of High Performance Bulk Materials and Coatings, *Naval Research Reviews*, 4:4–13 (1994)
3. Kear, B. H., Skandan, G., and Sadangi, R., High Pressure Synthesis of Nanophase WC/Co/Diamond Powders: Implications for Thermal Spraying, *J. Thermal Spray Tech.*, 7:412 (1998)
4. Lavernia, E. J., Lau, M. L., and Jiang, H. G., *Thermal Spray Processing of Nanocrystalline Materials*, in: *Proceedings of the NATO Advanced Study Institute on Nanostructured Materials: Science and Technology*, (G. Chow, and N. I. Noskova, eds.), pp. 283–302, Kluwer Academic Publishers, Dordrecht, The Netherlands (1998)
5. Suryanarayana, C., Nanocrystalline Materials, *Int. Mat. Rev.*, 40:41–64 (1995)

6. Lau, M. L., Jiang, H. G., Nuchter, W., and Lavernia, E. J., Thermal Spraying of Nanocrystalline Ni Coatings, *Phys. Stat. Sol. (A)*, 166:257–268 (1998)
7. Rittner, M. N., and Abraham, T., Nanostructured Materials: An Overview and Commercial Analysis, *Int. J. Powder Metall.*, 34:33–36 (1998)
8. Gleiter, H., Nanostructured Materials: State of the Art and Perspectives, *NanoStruct. Mat.*, 6:3–14 (1995)
9. Rawers, J. C., Thermal Stability of Nanostructured Metal Alloys, *J. of Thermal Spray Technology*, 7:427 (1998)
10. Murty, B. S., and Ranganathan, S., Novel Materials Synthesis By Mechanical Alloying/Milling, *Int. Mat. Rev.*, 43:101–141 (1999)
11. Gilman, P. S., and Benjamin, J. S., *Mechanical Alloying*, in: *Annual Review of Materials Science*, (R. A. Huggins, R. H. Bube, and D. A. Vermilyea, eds.) pp. 279–300, Annual Reviews, Palo Alto, CA (1983)
12. Benjamin, J. S., and Volin, T. E., The Mechanism of Mechanical Alloying, *Metall. Trans.*, 5:1929–1934 (1974)
13. Luton, M. J., Jayanth, C. S., Disko, M. M., Matras, S., and Vallone, J., Cryomilling of Nano-Phase Dispersion Strengthened Aluminum, *Mat. Res. Soc. Symp. Proc.*, 132:79–86 (1989)
14. Johnson, P. K., Nanodyne, Incorporated, *Int'l. J. of Powder Metall.*, 34:8–10 (1998)
15. He, J., Ice, M., and Lavernia, E. J., Synthesis of Nanostructured WC-12%Co Coating Using Mechanical Milling and HVOF Thermal Spraying, *Metall. Mater. Trans.*, 31A:541–553 (2000)
16. Karthikeyan, J., Berndt, C. C., Tikkanen, J., Reddy, S., and Herman, H., Plasma Spray Synthesis of Nanomaterial Powders and Deposits, *Mat. Sci. Eng. A*, 238:275–286 (1997)
17. Pawlowski, L., *The Science and Engineering of Thermal Spray Coatings*, John Wiley & Sons, England (1995)
18. He, J., Ice, M., and Lavernia, E. J., Synthesis and Characterization of Nanostructured Cr_3C_2-NiCr, *NanoStruct. Mat.*, 10:1271–1283 (1998)
19. van den Berge, F. M. J., Thermal Spray Processes: An Overview, *Advanced Materials & Processes*, 154:31–34 (1998)
20. Lau, M. L., Gupta, V. V., and Lavernia, E. J., Particle Behavior of Nanocrystalline 316-Stainless Steel During High Velocity Oxy-Fuel Thermal Spray, *NanoStruct. Mat.*, 12:319–322 (1999)
21. Lau, M. L., Jiang, H. G., and Lavernia, E. J., *Synthesis and Characterization of Nanocrystalline 316-Stainless Steel Coatings by High Velocity Oxygen-Fuel (HVOF) Spraying*, in: *Thermal Spray: Meeting the Challenges of the 21st Century: Proceedings of the 15th International Thermal Spray Conference*, (C. Coddet, ed.), pp. 379–384, ASM International, Nice, France (1998)

22. Sobolev, V. V., Guilemany, J. M., and Martin, A. J., *Influence of Mechanical and Thermal Behaviour of Stainless Steel Powder Particles During High Velocity Oxy-Fuel (HVOF) Spraying on Properties of Coatings*, in: *Proceedings of the 15th International Thermal Spray Conference*, (C. Coddet, ed.), pp. 503–510, ASM International, Nice, France (1998)

23. Scattergood, R. O., and Koch, C. C., A Modified Model for Hall-Petch Behavior in Nanocrystalline Materials, *Scr. Metall.*, 27:1195–1200 (1992)

24. Stewart, D. A., Dent, A. H., Harris, S. J., Horlock, A. J., McCartney, D. G., and Shipway, P. H., Novel Engineering Coatings with Nanocrystalline and Nanocomposite Structures by HVOF Spraying, *J. Thermal Spray Tech.*, 7:422 (1998)

25. Stewart, D. A., Shipway, P. H., and McCartney, D. G., Abrasive Wear Behaviour of Conventional and Nanocomposite HVOF-sprayed WC-Co coatings, *Wear*, 225–229:789–798 (1999)

26. Verdon, C., Karimi, A., and Martin, J.-L., A Study of High Velocity Oxy-Fuel Thermally Sprayed Tungsten Carbide Based Coatings. Part 1: Microstructures, *Mat. Sci. & Eng. A*, 246:11–24 (1998)

27. Cetegen, B. M., and Yu, W., In-situ Particle Temperature, Velocity, and Size Measurements in DC Arc Plasma Thermal Sprays, *J. Thermal Spray Tech.*, 8:57–67 (1999)

28. Bhola, R. and Chandra, S., *Splat Solidification of Tin Droplets*, in: *Thermal Spray: Practical Solution for Engineering Problems*, (C. C. Berndt, ed.), pp. 657–663, ASM International, Materials Park, Ohio (1996)

29. Eidelman, S., and Yang, X., *Optimization of Thermal Spray Guns and Coating Processes Using Numerical Simulations*, in: *Elevated Temperature Coatings: Science and Technology II*, (N. B. Dahotre and J. M. Hampikian, eds.), p. 47, TMS, Warrendale, PA (1996)

30. Eidelman, S., and Yang, X., Three Dimensional Simulation of HVOF Spray Deposition of Nanoscale Materials, *NanoStruct. Mat.*, 9(1–8):79–84 (1997)

31. Sobolev, V. V., and Guilemany, M., Dynamic Processes During High Velocity Oxyfuel Spraying, *Int. Mat. Rev.*, 41:13–32 (1996)

32. Yang, X., and Eidelman, S., Numerical Analysis of a High-Velocity Oxygen-Fuel Thermal Spray System, *J. Thermal Spray Tech.*, 5:175–184 (1996)

33. Knotek, O., and Schnaut, U., *Process Modeling of HVOF Thermal Spraying Systems*, in: *Thermal Spray: International Advances in Coatings Technology*, (C. C. Berndt, , ed.), pp. 811–816, ASM International, Materials Park, Ohio (1992)

34. Knotek, O., Lugscheider, E., Jokiel, P., Schnaut, U., and Wiemers, A., *Chromium Coatings by HVOF Thermal Spraying: Simulation and Practical Results*, in: *Thermal Spray Industrial Applications*, (C. C. Berndt, and S. Sampath, eds.), pp. 179–184, ASM International, Materials Park, Ohio (1994)
35. Clift, R., Grace, J. R., and Weber, M. E., *Bubbles, Drops, and Particles,* Academic Press, Inc., New York (1978)
36. Ranz, W. E., and Marshall, W. R., *Chemical Engineering Progress*, 48:141 (1952)
37. Varacalle, D. J., Ortiz, M. G., Miller, C. S., Steeper, T. J., Rotolico, A. J., Nerz, J., and Riggs , W. L., II, *HVOF Combustion Spraying of Inconel Powder*, in: *Thermal Spray: International Advances in Coatings Technology*, (C. C. Berndt, ed.), pp. 181–187, ASM International, Materials Park, Ohio (1992)
38. Lau, M. L., and Lavernia, E. J., Microstructural Evolution and Oxidation Behavior of Nanocrystalline 316-Stainless Steel Coatings Produced by High Velocity Oxygen-Fuel Spraying, *Mat. Sci. Eng. A.* 272:222–229 (1999)
39. *Metals Handbook: Properties and Selection: Irons, Steels, and High-Performance Alloys*, 1:871, ASM International, Materials Park, OH (1990)
40. Smith, M. F., Dykhuizen, R. C., and Neiser, R. A., *Oxidation in HVOF-Sprayed Steel*, in: *Thermal Spray: A United Forum for Scientific and Technological Advances*, (C. C. Berndt, ed.), p. 885, ASM International, Materials Park, Ohio (1997)
41. Sobolev, V. V., Guilemany, J. M., Nutting, J., and Miquel, J. R., Development of Substrate-Coating Adhesion in Thermal Spraying, *Int. Mat. Rev.*, 42:117–136 (1997)
42. Birks, N., and Meier, G. H., *Introduction to High Temperature Oxidation of Metals,* Edward Arnold, London (1983)
43. Morris, L. A., *Resistance to Corrosion in Gaseous Atmospheres*, in: *Handbook of Stainless Steel*, (D. Peckner, and I. M. Bernstein, eds.), p. 17, McGraw-Hill Book Co., New York (1977)
44. Themelis, N. J., *Transport and Chemical Rate Phenomena*, Gordon Branch Pub., London (1995)
45. Vardelle, A., Fauchais, P., and Themelis, N. J., *Oxidation of Metal Droplets in Plasma Sprays*, in: *Advances in Thermal Spray Science & Technology*, (C. C. Berndt, and S. Sampath, eds.), pp. 175–180, ASM International, Materials Park, Ohio (1995)

3

Nanostructured Materials and Composites Prepared by Solid State Processing

Hans J. Fecht

1.0 INTRODUCTION AND BACKGROUND

Mechanical deformation under shear conditions and high strain rates (~ 10^1–10^4 s^{-1}) leads to the formation of nanostructures within powder particles, thin foils or, at the surface of metals and alloys exposed to friction-induced wear conditions. For example, mechanical attrition and mechanical alloying of powder particles has been developed as a versatile alternative to other processing routes in preparing nanoscaled materials with a broad range of chemical composition and atomic structures.[1]–[3] In this process, lattice defects are produced by "pumping" energy into initially single-crystalline powder particles of typically 50 μm particle diameter. This internal refining process with a reduction of the average grain size by a factor of 10^3–10^4 results from the creation and self-organization of small-angle and high-angle grain boundaries within the powder particles during the mechanical deformation process. As a consequence, a change of the thermodynamic, mechanical and chemical properties of these materials has been observed with the properties of nanophase materials becoming controlled by the grain size distribution and the specific atomic structure and cohesive energy of the grain or interphase boundaries.

In the 1970s, the method of mechanical attrition (MA) of powder particles followed by high temperature sintering was developed as an industrial process to successfully produce new alloys and phase mixtures. For example, this powder metallurgical process allows the preparation of alloys and composites which cannot be synthesized via conventional casting routes. This method can yield:[4]

a. Uniform dispersions of ceramic particles in a metallic matrix (superalloys) for use in gas turbines.

b. Alloys with different compositions than alloys processed from the liquid.

c. Alloys of metals with quite different melting points with the goal of improved strength and corrosion resistance.

In the 1980s, the method of high-energy milling gained a lot of attention as a non-equilibrium solid-state process resulting in materials with nanoscale microstructures. The formation of nanocrystals within initially single crystalline powder samples has been first studied systematically in pure metals and intermetallic compounds.[5] Moreover, solid-state (mechanical) alloying beyond the thermodynamic equilibrium solubility limit can lead to the formation of amorphous metallic materials as observed for a broad range of alloys.[6]–[8] The amorphous phase formation occurs by intermixing the atomic species on an atomic scale, thus softening and destabilizing the crystalline lattice[9] and driving the crystalline solid solution outside of its stability range against "melting" or amorphization.[10] This process is considered as a result of both mechanical alloying[11][12] and the incorporation of lattice defects into the crystal lattice.[13] More recent investigations demonstrate that the nanostructure formation also can occur for several unexpected cases, such as brittle ceramics, ceramic-phase mixtures, polymer blends and metal/ceramic nanocomposites.

2.0 PHENOMENOLOGY OF NANOSTRUCTURE FORMATION

The milling of materials is of prime interest in the mineral, ceramic processing, and powder metallurgy industries.[14] Typical objectives of the milling process include particle size reduction (comminution), solid-state alloying, mixing or blending, and particle shape changes. These industrial processes are mostly restricted to relatively hard, brittle materials which fracture, deform, and cold-weld during the milling operation. While oxide dispersion strengthened superalloys have been the primary application of mechanical alloying, the technique has been extended to produce a variety of non-equilibrium structures including nanocrystalline,[15] amorphous[6] and quasicrystalline[16] materials (for a review, see Ref. 17).

A variety of ball mills have been developed for different purposes including tumbler mills, attrition mills, shaker mills, vibratory mills, planetary mills, etc.[18] The basic process of mechanical attrition is illustrated in Fig. 1. Powder particles with typical particle diameters of about 50 μm are placed together with a number of hardened steel or WC coated balls in a sealed container which is shaken or violently agitated.

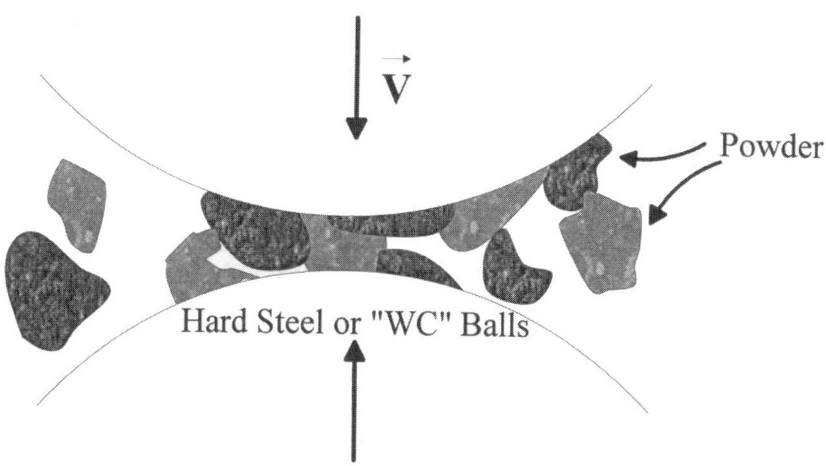

Figure 1. Schematic sketch of the process of mechanical attrition of metal powders.

High-energy milling forces can be obtained by using high frequencies and small amplitudes of vibration. Ball mills (e.g., SPEX model 8000) which are preferable for small batches of powder (i.e., ≈10 cc, sufficient for research purposes) are highly energetic, and reactions can take place by one order of magnitude faster compared with other types of mills. Since the kinetic energy of the balls is a function of their mass and velocity, dense materials (steel or tungsten carbide) are preferable to ceramic balls. Due to the continuous severe plastic deformation, a continuous refinement of the internal structure of the powder particles to nanometer scales occurs during high-energy mechanical attrition. The temperature rise during this process is modest and is generally estimated to be≤ 100 to 200°C. The collision time generally corresponds to about 2 μs.

For all nanocrystalline materials surface and interface contamination constitute major problems. During mechanical attrition, contamination by the milling tools (Fe) and atmosphere (trace elements of O_2, N_2 in rare gas) can be problematic. By minimizing the milling time and using the purest, most ductile metal powders available, a thin coating of the milling tools by the respective powder material can be obtained which reduces Fe-contamination tremendously. Atmospheric contamination can be minimized or eliminated by sealing the vial with a flexible "O"-ring after the powder has been loaded in an inert gas glove box. Small experimental ball mills can also be enclosed completely in an inert gas glove box. As a consequence, the contamination with Fe-based wear debris can generally be reduced to less than 1–2 at% and oxygen and nitrogen contamination to less than 300 ppm. In this respect, the nanoscaled powder material obtained has often a higher purity than materials synthesized by alternative methods, such as chemical processes or inert gas evaporation and condensation. In addition, the structural contributions of small pores occurring during compaction of small clusters can be safely neglected.[19]

Milling of refractory metals (e.g., tungsten) in a shaker or planetary mill for extended periods of time (> 30 hours) can result in levels of Fe-contamination of more than 10 at% if high vibrational or rotational frequencies are employed. On the other hand, contamination through the milling atmosphere can have a positive impact on the milling conditions if one intends to prepare metal/ceramic nanocomposites with one of the metallic elements being chemically highly reactive with the gas (or fluid) environment.

The deformation processes within the powder samples are important for fundamental studies of extreme mechanical deformation and the development of nanostructured states of matter with particular physical and chemical properties. Similar processes control the deformation of technologically relevant surfaces. For example, the effects of work-hardening, material transfer, and erosion during wear situations result in microstructures of wear surfaces comparable to those observed during mechanical attrition.[20] In particular, during sliding wear, large plastic strains and strain-gradients are created near the surface.[21] Similar to mechanical attrition of powder particles, this is the consequence of the formation of dislocation cell networks, subgrains, and grain boundaries with the subgrains becoming smaller near the surface.

3.0 HIGH-ENERGY BALL MILLING AND MECHANICAL ATTRITION

3.1 Examples

Metallic Elements and Intermetallics. During mechanical attrition, the metal powder particles are subjected to severe plastic deformation from collisions with the milling tools. Consequently, plastic deformation at high strain rates (~ 10^3–10^4 s^{-1}) occurs within the particles and the average grain size can be reduced to a few nanometers after extended milling. As the metal particle is plastically deformed, most of the mechanical energy expended in the deformation process is converted into heat but the remainder is stored in the metals, thereby raising the internal energy.[22][23] This was first investigated in detail for a number of high melting metals with bcc and hcp crystal structures.[5][15] Metals with fcc structure are inherently more ductile and often exhibit a stronger tendency to adhere to the container walls and to sinter into larger particles, often several millimeters in diameter, during the milling process. In a detailed study, the successful preparation of nanocrystalline fcc metals has been described.[24]

Since contamination from the milling devices (hardened steel) can occur, the experimental results discussed in the following are mainly concentrated on iron powder as a model system for mechanical attrition. The microstructural changes as a result of mechanical attrition can be followed by x-ray diffraction methods averaged over the sample volume. The x-ray diffraction patterns exhibits increasing broadening of the crystalline peaks as a function of milling time.

The peak broadening is caused by size as well as internal strain effects.[25][26] The average coherently diffracting domain size (grain or crystal size) and the microstrain as function of milling time are obtained from the integral peak widths assuming Gaussian peak shapes (Fig. 2).[27]

After corrections for K_α and instrumental broadening, the line broadening due to the small crystal size is constant in K-space and is given by $\Delta K = 0.9(2\pi/d)$ where d is the average domain or grain diameter. The strain broadening corresponds to $\Delta K = A <e^2>^{1/2} K$ with A being a constant depending on the strain distribution ($A \approx 1$ for a random distribution of dislocations)[28] and $<e^2>^{1/2}$ being the rms strain. Additional defects which might contribute to the peak broadening, such as stacking faults, can be safely neglected in all cases discussed here. However, for some metals with very small stacking fault energies, e.g., Co, the contribution of stacking faults to the peak broadening is considerable.[29]

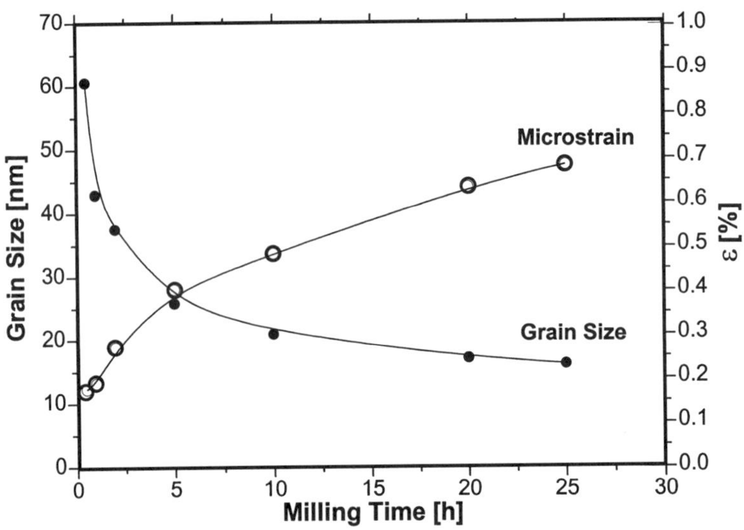

Figure 2. The average grain size and microstrains as a function of milling time for iron powder determined from x-ray line broadening.

In the very beginning, mechanical attrition leads to a fast decrease of the average grain size to 40–50 nm. Further refinement occurs slowly to about 15 nm after extended milling. The average atomic level strain reaches values up to 0.7%. Based on the method of x-ray analysis applied (Scherrer formula,[30] Williamson and Hall method at full width at half maximum or integral peak width at half maximum,[25] Warren-Averbach analysis,[31] etc.), the weighting of the grain size distribution is different and, therefore, the average grain size can vary by a factor of two.

Basically, all metals and compounds investigated so far exhibited similar behavior in terms of grain size reduction and increases of atomic level strains. Typical values for the average grain sizes of fcc, bcc, and hcp metals vary between 10 and 20 nm whereas the atomic-level strains can increase up to about 1%.[32]

It is furthermore interesting to note that by mechanical attrition, plastic deformation can be introduced in nominally brittle materials. The minimum grain or domain size for intermetallic compounds with CsCl structure has been found to vary between 12 nm for CuEr[33] and 2 nm (amorphous) for NiTi.[34] Furthermore, in the ordered intermetallic compounds, anti-site disorder is introduced during mechanical attrition leading to atomic-level strains up to 3%. Whereas, for the CsCl compounds, the reduction of the long-range chemical order parameter saturates at about 0.7, other intermetallic compounds exhibit complete disordering together with the formation of a nanocrystalline solid solution. For example, the A-15 type compounds Nb_3Al,[35] V_3Ga,[36] and Nb_3Au[37] transform after extended milling to a bcc solid solution with nanometer-sized grains.

Non-Equilibrium Crystalline and Amorphous Solid Solutions. The mechanical alloying of powder mixtures generally results in the formation of solid solutions extended in composition far beyond their equilibrium solubility limit (up to a factor of 10) and is caused by mechanically driven enhanced interdiffusion. The phenomenon of stress induced diffusion is typical if large potential gradients prevail, which lead to high-rate diffusion processes in the vicinity of a dislocation even at temperatures where self-diffusion is not possible.[38] At large deformations, layered structures or precipitates can, therefore, be resolved.

The action of deformation (i.e., driven system), especially shearing processes, in causing atomic-scale mixing, have been further clarified recently by computer simulation.[39][40] The mechanical alloying process is considered an athermal process which yields a high level of homogenization of the component atoms provided that sufficient deformation is applied. In fact, Monte Carlo simulations indicate that deformation can yield a solid solution even in alloys with a positive enthalpy of mixing. In this case, the driven system action, characterized by the forcing parameter (as the frequency ratio between forced and thermally activated jumps), results in a behavior with an enhanced diffusivity characteristic of a high-temperature high-entropy state with extended solubilities.[41]–[43]

Positive Enthalpy of Mixing. Surprisingly, mechanical alloying is achieved for powder mixtures having a positive enthalpy of mixing. Though in some cases, such as Ag-Fe, an intimate phase mixture of nanostructured Ag and Fe particles is produced,[44] in other cases real miscibility on an atomic level can be obtained, e.g., for Cu-Fe,[45] Cu-W,[46] Cu-Ta,[47] and Cu-V.[48] This apparent violation of the rules of equilibrium thermodynamics is a vital example of the potential of MA in synthesizing new materials under non-equilibrium conditions. For example, mechanical alloying can lead to the formation of nanocrystalline single-phase solid solutions of up to 60 at% Fe in Cu and 20 at% Cu in Fe.[49][50] The steady-state grain sizes range from 20 nm for Cu to 8–10 nm for Fe-rich alloys as determined by x-ray line broadening (Fig. 3).

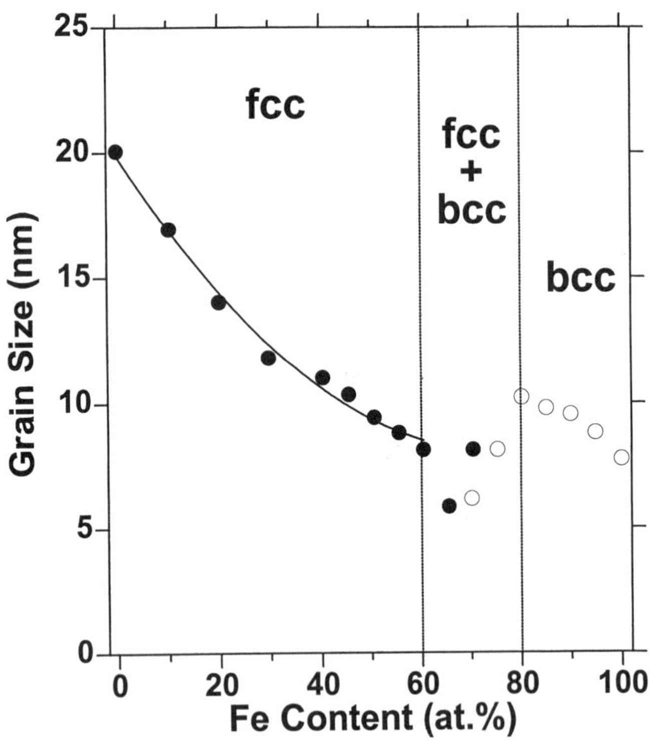

Figure 3. Average grain size for Fe_xCu_{100-x} powders after 24 hours of milling vs Fe content. *(After Ref. 50.)*

The enhanced solubility of alloys exhibiting spinodal behavior in coarse-grained systems has been attributed to the capillary pressure of the nanosized grains on the free energy due to their small radii of curvature. It has been found that during mechanical attrition of Fe-Cu powder mixtures, agglomerates of multilayers are formed leading to microstructures very similar to those obtained by cold-rolling.[51]

Furthermore, mechanical attrition can also produce ultrafine-scaled phase mixtures if a brittle material is milled together with a more ductile material. For example, 10 nm sized Ge particles can be embedded in a ductile matrix of Sn or Pb.[52] Similarly, very fine dispersions at the nanometer scale have been found, for example, in TiNi-C[53] and Ag-Fe.[54] By additional Mössbauer-studies, mutual solubility Ag in Fe domains (and Fe in Ag-domains, respectively) could be shown. As such, the resulting microstructure and chemical arrangement is very similar to nanocrystalline phase mixtures prepared by gas condensation methods.

Negative Enthalpy of Mixing/Glass Formation. Extended solid solutions far beyond the thermodynamic equilibrium have also been noted in course of mechanical milling for alloys with negative enthalpies of mixing.[55] For phase mixtures with large (> 15%) differences of atomic radii, the formation of an amorphous structure has often been observed.[56] During this process, long-range solute diffusion and, therefore, solute partitioning are suppressed.

For example, during mechanical alloying of 75a/o Zr and 25a/o Al, the formation of a supersaturated hcp (α-Zr) solid solution was observed prior to the solid-state-amorphization reaction.[57] However, in all cases of binary alloys, it remained unclear whether indeed a metallic glass has been formed or just an x-ray amorphous structure. More recently, a similar phase transformation sequence was found in a mechanically alloyed multicomponent elemental $Zr_{60}Al_{10}Ni_9Cu_{18}Co_3$ powder mixture with a bulk glass forming composition (Fig. 4).[58] The x-ray spectra at different stages of the milling process are characterized by the successive disappearance of the elemental Al, Co, Cu, and Ni peaks and a simultaneous shift of the Zr-peaks to higher-scattering angle, corresponding to a decrease in the lattice constant of the hcp-Zr as a result of the rapid dissolution of the smaller atoms, such as Cu, Ni, Co, and Al in the (α-Zr) matrix.[59]

Calorimetric investigations further revealed that a glass transition indeed occurred when the amorphous material was heated into the undercooled liquid state without prior crystallization. Figure 5 shows the corresponding increase of the specific heat capacity when the metallic glass sample, prepared in the solid state, was heated through the glass transition. Similar observations concerning the crystal-to-glass transition by mechanical attrition have also been observed for the equilibrium intermetallic phase mixture.[60] These rather surprising results confirm that a metallic glass can really be formed in the solid state (without cooling from the liquid state) by destabilizing a crystal due to the incorporation of static disorder causing high elastic stresses.[61]

Ceramics. Ductile materials can be deformed as described above, but it is surprising that nominally brittle materials, such as ceramics, also show a refined microstructure after extended mechanical attrition. For example, ZrO_2 has been milled for up to forty hours resulting in a grain size reduction to several nanometers. At a grain size of less than about 12 nm, a transition of the most stable monoclinic structure to a metastable orthorhombic modification has been observed.[62] A simple estimate shows that below a critical grain size of about 30 nm, the high-temperature phase becomes more stable than the low-temperature phase due to the defects incorporated.

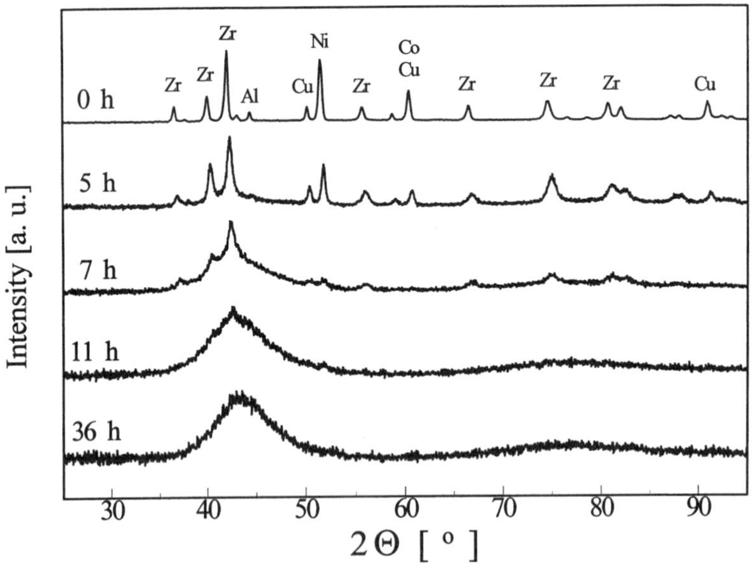

Figure 4. X-ray spectra for mechanically alloyed $Zr_{60}Al_{10}Ni_9Cu_{18}Co_3$ powder samples exhibiting a transition from the initially crystalline powder mixture to an amorphous glass-like structure.

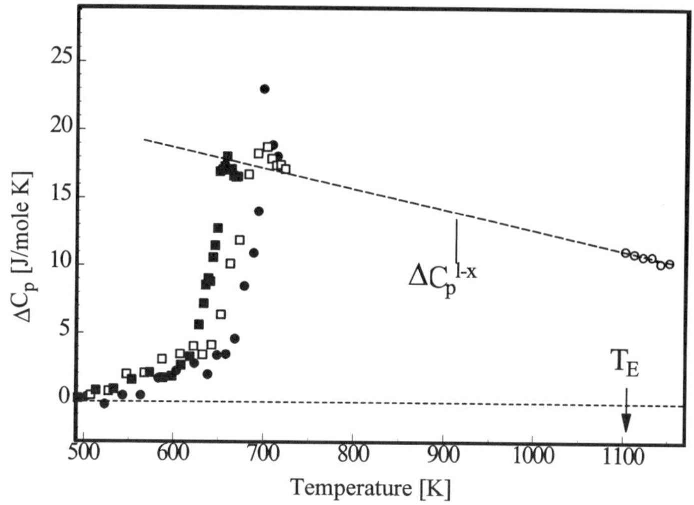

Figure 5. Thermal analysis exhibiting the difference in heat capacity of the sample material $Zr_{60}Al_{10}Ni_9Cu_{18}Co_3$ in comparison with the thermodynamically stable crystalline configuration for amorphous mechanically alloyed powder, bulk metallic glass, and foil stacks.

Mechanical alloying does occur for ceramic powder mixtures as well. For example, $(Fe,Cr)_2O_3$ solid solutions can be obtained from a Fe_2O_3/Cr_2O_3 powder mixture with an average grain size of about 20 nm.[63] Similar observations have been reported for ZrO_2/Y_2O_3 powder mixtures.[64] Also, chemical processes can be induced by milling ceramic materials. For example, extended periods of mechanical milling can lead to the deoxidation of oxides when appropriate materials are added. Mechanochemical reactions have been reported for α-Fe_2O_3/Ti, Ag_2O/C,[65] and CuO/Ca[66] mixtures.

Polymer Blends. Mechanical alloying of polymeric materials has been developed during the past decade as well. Similar to metallic materials, mechanical attrition leads to an increase of the internal energy. For example, polyamide (PA), polyethylene (PE), acrylonitrile-butadiene-styrene (ABS), polypropylene (PP), and polystyrene (PS) have been investigated in detail.[67] In order to fracture the polymer chains, the milling process is conducted below the respective glass transition temperatures. As a result, the crystallinity of the powder material can be decreased considerably by mechanical milling. The corresponding storage of energy allows consolidation of the powder to bulk samples at lower temperatures than conventionally processed material. In addition, the achieved mechanical properties are considerably improved.

The milling process also allows mechanical alloying of polymer mixtures (PA/PE, PA/ABS)[68] as well as mechanical alloying of polymers with ceramic (PP/SiC) and metal powder (PS/Sn, PP/Al, PE/Cu).[69] By mechanical milling, the normal compatibility restrictions for polymer formation from regular polymer science principles have been removed. This unique opportunity allows synthesizing of new materials and materials combinations with enhanced properties, which can not be achieved by any other method.

Nanocomposites. Mechanical attrition is also a very versatile process for preparing nanocomposites. Since this process is sensitive to contamination resulting from the milling environment, atmospheric control can be used to purposely induce chemical reactions between the milled powders and their environment. By a proper choice of a reactive gas atmosphere (O_2, N_2, etc.), or a milling fluid (organic fluids), the metal powder can be intentionally modified by reactive milling to a nanocrystalline metal-ceramic composite,[70] or fully reacted to a nanocrystalline ceramic, for example, a metal nitride.[71]

The metal powders (Ti, Fe, V, Zr, W, Hf, Ta, and Mo)[72]–[74] transform to a nanocrystalline nitride by high-energy ball milling under nitrogen gas flow. This solid-state interdiffusion reaction during reactive ball milling is triggered by fragmentation of the starting powder, thus creating new surfaces. These freshly created surfaces react with the flowing nitrogen gas to form a nitride surface layer over the unreacted core particle. With further milling, this reaction continues and a homogeneous nitride phase is formed and the unreacted core of metal disappears resulting in a nanostructured (often metastable) metal-nitride with a typical grain size of 5 nm.

By ball milling in organic fluids such as surfactants which are sometimes used to prevent contamination by the milling tools, chemical reactions can be induced leading to the formation of fine carbides. For example, by milling Al (-Ti, -Zr, or -Hf) alloys in hexane, an average grain size of 9 nm can be achieved with carbon dissolved in the matrix.[75] During dynamic compaction at about 1300 K, grain growth occurs up to about 44 nm together with precipitation of ZrC particles, 7 nm in size. Such ultrafine grained composites are expected to exhibit considerably improved strength and ductility.[76]

More recently, metallic glass/ceramic composites were obtained by mechanical alloying of multicomponent Zr-based elemental metallic powders together with SiC particles. [77][78] A secondary electron microscopy (SEM) image of such a $Zr_{65}Al_{7.5}Cu_{17.5}Ni_{10}$ + 10 vol% SiC metallic glass/ceramic composite after a milling time of thirty hours (Fig. 6) reveals a uniform distribution of fine SiC particles in the metallic glass powder matrix, as proven by further x-ray diffraction and EDX analysis. The size distribution of the SiC particles ranges from 1 μm down to values below 50 nm. It is further interesting to note that the SiC particles do not act as potent heterogeneous nucleation sites when the composite is heated to the crystallization temperature above the glass transition temperature, making mechanical alloying a convenient method for achieving dispersion-strengthened amorphous alloys with considerably improved strength and wear resistance by a powder metallurgical pathway.

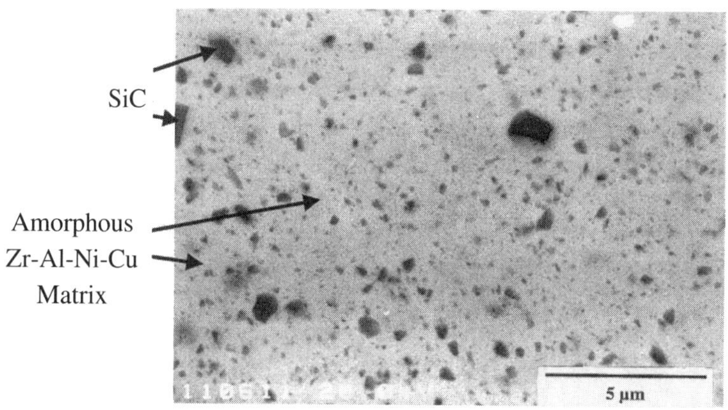

Figure 6. Scanning electron microscope image of a metal/ceramic composite of 10 vol% SiC particles in an amorphous matrix ($Zr_{65}Al_{7.5}Cu_{17.5}Ni_{10}$) prepared by mechanical alloying.

3.2 Mechanism of Grain Size Reduction

From wide angle x-ray spectra, the information about lattice defects (grain boundaries, dislocations, etc.) is obtained via their disturbing influence on the coherent superposition of radiation diffracted at the atomic lattice sites which causes the broadening of Bragg peaks. In small angle neutron scattering experiments (SANS), the lattice defects themselves give rise to a scattering contrast because of the (scattering length) density fluctuations associated with them.

Ball-milled Fe powder samples were measured in quartz cuvettes at a neutron wavelength ℓ of 0.60 nm. The cuvettes were filled up with D_2O resulting in a 79% reduction of scattering contrast at the wetted surfaces of the powder particles.

Different positions between the sample and the area-sensitive detector were chosen covering a range of momentum transfer $q = (4p/\ell) \sin(q)$ from 0.045 nm^{-1} to 0.85 nm^{-1} ($2q$: scattering angle). During the measurements a homogeneous magnetic field of 0.7 T was applied to the sample in the horizontal direction perpendicular to the incoming neutron beam.

For a magnetically saturated sample, the scattered intensity as a function of the vector of momentum transfer, q, can be written as $I(q) = I_N(q) + I_M(q) \sin^2 \alpha$ where $I_N(q)$ and $I_M(q)$ represent the structure function of the nuclear and magnetic scattering respectively,[79] and α is the azimuthal angle between q and the magnetic field projected on the area perpendicular to the incoming beam. The anisotropic intensity distributions were analyzed by radial averaging over angular sectors parallel and perpendicular to the direction of the applied magnetic field. While in the first case, $\alpha = 0°$, the spectra thereby obtained represent the nuclear scattering contribution, the perpendicular averaging, $\alpha = 90°$, yields a linear combination of nuclear and magnetic scattering. Radial distribution functions (RDF) were calculated from the SANS spectra by the indirect Fourier-transformation method.[80]

Figure 7 shows the SANS spectra of Fe powder samples before milling and after milling for 0.5 hours and 30 hours averaged parallel to the applied magnetic field. Compared to the spectrum of the unmilled sample, an increase of scattering intensity occurs after 0.5 hours milling over the entire q-range covered by the measurements. This increase may be explained by the refinement of microstructure in the early stages of the milling process (i.e., the scattering contribution of grain boundaries, dislocations, and triple junctions) which is proved by the x-ray peak broadening. It should be noted that the increase of scattering intensity extends to the high q range implying that structural inhomogeneities on a small length scale of a few nanometers are present after short milling times. This observation underlines that the structural refinement is strongly inhomogeneous and might occur in shear bands of high dislocation density surrounded by less deformed sample regions.

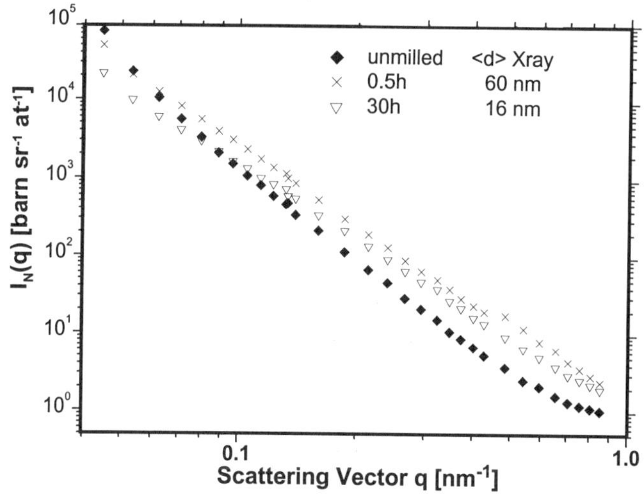

Figure 7. SANS spectra of ball-milled Fe powder after different milling times (0 hours, 0.5 hours, and 30 hours).

After 30 hours milling time, a drop of the scattering intensity is observed over the entire q-range compared to the 0.5 hours sample. The drop at high q values is especially surprising since, from x-ray measurements, it is known that the average grain size is further reduced to a volume average of about 16 nm, which is in good agreement with the average grain size derived from the RDF of about 15 nm after 30 hours milling. Furthermore, the RDF also shows a decrease in magnitude without any significant shift of the maxima (Fig. 8).

Obviously, the observed change of the SANS intensity cannot be solely explained by a shift of crystallite size distribution to smaller distances in real space during the milling process. Instead, it is believed that the scattering contrast due to dislocations plays an important role for the interpretation of the measured data. Since the volume dilatation in the vicinity of dislocations is small, their nuclear scattering contrast is small and disappears for pure screw dislocations.[81] Their magnetic scattering contrast due to orientation fluctuations of the magnetic moments of Fe atoms may exceed their contribution to nuclear scattering by factors up to 10–100.[82] These fluctuations are caused by magnetoelastic coupling between the magnetic moments and the dislocation strain field.

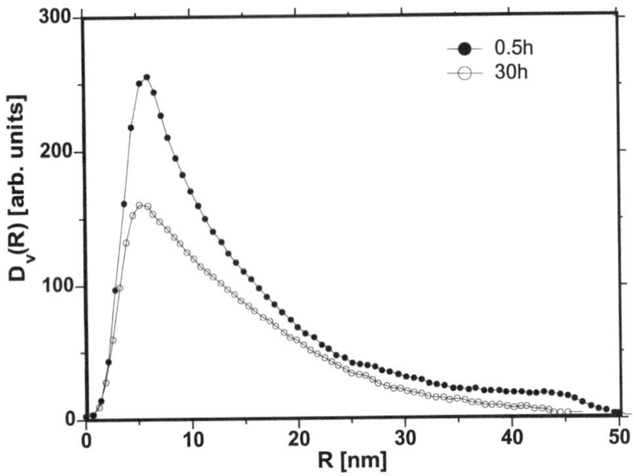

Figure 8. Radial distribution functions (volume weighted) calculated from spectra in Fig. 7 (milling times 0.5 hours, and 30 hours).

The SANS data may be interpreted with respect to the magnetic scattering contribution of dislocations as a result of a changing distribution of dislocations in the deformed material. According to this model, the increase of scattering intensity at high q after 0.5 hours milling is caused by an increasing dislocation density with an average dislocation distance of a few nanometers which is in the range of the final grain size after long milling times. The subsequent decrease of scattering intensity after 30 hours milling time may be caused by the rearrangement of dislocations to form grain boundaries and the absorption of dislocations as secondary grain boundary dislocations.

Furthermore, the grain refinement which can be followed only in an integral way by x-ray or neutron scattering methods has been investigated in more detail by TEM on intermetallic AlRu powder samples.[83] Figure 9 shows a series of TEM micrographs of an AlRu particle after mechanical attrition. It can be seen that the crystal is heavily strained and the deformation occurs in a rather inhomogeneous way. The arrows in Fig. 9a indicate a highly deformed region of a width of about 1 μm which extends throughout the entire particle. These shear bands have been observed in rolled metals and are typical for deformation mechanisms that occur at high strain rates in contrast to slip and twinning mechanisms at low and moderate strain rates.[84]

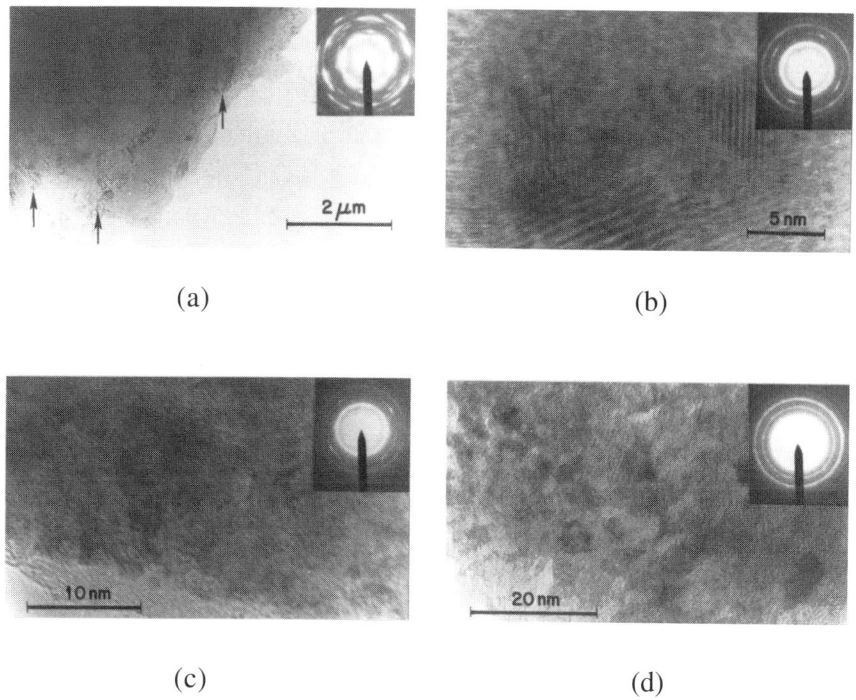

Figure 9. (*a*) TEM bright field images at relatively low magnification of an AlRu powder particle after 10 minutes of mechanical attrition. The arrows point to highly deformed regions (shear bands). The inset shows the corresponding diffraction pattern demonstrating the gradual smearing out of the initially sharp diffraction spots. (*b*) TEM high-resolution bright-field image with its corresponding diffraction patterns of AlRu after 10 minutes. (*c*) Two hours and (*d*) 64 hours of mechanical attrition.

The observed shear bands are separated by areas of similar lateral dimensions in the micrometer range having low defect densities. High resolution imaging of areas in the shear bands reveal a microstructure consisting of individual grains with a diameter of approximately 20 nm which are slightly rotated with respect to each other at a rotation angle of less than 20° as shown in Fig. 9b. With longer durations of mechanical attrition, the shear bands grow over larger areas and eventually (Fig. 9c) the entire sample disintegrates into subgrains with a final grain size of 5–7 nm for AlRu after 64 hours (Fig. 9d), thus ductilizing the originally brittle intermetallic compound.

The elemental processes leading to the grain size refinement include three basic stages:

i. Initially, the deformation is localized in shear bands consisting of an array of dislocations with high density.

ii. At a certain strain level, these dislocations annihilate and recombine to small angle grain boundaries separating the individual grains. The subgrains formed via this route are already in the nanometer size range (about 20–30 nm).

iii. The orientations of the single-crystalline grains with respect to their neighboring grains become completely random.

This can be understood in the following way. The yield stress, σ, required to deform a polycrystalline material by dislocation movement is related to the average grain size, *d,* by $\sigma = \sigma_0 + kd^{-1/2}$, where σ_0 and k are constants (Hall-Petch relationship).[85][86] An extrapolation to nanocrystalline dimensions shows that very high stresses are required to maintain plastic deformation. Experimental values for k and σ_0 are typically $k = 0.5$ MNm$^{-3/2}$ and $\sigma_0 = 50$ MPa.[87] For a grain size of 10 nm, the minimum yield stress is of the order of 5 GPa corresponding to 15% of the theoretical shear stress, which sets a limit to the grain size reduction achieved by plastic deformation during ball milling. Therefore, the reduction of grain size to a few nanometers is limited by the stresses applied during ball milling as long as no dramatic elastic softening of the crystal lattice occurs.

Further energy storage by mechanical deformation is only possible by an alternative mechanism. Grain boundary sliding has been observed in many cases at high temperatures leading to superplastic behavior. Alternatively, grain boundary sliding can also be achieved at a very small grain size and low temperature by diffusional flow of atoms along the intercrystalline interfaces which allows the synthesis of ductile ceramics.[88] This provides a mechanism for the self-organization and rotation of the grains observed here, thus increasing the energy of the grain boundaries proportional to their misorientation angle and excess volume.

This behavior is typical for deformation processes of bcc metals and intermetallic compounds at high strain rates. However, it is surprising that nominally brittle materials, such as intermetallics, develop considerable ductility under shear conditions.

3.3 Property—Microstructure Relationships

Decreasing the grain size of a material to the nanometer range leads to a drastic increase of the number of grain boundaries, reaching typical densities of 10^{19} interfaces per cm^3. The large concentration of atoms located in the grain boundaries, in comparison with the crystalline part, scales roughly with a reciprocal grain size dependence $1/d$.

Consequently, due to their excess free volume, the grain boundaries in nanocrystalline materials can cause considerable differences in the physical properties compared to the conventional polycrystalline counterpart.

Thermal Properties. As a result of the cold work, energy has been stored in the powder particles. During heating in a Differential Scanning Calorimeter (DSC), a broad exothermic reaction is generally observed. Integrating the exothermal signals gives the energy release, ΔH, during heating of the sample.

For example, the stored enthalpy reaches values up to 7.4 kJ/mol (after 24 hours) and 10 kJ/mol (after 32 hours) for Ru, which corresponds to 30–40% of the heat of fusion, ΔH_f. One would expect that the recovery rates during the milling process correlate with the melting point of the specific metal. With the exceptions of Co (due to a large number of stacking faults) and Hf, Nb and W (possibly due to an increased level of Fe-impurities from the milling tools stabilizing the nanostructure) such a relationship is indeed observed. Similar results have been obtained for metals with fcc structure as well.[89] Consequently, most effective energy storage occurs for metals with melting points above 1500 K resulting in average grain sizes between 6 nm (Ir) and 13 nm (Zr). For the compound phases, similar high values for the stored energies are found ranging from 5 to 10 kJ/mol and corresponding to values between 18 and 39% of the heat of fusion for grain sizes between 5 and 12 nm.

The final energies stored during mechanical attrition largely exceed those resulting from conventional cold working of metals and alloys (cold rolling, extrusion, etc.). During conventional deformation, the excess energy is rarely found to exceed 1–2 kJ/mol and, therefore, is never more than a small fraction of the heat of fusion.[90] In the case of mechanical attrition, however, the energy can reach values typical for crystallization enthalpies of metallic glasses, corresponding to about 40% ΔH_f.

A simple estimate demonstrates that these energy levels can not be achieved by the incorporation of defects which are found during conventional processing. In the case of pure metals, the contribution of point defects (vacancies, interstitials) can be safely neglected because of the high recovery rate at the actual processing temperature. Even taking non-equilibrium vacancies into account, which can form as a consequence of dislocation annihilation up to concentrations of 10^{-3},[91] such contributions are energetically negligible in comparison. On the other hand, for intermetallics, point defects are relevant in order to describe the stability of the material.

The maximum dislocation densities which can be reached in heavily deformed metals are less than 10^{16} m^{-2} which would correspond to an energy of less than 1 kJ/mol. Therefore, it is assumed that the major energy contribution is stored in the form of grain boundaries, and related strains within the nanocrystalline grains which are induced through grain boundary stresses.

Large differences generally also arise in the specific heat, c_p, at constant pressure. The specific heat of the heavily deformed powder particles was measured in the range from 130 K to 300 K, i.e., at temperatures low enough to prevent the recovery processes from taking place. For all samples, a considerable increase in c_p has been found experimentally after 24 hours milling, reaching values up to 15% for Ru. For pure metals, a linear correlation between the increase of the heat capacity Δc_p at 300 K, and the stored enthalpy, ΔH, given as a percentage of the heat of fusion ($\Delta H/\Delta H_f$) after extended mechanical attrition is observed (Fig. 10). Such a relationship is also predicted by the free volume model for grain boundaries.[92]

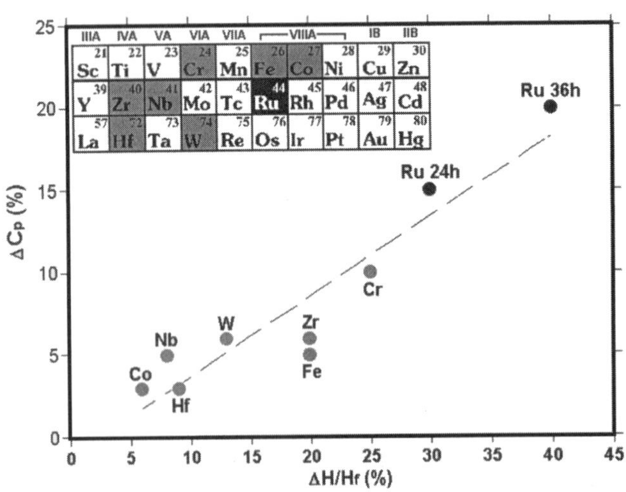

Figure 10. Specific heat increase Δc_p (%) in comparison to the unmilled state at room temperature as function of the stored enthalpy ΔH (given as percentage of ΔH_f) after 24 hours ball milling of pure elemental powder samples.

This energetic microstructure-property relationship is further emphasized by Fig. 11. Here the stored enthalpy, ΔH, in attrited Fe powder is shown as a function of average reciprocal grain size, $1/d$, since $1/d$ scales also with the volume density of grain boundaries in the nanocrystalline material ($\sim 3\delta/d$ where δ is the thickness of the grain boundary).[93] Two different regimes can be clearly distinguished: For small grain size reductions at the early stages of mechanical attrition, i.e., stage (*i*), the stored enthalpy shows only a weak grain size dependence typical for dislocation controlled deformation processes. After the average domain (grain) size is reduced below $d^* = 30–40$ nm, energy storage becomes more efficient. The critical grain size, d^*, corresponds to the size of nanograins which are formed within the shear bands. Therefore, for $d < d^*$, a regime can be identified where deformation is controlled by the properties of the small angle and, later, high angle grain boundaries which are developing in stages (*ii*) and (*iii*). The slope of the corresponding strain versus $1/d$ relationship corresponds in the grain boundary regime to 0.1 nm, typical for atomic level strains.[52]

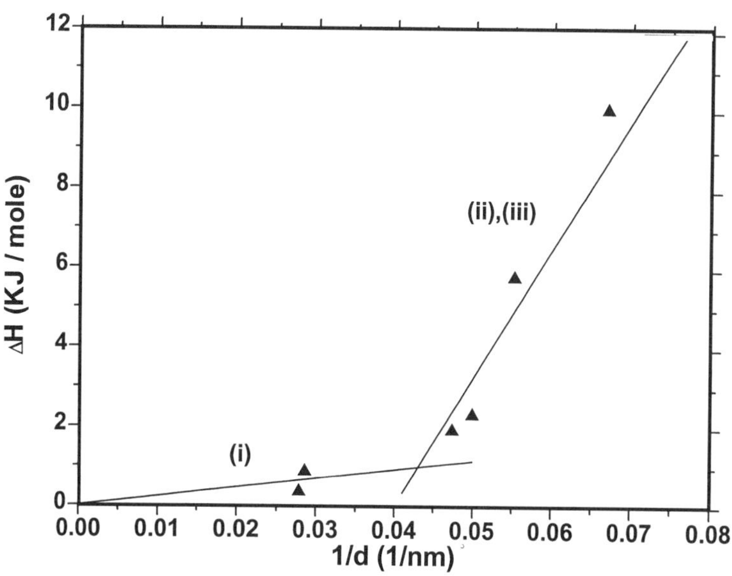

Figure 11. The stored enthalpy ΔH as function of reciprocal grain size $1/d$ of Fe at different levels of mechanical attrition. Two distinctively different stages can be observed: stage (*i*) which is dislocation-controlled for $d > 40$ nm and stages (*ii*),(*iii*) for $d < 40$ nm where deformation becomes controlled by grain boundaries.

Mechanical Properties. As a further consequence of the grain size reduction, a drastic change in the mechanical properties has been observed (for a review, see Refs. 94 and 95). In general, nanocrystalline materials exhibit very similar mechanical behavior to that of amorphous materials due to shear banding as the prevalent mechanism of deformation.[96] Strain hardening is not observed and thus conventional dislocation mechanisms are not operating. The lack of dislocations is the result of the image forces which act on dislocations near grain boundaries.

Local mechanical properties can be measured by nano-indentation methods. Here, the load as well as the indentation depth is monitored continuously during the loading and unloading process (Fig. 12). Typical results for nanocrystalline Fe powder samples exhibit an increase in hardness by a factor of 7 (9 GPa for *nx*-Fe with *d* about 16 nm versus 1.3 GPa for annealed *px*-Fe). In general, the hardness follows a trend similar to the Hall-Petch relationship, though the dislocation-based deformation mechanism in the nanocrystalline regime certainly does not apply as shown in Fig. 13. The Young's modulus can be measured by this method as well, and typically shows a decrease of about 10% compared to the polycrystal.

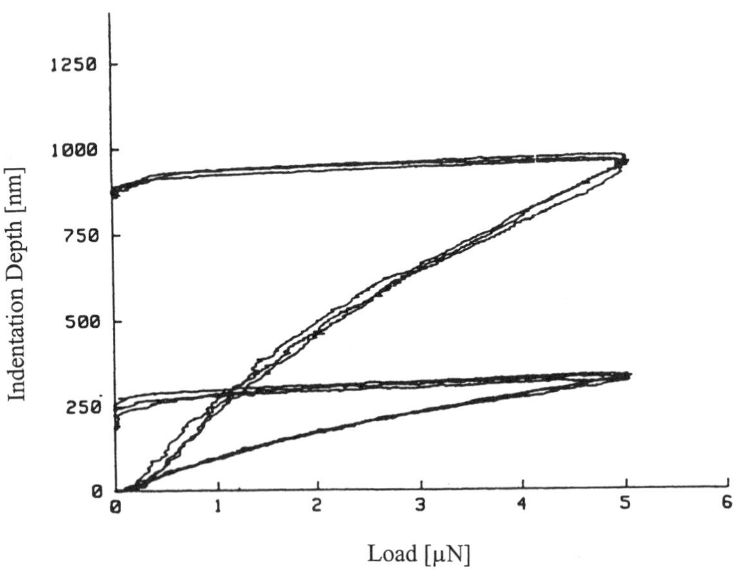

Figure 12. Hardness measurement using a nano-indentation device on polycrystalline *(upper curves)* and nanocrystalline *(lower curves)* attrited iron powder samples.

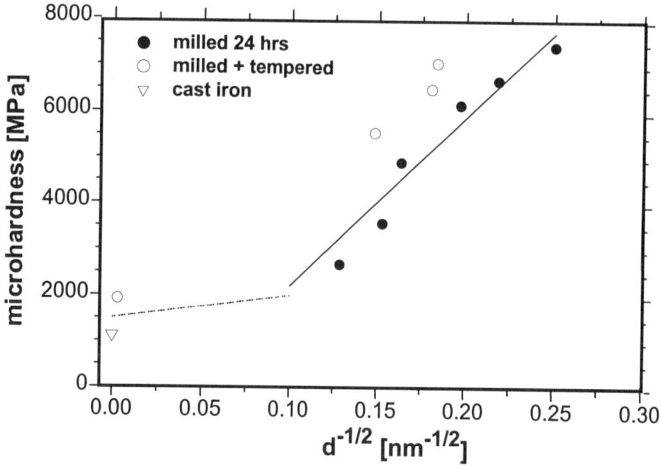

Figure 13. Hall Petch relationship for the hardness of *nx*- and *px* iron.

It is suggested that the mechanical properties of nanophase materials prepared by mechanical attrition after extended periods of milling are not being controlled by the plasticity of the crystal due to dislocation movement anymore but, rather, by the cohesion of the nanocrystalline material across its grain boundaries. From the considerable increase of hardness and the principle changes of the deformation mechanisms, improved mechanical properties can be expected as attractive features for the design of advanced materials as bulk or layered materials.

4.0 PHASE STABILITY AT ELEVATED TEMPERATURES

As a result of the cold work, considerable energy has been stored in the powder particles (see Figs. 10 and 11). Therefore, thermodynamically, these materials are far removed from their equilibrium configuration, and a large driving force towards equilibrium exists. On the other hand, for a number of applications, the nanocrystalline powder has to be compacted to a uniform bulk sample. Since cold compaction is usually not sufficient, Hot Isothermal Pressing (HIP) processes[97][98] are usually applied; therefore, it is important to understand the atomic processes occurring during annealing over extended periods of time.

The stored energy is released during heating to elevated temperatures due to recovery-relaxation processes within the boundaries, and grain growth. As a consequence, during annealing at elevated temperatures, relaxation and grain growth processes will occur, leading to a concomitant increase of the grain size.

This behavior has been investigated for iron in detail.[99][100] For extended periods of milling time, a decrease of the average grain size to nanometer dimensions is observed with a stationary average grain size d = 16 nm and 0.7% microstrain as shown in Fig. 2. The enthalpy release during a DSC heating experiment spreads over the entire temperature range of the scan as shown in Fig. 14. The very broad signal does not exhibit any distinct events but a further increase of the exothermic signal for $T > 250$–300°C.

X-ray diffraction of powder samples annealed for 80 minutes at each temperature revealed the evolution of grain size and strain as function of annealing temperature as shown in Fig. 15. The microstrain decreases rapidly below 200°C while the grain size remains nearly constant, so the enthalpy release during the first exotherm in Fig. 14 is only related to relaxation and not to grain growth. Grain growth starts to become significant above about 300°C. Furthermore, it has been found that, after a fast increase at early times, the average grain size d changes from 16 nm to about 30–40 nm. The average grain size remains constant for $t \geq 2400$ sec and reaches values of 100–200 nm at temperatures about 600°C.

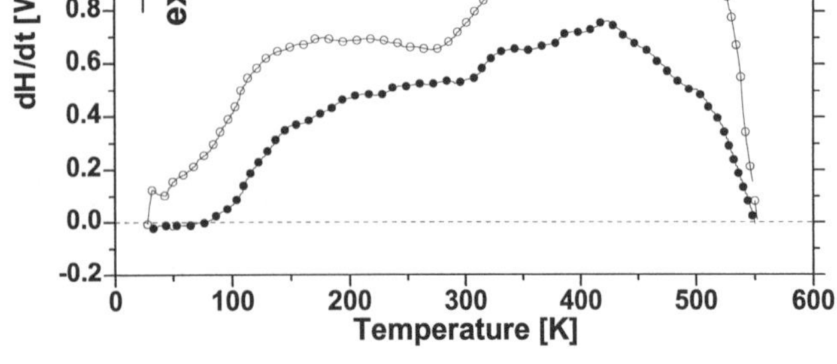

Figure 14. DSC heating scan at 10 K/min of iron powder after mechanical attrition for 5 and 25 hours.

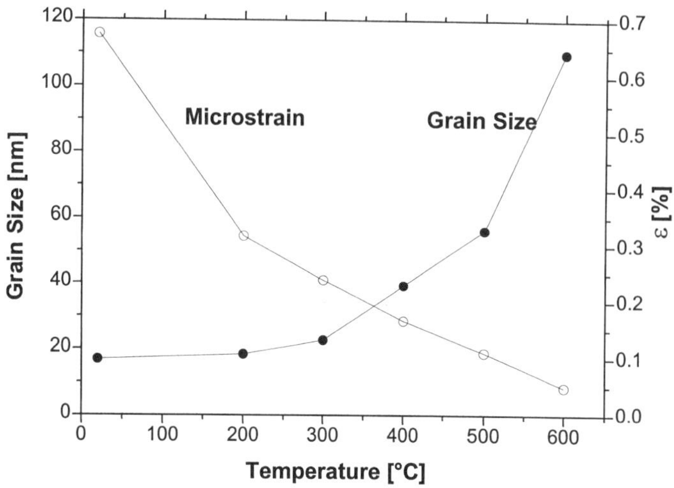

Figure 15. Dependence of stationary grain size and microstrain on annealing temperature.

Two regimes with and without grain growth can be distinguished. Since the influence of lattice point defects and lattice dislocations is negligible, the enthalpy release can be clearly assigned to the existence of grain boundaries. The reduction of the microstrains is probably caused by grain boundary relaxation and annihilation of secondary grain boundary dislocations. Based on elastic theory, it is estimated that this contribution to the overall energy is less than about 5%. On this basis, the grain boundary energy can be estimated.

By simple geometric considerations[101][102] the specific grain boundary excess enthalpy is estimated to be about 2.1 J/m^2. This would correspond to a value for non-equilibrium, unrelaxed, grain boundaries, whereas after relaxation, the grain boundary energy is reduced to 1.5 J/m^2. Values resulting from computer simulations suggest excess enthalpies between 1.2 and 1.8 J/m^2.[103] Therefore, we conclude that grain boundaries in the as-prepared state are characterized by increased values of about 25% due to their unrelaxed atomic structure or secondary grain boundary dislocations.

Isothermal DSC measurements allow further analysis of grain growth processes in nanocrystalline Fe. For example, the isothermal DSC curve shown in the upper part of Fig. 16 was measured at 500°C after annealing the sample at 400°C and heating to 500°C at a rate of 50°C/min. A monotonically decreasing signal typical for grain growth is observed. Similar signals are obtained at 200°C, 300°C, and 400°C and clearly differ from those measured in isothermal recrystallization processes controlled by nucleation and growth in conventional polycrystalline metals which are described by Johnson Mehl Avrami type models. Figure 16 does not exhibit the expected maximum related to an incubation time for nucleation, but shows only a decrease in the signal.

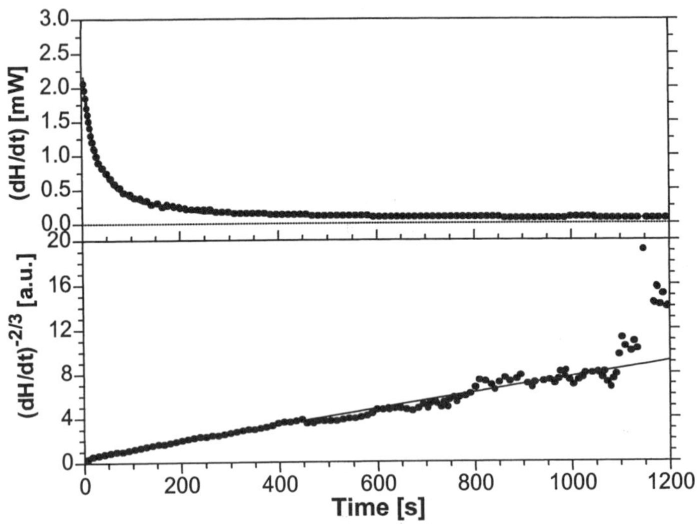

Figure 16. Isothermal exothermic DSC curve at 500°C of nanocrystalline iron *(upper part)* and plot of $(\Delta H/dt)^{-2/3}$ versus time *t (lower part).*

Furthermore, $(dH/dt)^{-2/3}$ should scale linearly with time if normal parabolic grain growth behavior is assumed. This assumption is well approximated for $t < 1200$ sec as shown in the lower part of Fig. 16. The upper part of Fig. 16 includes a fit to the measured DSC signal assuming parabolic grain growth. Based on the available data, an activation energy of 178 kJ/mol has been determined for grain growth in nanocrystalline Fe, which is comparable to the activation energy for grain boundary self diffusion in Fe.[104]

5.0 SEVERE PLASTIC DEFORMATION

5.1 General

Similar observations regarding the deformation mechanisms during mechanical attrition have been reported in chips removed during machining,[105] and simple metal filings,[106][107] as well as during extreme deformation of bulk materials.[108][109] Analogous to the mechanically attrited powder at the early stage, large inhomogeneities have been observed in filings with the deformation process leading to the formation of small angle grain boundaries. Here, the dislocation cell size dimensions are basically a function of the acting shear stress, τ, resulting in an average cell size dimension L of $L = 10\ Gb/\tau$ where G is the shear modulus and b is the Burgers vector.[110]

More detailed studies are obtained from cold rolling and torsion,[111] wire drawing,[112] and cyclic deformation[113] processes resulting in an asymptotic saturation of the flow stresses. This is considered to be a result of the simultaneous occurrence of dislocation multiplication and annihilation leading to a saturation of the dislocation density. In particular, under cyclic deformation of, for example, Cu, at amplitudes above $\gamma_{pl} \sim 10^{-4}$, slip becomes highly localized in so-called "persistent slip bands" (shear bands). These lie parallel to the primary glide plane and are separated by regions containing the original matrix structure. These bands consist of dense walls of dislocations; largely, screw dislocations having a density ~ 10^{13} m^{-2}. The closest spacing between screw dislocations of opposite sign is ~ 50 nm, the minimum distance before annihilation occurs. For edge dislocations, which are more relevant for the deformation of fcc crystals, this critical annihilation length is found to be 1.6 nm for Cu. It has been concluded that the annihilation of dislocations can set a natural limit to the dislocation densities which can be achieved by plastic deformation (typically less than 10^{13} m^{-2} for screw dislocations and 10^{16} m^{-2} for edge dislocations). Steady-state deformation is observed when the dislocation multiplication rate is balanced by the annihilation rate. This situation corresponds to the transition of stage (*i*) to stages (*ii*) and (*iii*) as described above. In this stage, the role of dislocations becomes negligible and further deformation occurs via grain boundaries. It is expected that the shear modulus of the grain boundary regions lowers by about 40% when the "volume-fraction" of the grain boundaries becomes comparable to that of the crystals.[114][115] Localized deformation then proceeds by the dilatation of the grain-boundary layers similar to superplastic behavior[116] with the undeformed crystallites moving in a "sea" of dilated grain-boundaries. Furthermore, the relative motion of the crystalline grains within the shear band leads to impingement which should give rise to large, locally inhomogeneous elastic stresses. As a consequence, in order to relax these strains, formation of nanovoids about 1 nm in diameter is expected to occur which inevitably leads to crack formation under tensile stress.[117] Such a deformation mode basically also provides a mechanism for the repeated fracturing and rewelding of the fresh surfaces during mechanical attrition leading to a steady-state particle size.

5.2 Cold Rolling of Thin Sheets

An alternate route to producing samples with high levels of both plastic deformation and interfacial area is by cold-rolling layered elemental sheets which are folded between each deformation cycle. In this case, the large increase in interfacial area is created internally with absolutely negligible contamination. Similarly, in contrast to mechanical attrition, the uncertainty in the temperature during processing is removed since the sample is in firm contact with the massive rolls, and deformation can be performed at a low strain rate to maintain ambient temperatures. This approach has been used to examine amorphous phase formation in several binary alloys such as Zr-Ni,[118] Cu-Er,[119] and Al-Pt[120] and also for the preparation of bulk Fe/Ag nano-multilayers with giant magnetoresistance.[121]

In the prior work on amorphous phase formation, deformation rates in excess of 1 sec^{-1} were employed and some annealing was needed to complete the amorphization reaction. Recently, fully amorphous foils of a multicomponent $Zr_{65}Al_{7.5}Cu_{17.5}Ni_{10}$ alloy have been synthesized at ambient temperatures from a layered array of individual elemental sheets by repeated low-strain-rate (0.1 sec^{-1}) cold rolling.[122] Figure 17 shows x-ray diffraction patterns from the Zr-Al-Ni-Cu foils taken after 10, 40, 80, and 120 deformation cycles. Each deformation cycle consisted of rolling the multilayer sandwich to a thickness of approximately 80 μm and subsequent folding. High resolution TEM analysis exhibits further evidence that a true amorphous phase has been formed as shown in Fig. 18.

The x-ray spectrum of the amorphous sample is basically identical to that of a metallic glass produced by liquid quenching with the same composition. Thermal analysis of the cold rolled amorphous sample with

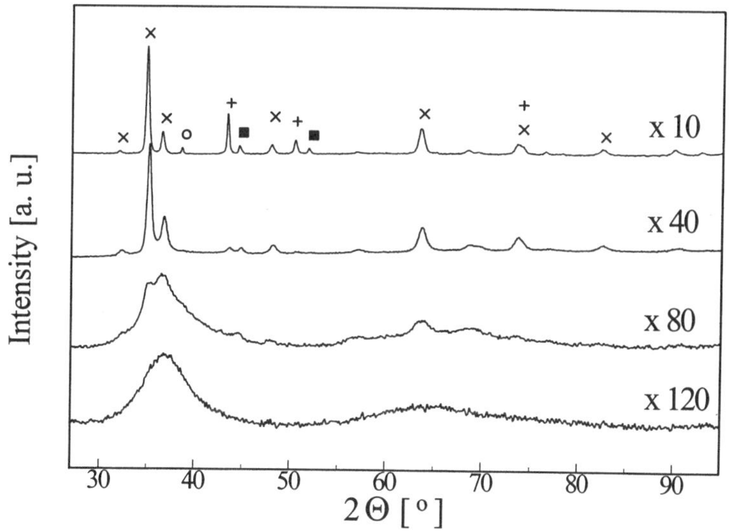

Figure 17. XRD of cold rolled thin foils of composition $Zr_{65}Al_{7.5}Cu_{17.5}Ni_{10}$.

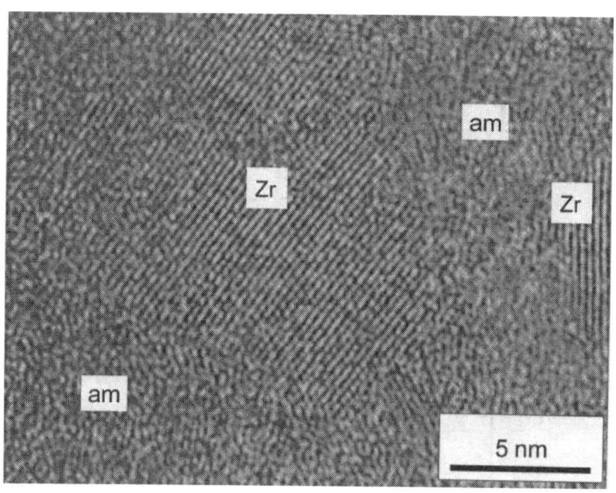

Figure 18. High resolution TEM image of nanocrystalline/glass transition during cold rolling of thin foils of composition $Zr_{65}Al_{7.5}Cu_{17.5}Ni_{10}$.

a DSC reveals a distinct glass transition at $T_g = 647$ K followed by a sharp exothermic crystallization peak at 745 K. Note that very similar to amorphization reactions observed in mechanically alloyed Zr-based powder mixtures of similar composition (see "Non-Equilibrium Crystalline and Amorphous Solid Solutions" in Sec. 3.1), the initial stage of cold rolling is characterized by the dissolution of solute into Zr along with a reduction in grain size to about 30 nm before the onset of the crystal-to-glass transition. The formation of similar amorphous phases from two inherently different initial states, i.e., the solid and the liquid state, suggest that compositionally induced static disorder in a mechanically driven system can lead to the same final glass state which is conventionally derived from freezing the dynamic disorder of a liquid out to a glass.

The present examples demonstrate that cold rolling for the synthesis of nanostructured materials or multicomponent metallic glasses is an attractive alternative to more traditional techniques such as mechanical attrition or liquid undercooling. Due to the relatively simple experimental conditions, size restrictions do not limit the formation of bulk samples.

5.3 Friction-Induced Surface Modifications

Many microscopic processes occur during mechanical attrition and mechanical alloying of powder particles which exhibit features in common with processes relevant in tribology and wear. For example, the effects of work hardening, material transfer, and erosion during wear situations result in microstructures of the wear surface which are similar to those observed during mechanical attrition.[123][124] In particular, during sliding wear, large plastic strains and strain-gradients are created near the surface. Typical plastic shear strain rates can correspond here to several 10^3 sec^{-1}.

Close to the surface of wear scars, as well as in the wear debris of Cu, nanocrystalline structures have been observed by high-resolution electron microscopy with an average grain size of 4–5 nm.[125] Within the interiors of the grains, no defects were observed, suggesting that most of the defects are absorbed by the grain boundaries due to their proximity. However, this type of plastic deformation at high strain rates does not seem to be limited to metals and alloys,[126] but has been observed in ceramics[127] and diamond[128] as well.

During sliding wear, a special tribological layer develops on the surface of a sliding component being subjected to large plastic strains. This surface layer often is called the *Beilby layer,* which, for a long time, was thought to be amorphous because its microstructure could not be resolved with the instruments commonly used.[129] There are indeed some systems in which truly amorphous layers are produced by sliding[130] but, in most cases, the sub-surface layer with a thickness of several micrometers has a nanocrystalline structure.

For example, during ultrasonic shot-peening, the formation of nanocrystalline Fe-surfaces has been observed where the initially coarse-grained structure in the surface layer was refined into equiaxed ultrafine grains (about 10 nm) with random crystallographic orientation as shown in Fig. 19.[131][132]

As a further example of technical relevance, the development of high speed trains reaching velocities higher than 300 km/h is also a materials challenge concerning the mechanical integrity and safety required for the railway tracks.[133][134] In particular, the interaction and slip between wheel and rail has been optimized and is controlled by sophisticated electronics whereas the materials for the rail have not been changed in two decades. In particular, on the steel surface (Fe-0.8 at% C-1.3 at% Mn) where the local pressure typically exceeds 1.0–1.5 GPa, solid state transformations have been observed which are caused by friction-induced shear forces and have strong similarities to mechanical attrition of powder samples.

Corresponding x-ray diffraction and TEM results indicate that the average grain size of the extremely deformed surface layer corresponds to about 20 nm, whereas a gradient in grain size is observed further away from the surface, reaching values up to 200 nm.[135] For example, Fig. 20 exhibits TEM photographs and the corresponding diffraction patterns of the initial pearlitic structure (a) in comparison with the nanocrystalline layer near the surface (b).

As a consequence, hardness measurements have been performed using a nanoindenter at small loads as shown in Fig. 21. Steep hardness gradients have been found in cross section with a lateral resolution of a few micrometers which are clearly correlated with the change in microstructure. As a result, the hardness is increased from typical values for the pearlitic steel of approximately 2.5 GPa to 13 GPa next to the surface.

This remarkable increase in hardness and mechanical strength of regions near the surface is clearly related to the fact that the average grain size is considerably decreased by the continuous deformation process. Similar results have been obtained for mechanically attrited α-Fe and α-Fe-C-powder processed in a ball-milling device. Here also, an increase of hardness by a factor of five has been observed.[136] During mechanical attrition of rail filings (identical composition as rail), a decrease in grain size to 7 nm after 50 hours of milling time has been observed together with a dissolution of the carbides. The same observation holds for the highly deformed rail surface, i.e., a dissolution of carbides and supersaturation of the α-Fe with carbon. However, due to the powder milling experiments, it is obvious that the main contribution to the hardness increase results from the grain size reduction and only partially from the highly strained martensite-like structure of the bcc iron supersaturated with carbon.

Moreover, the wear resistance of the nanostructured areas is increased by a factor of two. Fretting wear measurements typical for the type of wear in wheel-rail contact reveal wear rates of 1.55×10^{-5} mm^3m^{-1} for the nanostructured layer and 3.77×10^{-5} mm^3 m^{-1} for the undeformed surface. The large improvement of the mechanical properties clearly shows the importance of the formation of nanostructures for technologically relevant wear problems.

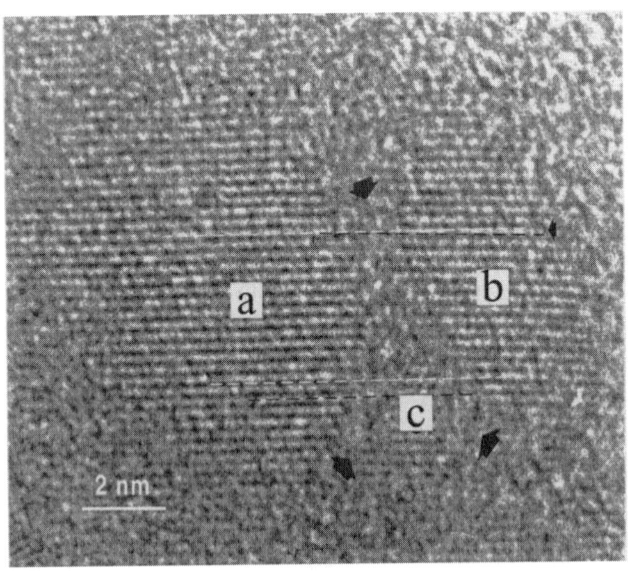

Figure 19. Nanocrystalline iron produced by means of ultrasonic shot peening. *(Courtesy of K. Lu.)*

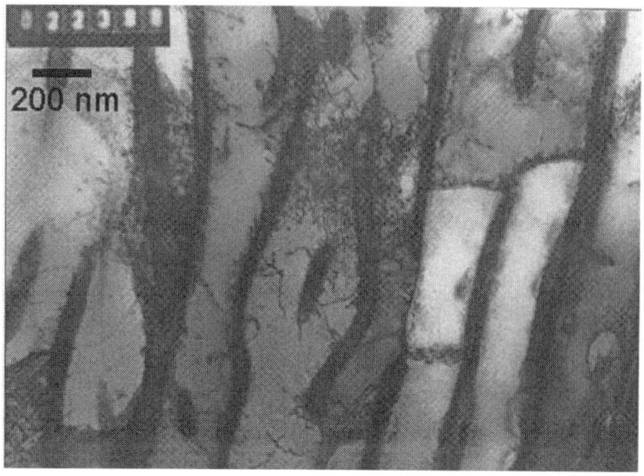

(a)

(b)

Figure 20. TEM images of (*a*) the pearlitic structure of a low alloyed Fe-C-Mn steel and (*b*) the nanocrystalline structure of an extremely worn surface of a high speed railway track.

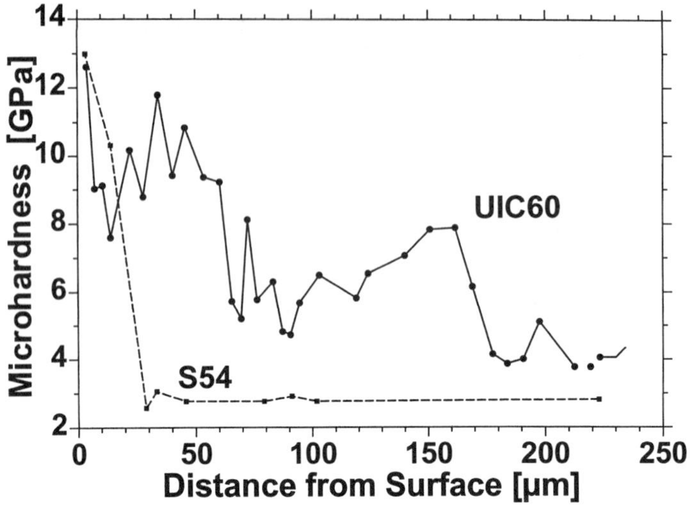

Figure 21. Hardness vs depth (distance from surface) for a high speed railway track with a nanostructured surface due to extreme wear conditions.

6.0 SUMMARY AND OUTLOOK

The solid-state processing methods of mechanical attrition and mechanical alloying have been developed as a versatile alternative to other processing routes in preparing nanoscaled materials with a broad range of chemical compositions and atomic structures. In this process, lattice defects are produced within the initially single-crystalline powder particles. The internal refining process with a reduction of the average grain size by a factor of $10^3 - 10^4$ results from the creation and self-organization of dislocation cell networks and the subsequent formation of small-angle and high-angle grain boundaries within the powder particles during the mechanical deformation process. As a consequence, a change of the thermodynamic, mechanical, and chemical properties of these materials has been observed with the properties of nanophase materials becoming controlled by the grain size distribution and the specific atomic structure and cohesive energy of the grain or interphase boundaries. Such a transition from dislocation controlled properties to grain boundary controlled properties is expected for nanocrystalline materials synthesized by other methods as well.[137]

Mechanical attrition offers interesting perspectives in preparing nanostructured powders with a number of different interface types in terms of structure (crystalline/crystalline, crystalline/amorphous) as well as atomic bonding (metal/metal, metal/semiconductor, metal/ceramic etc.). Due to the broad range of possible atomic structures, very different properties in comparison with conventional materials are obtained. For example, nano-structured particles prepared by mechanical attrition can exhibit unusually high values in hardness,[138] enhanced hydrogen solubility,[139][140] enhanced catalytic properties,[141] magnetic spin-glass behavior,[142] etc. This opens exciting possibilities for the preparation of advanced materials with particular grain- or interphase-boundary design.

It is expected that the study of mechanical attrition and alloying processes in the future not only opens new processing routes for a variety of advanced nanostructured materials but also improves the understanding of technologically relevant deformation processes on a nanoscale level.

ACKNOWLEDGEMENTS

The continuous financial support by the Deutsche Forschungsgemeinschaft (G. W. Leibniz program) and BMBF (OPTIKON, contract 03N3050G7) is gratefully acknowledged. The author would like to thank all the colleagues who have contributed to this topic over the past ten years for the collaboration and stimulating discussions, in particular Drs. W. L. Johnson (Caltech USA), J. H. Perepezko (UW-Madison, USA), H. Gleiter (Karlsruhe, Germany), A. Sagel (DaimlerChrysler AG, Germany), C. Ettl (Ulm, Germany), and M. Djahanbaksh (Ulm, Germany).

REFERENCES

1. Fecht, H. J., *Nanomaterials; Synthesis, Properties and Applications,* (A. S. Edelstein and R. C. Cammarata, eds.), Institute of Physics Publ., Bristol and Philadelphia (1966)
2. Gleiter, H., *Prog. Mat. Sci.*, 33:223 (1989)
3. *Nanophase Materials*, (G. C. Hadjipanayis and R. W. Siegel, eds), Kluwer Acad. Publ. (1994)

4. Benjamin, J. S., *Sci. Amer.*, 234:40 (1976)
5. Fecht, H. J., Hellstem, E., Fu, Z., and Johnson, W. L., *Adv. Powder Metallurgy*, 1:111 (1989)
6. Koch, C. C., Cavin, O. B., McKamey, C. G., and Scarbrough, J. O., *Appl. Phys. Lett.*, 43:1017 (1983)
7. Schwarz, R. B., Petrich, R. R., and Saw, C. K, *J. Non-Cryst. Solids*, 76:281 (1985)
8. Hellstern, E., and Schultz, L., *Appl. Phys. Lett.*, 48:124 (1986)
9. Ettl, C., and Samwer, K., *Mat. Sci. and Eng A*, 178:245 (1994)
10. Johnson, W. L., *Prog. Mat. Sci.*, 30:81 (1986)
11. Fecht, H. J., and Johnson, W. L., *Nature*, 334:50 (1988)
12. Shingu, P. H., (ed.), Mechanical Alloying, *Mat. Sci. Forum*, pp. 88–90 (1992)
13. Fecht, H. J., *Nature*, 356:133 (1992)
14. Benjamin, J. S., *Mat. Sci. Forum*, 88:1 (1992)
15. Fecht, H. J., Hellstem, E., Fu, Z., and Johnson, W. L., *Metall. Trans.A*, 21:2333 (1990)
16. Eckert, J., Schultz, L., and Urban, K., *Appl. Phys. Lett.*, 55:117 (1989)
17. Koch, C. C., *Mechanical Milling and Alloying in Materials Science and Technology*, (R. W. Cahn, P. Haasen, and E. J. Kramer, eds.), *Processing of Metals and Alloys,* 15:193, VCH, Weinheim (1991)
18. Kuhn, W. E., Friedman, I. L., Summers, W., and Szegvari, A., Powder Metallurgy, *ASM Metals Handbook,* 7:56, Metals Park, OH (1985)
19. Schaefer, H. E., Würschum, R., Gessmann, T., Stöckl, G., Scharwaechter, P., Frank, W., Valiev, R. Z., Fecht, H. J., and Moelle, C., *Nanostructured Materials,* 6:869 (1995)
20. Rigney, D. A., Chen, L. H., Naylor, M. G. S., and Rosenfield, A. R., *Wear,* 100:195 (1984)
21. Rigney, D. A., *Ann. Rev. Mater. Sci.*, 18:41 (1988)
22. Bever, M. B., Holt, D. L., and Titchener, A. L., *Prog. Mat. Sci.*, 17:5 (1973)
23. Miani, F., and Fecht, H. J., *Int. J. of Refract. Met. and Hard Mater.*, 17:133 (1999)
24. Eckert, J., Holzer, J. C., Krill, C. E., III, and Johnson, W. L., *J. Mat. Res.*, 7:1751 (1992)
25. Williamson, G. K., and Hall, W. H., *Acta Metall.*, 1:22 (1953)
26. Wagner, C. N. J., and Boldrick, M. S., *J. Mat. Sci. Eng. A*, 133:26 (1991)
27. Moelle, C., Ph.D. Thesis, Technical University, Berlin (1995)

28. Friedel, J., *Dislocations*, p. 418, Pergamon Press, Oxford (1964)
29. Moelle, C., Schmauss, M., and Fecht, H. J., unpublished research (1992)
30. Guinier, A., *X-ray Diffraction*, p. 121, W. H. Freemand and Co., San Francisco (1963)
31. Warren, B. E., and Averbach, B. L., *J. Appl. Phys.*, 21:595 (1950)
32. Koch, C. C., *Nanostructured Materials*, 2:109 (1993)
33. Hellstern, E., Fecht, H. J., Fu, Z., and Johnson, W. L., *J. Mater. Res.*, 4:1292 (1989)
34. Yamada, K., and Koch, C. C., *J. Mat. Res.*, 8:1317 (1993)
35. Oehring, M, and Bormann, R., *J. de Phys.*, (Paris), Colloq., 51(C-4):169 (1990)
36. Bakker, H., Di, L. M., and Lo Cascio, D. M. R., *Sol. State Phen.*, 23–24:253 (1992)
37. Di, L. M., and Bakker, H., *J. Appl. Phys.*, 71:5650 (1992)
38. Gleiter, H., *Acta Metallurgica*, 16:455 (1968)
39. Bellon, P., and Averback, R. S., *Phys. Rev. Lett.*, 74:1819 (1995)
40. Martin, G., and Bellon, P., *Solid State Physics*, 50:688 (1997)
41. Pochet, P., Bellon, P., Boulanger, L., Chaffron, L., and Martin, G., *Materials Science Forum*, 269–272:655 (1998)
42. Ma, E., He, J. H., and Schilling, P. J., *Phys. Rev. B*, 55:5542 (1997)
43. Pabi, S. K., Das, D., Mahapatra, T. K., and Manna, I., *Acta Met. Mat.* (1998)
44. Shingu, P. H., Huang, B., Niskitani, S. R., and Nasu, S., *Suppl. to Trans. JIM*, 29:3 (1988)
45. Yavari, A. R., Desré, P. J., and Benameur, T., *Phys. Rev. Lett.*, 68:2235 (1992)
46. Gaffet, E., Louison, C., Harmelin, M., and Faudet., F., *J. Mat. Sci. Eng. A*, 134:1380 (1991)
47. Veltl, G., Scholz, B., and Kunze, H. D., *J. Mat. Sci. Eng. A*, 134:1410 (1991)
48. Fukunaga, T., Mori, M., Inou, K., and Mizutani, U., *J. Mat. Sci. Eng. A*, 134:863 (1991)
49. Yavari, A. R., and Desré, P. J., in: *Ordering and Disordering in Alloys*, (A. R. Yavari, ed.), p. 414, Elsevier (1992)
50. Eckert, J., Birringer, R., Holzer, J. C., Krill, C. E., III, and Johnson, W. L., *J. Materials Research* (1993)
51. Bordeaux, F., and Yavari, A. R., *Z. Metallkd.*, 81:130 (1990)

52. Yang, J. S. C., and Koch, C. C., *J. Mat. Res.*, 5:325(1990)

53. Schlump, W., and Grewe, H., in: *New Materials by Mechanical Alloying Techniques*, (E. Arzt, and L. Schultz, eds.), p. 307, DGM-Oberursel (1989)

54. Shingu, P. H., Ishihara, K. N., and Kuyama, J., *Proc. of the 34th Japan Congress on Materials Research*, p. 19, The Society of Materials Science, Kyoto, Japan (1991)

55. Schwarz, R. B., *Materials Science Forum*, 269–272:269 (1998)

56. Schwarz, R. B., and Johnson, W. L., (eds.), Solid-State Amorphization Transformation, *Less Comm. Metals*, p. 140 (1988)

57. Fecht, H. J., Han, G., Fu, Z., and Johnson, W. L., *J. Appl. Phys.*, 67:1744 (1990)

58. Sagel, A., Wunderlich, R. K., Perepezko, J. H., and Fecht, H. J., *Appl. Phys. Lett.*, 70:580 (1997)

59. Sagel, A., Wanderka, N., Wunderlich, R. K., Schubert-Bischoff, N., and Fecht, H. J., *Scipta Mater.*, 38:163 (1998)

60. Sagel, A., Wunderlich, R. K., and Fecht, H. J., *Materials Letters,* 33:123 (1997)

61. Fecht, H. J., *Mat. Sci. and Eng. A*, 179/180:491 (1994)

62. Qi, V., and Fecht, H. J., *Materials Science Forum*, 269–272:187 (1998)

63. Michel, D., Mazerolles, L., and Chichery, E., *Materials Science Forum*, 269–272:99 (1998)

64. Tonejc, A. M., and Tonejc, A., *Materials Science Forum*, 225–227:497 (1996)

65. Tokumitsu, K., Nasu, T., Suzuki, K., and Greer, A. L., *Materials Science Forum*, 269–272:181 (1998)

66. Schaffer, G. B., and McCormick, P. G., *Appl. Phys. Lett.*, 55:45 (1989)

67. Shaw, W. J. D., *Materials Science Forum*, 269–272:19 (1998)

68. Pan, J., and Shaw, W. J. D., *J. of Appl. Poly. Sci.*, 56:57 (1995)

69. Karttunen, M., and Ruuskanen, P., *Mater. Sci. Forum*, 269–272:849 (1998)

70. Fu, Z., Fecht, H. J., and Johnson, W. L., *Mat. Res. Soc. Symp. Proc.*, 186:169 (1991)

71. Chalka, A., and Williams, J. S., *Mat. Sci. Forum*, 88–90:787 (1992)

72. El-Eskandarany, M. S., Sumiyama, K., Aoki, K., and Suzuki, K., *Mat. Sci. Forum*, 88–90:801 (1992)

73. Calka, A., *Appl. Phys. Lett.*, 59:1568 (1991)

74. Calka, A., and Williams, J. S, *Mat. Sci. Forum*, 88–90:787 (1992)

75. Schwarz, R. S., Desch, P. B., Srinivasan, S., and Nash, P., *Nanostructured Materials*, 1:37 (1992)
76. Schulson, E. M., and Barker, D. R., *Scripta Met.*, 17:519 (1983)
77. Lu, I. R., Moelle, C., Sagel, A., Wunderlich, R. K., Perepezko, J. H., and Fecht, H. J., *Materials Letters*, 35:297 (1998)
78. Moelle, C., Lu, I. R., Sagel, A., Wunderlich, R. K., Perepezko, J. H., and Fecht, H. J., *Mater. Sci. Forum*, 269–272:47 (1998)
79. Hayter, J. B., *J. Appl. Phys.*, 21:737 (1988)
80. Glatter, O., *J. Appl. Cryst.*, 10:415(1977)
81. Atkinson, H. H., and Hirsch, P. B., *Phil. Mag.*, 3:213 (1958)
82. Kronmüller, H., Seeger, A., and Wilkens, M., *Zeitschrift für Physik,* 171:291 (1963)
83. Hellstern, E., Fecht, H. J., Garland, C., and Johnson, W. L., *Mat. Res. Soc. Symp. Proc.*, 132:137 (1989)
84. Cottrell, A. H., in: *Dislocations and Plastic Flow in Crystals*, p. 162, Caledron Press, Oxford (1972)
85. Hall, E. O., *Proc. Phys. Soc.,* London, 643:747 (1951)
86. Petch, N. J., *J. Iron Steel Inst.*, London, 173:25 (1953)
87. Meyers, M. A., and Chawla, K. K., *Mechanical Metallurgy*, p. 494, Prentice-Hall, Englewood Cliffs, NJ (1984)
88. Karch, J., Birringer, R., and Gleiter, H., *Nature*, 330:556 (1987)
89. Eckert, J., Holzer, J. C., Krill, C. E., III, and Johnson, W. L., *J. Mat. Res.*, 7:1751 (1992)
90. Bever, M. B., Holt, D. L., and Titchener, A. L., *Prog. Mater. Sci.*, 15:5 (1973)
91. Essmann, U., and Mughrabi, H., *Phil. Mag. A*, 40:40 (1979)
92. Fecht, H. J., *Phys. Rev. Lett.*, 65:610 (1990)
93. Kuwano, H., Ouyang, H., and Fultz, B, *Nanostructured Materials*, 1:143 (1992)
94. Morris, D. G., *Mechanical Behaviour of Nanostructured Materials*, Trans Tech Publications Ltd., Switzerland (1998)
95. Koch, C. C., *J. of Metastable and Nanocrystalline Materials*, 2–6:565 (1999)
96. Fecht, H. J., *Phil. Magazine B*, 76:495 (1997)
97. Porat, R., Berger, S., and Rosen, A., *Mat. Sci. Forum*, 225–227:629 (1996)
98. Pfullmann, T., Oehring, M., Bohn, R., and Bormann, R., *Mat. Sci. Forum*, 225–227:757 (1996)

99. Moelle, C., and Fecht, H. J., *Nanostructured Materials*, 6:421 (1995)

100. Fecht, H.-J., and Moelle, C., *Materials Research Society Symposium Proceedings,* 457:113, *Nanophase and Nanocomposite Materials II,* (S. Komarneni, J. C. Parker, and H. J. Wollenberger, eds.) (1997)

101. Tschöpe, A., Birringer, R., and Gleiter, H., *J. Appl. Phys.*, 71:5391 (1992)

102. Chen, L. C., and Spaepen, F., *J. Appl. Phys.*, 69:679 (1991)

103. Wolf, D., *Phil. Mag. A*, 62:447 (1990)

104. Malow, T. R., and Koch, C. C., *Materials Science Forum*, 225–227:595 (1996)

105. Turley, D., *J. Inst. Metals*, 99:271 (1971)

106. Wagner, C. N. J., *Acta Met.*, 5:477 (1957)

107. Warren, B. E., *X-ray Studies of Deformed Metals in Prog. Metals Phys.*, 3:147 (1956)

108. Valiev, R. Z., Korznikov, A. V., and Mulyukov, R. R.,*Mat. Sci. and Eng. A*, 168:141 (1993)

109. Valiev, R. Z., Islamgaliev, R. K., Stolyarov, V. V., Mishra, R. S., and Mukherjee A. K., *Mater. Sci. Forum*, 269–272:969 (1998)

110. Kuhlmann-Wilsdorf, D., and Van der Merwe, J. H., *J. Mat. Sci. Eng.*, 55:79 (1982)

111. Haessner, F., and Hemminger, W., *Z. Metallk*, 69:53 (1978)

112. Lloyd, D. J., and Kenny, D., *Scripta Metall.*, 12:903 (1978)

113. Grosskreutz, J. C., and Mughrabi, H., *Constitutive Equations in Plasticity*, (A.S. Argon, ed.), p. 251, Cambridge, Massachusetts: MIT Press

114. Gilman, J. J., *J. Appl. Phys.*, 46:1625 (1975)

115. Donovan, P. E., and Stobbs, W. M., *Acta Metall.*, 31:1 (1983)

116. Hatherly, M., and Malin, A. S., *Scripta Metall.*, 18:449 (1984)

117. Goods, S. M., and Brown, L. M., *Acta Metall.*, 27:1 (1979)

118. Atzmon, M., Verhoeven, J D., Gibson, E. D., and Johnson, W. L., *Appl. Phys. Lett.*, 45:1052 (1984)

119. Atzmon, M., Unruh, K. M., and Johnson, W. L., *J. Appl. Phys.*, 58:3865 (1985)

120. Bourdeaux, F., and Yavari, A. R., *J. Appl. Phys.*, 67:2385 (1990)

121. Yasuna, K., Terauchi, M., Otsuki, A., Ishihara, K. N., and Shingu, P. H., *App. Phys. Lett.*, 82:2435 (1997)

122. Sagel, A., Sieber, H., Fecht, H. J., and Perepezko, J. H., *Phil. Mag. Lett.*, 77:109 (1998)

123. Ivanisenko, Y. V., Baumann, G., Fecht, H. J., Safarov, I. M.,. Korznikov, A. V., and Valiev, R. Z., *The Physics of Metals and Metallography*, 83:303 (1997)

124. Fecht, H. J., *Nanostructured Materials*, 6:33 (1995)

125. Ganapathi, S. K., and Rigney, D. A., *Scripta Metall.*, 24:1675 (1990)

126. Doyle, F. D., and Aghan, R. L., *Metall. Trans. B*, 6:143(1975)

127. Mehrotra, P. K., *Proc. Int. Conf. on Wear of Materials*, p. 194, Reston, VA, ASME, New York (1983)

128. Humble, P., and Hannink, R. H. J., *Nature*, 273:37 (1978)

129. Beilby, G., *Aggregation and Flow of Solids*, Macmillan (London) (1921)

130. Askenasy, P., Ph.D. Thesis, California Institute of Technology (1992)

131. Tao, N. R., Sui, M. L., Lu, J., and Lu, K., *Nanostructured Materials*, 11:433 (1999)

132. Lu, K., private correspondence (1999)

133. Baumann, G., Fecht, H. J., and Liebelt, S., *Wear*, 191:133 (1996)

134. Baumann, G., and Fecht, H. J., *Nanostr. Mat.*, 7(1/2):237 (1996)

135. Bürkle, G., Thesis, Ulm University (1999)

136. Siegel, R. W., and Fougere, G. E., in: *Nanophase Materials*, (G. C. Hadyipanayis and R. W. Siegel, eds.), p. 233, Kluwer Academic Press (1994)

137. Nieman, G. W., Weertman, J. R., and Siegel, R. W., *J. Mater. Res.*, 6:1012 (1991)

138. Kehrel, A., Moelle, C., and Fecht, H. J., in: Ref. 3.

139. Moelle, C., and Fecht, H. J., *Nanostructured Materials*, 3:93 (1993)

140. Ram, S., Fecht, H. J., Haldar, S., Ramachandrarao, P., and Banerjee, H. D., *Phys. Rev. B*, 56:1 (1997)

141. Zaluski, L., Zaluska, A., Tessier, P. Ström-Olsen, J. O., and Schulz, R., *Mater. Sci. Forum*, 225–227:853 (1996)

142. Zhou, G. F., and Bakker, H., *Phys. Rev. Lett.*, 72:2290 (1994)

4

Nanocrystalline Powder Consolidation Methods

Joanna R. Groza

1.0 INTRODUCTION

To take advantage of the unique properties of bulk nanocrystalline materials, the nanometer range powders have to be densified into parts of certain properties, geometry, and size. The key to the nanopowder consolidation process is to achieve densification with minimal microstructural coarsening and/or undesirable microstructural transformations. In addition, the fully dense specimen must be of sufficient size for reliable testing of final properties or a useful final product. Attempts to produce and densify nanopowders started as early as 1968.[1] These efforts were related to sintering MgO to achieve superplastic behavior. In the 80s, when nanopowder production was initiated on a larger scale, attention was directed to nanopowder processing, as well. However, densification commonly resulted in either grain coarsening, or unacceptably small specimen size and insufficient bonding. This severely limited the assessment of nanomaterial properties, particularly mechanical. Uncertain processing of nanopowders produced artifacts that generated at least some of the controversies on nanomaterial properties. For instance, lower Young's modulus and ductility values are now attributed to the remaining pores, oxides, or incomplete particle bonding from nanopowder processing.[2][3] The early 90s emphasized the need to develop reproducible processing methods for manufacturing nanopowders into sizable parts that retain nanometer features. The past half-decade has brought significant advances in the practice and theory of nanosintering which consequently resulted in the production of fully dense parts with nanometer grain size (considered throughout this chapter as ≤ 100 nm).

The densification process for conventional powders is well known, both theoretically and practically. However, the densification of nanopowders poses significant additional challenges. Powder agglomeration, high reactivity and, therefore, contamination, grain coarsening, and ultimate loss of the nanofeatures, and inability to fabricate large and dense parts are among the main problems. The lower temperatures for minimizing grain growth may hinder good intergranular bonding, thus compromising the expected high mechanical strength and ductility. Such low sintering temperatures may also interfere with the thermochemical oxide reduction on particle surfaces necessary for subsequent sinter bonding. The most recent efforts have been very fruitful in overcoming some of these problems (e.g., agglomeration and grain size control). This has been accomplished by major improvements in the nanopowder synthesis methods and understanding of the densification processes such as pore effects in nanosintering. For nanopowder consolidation, an intriguing question is whether the sintering mechanisms scale with grain size or are there changes in these mechanisms when nanoscale is reached? This review has been written with this question in mind. More specifically, differences between the sintering of regular and nanosize powders are highlighted. It is shown that when smaller scales are approached, atomic mechanisms become more obvious. For instance, the reorientation during early sintering stages becomes evident only when particles become very small. Alternatively, the possibility of new sintering mechanisms is examined.

A number of reviews specifically on nanopowder processing[4]–[6] and general reviews addressing sintering issues have been published.[7]–[11] This chapter first covers some thermodynamic and kinetic aspects of nanopowder densification: driving force, surface energy, sintering mechanisms, activation energies, and scaling laws. Due to the critical effect of surface contamination on small particles, the impurity role in sintering is described separately. Next, cold compaction with resultant pore size and distribution and their effects on sintering and grain coarsening are addressed. The presentation of the sintering process of nanoparticles is divided into pressureless (conventional) and pressure-assisted sintering.

Finally, methods for full densification of nanopowders and their ability to maintain the nanosize features are presented.

2.0 SPECIFIC ISSUES IN THE DENSIFICATION OF NANOCRYSTALLINE POWDERS

2.1 Thermodynamic and Kinetic Effects

Thermodynamically, nanopowders are highly unstable, sometimes by as much as 10–50 kJ/mol. The sintering process is driven by the tendency to reduce the excessively large surface area per unit volume. The extra energy of a surface with a radius of curvature, R, may be calculated as a stress (σ) in a Laplace equation:

Eq. (1) $$\sigma = \gamma / R$$

where γ is the surface energy. In nanomaterials, this sintering stress may reach very high values. For instance, the sintering stress may be as large as 300 MPa in 10 nm particles compared to only 3 MPa for 1 μm particles, if γ has a typical value of 1.5 J/m^2.[12]

For traditional sintering studies, the surface energy is assumed to be isotropic. The implications and difficulties brought about by this simplifying assumption in conventional powders, particularly for ceramics which are more anisotropic than metals, have been addressed.[13] For nanocrystals with significant surface area, the anisotropy problem becomes even more critical. First, sintering starts at a lower temperature. In this case, the effects of surface energy anisotropy are more pronounced. Usually, if the temperature is sufficiently high, the surface energy anisotropy may be reduced. Second, direct TEM studies of γ-Al_2O_3,[14] ZrO_2,[15][16] and CeO_2[17] indeed showed that nanoparticles have a faceted appearance with anisotropic surface energies. Even for metals, a departure from the ideal spherical particle concept and, therefore, isotropy is accentuated in the nanosize range. Theory and experiments have indicated that clean fcc particles, with diameters less than 4–7 nm, have an icosahedral configuration with multiple twins in their equilibrium condition.[18] In metals, HREM studies revealed faceted nanoparticles which are ascribed to surfaces with minimum energies.[19] The highly anisotropic behavior of nanoparticle surfaces may be responsible for the crystallographic alignment, which has been often observed. As detailed later, both in-situ TEM sintering studies and Molecular Dynamics simulations indicated that nanoparticles rotate and align when necking occurs between loose particles.[15][20] Agglomeration of β-SiC nanoparticles into open strings has been noticed by Conder, et al.[21] An atomic force microscopy study showed that nano-TiO_2 powders obtained by gas condensation have a preferred alignment in a chain-like structure for specimens compacted at room temperature.[22] Although not specifically shown, some crystallographic alignment may be implied.

The different local atomic arrangement at the surface of a nanocrystal may result in a different surface energy value than in conventional powders. This may be the case of amorphous layers which have been observed on Al nanoparticle surfaces.[23] As Cahn infers, the different nature of the surface oxide on nanopowders may be an indication of a different surface structure.[24] Experimentally, Trudeau, et al., showed highly non-stoichiometric surface oxides on nanocrystalline catalysts.[25] $BaTiO_3$ has a cubic surface layer on the tetragonal bulk nanocrystal.[26] Although some caution is in order, Payne also showed that interfaces display electronic states and, therefore, properties different from bulk crystals.[27] The interfacial energy may be modified by any distortion of a surface structure such as segregation or adsorption of impurities. Impurities dictate the thermodynamics of surface and surface behavior of nanoparticles. In the Al_2O_3 system, calorimetric studies by Navrotski and co-workers indicated that hydration at surfaces of nanocrystals results in equal surface energies of α and γ polymorphs, despite the lower surface energy of anhydrous γ-Al_2O_3.[28] This lower surface energy induces the stabilization of γ-Al_2O_3 in nano-aluminas instead of the α-Al_2O_3 which is stable in conventional crystals.[29] Surface hydroxyls on γ-Al_2O_3 surfaces which are present up to high temperatures induce sintering difficulties of this ceramic.[28] Macroscopically, stabilization of particle surfaces, i.e., lower surface energy values when oxides are formed, results in a decrease of the driving force for sintering. This has been observed in nanosize particles which typically contain large amounts of adsorbates (e.g., high oxygen in TiN[30]).

In addition to the inherent morphological metastability related to fine grain size, nanocrystalline materials often display topological (e.g., alternate crystal structures than at equilibrium) or compositional (e.g., extended solid solubility or amorphous phase) metastabilities.[5][31] Alternate crystal structures are found in materials with high pressure polymorphism, such as cubic

$BaTiO_3$,[32] monoclinic Y_2O_3,[33] and tetragonal ZrO_2[34] owing to Gibbs-Thompson effect. Alternate crystal structures due to interfacial energy effects at large undercoolings were shown by Krauss and Birringer in nanocrystalline refractory metals (the A15 modification of β-W and β-Ta).[35] These metastable phases are obtained only in powder form. When films of W and Ta have been synthesized with same grain size, only the equilibrium structures have formed. In contrast to high pressure ceramic phases, the latter metal structures have a lower density than the equilibrium crystals. The retention of initial powder metastability into the bulk parts is relevant to their sintering. To retain the unique properties of nanocrystals throughout sintering, grain size must remain below the critical value at which property conversion occurs. Some critical grain sizes at which nanomaterial structure or physical properties revert to those of conventional materials are compiled in Table 1. For some other properties, i.e., mechanical, such a threshold value has not been proved or this value is property dependent. Moreover, there is not yet consensus on the strengthening benefits in the nanocrystalline range, but the current efforts are to push the limit towards increasingly smaller grain sizes. Therefore, an arbitrary limit for the maximum grain size in the nanocrystalline materials is often considered at $\leq$ 100 nm.

Table 1. Critical Grain Size of Nanocrystals Below Which Structure/Property Characteristics Are Different From Conventional Materials.

Critical Grain Size (nm)	Structure/Property		Material	Reference
	Nano	Conventional		
≈13	monoclinic	cubic	Y_2O_3	
8-26	tetragonal	monoclinic	ZrO_2	34
24-30	reduced thermal conductivity	regular	ZrO_2 (YSZ)	36
50	anatase	rutile	TiO_2	37
>80	superparamagnetic	antiferromagnetic	Cr_2O_3	38
120	cubic	tetragonal	$BaTiO_3$	32
120	variable T_{Curie}	constant T_{Curie}	$BaTiO_3$	26

Significantly enhanced kinetics are expected for processes that display a direct grain size dependence. For sintering, this dependence may be illustrated using the equation for the densification rate (*dL/Ldt*) developed by Johnson and co-workers for all stages of sintering:[39]

Eq. (2)
$$-\frac{dL}{L\,dt}=\frac{\gamma\Omega}{kT}\left(\frac{\delta D_b\Gamma_b}{d^4}+\frac{D_v\Gamma_v}{d^3}\right)$$

where γ is the surface energy, Ω is the atomic volume, δ is the grain boundary width, D_b and D_v are the grain boundary and bulk diffusivities, Γ_b and Γ_v are functions of density, kT has the usual meaning, and d is the grain size. From this equation, it is seen that decreasing grain size by three orders of magnitude (e.g., from µm to nm) could enhance sintering rates by up to 12 orders of magnitude. Consequently, sintering of nanopowders may be accomplished at significantly lower temperatures and shorter durations than conventional powders. This has been noticed for numerous real nanoparticle sintering, as detailed later in this section. A more dramatic effect of the sintering temperature depression in nanopowders is seen in the liquid phase sintering systems, such as WC-Co. Due to the accelerated sintering at lower temperatures in the nanoregime, full densification may take place entirely during solid-state sintering.[40]

2.2 Sintering Mechanisms

The densification process consists of solid particle bonding or neck formation followed by continuous closing of pores from a largely open porosity to essentially a pore-free body. For simplicity, solid-state densification is considered to be accomplished through three stages: initial, intermediate and final. Multiple mechanisms are involved throughout these stages, namely, evaporation-condensation, surface diffusion, grain boundary diffusion, bulk diffusion, viscous flow, and plastic deformation. Each transport process exhibits a particular dependence on the particle/grain size and defect density. The highest sensitivity on particle size is that of surface and grain boundary diffusion. Although simultaneous mechanisms participate, the common sintering models attribute a predominant mass transport path to a specific sintering stage. For instance, surface diffusion is considered the principle mechanism during the initial stage when the main event is the neck formation. So far, a similar partition of the sintering process is accepted in the existing sintering theories applied to nanopowders.[4]

For nanoparticles with excessive surface area and highly curved surfaces, surface diffusion is expected to be extremely rapid in early sintering stages. Evidence of the surface diffusion contribution to the neck formation in nanopowder sintering is given by many researchers. Thuenissen, et al., attributed the neck area increase to surface diffusion in nanoscale Y-TZP ceramics.[41] Hahn, et al., observed the elimination of small pores by rapid surface diffusion between 425 and 775 K in nano-TiO_2.[42] Surface diffusion was the main mechanism for neck growth noticed by in-situ TEM studies of alumina[14] and zirconia.[15][43] Bonevich and Marks studied neck formation between two alumina grains which retained their orientation.[14] The fast neck formation and atomically sharp neck interfaces prompted them to conclude that surface diffusion is the most probable sintering mechanism. In their study of neck formation and growth in nano-zirconia, Rankin and Sheldon considered the lack of change in the particle center spacing as an indication of a surface diffusion mechanism.[15] Surface fluctuations involving clusters of atoms were visible in a way similar to formation and dissolution of ledges on atomically smooth surfaces.[15][43] Surface diffusion is known to occur by this terrace-ledge-kink mechanism. This mechanism becomes more evident for a faceted, anisotropic surface structure such as in ultrafine alumina particles.[14] This is an illustration that atomic mechanisms may be more easily unveiled when smaller scales are approached.

For conventional powders, it is known that surface diffusion does not lead to densification but to grain coarsening.[44] As already shown in this section, surface diffusion mechanisms are most sensitive to particle size. Therefore, enhanced surface diffusion with reduced low-temperature densification should be observed in sintering nanoparticles. Abnormal grain growth and pore coarsening without densification were noticed at low sintering temperatures in ultrafine silicon.[45] However, this is more of an exception, since for nanoparticles, enhanced low temperature densification has been typically observed. Whereas detailed later, a substantial reduction of the sintering onset temperature was experimentally reported.[4]–[6] Molecular Dynamics simulations also indicate extremely fast sintering behavior of nanoparticles.[20][46] Surface diffusion cannot explain this behavior. Furthermore, the usual approach of using high heating rates to avoid deleterious surface diffusion does not seem to always apply to small powders. In some cases, slow heating rates induced surprisingly high densities.[47] In other cases, no effect of the heating rates on densification or grain growth has been observed.[48] Although in the latter, the heating rates used (2–200 K/min) may be too slow to suppress grain coarsening, the contribution of surface diffusion to nanosintering is not clear. The ratio between bulk (and/or grain boundary) and surface diffusion contribution to overall densification is not known and may be different compared to conventional materials. Grain boundary diffusion is certainly enhanced in nanopowders compared to conventional sintering. Either a major shift in the contribution of various diffusion mechanisms or a change in the sintering mechanisms is possible, at least in the initial densification stage.

Some other mechanisms for neck formation in nanopowders have been suggested: grain boundary slip, dislocation motion, grain rotation, viscous flow and even grain boundary melting. Thuenissen, et al., implied that grain boundary sliding assisted by surface diffusion is the main contributor to early densification in Y-TZP.[41] For the particular grain and pore size conditions, the calculated sintering stress (70 MPa) is considered sufficient to cause grain boundary sliding and particle rearrangement. A low calculated activation energy, smaller than for any known diffusional mechanism in the context of large porosity and small grain size, may be compatible with grain boundary slip assisted by surface diffusion. As shown later in this section, if grain rotation is restricted, grain boundary sliding may become active.[46] This grain boundary sliding, similar to superplasticity, may be effective in the densification of nanoparticles by severe plastic deformation.[49]

Dislocation motion, viscous flow, and grain rotation mechanisms were identified by Averback and co-workers using Molecular Dynamics (MD) calculations to simulate nanoparticle sintering.[20][46][50][51] They showed sintering of a nanoparticle pair taking place in picoseconds. Calculations for the observed shrinkage based on surface or grain boundary diffusion cannot explain this rapid sintering.[46] To account for the shrinkage, the grain boundary diffusivity in Cu at 700 K has to be 10^{-2} cm^2/s, a value which is two orders of magnitude larger than liquid diffusivity in Cu. They attributed this rapid shrinkage simultaneous with neck formation to fast dislocation activity driven by the contact Hertzian stresses that exceed the ideal shear strength.[50] An experimental confirmation of such high stresses has been recently provided by dipole stress fringes observed at the contact of two Co nanoparticles.[52] In this case, dislocations are generated in the neck and move along the usual slip systems in the fcc crystals. This dislocation motion was sufficient to explain a shrinkage of ~10.6% in the sintering of two copper spheres. After the neck forms by dislocation slip, the adjacent particles rotate to achieve a minimum grain boundary energy.[20][50] TEM studies of ceria nanoparticles indicate some particle coalescence which may be attributed to such a rotation process.[17] Similarly, copper nanoparticles were shown to rotate to take the orientation of a clean copper substrate after deposition and modest heating to $0.35T_m$.[53] The reorientation that induces coalescence occurs when grain boundary motion is faster than surface diffusion. This rotation may at least partially explain twin formation in nanosize metals. Twins have been commonly observed in fcc nanoparticle pairs.[18][54][55] Particle rotation was actually observed by in-situ TEM of ZrO_2 nanoparticles.[15] An epitaxial regrowth process of one grain that initiated at the neck was also reported in δ-Al_2O_3.[14] A simulation of seven initial cylindrical particles further indicated grain rotation to finally form one single grain.[46][50] Experimentally, Thuenissen, et al., documented the formation of single crystal aggregates (14 nm) from several initial 8 nm particles.[41] This occurred in early densification stages with no visible macroscopic shrinkage, typical when surface diffusion operates.

For materials with difficult dislocation motion, such as intermetallic compounds, MD simulations indicate that viscous flow is the main mechanism for neck formation and growth.[46] When three or more particle sintering is simulated, the contributions of these mechanisms is different. Dislocation motion cannot provide full densification, and diffusion mechanisms have to contribute to densification. Grain rotation is also restricted. This restriction creates large residual stresses. Further densification is accomplished by grain boundary sliding.[46] Alternatively, the large grain boundary stresses may induce melting or amorphization that also results in enhanced densification.[50][51] On a different theoretical basis, Trusov, et al., also claim that viscous flow is the predominant mechanism in the early stages of sintering nanoparticles.[56] They demonstrate that vacancies generated at grain boundaries and pores may reach pre-melting levels in a very short time. Dominguez, et al., also describe sintering of nanosize Cu and Fe based on grain boundary amorphization or melting mechanisms.[57]

While there is some theoretical and experimental evidence to support these new mechanisms involved in the nanopowder sintering, there is still much uncertainty about their individual or collective contribution to the overall sintering process. Whether these new mechanisms are generally applicable or system specific is largely unknown. If new mechanisms contribute to nanosintering, a simple scaling down of the conventional sintering equations is not applicable. Careful re-inspection of the simplifying assumptions used to derive the sintering equations is required. Particularly, surface effects and particle anisotropy should be reconsidered. Sintering diagrams indicating different mechanisms as a function of sintering stage and powder size would be most helpful.

Activation Energy. Commonly, the calculation of activation energy values provides insight into the microscopic process mechanisms. As already shown in this section, sintering is accomplished by multiple mechanisms. The identification of the specific mechanisms is complicated since they change while sintering progresses. As a result, the interpretation of the Arrhenius plots in sintering is difficult, as emphasized in sintering literature.[44][58] This is even more true for nanoparticles as seen in only a few activation energy values available. In addition to the difficulty of identifying a dominant mechanism, sintering mechanisms change with particle size. The largest departure from conventional sintering is seen in the initial sintering stages. In these stages, lower activation energies than for conventional diffusion have been reported. For instance, Trusov, et al., report a lower activation energy (134 kJ/mol) than for surface diffusion (~ 300 kJ/mol[44]) in 40 nm tungsten.[56] Vergnon, et al., cite activation energies of 234 kJ/mol for 13.5 nm Al_2O_3 and 96 kJ/mol for 11.5 nm TiO_2.[59] As mentioned earlier, Thunissen, et al., also found an atypically low activation energy (100 kJ/mol) in the early sintering of 15 nm Y-TZP.[41] In contrast, when the particle size was 50 nm, the activation energy had the usual value of 275 kJ/mol. An activation energy comparable to liquid diffusion was reported for nanoparticle iron and copper sintered in hydrogen by Dominguez, et al.,[57] Some of these low activation energy values can not unambiguously be assigned to surface diffusion due to the lack of comparable literature data, particularly for ceramics. In addition, a more valid comparison should use nanomaterial diffusion data rather than diffusion data on regular materials. As already mentioned, the low activation energy values in early sintering stages may be indicative of some different mechanisms that occur in sintering nanopowders.

For late sintering stages, activation energies are in better agreement with values determined for sintering of conventional powders.[41][60][61] For instance, Messing and Kumagai found activation energies that correspond to volume diffusion for sintering α-Al_2O_3 at 90% density.[60] By late sintering conditions, the initial nanopowders have already lost some of the more conspicuous initial nanofeatures and, therefore, their densification behavior converges towards that of regularly grained materials. Only slight changes, such as the predominance of grain boundary diffusion instead of volume diffusion, have been noticed in the densification by power law creep of Fe-10 wt% Cu during sinter forging.[61] This shift to grain boundary diffusion is not unreasonable since sintering temperatures are lower than in large grain powders.

Scaling Law. As already mentioned, the onset of sintering is consistently reported to be significantly lower in nanoparticulates compared to conventional materials.[30][57][61]–[69] The sintering of both metal and ceramic nanoparticles was found to start at temperatures of 0.2–0.4 T_m as compared to 0.5–0.8 T_m for conventional powders. Some examples of the sintering onset variation with particle size are shown in Table 2. Densification of nano-TiN starts at 1170 K and is completed below 1823 K, whereas, the microcrystalline powder is only 63% dense at 1823 K.[30] Full densification of nanopowders is completed at temperatures lower than that for conventional powders, as well. In Si_3N_4, Pechenik, et al., fully sintered nanopowders at 700 K lower than the coarse grained analogs.[66][67] TiC powder of 140–170 nm particle size sintered to 91% density at 1900 K as compared to 5 µm powder that achieved the same density at 3070 K.[68] Sintering of nano-ZrO_2 is completed at 1745 K, while commercial powders sinter at > 1945 K.[63] Nanometer size TiO_2 (12–14 nm) completes sintering at ~1300 K compared to more than 1670 K for 1.3 µm TiO_2.[42] For gel-derived TiO_2 powders, near full density is achieved at 1075 K.[69] Presently, the lower sintering temperatures of nanopowders are well recognized and fully exploited in numerous applications. Technologically, there are significant benefits from the lower sintering temperatures of nanopowders: possibility of avoiding sintering aids, phase decomposition, deleterious interfacial interactions, and undesirable phase transformations (e.g., the monoclinic to tetragonal ZrO_2 transformation that inevitably results in cracking).[63][70]

Table 2. Sintering Onset For Nanoparticles

Material	Particle Size (nm)	T, K (T/T_m)	Ref.
TiO_2	40	950 (0.46)	62
TiO_2	13	823 (0.4)	42
Y_2O_3	4	600 (<0.25)	63
ZrO_2	70	1370 (~0.5)	64
ZrO_2	8-9	870-920 (~0.3)	63
W	9	900 (0.24)	65
Fe	2000	~900 (0.5)	57
Fe	30	393 (0.21)	57

To rationalize this decrease in the sintering temperature, different scaling laws have been applied. For particles under 100 nm, Alymov, et al., developed an empirical relationship for the dependence of the sintering onset temperature, T_s, on the mean particle size, d:[65]

Eq. (3) $\ln T_s = 1/d$

They calculated a ratio of sintering onset to melting temperature which is in good agreement with experimental data for sintering of nanometals. In 1950, Herring developed the classical scaling rule for the effect of the particle size on sintering time.[71] The time, t, to achieve the same sintering condition is related to the powder particle size, d, if the same sintering mechanisms operate, by the equation:

Eq. (4) $t_1/t_2 = (d_1/d_2)^n$

where n is the exponent which has a value of three if the main sintering mechanism is volume diffusion and a value of four if the main sintering mechanism is grain boundary diffusion. Considering the Arrhenius expression for temperature, the sintering temperature dependence on the particle size becomes:

Eq. (4a) $n \ln(d_1/d_2) = Q/R[(1/T_1)-(1/T_2)]$

where Q is the activation energy for the predominant sintering mechanism, R is the gas constant, d_1 and d_2 are the different powder particle sizes and T_1 and T_2 are their respective sintering temperatures. Reasonable agreement of experimental and calculated data was found in TiO_2 and Al_2O_3 assuming certain diffusion mechanisms.[60][62] For instance, in TiO_2 the quantitative agreement was better when grain boundary diffusion was considered the sintering mechanism.[62] The fitting value for the activation energy used (139 kJ/mol) is close to half of the activation energy for lattice diffusion in TiO_2 (251 kJ/mol), a typical ratio for grain boundary to lattice diffusion. For Al_2O_3 the calculated values are in closer agreement with the experimental values again if the activation energy for grain boundary diffusion is used.[60] But, in this system, Kumagai and Messing measured an activation energy for sintering close to volume diffusion (543 kJ/mol). When applying Herring's law to Andrievski's data[72] on sintering pure nickel particles of different sizes to 60% density, the calculated activation energy value (239 kJ/mol) was closer to volume diffusion (284 kJ/mol) than to grain boundary diffusion (115 kJ/mol) in nickel. However, volume diffusion is less likely to be the principle sintering mechanisms since temperatures are low and particles are fine. These discrepancies point to the uncertainties in using the scaling law to determine the activation energy values and thus, deduce the sintering mechanisms. Conversely, the predictions of sintering temperatures as a function of particle size may not be accurate, giving primarily qualitative estimates. The main concern in using Herring's scaling law is related to a possible change of sintering mechanisms from conventional size to nanopowders and sintering temperature. Sintering at lower temperatures may induce a mechanism shift even if particles are conventional size. For nanoparticles, indications of possible changes in the sintering mechanisms have been already shown. Furthermore, in using Eq. (4), the initial grain size is considered despite grain coarsening that always occurs starting with the intermediate sintering stage. Usually, sintering models also neglect grain growth. Finally, as in any extrapolation, a careful inspection of the initial assumptions is in order. Herring cautioned that the degree of contamination on the surface should be kept the same for all particle sizes. This is definitely questionable when conventional and nanosize particles are compared. Not only have higher contamination levels been reported,[72] but also different types of oxides have been identified on nanoparticles.[23][25] In addition, when sintering occurs at lower temperature, the complete reduction of oxides or hydroxides may be hindered. The specific effect of the surface contamination on sintering is discussed in the next section.

2.3 Impurity Role

As a surface controlled process, sintering is critically dependent on particle surface condition. Surface properties become even more dominant in the sintering of nanoparticles. Studies of this topic abound since particle surface properties affect not only individual particle properties and sintering behavior, but also final mechanical properties. This explains the precautions to eliminate contamination by in-situ consolidation of metal nanopowders produced by inert gas condensation for careful mechanical property characterization.[2][73] In ceramics, sinterability and mechanical properties are also highly affected by surface impurities. For instance, when the inherent surface SiO_2 is present on Si_3N_4 particles, sintering is enhanced, but creep strength and toughness are impaired during Si_3N_4 part service.[74]

The excessive surface area of nanoparticles justifies the focus on powder contamination. Using photoacoustic infrared spectroscopy, Ying documented larger amounts of adsorbates on nanoclusters than on conventional γ-Al_2O_3.[75] Andrievski measured about 120 cm^3/g of absorbed gases such as N_2, H_2 and H_2O in 50 nm pure Ni powders compared to 40 and 8 cm^3/g for 5 and 50 μm powders, respectively.[72] The impurity amount is highly dependent on the processing method. Generally, the processing path has a significant effect on nanocrystalline microstructures and, hence, sinterability.[4][8][60][76] For instance, powders obtained by mechanical attrition are inevitably more contaminated than those produced by other processes (e.g., by inert gas condensation). Oxides, nitrides, and other compounds are often found in consolidated parts made of attrition-milled nanopowders.[77][78] Although products of contamination, these compounds may be used to advantage to prevent grain coarsening[78] when present as fine dispersions. Conversely, a fine grain size was more difficult to retain in high purity powders or during consolidation in high purity conditions.[73]

The influence of contamination on sintering was studied by in-situ TEM under controlled oxygen levels compared to ultrahigh vacuum conditions.[18] Studies of the early sintering stages in ultraclean conditions have been performed for both metal and ceramic nanopowders.[14][15][79] All results indicate that ultraclean nanoparticles sinter very rapidly even at room temperature. This rapid sintering is in agreement with MD simulations.[20][46][51] As shown in Sec. 2.2, neck growth occurs by surface diffusion, dislocation motion, and grain rotation. In contrast, when the sintering of nanoparticles takes place in the presence of controlled oxygen traces, very little or no neck growth is observed.[18] This oxygen contamination decreases the surface energy and slows down the sintering kinetics. More specifically, TEM studies showed that sintering of Fe-Ni nanoparticles starts only when oxide layers are reduced.[80] Sintering occurred above 500 K in hydrogen. When vacuum was used, no sintering took place up to 725 K. As already mentioned, reducing the amount of oxygen in TiN nanopowders decreases the sintering temperature.[30]

The sintering environment appears to have a similar effect in nanopowders as in conventional ones. Dominguez, et al., calculated a lower activation energy for the sintering nano-iron in hydrogen compared to vacuum.[57] In these results, the former activation energy corresponds to liquid-like diffusion, while the latter is for more conventional grain boundary diffusion. Consistently, ceramics sintered better—higher densities at same temperatures or similar densities at lower temperatures—in vacuum than in air.[63][64]

To avoid further contamination, the as-produced nanopowders may be either consolidated in-situ or handled in controlled environment prior to sintering. Both approaches add large costs to the actual component production. In a manufacturing environment, powders are consolidated in a place remote from the powder producing area. Therefore, contamination of small particles is likely to occur. The oxygen role in nano-iron sintering was addressed by Bourell and Kaysser.[81] The powders were produced by an evaporation-condensation method. Although sintering was carried out in a reducing hydrogen atmosphere, oxygen was found as fine oxides in the densified product. Alternatively, consolidation processes that are less susceptible to the contamination levels are recommended (e.g., high pressure consolidation and severe plastic deformation), although not always practical. Some electrical field assisted sintering processes that claim an initial oxide reduction may be also helpful (e.g., plasma activated sintering).

2.4 Green Density of Nanopowders

The initial step in most densification processes is to compact the powders at room temperature, or cold compaction, to form a green body. Final sintering results are largely dictated by the green compact microstructure. A larger number of initial point contacts, smaller pores in a high green density compact, and a uniform pore distribution favor higher final density. Sintering times may be shorter and lower sintering temperatures may be used. Conversely, many sintering defects may be traced to the green compact structure. Inhomogeneities in density, packing, and particle size in green compact will limit the final sintered density. Such an example is the crack generation in ceramics upon sintering inhomogeneous cold compacts. Generally, nanocrystalline powders are less forgiving of defects in green compacts compared to conventional powders. The most common problem is the elimination of large pores that originate from the green compact. This elimination requires high temperatures upon subsequent sintering, thereby, promoting unwanted grain growth and losing the desired nanosize features.

Cold compaction comprises specific stages that involve sliding and rearrangement of particles, elastic compression at particle contact points, plastic yielding for metals or fragmentation for brittle materials. Similar steps are seen in nanopowder compaction. The principle differences in nanoparticles arise from the specifics of small particle sliding and friction, as well as from the changes in dislocation-assisted deformation behavior. Sliding and rearrangement in nanopowders are severely restricted owing to large frictional forces among powder particles.[4][72][82] These forces are a result of mechanical, electrostatic, van der Waals, and surface adsorption phenomena that become much more significant as particle size decreases. Mechanical friction resistance is substantial due to numerous interparticle contact points. Irregular particle boundaries favor agglomerate formation. As a result, particle rearrangement is hindered and lower green densities are likely to be achieved compared to conventional micron size powders (Fig. 1). For example, Bourell, et al., achieved 45% density for 70 nm Y-TZP and only 29% for 40 nm.[84] Conversely, when particle sliding is facilitated, high green values are obtained. This may be accomplished by using lubricants or coatings. The latter technique has been used predominantly in ceramic compaction to isolate individual particles by encapsulation or sheathing.[85][86]

As an example, rare earth carbide nanocrystals have been carbon-coated to prevent degradation by oxidation or hydrolysis.[85] Similarly, carbon coating on ferromagnetic powders acted as an effective oxidation barrier.[86] Although no results on packing or consolidation behavior of coated particles are presented, the carbon surface layer and the spherical morphology thus obtained should improve powder flow and alleviate agglomeration problems. For both metals and ceramics, the lubrication action of liquid nitrogen during low temperature compaction has been shown to improve rearrangement, thus providing higher green density values (note the cryogenic compaction data point in Fig. 1).[82] As a natural extension, wet processing provides significantly improved packing uniformity of ceramic nanopowders.[4] For instance, a green density (74%) close to the ideal value that results by packing monosize spheres was achieved in 12.5 nm Y_2O_3-doped ZrO_2 by centrifugation from a slip suspension.[87]

A better initial particle rearrangement facilitated by good wetting enhances sinterability even without increasing green density. For instance, Yan and Rhodes achieved 99% density by pressureless sintering of only 41% dense green compact of TiO_2 nanopowders prepared by the centrifugation method.[62] Similar full densification with 80 nm final grain size after wet compaction is reported by Mayo in ZrO_2-3 mol% Y_2O_3.[4] These results are to be contrasted to dry compaction of the same powders that requires 0.5–1 GPa pressure to achieve similar densification characteristics. Therefore, techniques that take advantage of the wet processing of ceramics such as centrifugation, tape and slip casting, and osmotic and pressure filtration are on the rise. Osmotic compaction has been recently applied to ceramic powder compaction based on the osmotic chemical potential action. Equal or greater green densities than in physical pressure applications have been achieved with no mechanical breakage that external forces may produce. In 8 nm zirconia, Miller and Zukovski obtained about 47% dense samples by an osmotic pressure equivalent to 12 MPa.[88] Although final sintering of osmotic specimens has not yet been performed, the expectation is for high density values due to homogeneous packing similar to wet compaction. More results on cold compaction of ceramics are discussed in a review by Mayo.[4]

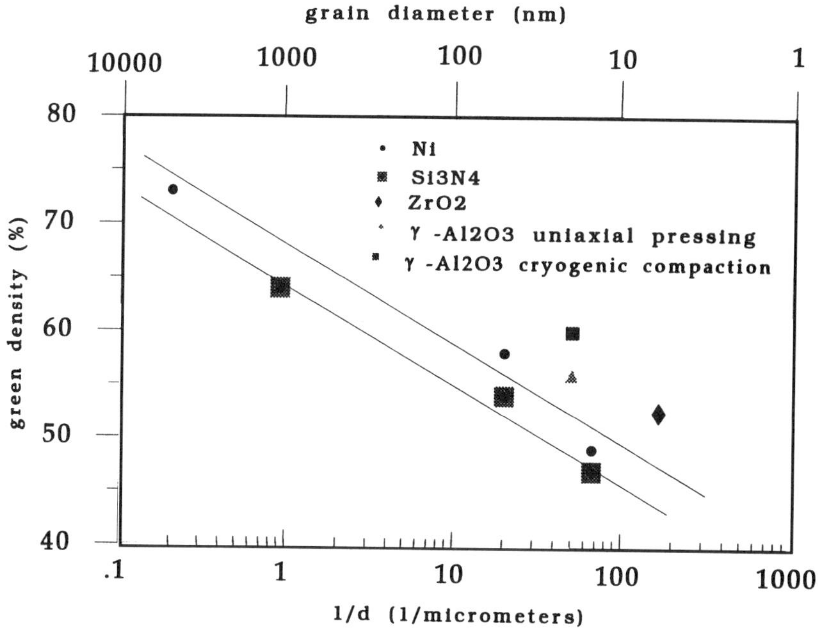

Figure 1. The effect of particle size on green density upon conventional dry compaction (uniaxial pressing) for 50 nm Ni and 50 nm Si_3N_4.[71] Also shown are data points for dry (triangle) and cryogenic compaction (square) of 20 nm γ-Al_2O_3[82][83] and 8 nm ZrO_2 (diamond).[64] All compactions performed at 1 GPa.

Cold Compaction of Nanocrystalline Ceramics. In dry compaction, ceramic powders are more difficult to compact than metals (compare Ni and Si_3N_4 compaction lines in Fig. 1). An equation for calculating the pressure necessary to achieve a certain density based on volume changes with pressure was developed by Chen and Malghan for cold compaction.[83] Experimentally, the literature reports indicate room temperature densities reaching >95% for nanometals but only 75–90% for nanoceramics.[9][89] However, there are a few cases when ceramics pack better than metals by compaction at room temperature (see Al_2O_3 and ZrO_2 data points in Fig. 1). Note that all these high densities were achieved by applying a very high pressure of 1 GPa. Indeed, green density increases with pressure applied. For instance, results for 70–80 nm TiN nanopowders show that green densities up to 65% may be reached by conventional cold compaction. At very high pressures (7 GPa) the density obtained was close to 80%.[90] In cold isostatic pressing at very high pressures (up to 5.6 GPa), densities in excess of 90% for nanocrystalline Al_2O_3 and 80% for SiO_2 have been achieved by Gallas, et al.,[89] A major inconvenience in applying large pressures, other than equipment limitation, is the high level of residual stresses in ceramic powders that may result in fracturing upon subsequent handling.[4]

At lower pressure values, the hard agglomerates formed in ceramics cannot be fractured. To break these agglomerates, a critical pressure is necessary (Fig. 2). On a density-pressure plot, the transition point, P_y, where the slope change occurs, is interpreted as the strength of the agglomerates. Above P_y, the weak agglomerates break (Curve 1, Fig. 2). When agglomerates are strong and dense, they do not break and, consequently, no transition point is seen (Curve 2, Fig 2). This is the case of the agglomerates in calcined water-washed yttria-stabilized zirconia that survive cold isostatic pressing at 400 MPa.[91] If powders contain no or weak agglomerates, the final result of the pressure application is a uniform distribution of small pores with favorable consequences for final sintering. Such a gradual change from a bimodal pore distribution in agglomerated powders to only small pores by pressure application was documented by van de Graaf, et al., in 8 nm 17% Y_2O_3–ZrO_2 ceramics using a porosimetry technique.[92]

A high value of the green density may not always reflect a uniform pore size and distribution. When large agglomerates are present, the green density may be larger than in non-agglomerated powders, similar to bimodal size powder packing. For instance, a high density of 75% in as-compacted nano-TiO_2 at 425 K under 2 GPa pressure for 2 hours was explained by easier packing of large agglomerates (50 nm) rather than individual smaller particles (14 nm).[42] For practical purposes, transparency of green ceramics may be used as a quick check of the compaction results.[4][89] A translucent green compact is an indication that pores are smaller than the visible light wavelength and consequently may reach high densities upon further densification.

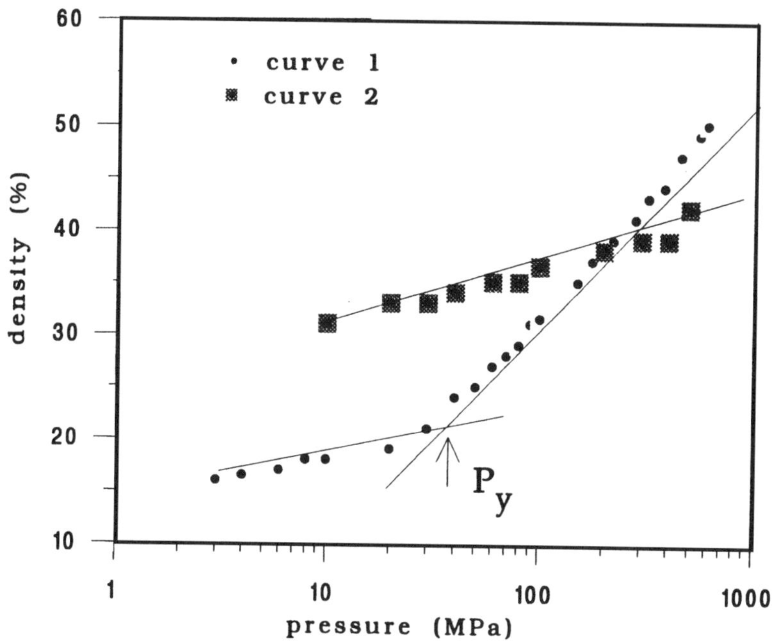

Figure 2. Compaction behavior of ceramic powders (from Refs. 91 and 92). *Curve 1*: Weakly agglomerated ZrO_2-17% Y_2O_3. *Curve 2*: Hard agglomerates in 5%Y-TZP.

Cold Compaction of Metals. To achieve high green densities of metal nanopowders, plastic yielding is necessary which, in turn, requires high compaction stresses. These high stresses are rationalized based on the Hall-Petch dependence of yield stress, σ, upon grain size, d, (i.e., $\sigma \propto d^{-1/2}$).[93] According to the pressure densification models, plastic yielding takes place when the effective pressure is several times the yield strength of the particle material.[8][61] Such high stresses may be commonly achieved using high pressure equipment such as a diamond or WC cubic anvil cell. Gutmanas, et al., obtained high densities up to 97% at room temperature and 99% at 575 K in Ni (65 nm) and Fe (30 nm) at 3 GPa in a high pressure cell made of high speed steel.[94] Historically, high pressure consolidation has been the preferred in-situ consolidation method for classical nanopowders obtained by the inert gas condensation (IGC) method.[9][95]–[98] Specimen size is usually small, not exceeding 1 mm in thickness. Sometimes, densities close to the theoretical have been achieved by low temperature sintering (e.g., 575 K) after cold compaction under high-pressure of nanocrystalline metals (Ni, Fe, Al).[94] For a more detailed theoretical and experimental treatment of high pressure cold compaction of nanopowders, the reader is referred to Gutmanas' review.[7]

The usual dependence of green density of nanometals as a function of applied pressure has two regions (Fig. 3). At lower pressures, there is a linear increase of density with pressure. At larger pressure values, the slope is considerably less. This region is attributed to work hardening.[7][93] As seen in Fig. 3, little work hardening is experimentally observed (Note the flattened compaction curves at high pressures in 50 nm Ni, Fig. 3a). This is in contrast to submicron or micron size powders for which strain hardening of 200–300 MPa has been observed in cold sintering.[7] The difficulty for strain hardening in nanomaterials is explained by restricted dislocation generation and flow. A more detailed discussion of the role of strain hardening in nanopowder consolidation may be found in Gutmanas.[7]

A particular case is represented by mechanically alloyed powders which most commonly consist of micron size particles with multi-grains in the nanometer range. In this case, the particle size dictates the cold compaction behavior and high green densities are, therefore, achieved. For instance, a green density of ~90% has been reached under 1 GPa pressure in mechanically alloyed Fe_3Al powders.[99]

As shown in Sec. 2.1, the particle surface state becomes increasingly important in nanopowders due to the dramatic increase in the surface area. Literature examples illustrate the interdependence of green density value on surface chemistries, adsorption characteristics, and reactivity. Andrievski measured high absorbed-gas levels on 50 nm Ni powders and, consequently, attained low final green densities even at high pressures.[72] As seen in Fig. 3a, the green density values reached in these powders at 1 GPa pressure by uniaxial pressing are only ~58%. Even lower green density values were obtained for smaller Ni powders (e.g., 15 nm Ni in Fig. 3a). For comparison, 50 µm Ni powders reached 82% density in uniaxial pressing and 86% in cold isostatic conditions. The degradation of surface condition by oxidation of metal nanopowders due to storage further limits the green density values obtained. Andrievski showed such a decrease of the final green density of metal nanopowders after storage (see the two data points in Fig. 3b). Noteworthy also is the minor density difference after storage in air and argon. In contrast, adsorbed humidity in ceramics is at significantly lower levels. The adsorbates on Si_3N_4 powders were about twenty fold less than to Ni powders of same grain size (50 nm).[71] As a result, most ceramics are far less sensitive to storage in humid conditions.[4] However, the level of adsorbed gases such as CO, CO_2, N_2 and H_2 in ceramic nanopowders may be quite high.[72][90] Even in very controlled preparation conditions such as in inert gas condensation, nanopowders may retain high levels of adsorbates.[73]

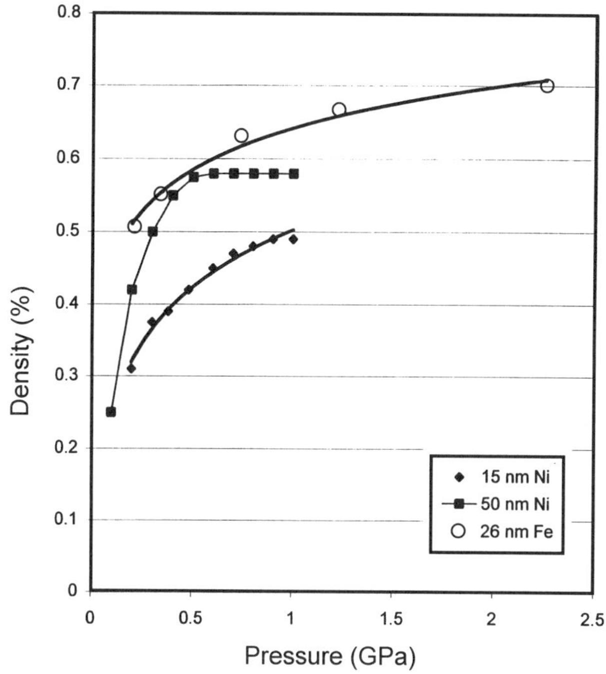

(a)

Figure 3. Pressure effects on the cold compaction of 15 and 50 nm Ni powders and 26 nm Fe powders (from Refs. 72 and 93); (*a*) uniaxial pressing, (*b*) cold isostatic pressing (CIP). Note slightly higher density values for CIP than for uniaxial compaction. The two data points in (*b*) are for air stored (square) and argon stored (full triangle) Ni powders.[72]

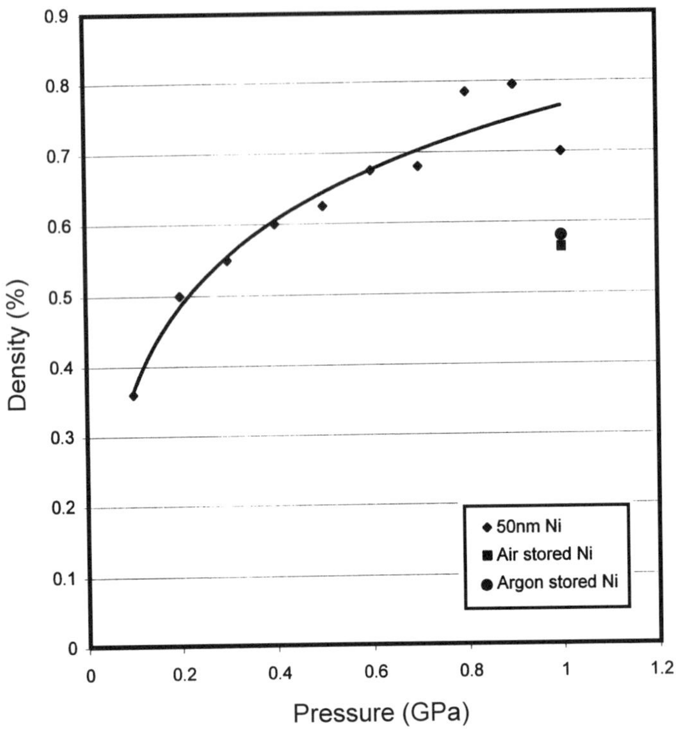

(b)

Figure 3. *(Cont'd.)*

Warm Compaction. To eliminate adsorbates and enhance interparticle bonding, warm compaction at temperatures up to 675 K has been largely applied to nanopowders. For most metal nanopowders, these temperatures are still in the cold working regime. Selection of such temperatures below recrystallization is purposely sought to prevent grain growth.[90][94] Only for low melting point materials, compaction at room temperature may induce recrystallization and sintering effects. For instance, in Bi-Sb system, cold pressing at 350 MPa yielded an 89% dense specimen while grain size doubled.[100] Significant work in warm compaction has been carried out for metal nanopowders by Weertman and coworkers.[73][101]–[103] The IGC processed powders were densified under high pressure application (1.4 GPa) for 5–900 min at temperatures between 375 and 575 K.[101][102] Densities exceeded 95%. Higher densities were obtained when volatile contaminants were carefully eliminated based on studies of gas desorption temperatures and kinetics.[73][103] The main purpose of such studies was to minimize the consolidation artifacts which hindered the accurate characterization of the mechanical properties of nanomaterials. High pressure compaction (1 GPa) at 525 K also resulted in high densities of TiAl intermetallics.[104] Ceramic nanoparticles have also been densified under high pressures at low temperatures. Warm compaction at 675 K of 70–80 nm TiN powders resulted in densities exceeding 90%.[90] Similar to metals, the degassing of ceramic powder contributes to high densities.[45][68] The key to achieving such high densities by cold or warm compaction is clean particle boundaries that allow interatomic forces to come into play, thus, providing better intergranular bonding. Higher compaction temperatures and, consequently, slightly larger grain sizes may also allow plastic deformation to contribute to nanometal densification.[7][102]

2.5 Pore Size and Its Effects on the Densification Behavior

Similar to conventional powders, full density or rapid sintering of nanosize powders is achieved when the green structure contains a narrow pore size distribution. Conversely, densification is retarded or inhibited when pore distribution is wide. In this case, big pores become larger and only small pores shrink. The removal of large pores is a lengthy process and requires higher temperatures. This way, the overall effect of large pores is to slow densification and induce undesirable coarsening.

A thermodynamic model of the pore shrinkage was developed by Kingery.[105] Based on curvature considerations, he predicted the pore size for the transition between pore shrinkage and pore growth. A complicating event is the separation of large pores from the grain boundaries that occurs in final sintering stages. Brook developed pore–size-grain size maps to designate the regions where such pore separation from grain boundary occurs.[106] When pore–grain-boundary breakaway takes place, the detached pores will no longer benefit from easy transport paths such as grain boundaries. Instead, transport is by slower volume diffusion and, consequently, densification occurs at a slower rate. This is the stage when open porosity breaks down and pores become closed. Since grain boundary migration is no longer restricted by pores, grain coarsening takes place, or pinning of grain boundary migration due to residual pores is no longer effective. Therefore, in the final sintering stage, densification is hindered and significant grain growth occurs. Mayo, et al., extended Kingery's treatment to second stage sintering of nanoceramics and demonstrated that they also obey Kingery's critical ratio for pore shrinkage.[107][108] Full densification is likely to occur when all pores are smaller than the critical pore-to-grain size ratio. Large pores are more likely to undergo the pore-boundary separation. This prevents high densities from being achieved. The effect of large pores from a non-uniform green structure on final sintering of nanopowders is even less forgiving than for normal grained powders. If coarsening occurs in late sintering stages, the overall notion ofnanograin size is compromised. Experimentally, a strong correlation between the closure of open porosity and the onset of exaggerated grain growth was noted in both ceramic and metal nanopowders.[42][63][72][107][109]–[113] As seen in Fig. 4a for Ni and TiO_2, accelerated grain growth occurs at densities above 90% or when the pores become closed. Often, the final grain size upon full densification is close to 1 µm,or the materials are no longer nanocrystalline. In contrast, no exaggerated grain growth takes place upon densification of non-agglomerated Al_2O_3 powders (Fig. 4a).

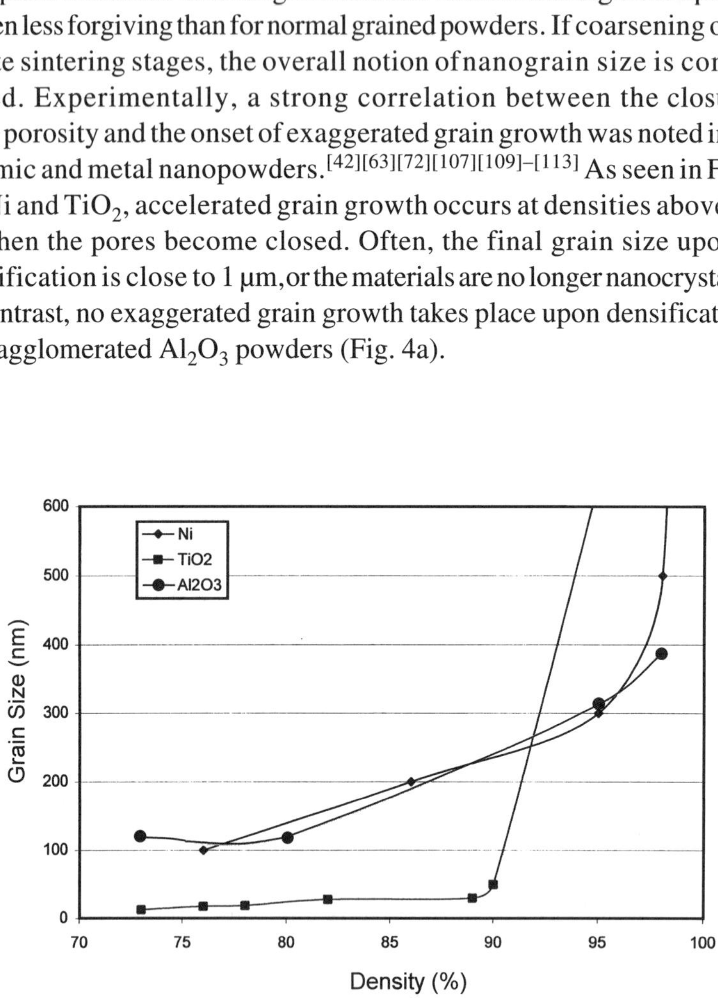

(a)

Figure 4. (*a*) Exaggerated grain growth in Ni and TiO_2 when porosity becomes of closed type.[72][104] The densification behavior for non-agglomerated powders is shown for Al_2O_3.[60] (*b*) The effect of dopants on grain growth of TiO_2.[113]

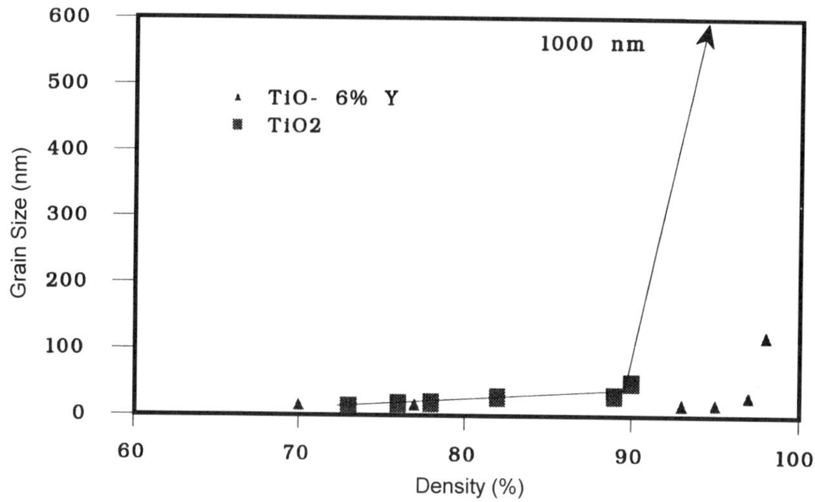

(b)

Figure 4. *(Cont'd.)*

Large pores usually originate from agglomerated powders. As shown in Sec. 2.4, nanoparticles are inevitably linked to a high agglomeration tendency. Although agglomerates are more often reported in ceramic nanopowders, they have been also noticed in metals.[56][114] For instance, Trusov, et al., gave experimental evidence that aggregates of ~1 μm may be found in 10 nm metallic powders.[56] Agglomerated powders have a bimodal pore distribution with small inter-agglomerate and large intra-agglomerate pores. The removal of large inter-agglomerate pores based on vacancy diffusion requires significantly higher temperatures and longer sintering times. Mayo, et al., compiled evidence for three TiO_2 powders showing that sintering temperature scales with agglomerate size, rather than with size of the individual powder particle (Fig. 5). Also noteworthy in Fig. 5 is that the smaller the particles, the larger the agglomerates. The non-agglomerated powders sinter at the lowest temperature, despite their largest nanoparticle size. See NA (<40 nm) data points in Fig. 5.

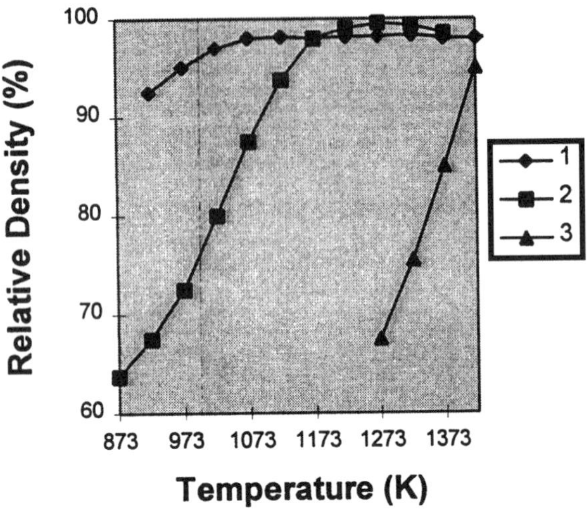

Figure 5. The effect of agglomerate size on the sintering temperature of nano-TiO_2.[107] Agglomerate size (particle size): *1*) Non-agglomerated (<40 nm), *2*) 80 nm (16 nm), *3*) 340 nm (8–10 nm). *(With permission from Elsevier Science.)*

Based on the observed dependence of the sintering process on pore size, Mayo developed a modified sintering law that directly accounts for pore size effects on the densification rate:[4]

Eq. (5) $$\frac{1}{\varrho(1-\varrho)}\frac{d\varrho}{dt} \propto \frac{1}{d_n}\frac{1}{r}\exp\left(\frac{-Q}{RT}\right)$$

where ρ is density, d is the particle size, n is a constant dependent on the sintering mechanism, r is the pore radius, Q is the activation energy, R is the gas constant, and T is the absolute sintering temperature. This equation predicts that the highest densification rate occurs for the finest pore size. It was found to hold throughout the sintering process and for a large range of pore sizes, at least when pore size distribution is uniform. This relationship has two implications. First, the pore size, in addition to grain size, should be controlled during sintering. Fast sintering kinetics result when pore sizes are small. Second, the densification rate is dictated by the instantaneous pore size, not only the initial pore size. Therefore, to maintain a fast sintering rate in late sintering stages, the pore should remain small even in late sintering stages. A small pore size throughout the sintering process is also critical in controlling the final grain size. For these purposes, a small and uniform pore population is desired in the green compact. Most often, such a pore distribution is associated with a high green density in non- or weakly agglomerated powders.[4][92][115][116] As early as 1981, Rhodes demonstrated that de-agglomeration of a commercial ZrO_2 may reduce sintering from 4 h at 1775 to 1 h at 1375 K.[87] In the past years, efforts to produce non-agglomerated powders have intensified. A good summary of the progress in producing non-agglomerated ceramic powders may be found in Mayo's review.[4] The benefits of a small pore size and narrow size distribution in reaching high densities and reduced grain coarsening have been shown by many researchers in non-agglomerated Al_2O_3[60][117] (see also Fig. 4a), Y_2O_3 and ZrO_2,[63][64][115] SnO_2,[116] and TiN[30] nanopowders.

2.6 Grain Growth

The success in nanopowder consolidation is intimately related to the control of the competition between densification and coarsening. As shown in Sec. 2.1, the driving force for densification is largely due to the extreme curvature of nanoparticles. As sintering progresses and grain boundaries are created, the driving force, Δp, for mass transport may be expressed as a Gibbs-Thomson effect:[118]

Eq.(6) $$\Delta p = \frac{2\gamma_b}{d} + \frac{2\gamma}{r}$$

where γ_b is the grain boundary energy, d is the average grain size, γ is the surface free energy, and r is the radius of curvature of pore surfaces. The first term is the grain coarsening tendency, while the second is the sintering driving force. When pores are small, sintering predominates or the pores control grain growth. When particles bond and grain sizes are small, the driving force for coarsening becomes important. The key to nanopowder consolidation is to control this coarsening which inevitably competes with sintering.

The thermal stability of nanocrystalline materials has been the subject of recent reviews[119][120] or covered by a number of general reviews.[8][10] Here the attention is mainly on grain growth during sintering. In nanomaterials, grain coarsening is described by an equation similar to that of conventional materials:

Eq. (7) $$d^n - d_0^n = kt$$

where d is the average grain diameter, d_0 is the initial grain diameter, n is the grain growth exponent, k is an Arrhenius type constant, and t is time. Koch and coworkers summarized the n values for the isothermal grain growth of nanocrystals and compared them to those for conventional polycrystalline materials.[120] At low temperatures, n values are high or grain growth is slow, similar to regular materials. Such increased n values are rationalized by restricted grain boundary mobility which is most commonly due to pore or solute effects. In nanocrystalline materials such as iron, Mallow and Koch found an average n value of 12 at low temperatures but the n value decreased with increasing temperatures in the same way as for the coarse grained materials.[121] In ZrO_2-3 mol% Y_2O_3 ceramics, n is larger than four at low temperatures and decreases to three at intermediate temperatures.[4] In other cases, constant n values were reported, e.g., $n = 4$ in grain growth during sintering of Al_2O_3[122] or $n \approx 3$ in TiO_2 and yttria stabilized ZrO_2.[4][41][123]

Activation energies for grain coarsening are generally difficult to assess particularly when the n exponent deviates from the ideal value of two. Malow and Koch compiled the existing activation energy data for grain coarsening in consolidated nanosize metals, intermetallic compounds, and ceramics.[120] In most cases, the activation energy is close to that of grain boundary diffusion, which is similar to large grained materials. Some exceptions were found for elemental nanometals with activation energies equal to that of lattice diffusion.[121][124] Two apparent activation energy values were calculated in nanocrystalline Fe: a low value for low temperatures (125 kJ/mol) and a high value (248 kJ/mol) corresponding to high temperatures.[121] The latter value was close to that of volume diffusion in iron and is similar to the value found in coarse grained iron. The low temperature value is less than that for grain boundary diffusion. A low activation energy value for grain coarsening in early sintering stages was also calculated by Thunissen, et al., in ZrO_2-5.8 mol% Y_2O_3, although some uncertainty may be due to an unknown n value.[41] Eastman also reports a low activation energy for grain growth (83 ± 40 kJ/mol) in nano-TiO_2.[123] However, any speculation for a new mechanism based on these low activation energy values is risky because of the reported uncertainties in impurity or porosity levels, as well as the lack of specific diffusion data in nanomaterials for direct comparison.

The influence of pores on grain coarsening has been well documented theoretically[4][125][126] and experimentally, particularly in ceramics.[42][107][127][128] As shown in Sec. 2.5, the presence of open pores in nanopowders inhibits grain growth similar to the way pores prevent grain coarsening in ordinary grain sized ceramics. As soon as the pore isolation occurs, grain growth is unrestricted (Fig. 4a). The pinning action of the pores is, however, difficult to predict. Gupta established an empirical direct relationship between density and grain size in the intermediate stage sintering.[129] On a better fundamental basis in a pore controlled grain growth model, Liu and Patterson found a linear relationship between the inverse of grain size and the pore surface area per unit volume.[130] This model, similar to Zener's drag effect, is valid only for an immobile pinning phase. Therefore, the experimental verification of this model was performed on materials that have either open pores (e.g., in UO_2 between densities of 62 and 86%) or dopants that suppress pore breakaway (e.g., MgO additions in Al_2O_3). This relationship holds experimentally for metals (Cu, W), as well. A further improvement in grain growth predictions may be accomplished if the modeling of pore closure is achieved.

The simplest way to control grain coarsening is to take advantage of the pinning effect of open pores and limit the density to ~90%, if this density is acceptable for the final product. To limit grain size beyond 90% density, a small and uniform pore size distribution in the green compact is desired. As shown in Eq. 5, small pores enhance densification because densification rates scales with the inverse of pore size, while a large number of pores (i.e., many small pores rather than a few large pores) pin the grain boundaries in an effective manner.[4][108][116][122] Other strategies to control grain growth after pore closure include kinetic and thermodynamic approaches.[10] The coarsening kinetics may be slowed down by restricting grain boundary mobility using a solute drag or particle pinning effect. Small particles or second phase precipitates stabilize a fine grain size upon heating at high temperatures, as long as particles do not dissolve or coarsen. The maximum grain size (d_{max}) is dependent on the particle size, r_p, and volume fraction, f, by a relationship given as Zener drag:

Eq. (8) $$d_{\max} = \frac{4r_p}{3f}$$

The solute drag effect is based on the decrease of grain boundary mobility and free energy when solute segregates at grain boundaries. Theoretically and experimentally, it was shown that grain boundaries in nanomaterials have increased solubilities for solute atoms.[10][131][132] The enrichment of the grain boundary with solute atoms has been shown to diminish or even reverse the free energy available for grain growth. Weissmuller demonstrated that segregation at grain boundaries reduces the grain boundary specific energy, σ, to zero or negative values.[10][132] In these cases, the nanocrystal becomes thermodynamically metastable with respect to grain growth. The experimental verification of the grain growth metastability theory was performed in the Y-Fe system and a good stability of grain size was found in alloys that contain up to 20 at% Fe.[132]

Experimentally, dopants in ceramics have been widely used to suppress grain growth by providing a drag force on the moving boundaries or forming secondary phase particles.[42][113][127][133]–[135] For instance, Hahn, et al., found that 6% Y addition to TiO_2 retarded grain coarsening to >95% density, thus retaining a grain size less than 100 nm[113] (see Fig. 4b). Terwilliger and Chiang showed that Ca and Sn are grain growth inhibitors in 95% dense nano-TiO_2.[133] Perhaps one of the most dramatic effects of dopants is the addition of 6% Ca to CeO_2 that retained a 30 nm grain size after full sintering by heating at 10 K/min to 1625 K.[135] In the same sintering conditions, undoped CeO_2 reached a grain size of 400 nm. Grain boundary chemical segregation of 5% Nb was very effective in suppressing grain growth of milled Fe_3Si.[136] Promotion of chemical ordering in intermetallics [Ni_3Al, Fe_3Si and $(Fe, Mn)_3Si$] also retards grain growth by reducing grain boundary mobility.[137]

If no dopants are desired because of their negative effect on the final properties (e.g., on dielectric[62] or mechanical[74] properties), sintering under pressure may be another effective means to suppress grain growth. This approach is discussed in Secs. 3.3 and 3.4.

3.0 METHODS FOR FULL DENSIFICATION OF NANOPOWDERS

3.1 Characterization of Nanomaterials Densification: Density and Grain Size Measurements

To investigate any consolidation results, two main characteristics are usually examined: density and grain size. When analyzing different nanopowder densification methods, caution is required when comparing reported density and grain size values. This is due to the diversity of measurement methods and their inherent experimental errors. Some issues related to these measurement techniques for nanoparticles and theoretical density values are briefly addressed below.

Density Measurements. The density may be displayed as a physical value or a percentage of the theoretical density. The theoretical density value of nanocrystalline materials is controversial. Some argue that a lower density value in nanocrystals may arise due to a large volume fraction of grain boundaries which, in turn, may have lower densities than adjacent crystals.[138]–[141] A more detailed discussion on the theoretical density value of nanocrystalline materials may be found in Ref. 142. Another source of theoretical density error is the uncontrolled presence of a secondary phase such as remaining oxides.[77][81]

Experimentally, density measurements are usually performed using Archimedes' principle. The errors associated with this method derive mainly from thermal and surface effects, pore presence and specimen size. The reproducibility of density values is within ~ 2%. To increase the accuracy of density measurements, some higher density immersion liquids (e.g., ethyl phtalate or methylene iodide) may be used. Larger specimens result in less density error. For specimens with interconnected porosity, prior impregnation is required.[143] The ultimate indication of pore elimination is by positron annihilation spectroscopy. This technique is very sensitive to ultra-fine pores, which are otherwise difficult to observe even in TEM. It has been used to study densification by monitoring the pore sizes as a function of sintering parameters.[144][145] Further information on nanopores and their distribution may be furnished by small angle x-ray or neutron scattering and BET (Brunauer, Emmett, Teller) nitrogen adsorption methods.[4][30][42][108][110][146] Similar to other porosimetry techniques, BET is suitable only for open porosity. Some more sophisticated pore size characterization has been performed using three dimensional scanning tunneling microscopy (STM) in cold compacted metals produced by the IGC method.[147]

Grain Size Measurements. This type of measurement is commonly carried out by x-ray diffraction (XRD) and electron microscopy techniques (TEM or high resolution SEM). In TEM and SEM, grain sizes are measured directly and grain size distribution may also be obtained. In contrast, XRD techniques provide only an average grain size. These techniques rely on the dependence of broadening of Braggs' reflections on grain size. Corrections for instrumental broadening must be made. Scherrer analysis is straightforward and assumes that the broadening of x-ray peaks is only due to the small crystallite size. The Warren-Averbach (WA) method is more sophisticated and accounts for broadening caused by internal stresses, as well. Furthermore, the WA analysismay distinguish between broadening due to grain size and to intrinsic faults and twins. Therefore, measurable differences may be found between Scherrer (larger values) and WA (smaller values) results and also when using different x-ray peaks for grain size calculations. The Scherrer method yields a volume-weighted average while the WA yields an area-weighted average grain size. Typically, these two averages differ by a factor of 2.[10] Nieman, et al., report a smaller error (10–25%) in grain size determination by the WA method for grain sizes up to 25 nm but thc error may be up to 100% when the grain size exceeds 50 nm.[96]

Generally, average TEM grain size and x-ray results are in good agreement, particularly for the WA analysis with strain and fault corrections.[10] As expected, the best agreement is found when the grain size distribution is narrow (e.g., Ref. 61). When it is large, the TEM method is preferable. In only in a few cases, large discrepancies are reported between microscope and XRD measured grain size.[42][148] For instance, Hahn, et al., reported SEM grain sizes to be four times larger than XRD line broadening data.[42] However, in this case, the SEM size was found to be that of agglomerates rather than that of individual nanoparticles. More details about grain size measurement methods in nanocrystalline materials may be found in Refs. 10 and 119.

3.2 Conventional Sintering

Both ceramic and metal nanopowders have been fully densified by conventional (pressureless) sintering. As pointed out in Sec. 2, sintering of nanoparticles is expected to take place at significantly lower temperatures due to an increase in driving force. These lower temperatures may minimize or even eliminate the use of sintering additives or grain growth inhibitors. For instance, Siegel, et al., sintered nanophase TiO_2 at a temperature 600 K lower than for conventional sintering, thus avoiding the need for sintering aids.[145] In some other systems such as WC-Co, the low sintering temperatures required to consolidate nanopowders may render the grain growth inhibitors unnecessary.[40] However, little research has been performed in understanding the interplay of dopant and pore effects to inhibit grain growth. A better understanding of the mechanism by which grain growth is inhibited may result in significant payoffs.

Ceramic Sintering. The benefits of fine grain sizes in ceramic sintering were reported more than 30 years ago when the sintering onset for very fine grained MgO was found as low as 625 K.[1] Low temperature sintering of ceramics has been achieved in fine grained ceramics in the

1980s.[62][72][87][149] Rhodes sintered the centrifugally cast ZrO_2-6.5 mol Y_2O_3 to 99.5% density at 1375 K to a final grain size of 200 nm.[87] Colloidally processed alumina sintered to 99.5% density at 1425 K.[149] Similar attempts to sinter ZrO_2 achieved 98% density at 1493 K when Y_2O_3 was the additive and 93% at 1208 K when the additive was Bi_2O_3.[72] In both latter cases, the final grain size was 600–700 nm. Lately, intensified efforts on sintering nanograined ceramic powders (ZrO_2, Al_2O_3, CdO, and TiO_2) by pressureless sintering resulted in fully dense ceramics with restrained grain growth. The sinterability of ceramic nanopowders has been greatly enhanced by improvements in nanopowder synthesis, particularly the control of nanopowder agglomeration.[4][60][150][151] Examples of dense ceramics that have retained nanosize grains up to 100 nm by conventional sintering are provided in Table 3. As seen in this table, full densification without grain coarsening has been achieved predominantly in the ZrO_2 system. This system has received substantial attention for two main reasons.[4][41][48][63][87] First, ZrO_2 has the slowest coarsening tendency among ceramics. Second, the pore size does not seem to affect grain coarsening.[4] Certainly, a better understanding of sintering and coarsening competition in other systems will improve our ability to fully densify them while retaining nanosize grains.

In the sol-gel titania, the final nano-range grain size is attributed to a transformation enhanced sintering.[69] This is in contrast to the results of Mayo who found that phase transformation during sintering of titania promoted grain growth.[109][153] A complicating factor in this comparison is powder agglomeration. The latter powders were agglomerated while the former had a low degree of aggregation. The exact role of phase transformation and powder agglomeration in pressureless sintering is certainly worth further attention. In conventional sintering of coarsely grained ceramics, high densities and smaller grain sizes may be achieved by using fast heating rates. The explanation is based on bypassing the low temperature range where the effect of surface diffusion is predominant and grain coarsening takes place. The theoretical predictions require heating rates on the order of 100 K/s to avoid surface diffusion contribution in early densification stages.[154] The only results consistent with this theory were achieved in sintering alumina and titania nanoparticles to high densities (95%) by using a heating rate of ~100 K/s.[59] Otherwise, the effect of heating rate on nanoceramic densification seems to be controversial. On one hand, no effect of heating rate in the range of 2–200 K/min on grain size and retarded densification have been noticed in 35 nm ZrO_2-3% Y_2O_3.[48][155] On the other hand, sintering with a similar heating rate (> 100 K/min) resulted in high densities and small grain sizes in Y-TZP.[156] A practical

Table 3. Density and Grain Sizes in Nanoceramics Consolidated by Conventional (Pressureless) Sintering

Material	Sintering Parameters:				Final Properties:		Ref.
	Init. Gr. Size (nm)	Temp. (K)	Time (hrs)	Atm.	Dens. (%)	Gr. Size (nm)	
CeO_2	10 to 15	1523	2	air	near full	~ 100	135
CeO_2-6 at.% Ca	10 to 15	1623	10 C/min.	air	full	30	135
ZrO_2	6 to 9	1400	1.3	air	full	80	64
ZrO_2	6 to 9	1250	.67	vac	near full	60	64
ZrO_2-3 mol.% Y_2O_3	6	1343	2.00	NR	near full	<100	152
ZrO_2-3 mol.% Y_2O_3[1]	<10	1373	1	NR[3]	99.9	80	4
ZrO_2-5.8 mol.% Y_2O_3	8	1273	100	air	>95	80	41
ZrO_2-5.8 mol.% Y_2O_3	8	1323	7	air	96	60	41
TiO_2	14	1173	15 to 20	oxygen	~ 90	50	42
TiO_2[2]	~ 6	873	NR	NR	99	<60	69
TiO_2-6% Y	~ 15	1173	15-20	oxygen	~98	30	113

1. Centrifugally consolidated. 2. Sol-gel prepared. 3. NR= not reported.

problem in the fast sintering of low thermal conductivity nanoceramics is the flaw formation as a consequence of thermal gradients.[48] These results, along with enhanced sintering at low temperatures, emphasize the complex and unclear role of surface diffusion in the sintering of nanopowders which certainly warrants further studies.

Metals. Conventional sintering has been less successful when applied to nanosize metals and intermetallics compared to nanoceramics. For instance, full densities have been achieved only in a few cases such as pure Ni and Fe-(Fe, $Mo)_6C$ powders.[157][158] Rate controlled sintering of pure Ni resulted in 99% dense specimen with 70–80 nm grain size.[157] However, for the other cases, the final grain size was 500–600 nm since sintering was carried out at regular sintering temperatures ($>0.5T_m$). Consequently, warm compaction has been the preferred way to combine high densities and grain sizes less than 100 nm in nanometals, as shown in Sec. 2.4. Nanograined WC-Co powders have been fully consolidated with minimal grain growth by conventional sintering, but the final grain size was on the order of 100–200 nm.[158]–[161] This final grain size is mainly attributed to the liquid phase mechanism which typically promotes grain growth. One approach in restricting grain growth in WC-Co is to use inhibitors.[159] Another promising way is to explore the sintering temperature depression in the nanoregime to possibly achieve densification only in the solid state and thus limit the grain growth.[40] TiAl intermetallics produced by IGC were sintered to >95% density by conventional sintering at low temperatures ($0.4T_m$) with grain sizes less than 15–20 nm.[162][163]

Nano-Composite Densification. Handwerker, et al., summarized the sintering behavior of ceramic composites.[149] Three different behaviors are identified in the sintering of viscous matrices. The common rule of mixture applies only when the volume fraction of the second phase is less than 10%. Scherrer's theory for viscous sintering with rigid inclusions applies between 10 and 12%. At higher volume fractions, deviations from both models are observed. For crystalline matrices, densification rates are lower than theoretical at all volume fractions, indicating that the second phase retards matrix densification. Sintering of nanocomposites seems to adhere to the same mechanisms. For instance, Prabhu and Bourell have observed that the experimental shrinkage for a nanocomposite of ZrO_2-3 mol% Y_2O_3 with 20 wt% Al_2O_3 was lower than for the zirconia matrix.[164] The densification rate of the nanocomposite is not a linear function of the densification rate of the two components and each phase in the mixture has its own densification and coarsening rate. The nanocomposite sintered to a density of 97% at 1573 K for 1 hour using a pressure of 30 MPa in hot pressing, while zirconia sintered to a density of 98% density at 1498 K. The grain size exceeded the nanometer range (200–300 nm). Similarly, the addition of SiC particles delayed the densification and grain growth rates of the Al_2O_3 matrix.[165] In an Al_2O_3 - 20% ZrO_2 composite, densities of 87–97% have been reached but the final grain size was between 400 and 700 nm.[166] A final density of 98% was achieved in Al_2O_3-SiC composite but grain size was in the micron range.[167] In a Cu-Nb composite, the Nb phase successfully restricted grain growth upon conventional sintering that achieved a high density and retained a grain size of 100 nm.[168]

3.3 Pressure Effects in Nanopowder Consolidation

The classical temperature-pressure trade-off for sintering brings distinct advantages to nanopowder densification due to the ability to restrict grain growth. Numerous examples of pressure-assisted sintering of nanopowders demonstrate enhanced densification with reduced grain growth compared to pressureless (conventional) sintering.[42][61][84][115][134][169]–[173] An illustration of the pressure effects on nanopowder densification is shown in Fig. 6.

An applied stress adds a new component to the curvature related driving force for densification. The applied mean stress, known as the hydrostatic stress, induces new plasticity-driven sintering mechanisms, as well as a stress-assisted diffusion mechanism. Both mechanisms enhance the densification rate. The shear component of the applied stress causes particle rearrangement and the collapse of large pores. The particle rearrangement and macroscopic deformation of pores increase the number of particle contacts. This different pore size and distribution is more amenable to pore removal than in pressureless sintering.[125][128] The main benefit is the elimination of large pores in agglomerated powder. As shown in Secs. 2.5 and 2.6, the presence of large pores is the cause of the exaggerated grain coarsening at densities above 90%. If large pores can be eliminated, grain growth is minimized. Generally, the capability to collapse the pores in pressure-assisted sintering scales with the shear stress level. This stress is minimal in *hot isostatic pressing* (HIP) and increases gradually from quasi-isostatic pressing to uniaxial pressing in a die (hot pressing), to uniaxial pressing without a die (sinter-forging) and finally to extrusion.[174] A shear stress is also beneficial for mechanical disruption of surface oxide layers which provides better interparticle bonding.

(a)

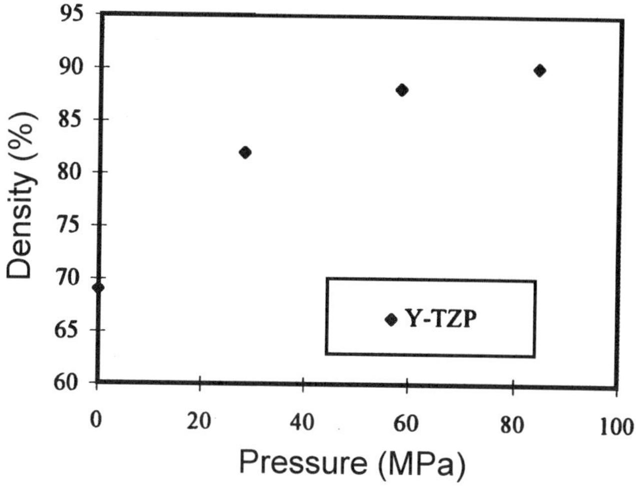

(b)

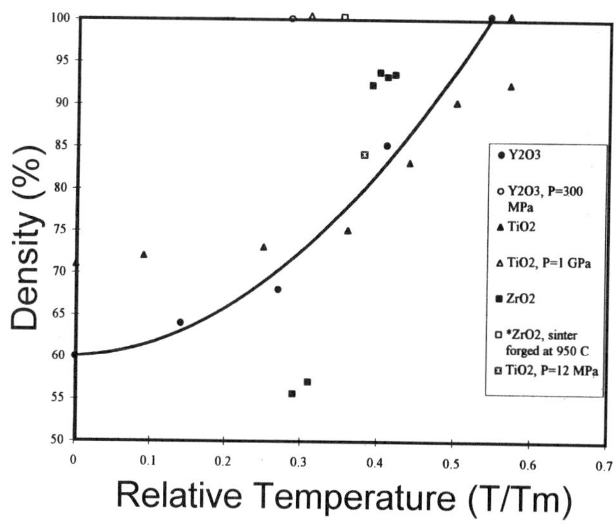

Figure 6. Pressure effect on the densification of nanoceramics. (*a*) Density improvement with the increase in the applied stress during sinter forging of Y-TZP at 1375 K for 14 min;[173] (*b*) Pressure application yields full density (open symbols) at lower temperatures than pressureless sintering (full symbols and fitting curve).[63][172] Note that a pressure of 12 MPa has no influence on TiO_2 sintering (open star).

In nanocrystalline powders, the effect of the applied stress is noticeable only if it exceeds the intrinsic curvature-driven sintering stress. Skandan, et al., analyzed the contributions of the intrinsic and external pressures.[64][115][175] The applied pressure is independent of the particle size. The intrinsic sintering pressure increases when the particle size is reduced and may reach high values as the particles become increasingly small. At this particle size, the applied pressure has to be higher than the curvature driven pressure. Consistent with this threshold effect, no external pressure effect was observed in the densification of ZrO_2,[64][104] TiO_2[107][109][176][177] (see Fig. 6b) or CdO[178] unless a certain pressure level was achieved. However, this stress level significantly departed from the threshold values calculated based on the particle curvature effects. As shown in Sec. 2.1, the intrinsic curvature stress in nanopowders may be on the order of hundreds of MPa. The experimentally observed threshold pressure in nanopowders is an order of magnitude less.[64][107][109][177] This discrepancy may suggest a wide particle size distribution with large particles which may give only a small contribution to the intrinsic sintering stress. Alternatively, grain coarsening may occur by the time the pressure is applied, thus lowering the intrinsic sintering stress. Many times, pressure application is deliberately carried out only at high temperatures where the resistance to deformation is less and the pores are more likely to collapse.[64][107][179][180] As shown in Sec. 2.4, large pores are more difficult to fracture, even at very high pressures, when the temperature is low. If pressure is applied at an elevated temperature, densification of the compact is much easier. For instance, Schwarz, et al., achieved full density and retained fine grain size (44 nm) in mechanically alloyed Al_5CuZr using a hot forging technique with pressure applied when the system achieved the sintering temperature.[179] Similarly, density improvements have been noticed when pressure was applied after heating vs application of pressure at room temperature in sintering mechanically alloyed iron powders.[180]

The classical pressure-assisted sintering diagrams developed by Ashby and co-workers[181][182] have been used to describe nanopowder consolidation.*[61][84][183] For the same pressure level, Bourell, et al., indicated a change in the sintering mechanisms from stress-assisted densification in coarse grained zirconia to curvature driven grain boundary diffusion in nanocrystalline Y-TZP (Fig. 7).[84] Similarly, McKimpson showed an enhanced boundary diffusion contribution when calculating Ashby maps

* A new and comprehensive model has been developed since the completion of this chapter. See Suryanarayan, R., and Sastry, S. M. L., Consolidation of Nanoparticles—Development of a Micromechanistic Model, *Acta Mater.*, 47:3079–3098 (1999).

for hot isostatic pressing of nanosize (Al, Cr)$_3$Ti, Cu, and Al_2O_3 powders.[183] However, comparison with experimental HIP results indicate that the calculated densification rates are consistently overestimated. For instance, 20 nm Al_2O_3 HIPped for 1 hour at 1625 K under 300 MPa pressure resulted in an only 84% dense sample. Among the possible reasons is the retarded densification due to large pores in agglomerated powders. HIP induces only local shearing stresses and at the lowest levels among the pressure enhanced sintering methods. However, even these small deviations from a purely hydrostatic stress in HIP have been shown to contribute to enhanced sintering rates.[184]

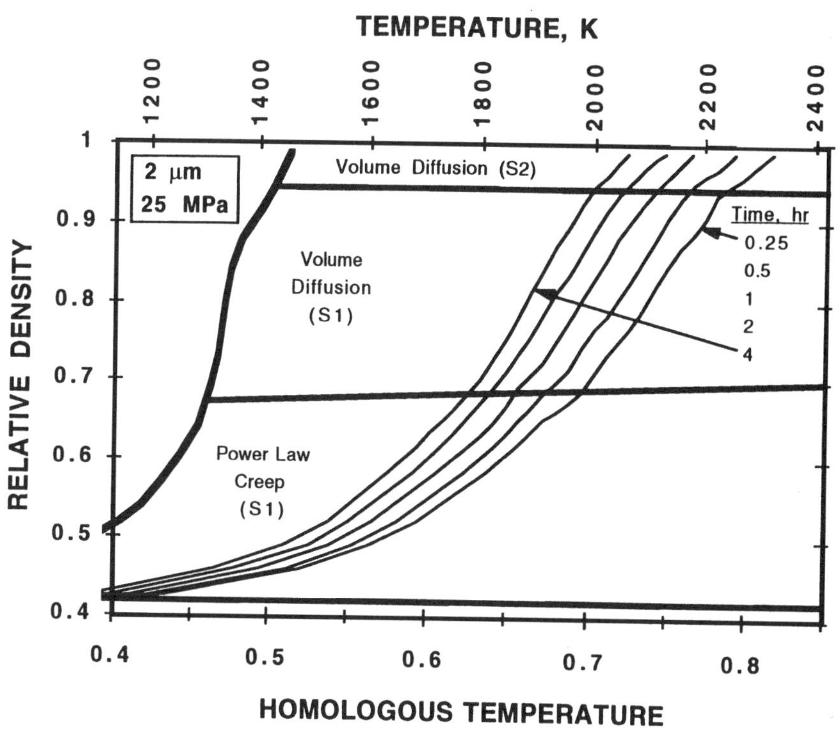

(a)

Figure 7. Densification maps of (*a*) regular size and (*b*) nanocrystalline Y-TZP powders. *(Courtesy of D. L. Bourell.)*

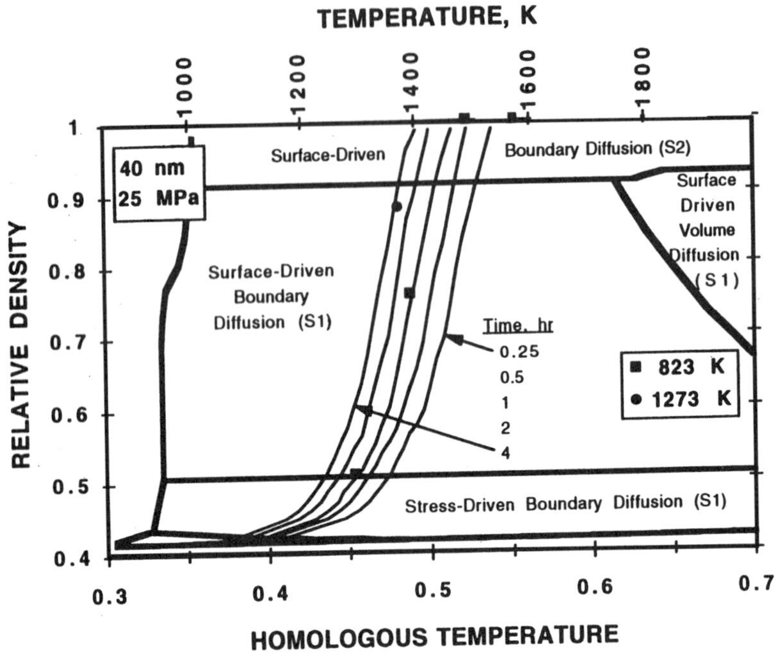

(b)

Figure 7. *(Cont'd.)*

Shaik and Milligan applied the Ashby model to rapid sinter forging of milled Fe-Cu nanopowders.[61] They used the experimental green density (0.8) compared to the theoretical value (0.64) used by Ashby, and also modified the stress state. Noteworthy is the good agreement with the experimental results when grain boundary diffusion was considered the mass transfer mechanism in nanograin powders within micron particle sizes. Creep was the dominant densification mechanism. The contributions of yielding and diffusion were significantly less with densification rates about two orders of magnitude lower than those for creep densification. When the applied pressure is high, such as in the piston-cylinder method or explosive forming, plastic deformation may contribute to nanocrystalline materials densification, despite their high yield stress value. Other mechanisms such as grain boundary sliding and grain rotation may also play a role in nanopowder deformation sintering such as in sinter forging.[104][185][186] Grain boundary sliding is assumed to be the main densification mechanism in severe plastic deformation consolidation.[49]

3.4 Pressure-Assisted Consolidation Methods

With only a few exceptions, all known pressure-assisted consolidation methods have been applied for nanopowder densification: hot pressing, sinter forging, HIP, extrusion, and high pressure techniques. As already shown in Sec. 3.3, hot pressing and sinter forging involve a uniaxially applied pressure in a die (hot pressing), or with no die (sinter forging). In HIP, a fluid transmits a hydrostatic stress. Variants of HIP with quasi-isostatic stress distribution may be obtained if a fluid-like pressure transmitting medium is used such as in the Ceracon process.[174][187]

Hot Pressing. Hot pressing gives distinct advantages in comparison to pressureless sintering in achieving full densities and minimal grain growth as shown, for example, in nanograined Fe-(Fe, Mo)$_6$C,[188][189] TiN,[5] ZrO_2-Al_2O_3,[164] or TiO_2.[42] Theoretical or near-theoretical densities and grain sizes less than 100 nm have been achieved by hot pressing mechanically alloyed Fe-2% Al,[77] Fe-10% Al,[78] Al-10% Ti,[190] Fe, Fe_3Al and Ni_3Al[191][192] and TiAl.[193] A grain size of only 16 nm has been retained in cryomilled Fe-10 wt% Al which was hot pressed at 823 K and heat treated for 1 hour at 1223 K.[78] This unusual stability is attributed to nanometer dispersoids of γ-Al_2O_3 and AlN particles.[78] Similar densification results with final nanosize structures have been reported by hot pressing W-Ti and metal-nitrides composites.[194][195] Hot pressing retained an amorphous structure in (Fe, Co, Ni)B alloys.[196] Hot pressing of nanoceramics such as ZrO_2,[64] TiO_2,[197][198] and CeO_2[199] or nanoceramic composites such as ZrO_2- Al_2O_3,[200] Si_3N_4/SiC,[201] and Al_2O_3/Ni[202] achieved full densities and grain sizes below 100 nm, as well. Generally, the pressures used in hot pressing span a large range from low (<100 MPa),[77][84][164] to moderate (100–500 MPa),[64][78][198] and high pressure levels (>0.5 GPa).[42][63][191][203] Increasing the pressure diminishes the final grain size. Some meaningful examples are given by Hahn who sintered nano-TiO_2 to its theoretical density at 725–825 K (~ 0.35 T_m) applying 1 GPa with no grain growth[63] and Araki, et al., who densified mechanically alloyed Al-10.7 at% Ti powders to 98% under 2 GPa at 573 K with virtually no grain growth and retention of the initial Al supersaturation.[203] A very high pressure application (>1 GPa) retained grain sizes less than 75 nm in nearly fully dense ceramics such as TiO_2 and Al_2O_3 sintered to 95% density[204][205] and aluminides (Fe_3Al and Ni_3Al) sintered to 91–95% at 775 K.[191]

Hot Isostatic Pressing (HIP). By design, HIP densification involves longer times and therefore grain growth is more likely to occur than in hot pressing. The final grain size in dense materials obtained by HIPping nanopowders may easily reach micron size.[206]–[208] However, a careful control of HIP parameters, particularly temperature, resulted in grain sizes of 100–300 nm in ball-milled TiAl and Ti_3Al,[209] Fe-10% Cu,[171] Si_3N_4-SiC nanocomposites,[210] and sometimes less than 100 nm in FeAl[211] and Cu-Nb.[212] The latter dense nanocrystalline materials were obtained at temperatures less than 0.3 T_m and pressures between 175 and 300 MPa. HIP was applied to nanoceramic densification in only a few cases where densities of 91–92% were reached such as in SiC at 2175 K[207] and Si_3N_4 at 1925 K.[206] Grain sizes well exceeded nanometer range. One limitation in using HIP for ceramic densification is the lack of adequate encapsulation materials with temperatures close to ceramic sintering temperatures.[4]

Sinter Forging. Generally, the stress levels required for densification by sinter forging are lower than in hot pressing or HIP. This method received substantial attention, both theoretically and experimentally, as applied to ceramic densification.[128][169][185][213][214] The most attractive benefit in using the sinter-forging technique is the capability of densifying green compacts with large interagglomerate pores. As shown in Sec. 3.3, the high shear stresses associated with uniaxial pressure application contribute to the closure of large pores that cannot otherwise be eliminated by only diffusion. Plastic strains necessary to close large pores are high. Beyond pore closure, the application of high plastic strains induces dynamic grain growth. Therefore, to eliminate large pores and prevent grain coarsening, large strains must be applied before small pores are eliminated by diffusion, i.e., during the intermediate sintering stage.[128] For this purpose, a two-stage sinter-forging (high, then low, strain rate) has been developed and resulted in the highest densities and smaller grain sizes.[115][177] Conversely, if large pores are not fractured, such as in some constant load sinter-forging cases, densification is not complete.[128][186] Sinter-forging has been extensively applied to nanoceramic particle consolidation such as zirconia,[64][128][134][173][175][186][215] alumina,[213][216] titania,[104][107][153][177] and zirconia-toughened alumina.[215] As an illustration of the efficiency of sinter forging in retaining nanometer grains, Mayo, et al., sinter-forged nano-TiO_2 with a grain size of 87 nm while pressureless sintering gave a 400 nm grain size for the same final density (91%).[107][109] The benefits of sinter forging in comparison to HIPping may be seen in Fe-10% Cu powders in which the former consolidation process achieved 45 nm grains at 800 K and

525 MPa whereas HIPping at 975 K and 170 MPa yielded a final grain size of 130 nm.[61][171] In numerous cases, sinter forging was very effective in achieving full or near full densities and grain sizes less than 100 nm in both ceramics[64][113][115][177][215] and metals.[61][170][191] Grain sizes even less than 50 nm have been often retained such as in nanocrystalline ZrO_2 at $0.4T_m$,[64] TiO_2 at 0.43 T_m,[177] Fe-Cu at 0.44-0.5 T_m.[170][171] Practically, precautions have to be considered to avoid ceramic specimen fracture at high strain rates at low temperatures.[128] In contrast, high strain rates may be used in sinter forging of metals.[61]

Extrusion. Hot extrusion usually involves high stresses that may be applied at relatively lower temperatures than in other pressure-assisted techniques, such as HIP. A good grain boundary strength is achieved only if sufficient diffusion and bonding occur at extrusion temperature. A compromise is usually sought between the strength, which requires high temperatures, and final grain size, for which low temperatures are desired. Hot extrusion has been primarily used to consolidate metal nanoparticles.[217]–[220] Grain sizes less than 100 nm were achieved at 1120 K using a 0.5 GPa stress in Ni and Fe,[217] Al-Ni-Zr with mischmetal additions under an extrusion ratio of 10:1,[218] and in Mg-5% Cu-10% Y alloys at 323–773 K with extrusion ratios of 5:1 to 10:1.[220] If the extrusion temperature is sufficiently low, an amorphous structure may be retained in the latter alloys.

Ultrahigh Pressure Sintering. Pressure levels up to 5–7 GPa have been attained using piston-anvil devices, cubic anvil cells, and toroidal- and belt-type apparatuses. Similarly high pressure levels may be achieved by a new technique known as *severe plastic deformation consolidation* (SPDC). Shockwave or dynamic consolidation may reach even higher pressure levels, up to tens of GPa. This latter technique will be detailed in the next section. Some examples of high pressure sintering have already been shown in the paragraph on hot pressing. Misra, et al., consolidated Al_2O_3 to near full densities (>99%) and a final grain size of 93 nm at 1223 K using a piston cylinder method to reach 1 GPa.[221] A pressure of 5.5 GPa consolidated Al_2O_3 to 95% density and 73 nm grain size at 1073 K in 15 minutes.[205] A bulk nanostructure with 22 nm grain size was achieved in Al-Fe (Mo, Si, B) alloys by high pressure consolidation (3 GPa) at 1073 K of alternate amorphous layers of constituent metals.[222] Fully dense Ni and Cu-50 wt% Ag specimens with $\leq$ 20 nm grain size were produced by SPDC.[223][224]

3.5 Non-Conventional Sintering Methods

A number of non-conventional consolidation methods have been applied to nanopowder densification: microwave sintering, shock or dynamic consolidation, and field assisted sintering. The main purpose in using these methods is to enhance densification, thus, reducing the sintering temperature or time with the ultimate benefit of preserving final fine grain sizes.

Microwave Sintering. The processing time in microwave sintering is reduced compared to conventional heating from external sources due to the direct energy coupling with electric dipoles within the heating body.[225][226] These shorter times bring about energy and final property benefits. Temperature gradients are reduced and an overall short sintering time minimizes grain growth. In addition, the rapid heating rate can bypass the low temperature region where the rate of grain growth is higher than the rate of densification. Since grain boundaries are the primary sites for electric dipoles, microwave sintering appears particularly attractive for the densification of nanocrystalline powders. The method has been applied to ceramic nanoparticles such as TiO_2[227]–[229] and Al_2O_3.[229]–[231] In these materials, only densities less than 95% have been achieved when grain size was less than 100 nm. For instance, microwave sintering had to be restricted to 1425 K to maintain a nanometer grain size in γ-Al_2O_3 which fully transformed to α-Al_2O_3.[231] The final density was 93%. Contrary to expectations, the density-grain size combinations in γ-Al_2O_3 were almost identical to those achieved in more conventional pressureless sintering experiments at the same temperature.[230] The sequence of phase transformation in Al_2O_3 was also similar to that in conventional sintering.

Field Assisted Sintering. Electric field application has been shown to enhance powder densification thereby achieving higher densities at lower temperatures or in shorter times.[232]–[235] Variants of field sintering which involve pulsed current discharge and resistance heating, sometimes termed *plasma activated sintering* (PAS), *field activated sintering technique* (FAST), *spark-plasma sintering,* or *pulse electro-discharge consolidation*, have been effectively applied to nanopowder consolidation. The enhanced densification in PAS is most noticeable at lower temperatures or when multiple discharges of current through the powder compact are applied.[180][232] For non-conductive ceramics, a tentative explanation of the enhanced densification by multiple discharges is related to the increase in the dielectric constant with the temperature level at which the discharge was applied.[74][232] Generally, it is believed that pulsed current application promotes the removal of adsorbates or surface oxides due to a possible plasma generation.[233] This enhances the sinter bonding of particulates.

Field sintering for three minutes at 725 K of mechanosynthesized Fe-85% Fe_3C achieved 99% density with a final grain size of 45 nm.[233] This is to be compared to HIPping of the same powders at 1025 K for sixty minutes which yielded the same density but final grain size of 87 nm. Kimura and Kobayashi fully sintered mechanically alloyed TiAl powders and retained nanosize grains by spark sintering at 1051–1312 K under modest pressures (29–147 MPa).[234] Similar to conventional sintering, they showed that the densification temperature inversely scaled with the applied pressure.

An alternate electric pulsed-power method induces short, high-pressure pulses that result in powder densification such as in *dynamic magnetic compaction* (DMC).[236][237] Ivanov, et al., obtained up to 5 GPa pressure and thus consolidated alumina powders to 83% density while retaining a nanocrystalline structure.[236] The DMC method was applied to sintering of large parts of nanocrystalline iron and alumina.[237]

Shockwave Consolidation. Dynamic or shockwave consolidation proceeds with the passage of a large-amplitude compressive stress generated by plate impact or explosion without any external heating.[187][238] The peak pressure values may be on the order of tens of GPa, thus providing densification by plastic yielding for both metals and ceramics.[238]–[242] Localized heating, possibly up to melting temperatures, due to particle interfriction occurs and enables good interparticle bonding. In nanosize powders, the heat may transfer throughout the entire particle, thus providing an advantage over coarser materials where the heating is only superficial. Best results are achieved when high temperatures are reached before the shockwave passes. If particles are heated, they may deform rather than fracture when the stress is applied. This very short, high-temperature exposure provides the best means to retain fine grain size or out-of-equilibrium conditions such as amorphous structures,[240] or supersaturated solid solutions.[242] The major drawback is the difficult coordination of these short stress and heat application events which result in frequent specimen fracture (cracking).

Shockwave consolidation has been applied to consolidation of both ceramic and metal nanoparticles. For instance, fully dense specimens with grain size of 20 nm were obtained in ball milled Fe-N solid solution.[242] In the Ti-Si system, shock consolidation yielded 30–40 nm grain size of crystalline $TiSi_2$ and Ti_5Si_3 phases.[240] Only limited grain coarsening took place upon subsequent annealing at 800°C for one hour. Full densification was also reported in mechanically alloyed TiAl specimens with final grain sizes of 15 nm.[239] These results are compared with HIPping that provided full consolidation at 1075 K, 207 MPa, 2 hours, but grain sizes were about 100 nm. Ceracon consolidation at 1225 K, 1240 MPa, 1 minute, resulted in grain sizes of 60 nm in TiAl and 120 nm in Ti_3Al.[187]

4.0 SUMMARY

The most distinctive features of the sintering process of nanosize powders are the high driving force and enhanced kinetics due to large curvature effects. Densification of nanopowders takes place at temperatures consistently below those of larger grained powders by up to several hundreds of degrees. There are numerous benefits from using lower sintering temperatures: small final grain sizes, elimination of sintering aids, avoiding undesirable phase transformations, and deleterious decomposition or interfacial reactions.

Analysis of the nanopowder sintering process provides new insights into less conspicuous features for regular sintering such as interface structure, anisotropy, or contaminant and defects role. The atomic mechanisms become more obvious when smaller scales are approached. When particle size decreases, new sintering mechanisms such as grain boundary slip, dislocation motion, grain rotation, viscous flow, and grain boundary melting may also become operative. Densification of nanopowders has lead to a new and better understanding of the role of pores in sintering and grain growth.

Densification behavior is largely dictated by pore size and distribution. A compact with uniform and fine pore structure sinters to full densities without grain coarsening. Most of the time, such a pore distribution is associated with a high green density in non- or weakly agglomerated powders. The recent efforts to synthesize non-agglomerated nanopowders have reflected positively on their final densification results. Better control of synthesis and processing enables fabrication of fully dense nanocrystalline parts, particularly ceramic, even by conventional sintering. Pressure-assisted densification applied to nanopowders has also shed light on new stress effects on sintering, such as the threshold effect. The science and practice of pressure application has significantly improved with positive results to achieve sound and dense nanocrystalline parts. This is particularly the case for sinter forging. Other pressure-assisted sintering techniques are still in the development stage.

Although some distinct differences in the densification of nano- vs micron-grained powders seem to emerge and a better understanding of nanosintering has been accomplished, the specific effect of densification variables on the final density and properties of nanomaterials is not well understood yet. The interdependence of the sintering behavior on size, chemistry, and structure of nanoscale particles is an area of fruitful future research. More fundamentally, there is a tremendous need for specific materials data in the nanoregime, such as diffusion coefficients and surface energies.

ACKNOWLEDGMENTS

The author's research on nanocrystalline materials densification has been supported by National Science Foundation, Chemical and Transport Systems (grant # CTS-9632280) and Manufacturing Processes and Equipment Program (grant # DMII–9532072). The author wishes to thank all those who provided her with unpublished results or preprints of their recent work. Irene Thuenissen is gratefully acknowledged for her help in the manuscript preparation.

REFERENCES

1. Morgan, P. E. D., *Superplasticity in Ceramics,* in: *Ultrafine-Grained Ceramics* (J. J. Burke, L. R. Norman, and V. Weiss, eds.), pp. 251–271, Syracuse University Press, Syracuse (1968)
2. Sanders, P. G., Eastman, J. A., and Weertman, J. R., Elastic and Tensile Behavior of Nanocrystalline Copper and Palladium, *Acta Mater.*, 45:4019–4025 (1997)
3. Malow, T. R., and Koch, C. C., Mechanical Properties in Tension of Mechanically Attrited Nanocrystalline Iron by the Use of the Miniaturized Disk Bend Test, *Acta Mater.*, 46:6459–6473 (1998)
4. Mayo, M., *P*rocessing of Nanocrystalline Ceramics from Ultrafine Particles, *Int. Mater. Rev.*, 41:85–115 (1996)
5. Bourell, D. L., and Groza, J. R., *Consolidation of Ultrafine and Nanocrystalline Powder,* in: *PowderMetal Technologies and Fabrication, ASM Handbook,* pp. 583–589, ASM International, Metals Park, OH (1998)

6. Andrievski, R. A., Nanocrystalline High Melting Point Compound-Based Materials, *J. Mat. Sci.*, 29:614–631 (1994)
7. Gutmanas, E. Y., Materials with Fine Microstructures by Advanced Powder Metallurgy, *Progr. Mat. Sci.*, 34:261–366 (1990)
8. Suryanarayana, C., Nanocrystalline Materials, *Int. Mater. Rev.*, 40:41–64 (1995)
9. Siegel, R. W., *Synthesis and Processing of Nanostructured Materials,* in: *Mechanical Properties and Deformation Behavior of Materials Having Ultra-Fine Microstructures,* (M. Nastasi, D. Markin, and H. Gleiter, eds.), pp. 509–538, Kluwer Academic Publ., Dordrecht, Netherlands(1993)
10. Weismuller, J., *Nanocrystalline Materials - An Overview,* in: *Synthesis and Processing of Nanocrystalline Powder,* (D. L. Bourell, ed.), pp. 3–19, TMS, Warrendale, PA (1996)
11. Lu, K., Nanocrystalline Metals Crystallized from Amorphous Solids: Nanocrystallization, Structure and Properties, *Mat. Sci. Eng.*, R16:161–221 (1996)
12. Porter, D. A., and Easterling, K. E., *Phase Transformations in Metals and Alloys*, 2nd ed., p. 113, Chapman and Hall, London (1992)
13. Johnson, D. L., *Solid State Sintering Models,* in: *Sintering Processes* (G. C. Kuczynski, ed.), pp. 97–106, MRS, Plenum Press, NY (1980)
14. Bonevics, J. E., and Marks, L. D., The Sintering Behavior of Ultrafine Alumina Particles, *J. Mater Res.*, 7:1489–1500 (1992)
15. Rankin, J., and Sheldon, B. W., In Situ TEM Sintering of Nano-Sized ZrO_2 Particles, *Mater. Sci. Eng. A*, 204:48–53 (1995)
16. Carim, A. H., *Microstructure in Nanocrystalline Zirconia Powders and Sintered Compacts,* in: *Nanophase Materials, Synthesis-Properties-Applications,* (G. C. Hadjinapanis and R. W. Siegel, eds.), pp. 283–286, Kluwer Academic Publ., Netherlands (1994)
17. Guillou, N., Nistor, L. C., Fuess, H., and Hahn, H., Microstructural Studies of Nanocrystalline CeO_2 Produced by Gas Condensation, *Nanostr. Mater.,* 8:545–557 (1997)
18. Olynick, D. L., Gibson, J. M., and Averbach, R. S., Trace Oxygen Effects on Copper Nanoparticle Size and Morphology, *Appl. Phys. Lett.*, 68:343–345 (1996)
19. Champion, Y., and Bigot, J., Synthesis and Structural Analysis of Al Nanocrystalline Powders, *Nanostr. Mater.,* 10:1097–1110 (1998)
20. Zhu, H., and Averback, R. S., Sintering Process of Two Nanoparticles: A Study by Molecular Dynamics Simulations, *Phil. Mag. Lett.*, 73:27–33 (1996)

21. Conder, R. J., Ponton, C. B., and Marquis, P. M., *Processing of Alumina/ Silicon Carbide Nanocomposites,* in: *Nanoceramics* (R. Freer, ed.), pp. 105–116, The Institute of Materials, London (1993)
22. Ying, J. Y, Chi, L. F., Fuchs, F., and Gleiter, H., Surface Morphology of Nanocrystalline Titanium Oxide by AFM, *Nanostruct. Mater.*, 3:273–281 (1993)
23. Luo, P., Nieh, T. G., Schwartz, A. J., and Lenk, T. J., Surface Characterization of Nanostructured Metal and Ceramic Particles, *Mat. Sci. Eng. A*, 204:59–64 (1995)
24. Cahn, R. W., Nanostructures Come of Age, *Nature*, 359:591–592 (1992)
25. Trudeau, M. L., Tschope, A., and Ying, J. Y., XPS Investigation of Surface Oxidation and Reduction in Nanocrystalline $CexLa_{1-x}O_{2-y}$, *Surface and Interf. Analysis,* 23:219–226 (1995)
26. Takeuchi, T., Ado, K., Asai, T., Kageyama, H., Saito, Y., Masquelier, C., and Nakamura, O., Thickness of Cubic Surface Phase on Barium Titanate Single-Crystalline Grains, *J. Am. Ceram. Soc.*, 77:1665–1668 (1994)
27. Payne, M. C., *Electron Theory Applied to Phases with Low Dimensions,* in: *Mechanical Properties and Deformation Behavior of Materials Having Ultra-Fine Microstructures*, (M. Nastasi, D. Markin, and H. Gleiter, eds.), pp. 37–52, Kluwer Academic Publ., Netherlands (1993)
28. McHale, J. M., Navrotsky, A., and Perrotta, A. J., Effects of Increased Surface Area and Chemisorbed H_2O on the Relative Stability of Nanocrystalline γ-Al_2O_3 and α-Al_2O_3, *J. Phys. Chem. B*, 101:603–613 (1997)
29. McHale, J. M., Auroux, A., Perrotta, A. J., and Navrotsky, A., Surface Energies and Thermodynamic Phase Stability in Nanocrystalline Aluminas, *Science,* 277:788–791 (1997)
30. Rabe, T., and Wasche, R., Sintering Behavior of Nanocrystalline Titanium Nitride Powders, *Nanostruct. Mater.*, 6:357–360 (1995)
31. Turnbull, D., Metastable Structures in Metallurgy, *Met. Trans.*, 12A:695–708 (1981)
32. Uchino K., Sadanaga E., and Hirose, T., Dependence of the Crystal Structure on Particle Size in Barium Titanate, *J. Am. Ceram. Soc.*, 72:1555–1558 (1989)
33. Skandan, G., Foster, C. M., Frase, H., Ali, M. N., Parker, J. C., and Hahn, H., Phase Characterization and Stabilization Due to Grain Size Effects of Nanostructured Y_2O_3, *Nanostr. Mater.*, 1:313–322 (1992)
34. Garvie, R. C., Stabilization of the Tetragonal Structure in Zirconia Microcrystals, *J. Phys. Chem.*, 82:218–223 (1978)

35. Krauss, W., and Birringer, R., Metastable Phases by Inert-Gas Condensation, *Nanostr. Mater.*, 9:109–112 (1997)

36. Eastman, J. A., Choi, U. S., Li, S., Soyez, G., Thompson, L. J., and DiMelfi, R. J., Novel Applications Exploiting the Thermal Properties of Nanostructured Materials, presentation: Fine, Ultrafine, and Nano Powders 98 Conf., New York (Nov. 8–10, 1998)

37. Hague, D. C., and Mayo, M. J., The Effect of Crystallization and a Phase Transformation on the Grain Growth of Nanocrystalline Titania, *Nanostr. Mater.*, 3:61–67 (1993)

38. Balachandran, U., Siegel, R. W., Liao, Y. X., and Askew, T. R., Synthesis, Sintering and Magnetic Properties of Nanophase Cr_2O_3, *Nanostr. Mater.*, 5:505–512 (1995)

39. Hansen, J. D., Rusin, R. P., Teng, M.H., and Johnson, D. L., Combined-Stage Sintering Model, *J. Am. Ceram. Soc.*, 75:1129–1135 (1992)

40. Porat, R., Berger, S., and Rosen, R., Dilatometric Study of the Sintering Mechanism of Nanocrystalline Cemented Carbides, *Nanostr. Mater.*, 7:429–436 (1996)

41. Thuenissen, G. S. A. M., Winnubst, A. J. A., and Burggraaf, A. J., Sintering Kinetics and Microstructure Development of Nanoscale Y-TZP Ceramics, *J. Europ. Ceram. Soc.*, 11:319–324 (1993)

42. Hahn, H., Logas, J., and Averbach, R. S., Sintering Characteristics of Nanocrystalline TiO_2, *J. Mater. Res.*, 5:609–614 (1990)

43. Kusunoki, M., Yonemitsu, K., Sasaki, Y., and Kubo, Y., In Situ Observation of Zirconia Particles at 1200°C by High-Resolution Electron Microscopy, *J. Am. Ceram. Soc.*, 76:763–765 (1993)

44. German, R. M., *Sintering Theory and Practice*, John Wiley & Sons, New York, pp. 86–88 (1996)

45. Hayashi, K., and Kihara, H., Sintering Behavior of Ultrafine Metal Powders, in: *Fundamentals of Diffusion Bonding* (I. Yshida, ed.), pp. 501–509, Elsevier, NY (1987)

46. Averback, R. S., Zhu, H., Tao, R., and Hofler, H. J., Sintering of Nanocrystalline Materials: Experiments and Computer Simulations, in: *Synthesis and Processing of Nanocrystalline Powder* (D. L. Bourell, ed.), pp. 203–216, TMS, Warrendale, PA (1996)

47. Yoo, Y. S., Kim, J. J., and Kim, D. Y., Effect of Heating Rate on the Microstructural Evolution During Sintering of $BaTiO_3$ Ceramics, *J. Am. Ceram. Soc.*, 70:C322–C324 (1987)

48. Chen, D. J., and Mayo, M., Densification and Grain Growth of Ultrafine 3 mol% Y_2O_3-ZrO_2 Ceramics, *Nanostr. Mater.*, 2:469–478 (1993)

49. Valiev, R. Z., private communication, UC Davis (1997)

50. Zhu, H., and Averback, R. S., Sintering of Nano-Particle Powders: Simulations and Experiments, *Mater. Manuf. Processes*, 11:905–923 (1996)

51. Zhu H., and Averback, R. S., Molecular Dynamics Simulations of Densification Processes in Nanocrystalline Materials, *Mat. Sci. Eng. A*, 204:96–100 (1995)

52. Yao, Y., and Tholen, A. R., Adhesion between Nanoparticles, *Nanostr. Mater.,* 12:661–664 (1999)

53. Yeadon, M., Yang, J. C., Ghaly, M., Olynick, D. L., Averbach, R. S., and Gibson, J. M., *Sintering of Sputtered Copper Nanoparticles on (001) Copper Substrates,* in: *Nanophase and Nanocomposite Materials ll* (S. Komarneni, J. C. Parker, and H. J. Wollenberger, eds.), pp. 179–184, MRS, Pittsburgh, PA (1997)

54. Champion, Y., and Bigot, J., Preparation and Characterization of Nanocrystalline Copper Powders, *Scr. Mater.*, 35:517–522 (1996)

55. Ross, F., Fendorf, M., and Groza, J. R., TEM Studies of Copper Nanoparticles, unpublished (1997)

56. Trusov, L. I., Lapovok, V. N., and Novikov, V. I., *Problems of Sintering Ultrafine Powders,* in: *Science of Sintering,* (D. P. Uskokovic, H. Plamour III, and R. M. Spriggs, eds.), pp. 185–192, Plenum Press, New York (1989)

57. Dominguez, O., Champion, Y., and Bigot, J., Liquidlike Sintering Behavior of Nanometric Fe and Cu Powders: Experimental Approach, *Met. Mater. Trans.,* 29A:2941–2949 (1998)

58. Pask, J. A., *Grain Boundaries in Sintering,* in: *Science of Sintering,* (D. P. Uskokovic, H. Plamour III, and R. M. Spriggs, eds.), pp. 381–393, Plenum Press, New York (1989)

59. Vergnon, P., Astier, M., and Teichner, S. J., *Initial Stage for Sintering of Ultrafine Particles (TiO_2 and Al_2O_3),* in: *Fine Particles* (W. E. Kuhn, ed.), pp. 299–307, The Electrochemical Society, Princeton (1974)

60. Messing, G. L., and Kumagai, M., Low-Temperature Sintering of α-Alumina Seeded Bohemite Gels, *Am. Ceram. Bull.*, 73:88–91 (1994)

61. Shaik, G. R., and Milligan, W. W., Consolidation of Nanostructured Metal Powders by Rapid Forging: Processing, Modeling, and Subsequent Mechanical Behavior, *Met. Mater. Trans.*, 28A:895–904 (1997)

62. Yan, M. F., and Rhodes, W. W., Low Temperature Sintering of TiO_2, *Mater Sci Eng.*, 61:59–66 (1983)

63. Hahn, H., Microstructure and Properties of Nanostructured Oxides, *Nanostr. Mater.*, 2:251–265 (1993)

64. Skandan, G., Processing of Nanostructured Zirconia Ceramics, *Nanostr. Mater.*, 5:111–126 (1995)
65. Alymov, M. I., Maltina, E. I., and Stepanov, Y. N., Model of Initial Stage of Ultrafine Metal Powder Sintering, *Nanostr. Mat.*, 4:737–742 (1992)
66. Pechenik, A., Piermarini, G. J., and Danforth, S. C., Fabrication of Transparent Silicon Nitrides from Nanosize Particles, *J. Am. Ceram. Soc.*, 75:3283–3288 (1992)
67. Pechenik, A., Piermarini, G. J., and Danforth, S. C., Low Temperature Densification of Silicon Nitride Nanoglasses, *Nanostr. Mater.*, 2:479–486 (1993)
68. Panfilov, S. A., Padalko, O. V., and Mitrenko, N. G., Behavior of Ultrafine Titanium Carbide Powders During Annealing and Sintering, *Poroshkovaya Metallurgiya,* 11:45–49 (1985)
69. Kumar, K.-N. P., Keizer, K, Burggraaf, A. J., Okubo, T., Nagamoto, H., and Morooka, S., Densification of Nanostructured Titania Assisted by a Phase Transformation, *Nature,* 358:48–51 (1992)
70. Winnubst, A. J. A., Keizer, K., and Burggraaf, A. J., Mechanical Properties and Fracture Behavior of ZrO_2-Y_2O_3 Ceramics, *J. Mater. Sci.*, 18:1958–1966 (1983)
71. Herring, C., Effect of Change of Scale on Sintering Phenomena, *J. Appl. Phys.*, 21:301–303 (1950)
72. Andrievski, R. A., Compaction and Sintering of Ultrafine Powders, *Intern. Powder Metall.*, 30:59–66 (1994)
73. Sanders, P. G., Fougere, G. E., Thompson, L. J, Eastman, J. A., and Weertman, J. R., Improvements in the Synthesis and Compaction of Nanocrystalline Materials, *Nanostr. Mater.*, 8:243–252 (1997)
74. Schneider, J. A., Processing and Properties of Silicon Nitrides Ceramics, PhD Thesis, UC Davis (1997)
75. Ying, J. Y., *Sintering of Nanocrystalline Al_2O_3 Study by Photoacoustic Infrared Spectroscopy,* in: *Mechanical Properties and Deformation Behavior of Materials Having Ultra-Fine Microstructures* (M. Nastasi, D. M. Parkin, and H. Gleiter, eds.), pp. 565–570, Kluwer Academic Press, Netherlands (1993)
76. Koch, C. C., *Processing of Nanophase Materials by High Energy Milling,* in: *Nanophases and Nanocrystalline Structures* (R. D. Shull and J. M. Sanchez, eds.), pp. 19–31, TMS, Warrendale, PA (1994)
77. Rawers, J., Slavens, G., Govier, D., Dogan, C., and Doan, R., Microstructure and Tensile Properties of Compacted, Mechanically Alloyed Nanocrystalline Fe-Al, *Met. Mater. Trans.*, 27A:3126–3134 (1996)

78. Perez, R. J., Huang, B., and Lavernia, E. J., Thermal Stability of Nanocrystalline Fe-10 wt%Al Produced by Cryogenic Mechanical Alloying, *Nanostr. Mater.*, 7:565–572 (1996)

79. Olynick, D. L., Gibson, J. M., and Averbach, R. S., In-Situ Ultra-High Vacuum Transmission Electron Microscopy Studies of Nanocrystalline Copper, *Mat. Sci. Eng. A*, 204:54–58(1995)

80. Eldrup, M., Bentzon, M. D., Pedersen, A. S., Linderoth, S., Pedersen, N. J., and Larsen, B., *Reduction of Surface Oxide on Ultrafine FeNi Particles,* in: *Mechanical Properties and Deformation Behavior of Materials Having Ultra-Fine Microstructures* (M. Nastasi, D. M. Parkin, and H. Gleiter, eds.), pp. 571–577, Kluwer Academic Press, Netherlands (1993)

81. Bourell, D. L., and Kaysser, W. A., Nanocrystalline Iron Sintering Behavior and Microstructural Development, *Met. Mater. Trans.*, 25A:677–685 (1994)

82. Chen, W., Pechenik, A., Dapkunas, S. J., Piermarini, G. J., and Malghan, S. G., Novel Equipment for the Study of Compaction of Fine Powders, *J. Am. Ceram. Soc.*, 77:1005–1010 (1994)

83. Chen, W., and Malghan, S. G., Investigations of Compaction Equations for Powders, *Powder Technology,* 81:75–81 (1994)

84. Bourell, D. L., Parimal, and Kaysser, W. A., Sol-Gel Synthesis of Nanophase Yttria-Stabilized Tetragonal Zirconia and Densification Behavior Below 1600 K, *J. Am. Ceram. Soc.*, 76:705–711 (1993)

85. Majetich, S. A., Scott, J. H., Brunsman, E. M., McHenry, M. E., and Nuhfer, N. T., Structure and Magnetism of Carbon-Coated Rare Earth Carbide Nanocrystals, in: *Fullerenes: Physics, Chemistry, and New Directions Vl*, (R. S. Ruoff and K. Kandish, eds.), pp. 1448–1462, The Electrochem. Society, Pennington, NJ (1994)

86. McHenry, M. E., Nakamura Y., Kirkpatrick, S., Johnson, F., Curtin, S., DeGraef, M., Nuhfer, N. T., Majetich, S. A., and Brunsman, E. M., *Carbon-Coated Fine Particle Ferromagnets Produced by a Carbon-Arc Technique,* in: *Fullerenes: Physics, Chemistry, and New Directions Vl*, (R. S. Ruoff and K. Kandish, eds.), pp. 1463–1478, The Electrochem. Society, Pennington, NJ (1994)

87. Rhodes, W. H., Agglomerate and Particle Size Effects on Sintering Yttria-Stabilized Zirconia, *J. Am. Ceram. Soc.*, 64:19–22 (1981)

88. Miller, K. T., and Zukovski, C. F, Osmotic Consolidation of Suspensions and Gels, *J. Am. Ceram. Soc.*, 77:2473–2478 (1994)

89. Gallas, M. R., Rosa, A. R., Costa, T. H., da Jornada, J. A. H., High Pressure Compaction of Nanosize Ceramic Powder, *J. Mater. Res.*, 12:764–768 (1997)

90. Andrievski, R. A., Torbov, V. I., and Kurkin, E. N., Nanocrystalline Titanium Nitride, *Proceedings of the 13th International Plansee Seminar*, (H. Bildstein and R. Eck, eds.), 3:649–655, Metallwerk Plansee, Reutte (1993)

91. Sagel-Ransijn, C. D., Winnbust, A. J. A., Kerkwijk, B., Burggraaf, A. J., and Verweij, H., Production of Defect-Poor Nanostructured Ceramics of Yttria-Zirconia, *J. Europ. Ceram. Soc.*, 17:831–841 (1997)

92. Van de Graaf, M. A. C. G., Termaat, J. H. H., and Burggraaf, A. J., Microstructure and Sintering Kinetics of Highly Reactive ZrO_2-Y_2O_3 Ceramics, *J. Mater. Sci.*, 20:1407–1418 (1985)

93. Dominguez, O., Phillippot, M., and Bigot, J., The Relationship Between Consolidation Behavior and Particle Size in Fe Nanometric Powders, *Scr. Met. Mater.*, 32:13–17 (1995)

94. Gutmanas, E. Y., Trusov, L. I., and Gotman, I., Consolidation, Microstructure and Mechanical Properties of Nanocrystalline Metal Powders, *Nanostr. Mater.*, 8:893–901 (1994)

95. Nieman, W., Weertman, J. R., and Siegel, R. W., Mechanical Behavior of Nanocrystalline Cu and Pd, *J. Mater. Res.*, 6:1012–1027 (1991)

96. Nieman, G. W., Weertman, J. R., and Siegel, R. W., Mechanical Behavior of Nanocrystalline Metals, *Nanostr. Mater.*, 1:185–190 (1992)

97. Kobelev, N. P., Soifer, Y. M., Andrievski, R. A., and Gunther, B., Microhardness and Elastic Properties of Nanocrystalline Silver, *Nanostr. Mater.*, 2:537–544 (1993)

98. Kumpmann, A., Gunther, B., and Kunze, H. D., Thermal Stability of Ultrafine-Grained Metals and Alloys, *Mat. Sci. Eng. A*, 168:165–169 (1993)

99. Bonetti, E., Valdre, G., Enzo, S., Cocco, G., and Soletta, I., Nanostructured Fe_3Al Intermetallic obtained by Mechanical Alloying and Thermal Aging, *Nanostr. Mater.*, 2:369–375 (1993)

100. Devaux, X., Brochin, F., Dauscher, A., Lenoir, B., Martin-Lopez, R., Scherrer, H., and Scherrer, S., Production of Ultrafine Powders of Bi-Sb Solid Solution, *Nanostr. Mater.*, 8:137–147 (1997)

101. Sanders, P. G., Witney, A. B., Weertman, J. R., Valiev, R. Z., and Siegel, R. W., Residual Stress, Strain and Faults in Nanocrystalline Palladium and Copper, *Mat. Sci. Eng. A*, 204:7–11 (1995)

102. Fougere, G. E., Weertman, J. R., and Siegel, R. W., Processing and Mechanical Behavior of Nanocrystalline Fe, *Nanostr. Mater.*, 5:127–134 (1995)

103. Elliott, B. R., and Weertman, J. R., Influence of Processing on Internal Structure of Nanocrystalline Ni, Presentation at TMS Conference, Orlando (1997)

104. Averback, R. S., Hofler, H. J., and Tao, R., Processing of Nano-Grained Materials, *Mater. Sci. Eng. A*, 166:169–177 (1993)

105. Kingery, W. D., Bowen, H. K., and Uhlmann, D. R., *Introduction to Ceramics*, 2nd ed., p. 486, Wiley, New York (1976)

106. Brook, R. J., Pore-Grain Boundary Interactions and Grain Growth, *J. Am. Ceram. Soc.*, 52:56–57 (1969)

107. Mayo, M., Hague, D. C., and Chen, D. J., Processing Nanocrystalline Ceramics for Applications in: Superplasticity, *Mater. Sci. Eng, A*, 166:145–159 (1993)

108. Mayo, M. J., and Hague, D. C., Porosity-Grain Growth Relationships in the Sintering of Nanocrystalline Ceramics, *Nanostruct. Mat.*, 3:43–52(1993)

109. Mayo, M. J., Superplasticity in Nanostructured Ceramics, in: *Mechanical Properties and Deformation Behavior of Materials Having Ultra-Fine Microstructures*, (M. Nastasi, D. M. Parkin, and H. Gleiter, eds.), pp. 361–380, Kluwer Academic Press, Dordrecht, Netherlands (1993)

110. Hackley, V. A., Anderson, M. A., and Spooner, S. A., Small-Angle X-ray Scattering Study of Microstructure Evolution During Sintering of Sol-Gel-Derived Porous Nanophase Titania, *J. Mat. Res.*, 7:2555–2571(1992)

111. Wen, T. L., Hebert, V., Vilminot, S., and Bernier, J. C., Preparation of Nanosized Yttria-Stabilized Zirconia Powders and Their Characterization, *J. Mater. Sci.*, 26:3787–3791 (1991)

112. Wu, J. M., and Wu, C. H., Sintering Behavior of Highly Agglomerated Ultrafine Zirconia Powders, *J. Mat. Sci.*, 23:3290–3299 (1988)

113. Hahn, H., Averback, R. S., Hofler, H. J., and Logas, J., *Sintering and Deformation of Nanocrystalline Ceramics,* in: *Clusters and Cluster-Assembled Materials*, (R. S. Averbach, J. Bernholc, and D. L. Nelson, eds.), pp. 569–580, MRS, Pittsburgh, PA (1991)

114. McCandlish, L. E., Kevorkian, V., Jia, K., and Fisher, T. E., *Nanostructured WC-Co Composite Powders,* in: *Advances in Powder Metallurgy and Particulate Materials*, (C. Lall and A. J. Neupaver, eds.), pp. 328–337, MPIF, Princeton, NJ (1994)

115. Skandan, G., Hahn, H, Kear, B. H., Roddy, M., and Cannon, W. R., *Processing of Nanostructured Zirconia Ceramics,* in: *Molecularly Designed Ultrafine/Nanostructure Materials*, (K. E. Gonsalves, G. M. Chow, T. D. Xiao, and R. C. Cammarata, eds.), pp. 207–212, MRS, Pittsburgh, PA (1994)

116. Ahn, J. P., Huh, M. Y., and Park, J. K., Effect of Green Density on Subsequent Densification and Grain Growth of Nanophase SnO_2 Powder During Isothermal Sintering, *Nanostr. Mater.*, 8:637–643 (1997)

117. Wu, S. J., and De Jonghe, L. C., Sintering of Nanophase γ- Al_2O_3 Powder, *J. Am. Ceram. Soc.*, 79:2207–2211 (1996)

118. Raj, R., Analysis of the Sintering Pressure, *J. Am. Ceram Soc.,* 70:C-210–C-211 (1987)

119. Malow, T. R., and Koch, C. C., Thermal Stability of Nanocrystalline Materials, *Mat. Sci. Forum,* 225–227:595–604 (1996)

120. Malow, T. R., and Koch, C. C., *Grain Growth of Nanocrystalline Materials - A review*, in: *Synthesis and Processing of Nanocrystalline Powder,* (D. L. Bourell, ed.), pp. 33–44, TMS, Warrendale, PA (1996)

121. Malow, T. R., and Koch, C. C., Grain Growth in Nanocrystalline Iron Prepared by Mechanical Attrition, *Acta Mater.*, 45:2177–2186 (1997)

122. Zhao, J., and Harmer, M. P., Sintering Kinetics for a Model Final-Stage Microstructure: A Study of Al_2O_3, *Phil. Mag. Lett.*, 63:7–14 (1991)

123. Eastman, J. A., Microstructural Development in Nanophase TiO_2 During Sintering, *J. Appl. Phys.*, 75:770–779 (1994)

124. Eckert, J., Holzer, J. C., Krill, C. E. III, and Johnson, W. L., Structural and Thermodynamic Properties of Nanocrystalline fcc Metals Prepared by Mechanical Attrition, *J. Mater. Res.*, 7:1751–1761 (1992)

125. Lange, F. F., Powder Processing Science and Technology for Increased Reliability, *J. Am. Ceram. Soc.*, 72:3–15 (1989)

126. Mortensen A., Kinetics of Densification by Solution-Reprecipitation, *Acta Met. Mater.*, 45:749–758 (1997)

127. Nieh, T. G., and Wadsworth, J., Dynamic Grain Growth During Superplastic Deformation of Yttria-Stabilized Tetragonal Zirconia Polycrystals, *J. Am. Ceram. Soc.*, 72:1469–1472 (1989)

128. Hague, D. C., and Mayo, M. J., Sinter-Forging of Nanocrystalline Zirconia: 1, Experimental, *J. Am. Ceram. Soc.*, 80:149–156 (1997)

129. Gupta, K., Possible Correlation Between Density and Grain Size During Sintering, *J. Am. Ceram. Soc.*, 55:276–277 (1972)

130. Liu Y., and Patterson, B. R., Grain Growth Inhibition by Porosity, *Acta Met. Mater.*, 41:2651–2656 (1993)

131. Weissmuller, J., Krauss, W., Haubold, T., Birringer. R., and Gleiter, H., Atomic Structure and Thermal Stability of Nanostructured Y-Fe Alloys, *Nanostr. Mater.*, 1:439–447 (1992)

132. Weissmuller, J., Alloy Effects in Nanostructures, *Nanostr. Mater.*, 3:261–272 (1993)

133. Terwilliger, C. D., and Chiang, Y. M., The Effect of Calcium Segregation on Grain Growth in Nanocrystalline TiO_2, *Nanostruct. Mater.*, 4:651–661 (1994)

134. Owen, D. M., and Chokshi, A. H., An Evaluation of the Densification Characteristics of Nanocrystalline Materials, *Nanostr. Mater.*, 2:181–187 (1993)

135. Rahaman, M. N., *Sintering and Grain Growth of Ultrafine CeO_2 Powders,* in: *Sintering Technology,* (R. M. German, G. L. Messing, and R. Cornwall, eds.), pp. 93–100, M. Dekker, NY (1996)

136. Gao, Z. Q., and Fultz, B., Inter-Dependence of Grain Growth, Nb Segregation and Chemical Ordering in Fe-Si-Nb Nanocrystals, *Nanostr. Mater.*, 4:939–947 (1994)

137. Bansal, C., Gao, Z. Q., and Fultz, B., Grain Growth and Chemical Ordering in $(Fe, Mn)_3Si$, *Nanostr. Mater.*, 5:327–336 (1995)

138. Nazarov, A. A., Romanov, A. E., and Valiev, R. Z., On the Nature of High Internal Stresses in Ultrafine Grained Materials, *Nanostr. Mater.*, 4:93–101 (1994)

139. Misra, R. S., Valiev, R. Z., and Mukherjee, A. K., Fully Dense Nanocrystalline Nickel by Severe Plastic Deformation Consolidation, *Mat. Sci. Forum,* 225–227:605–610 (1996)

140. Gleiter, H., Nanostructured Materials: State of the Art and Perspectives, *Zeitschr. Metallk.*, 86:78–83 (1995)

141. Eastman, J. A., and Fitzsimmons, M. R., On the Two-State Microstructure of Nanocrystalline Chromium, *J. Appl. Phys.*, 77:522–527 (1995)

142. Groza, J. R., and Dowding, R., Nanoparticulate Materials Densification, *Nanostr. Mater.*, 8:749–768 (1997)

143. ASTM B 328-92 Standard Test Method for Density, Oil Content, and Interconnected Porosity of Sintered Structural Parts and Oil-Impregnated Bearings.

144. Schaefer, H. E., Wurschum, R., Scheytt, M., Birringer, R., and Gleiter, H., Structure of Nanometer-Sized Polycrystalline Iron Investigated by Positron Lifetime Spectroscopy, *Mater. Sci. Forum,* 15–18:9545–9554 (1987)

145. Siegel, R. W., Ramasamy, S., Hahn, H., Li, Z., Lu, T., and Gronsky, R., Synthesis, Characterization, and Properties of Nanophase TiO_2, *J. Mater. Res.*, 3:1367–1372 (1988)

146. Allen, A. J., Krueger, S., Long, G. G., Kerch, H. M., Hahn, H., and Skandan, G., Small-Angle Neutron Scattering Studies of Ceramic Nanophase Materials, *Nanostr. Mater.,* 7:113–126 (1996)

147. Ying, J. Y., Wang, G. H., Fuchs, H., Laschinski, R., and Gleiter, H., STM AFM Study of Grain Boundary Migration in Nanostructured Solids, *Mater. Lett.*, 15:180–185 (1992)

148. Kats-Demyanets, A., and Chaim, R., Grain Growth Control in Nanocrystalline Oxide Ceramics, *Nanostr. Mater.*, 6:851–854 (1995)

149. Handwerker, C. A., Blendell, J. E., and Coble R. L, *Sintering of Ceramics,* in*: Science of Sintering*, (D. P. Uskokovics, H. Palmour, III, and R. M. Spriggs, eds.), pp. 3–37, Plenum Press, NY (1989)

150. Boutz, M. M. R., Olde Scholtenhuis, R. J. M., Winnubst, A. J. A., Burggraaf, A. J., A Hydrothermal Route for Production of Dense, Nanostructured Y-TZP, *Mat. Res. Bull.*, 29:31–40 (1994)

151. Murray, M. G. S., Ponton, C. B., and Marquis, P. M., *Processing of Ultrafine Grained Hydroxyapatite Bioceramics*, in: *Nanoceramics*, (R. Freer, ed.), pp. 25–30, Inst. of Materials, London (1993)

152. Duran, P., Villegas, M., Capel, F., Fernandez, J. F., and Moure, C., Nanostructured and Near Defect-Free Ceramics by Low-Temperature Pressureless Sintering of Nanosized Y-TZP Powders, *J. Mater. Sci.*, 32:4507–4512 (1997)

153. Hague, D. C., and Mayo, M. J., The Effect of Crystallization and a Phase Transformation on the Grain Growth of Nanocrystalline Titania, *Nanostr. Mater.*, 3:61–67 (1993)

154. Johnson, D. L., *Rapid Sintering of Ceramics,* in: *Advanced Ceramics II* (S. Somiya, ed.) pp. 1–6, Elsevier Applied Science, London (1986)

155. Chen, D. J., and Mayo, M. J., Rapid Rate Sintering of Nanocrystalline ZrO_2-3 mol% Y_2O_3, *J. Am. Ceram. Soc.,* 79:906–912 (1996)

156. Venkatachari, K. R., Huang, D., Ostrander, S. P., Schulze, W. A., and Stangle, G. C., Preparation of Nanocrystalline Yttria-Stabilized Zirconia, *J. Mater. Res.,* 10:756–761 (1995)

157. Ragulya, A. V., and Skorokhod, V. V., Rate-Controlled Sintering of Ultrafine Nickel Powder, *Nanostr. Mater.*, 5:835–843 (1995)

158. Sadangi, R. K., Kear, B. H., and McCandlish, L. E., *Synthesis and Processing of Nanograined Fe-(Fe,Mo)$_6$C Composite Powders,* in: *Molecularly Designed Ultrafine/Nanostructure Materials* (K. E. Gonsalves, G. M. Chow, T. D. Xiao, and R. C. Cammarata, eds.), pp. 219–225, MRS, Pittsburgh, PA (1994)

159. Wu, L., Kim, B. K., Kear, B. H., and McCandlish, L. E., *Grain Growth Inhibition in Sintering of Nanostructured WC-Co Alloys,* in: *Proc. 13th Intern. Plansee Seminar*, (H. Bildstein and R. Eck, eds.) 3:667–679, Metallwerk Plansee, Reute (1993)

160. Fang, Z., and Eason, J. W., Study of Nanostructured WC-Co Composites, *Int. J. Refractory & Hard Met.,* 13:297–303 (1995)

161. McCandlish, L. E., Kear, B. H., and Kim, B. K., Processing and Properties of Nanostructured WC-Co, *Nanostr. Mater.*, 1:119–124 (1992)

162. Chang, H., Hofler, H. J., Alstetter, C. J., and Averbach, R. S., Synthesis, Processing and Properties of Nanophase TiAl, *Scr. Metall. Mater.*, 25:1161–1166 (1991)

163. Chang, H., Hofler, J., Altstetter, C., and Averback, R. S., Synthesis, Processing and Properties of Nanophase Aluminide, *Mat. Sci. Eng. A*, 153:676–678 (1992)

164. Prabhu, G. B., and Bourell, D. L., Synthesis and Sintering Characteristics of Zirconia and Zirconia-Alumina Nanocomposites, *Nanostr. Mater.*, 6:361–364 (1995)

165. Nakahira, A., and Niihara, K., Sintering Behavior and Consolidation Process for Al_2O_3/SiC Nanocomposites, *Jpn. J. Ceram. Soc.*, 100:448–453 (1992)

166. Wittenauer, J., *Synthesis and Plasticity Characteristics of a Nanocrystalline Alumina-Zirconia Composite,* in: *Superplasticity and Superplastic Forming 1995,* (A. K. Ghosh, and T. R. Bieler, eds.), pp. 85–92, TMS, Warrendale (1996)

167. Anya, C. C., and Roberts, S. G., Pressureless Sintering and Elastic Constants of Al_2O_3-SiC Nanocomposites, *J. Eur. Ceram. Soc.,* 17:565–573 (1997)

168. Holtz, R. L., and Provenzano, V., Enhanced Microhardness of Copper-Niobium Nanocomposites, *Nanostr. Mater.*, 4:241–256 (1994)

169. Averback, R. S., Hofler, H. J., Hahn, H., and Logas, J. C., Sintering and Grain Growth in Nanocrystalline Ceramics, *Nanostr. Mater.*, 1:173–178 (1992)

170. He, L., and Ma, E., Processing and Microhardness of Bulk Cu-Fe Nanocomposites, *Nanostr. Mater.*, 7:327–339 (1996)

171. Carsley, J. E., Milligan, W. W., Hackney, S. A., and Aifantis, E. C., Glasslike Behavior in a Nanostructured Fe/Cu Alloy, *Met. Mater. Trans.*, 26A:2479–2481 (1995)

172. Skandan, G., Hahn, H., Roddy, M., and Cannon, R. W., Ultrafine-Grained Dense Monoclinic and Tetragonal Zirconia, *J. Am. Ceram. Soc.,* 77:1706–1710 (1994)

173. Boutz, M. M. R., Olde Scholtenhuis, R. J. M., Winnubst, A. J. A., and Burggraaf, A. J., *Preparation, Microstructural Control and Superplasticity of Nanostructured Yttria/Ceria Stabilized Tetragonal Zirconia Ceramics,* in: *Nanoceramics,* (R. Freer, ed.), pp. 75–86, The Institute of Materials, London (1993)

174. Groza, J. R., Non-Conventional Pressure-Assisted Powder Consolidation Processes, *J. Mat. Eng.Perform.,* 2:283–290 (1993); and *Powder Consolidation in Non-equilibrium Processing of Materials,* (C. Suryanarayana, ed), pp. 345–372, Pergamon, Elsevier Science (1999)

175. Skandan, G., Hahn, H., Kear B. H., Roddy, M., and Cannon, R. W., The Effect of Applied Stress on Densification of Nanostructured Zirconia during Sinter-Forging, *Matt. Lett.*, 20:305–309 (1994)

176. Hofler, H. J., and Averbach, R. S., *Sinter-Forging of Nanocrystalline TiO_2*, in: *Nanophase and Nanocomposite Materials,* (S. Komarneni, J. C. Parker, and G. J. Thomas, eds.), 286:9–14, MRS, Pittsburgh, PA (1993)

177. Uchic, M., Hofler, H. J., Flick, W. J., Tao, R., Kurath, P., and Averbach, R. S., Sinter-Forging of Nanophase TiO_2, *Scr. Metall. Mater.*, 26:791–796 (1992)

178. Rahaman, N. R., and De Jonghe, L. C., Sintering of CdO Under Low Applied Stress, *J. Am. Ceram. Soc.*, 67:C205–C207 (1984)

179. Schwarz, R. B., Desch, P. B., Srinivasan, S., and Nash, P., Synthesis and Properties of Trialuminides with Ultra-Fine Microstructures, *Nanostr. Mater.*, 1:37–42 (1992)

180. Yoo, S. H., Field Effects in Plasma Activated Sintering, MS Thesis, UC Davis (1996)

181. Arzt, E., Ashby, M. F., and Easterling, K. E., Practical Applications of Hot-Isostatic Pressing Diagrams: Four Case Studies, *Met. Trans.*, 14A:211–221 (1983)

182. Swinkels, F. B., Wilkinson, D. S., Arzt, E., and Ashby, M. F., Mechanisms of Hot-Isostatic Pressing, *Acta Metall.*, 31:1829–1840 (1983)

183. McKimpson, M. G., Densification Maps for Nanosized Powders, *Mat. Manuf. Proc.*, 11:935–949 (1996)

184. Duva, J. M., and Crow, P. D., The Densification of Powders by Power-Law Creep During Hot Isostatic Pressing, *Acta Metall. Mater.*, 40:31–35 (1992)

185. Hague, D. C., and Mayo, M. J., Modeling Densification During Sinter-Forging of Yttria-Partially-Stabilized Zirconia, *Mat. Sci. Eng. A*, 204:83–89 (1995)

186. Boutz, M. M. R., Winnubst, L., Burggraaf, A. J., Nauer, M., and Carry, C., Low-Temperature Sinter Forging of Nanostructured Y-TZP and Ce-TZP, *J. Am. Ceram Soc.*, 78:121–128 (1995)

187. Korth, G. E., and Williamson, R. L., Dynamic Consolidation of Metastable Nanocrystalline Powders, *Met. Mater. Trans.*, 26a:2571–2578 (1995)

188. Sadangi, R. K., Kear. B. H., and McCandlish, L. E., Hot Pressing of Nanograined Fe-$(Fe, Mo)_6C$ Composite Powders, *Nanostr. Mater.*, 2:563–569 (1993)

189. Sadangi, R. K., Kear, B. H., and McCandlish, L. E., Pressureless Sintering of Nanograined Fe-$(Fe,Mo)_6$ Composite Powders, *Powder Met.*, 37:277–282 (1994)

190. Guoxian, L., Qingchang, M., Zhichao, Li, and Erde, W., Consolidation of Nanocrystalline Al-Ti Alloy Powders Synthesized by Mechanical Alloying, *Nanostr. Mater.*, 5:673–678 (1995)

191. He, L., and Ma, E., Nanophase Metallic Alloys Consolidated from Powders Prepared by Mechanical Alloying, *Mat. Sci. Eng. A*, 204:240–245 (1995)

192. He, L., and Ma, E., Full-Density Nanocrystalline Fe-29Al-2Cr Intermetallic Consolidated from Mechanically Milled Powders, *J.Mat.Res.*, 11:72–80 (1996)

193. Aoki, K., Itoh, Y., Park, J., Kawamura, Y., Inoue, A., and Masumoto, T., Preparation of Nanocrystalline Bulk TiAl by a Combination of HDDR and Hot Pressing, *Intermetallics*, 4:S103–S111 (1996)

194. Axelbaum, R. L., Huertas, J. I., Lottes, C. R., Hariprasad, S., and Sastry, S. M., Nano-Phase W and W-Ti Composite via Gas-Phase Combustion Synthesis, *Mater. Manuf. Proc.*, 11:1043–1053 (1996)

195. Secondi, J., Drbohlav, O., and Yavari, R., Metal-Nitride Composites Obtained by Alloy Milling under Nitrogen Gas, *Mat. Sci. Forum*, 179:287–294 (1995)

196. Saida, J., Inoue, A., and Masumoto, T., The Formation of Nanostructured Materials by Consolidating Ultrafine Amorphous Alloy Particles Prepared by Chemical Reduction, *Mat. Sci. Eng. A*, 179–A180:577–581 (1994)

197. Hahn, H., and Averbach, R. S., High Temperature Mechanical Properties of Nanostructured Ceramics, *Nanostr. Mater.*, 1:95–100 (1992)

198. Terwilliger, C. D., and Chiang, Y. M., Characterization of Chemically-and Physically-Derived Nanophase Titanium Oxide, *Nanostr. Mater.*, 2:37–45 (1993)

199. Lavik, E. B., Chiang, Y. M., Kosacki, I., and Tuller, H. L., Enhanced Electrical Conductivity and Nonstoichiometry in Nanocrystalline CeO_{2-x}, in: *Metastable Phases and Microstructures* (R. Bormann, G. Mazzone, R. D. Shull, R. S. Averbach, and R. F. Ziolo, eds.), pp. 359–364, MRS, Pittsburgh (1996)

200. Bose, S., Application of Nanocrystalline Materials as Matrices of Composites: Processing and Performance Advantages, *Mat. Sci. Eng.*, A196:105–109 (1995)

201. Lences, Z., Bellosi, A., and Monteverde, F., Factors Influencing the Crystallization and the Densification of Ultrafine Si/N/C Powders, *Mat. Chemistry Phys.*, 41:46–54 (1995)

202. Sekino, T., Nakajima, T., and Niihara, K., Mechanical and Magnetic Properties of Nickel Dispersed Alumina-Based Nanocomposite, *Mat.Lett.*, 29:165–169 (1996)

203. Araki, H., Saji, S., Okabe, T., Minamino, Y., Yamane, T., and Miyamoto, Y., Consolidation of MA Al-10.7 at% Ti Powder at Low Temperature and High Pressure of 2GPa, *Mat. Trans. JIM*, 36:465–468 (1995)

204. Liao, S. C., Mayo, W. E., and Pae, K. D., Theory of High Pressure/Low Temperature Sintering of Bulk Nanocrystalline TiO_2, *Acta Mater.*, 45:4027–4040 (1997)

205. Liao, S. C., Chen, Y. J., Kear, B. H., and Mayo, W. E., High Pressure/Low Temperature Sintering of Nanocrystalline Ceramics: Processing Issues and Emerging Applications, *Fine, Ultrafine and Nano Powders '98 Conf.*, (Nov. 8–10, 1998)

206. Danforth, S. C., Symons, W., and Nilsen, K. J., *Processing of Silicon Nitride Powders,* in: *Advanced Ceramic Processing and Technology 1*, (J. G. P. Binner, ed.), pp. 39–71, Noyes, Park Ridge, NJ (1990)

207. Kaiser, A., Vassen, R., Stover, D., and Buchkremer, H. P., Heat Treatment of Ultrafine SiC Powders to Reduce Oxygen Sensitivity and Grain Growth, *Nanostr. Mat.*, 4:795–802 (1994)

208. Borner, I., and Eckert, J., Grain Size Effects and Consolidation in Ball Milled Nanocrystalline NiAl, *Mat. Sci. Forum*, 235–238:79–84 (1997)

209. Suryanarayana, C., Korth, G. E., Chen, G. H., Frefer, A., and Froes, F. H., Thermal Stability of Nanostructured Titatnium Aluminides, *Nanostr. Mater.,* 2:527–535 (1993)

210. Kaiser, A., Vassen, R., Stover, D., Buchkremer, H. P., Forster, J., and Uhlenbusch, J., Si_3N_4//SiC Composites Using Conventional and Nanosized Powders, *Nanostr. Mater.*, 6:917–920 (1995)

211. Senkov, O. N., Ovecoglu, M. L., Srisukhumbowornchai, N., and Froes, F. H., Stability of Nanocrystallyne TiAl-based Alloys, *Nanostr. Mater.,* 10:935–945 (1998)

212. Murphy, B. R., and Courtney, T. H., Synthesis of Cu-NbC Nanocomposites by Mechanical Alloying, *Nanostr. Mat.*, 4:365–369 (1994)

213. Kwon, O. H., Nordahl, C. S., and Messing, G. L., Submicrometer Transparent Alumina by Sinter-Forging Seeded γ-Al_2O_3 Powders, *J. Am. Ceram. Soc.*, 78:491–494 (1995)

214. Kellet, B. J., and Lange, F. F., Experiments on Pore Closure During Hot Isostatic Pressing and Forging, *J. Am. Ceram. Soc.*, 71:7–12 (1988)

215. Winnubst, A. J. A., Boutz, M. M. R., He, Y. J., Burggraaf, A. J., and Verweji, J. H., Plasticity of Nanocrystalline Zirconia Ceramics and Composites, *Ceramics Intern.,* 23:215–221 (1997)

216. Nordahl, C. S., and Messing, G. L., Transformation and Densification of θ-Alumina During Sinter Forging, *J. Am. Ceram. Soc.*, 79:3145–3154 (1996)

217. Alymov, M. I., and Leontieva, O. N., Synthesis of Nanoscale Ni and Fe Powders and Properties of Their Compacts, *Nanostr. Mater.*, 6:393–395 (1995)

218. Nagahama, H., Ohtera, K., and Higashi, K., Mechanical Properties of Rapidly Solidified Aluminium Alloys Extruded from Amorphous or Nanocrystalline Powders, *Phil. Mag. Lett.*, 67:225–230 (1993)

219. Morris, M. A., and Morris, D. G., Microstructural Refinement and Associated Strength of Copper Alloys Obtained by Mechanical Alloying, *Mater. Sci. Eng. A*, 111:115–127 (1989)

220. Kato, A., Suganuma, T., Horikiri, H., Kawamura, Y., Inoue, A., and Masumoto, T., Consolidation and Mechanical Properties of Atomized Mg-based Amorphous Powders, *Mat. Sci. Eng. A*, 179–A180:112–117 (1994)

221. Misra, R. S., Lesher, C. E., and Mukherjee, A. K., High-Pressure Sintering of Nanocrystalline γ-Al_2O_3, *J. Am. Ceram. Soc.*, 79:2989–2992 (1996)

222. Yao, B., Ding, B. Z., Wang, A. M, Li, D. J., and Hu, Z. Q., Preparation of a Single Phase Bulk Al-Fe(Mo,Si, B) Nanostructured Alloy Under High Pressure, *Matt. Lett.*, 22:81–86 (1995)

223. Valiev, R. Z., Misra, R. S., Groza, J. R., and Mukherjee, A. K., Processing of Nanostructured Nickel by Severe Plastic Deformation of Ball Milled Powder, *Scr. Met. Mater.*, 34:1443–1448 (1996)

224. Shen, H., Li, Z., Gunther, B., Korznikov, A. V., and Valiev, R. Z., Influence of Powder Consolidation Methods on the Structural and Thermal Properties of a Nanophase Cu-50wt-Percent Ag Alloy, *Nanostr. Mater.*, 6:385–388 (1995)

225. Clark, D. E., and Sutton, W. H., Microwave Processing of Materials, *Annual Rev. Mat. Sci.,* 26:299–331 (1996)

226. Katz, J. D., Blake, R. D., and Scherer C. P., Microwave Sintering of Titanium Diboride, *Ceram. Eng. and Sci. Proceed.*, 10:857–867 (1989)

227. Lewis, D., Rayne, R. J., Bender, B. A., Kurihara, L. K., Chow, G. M., Fliflet, A., Kincaid, A., and Bruce R., Conventional and High Frequency Microwave Processing of Nanophase Ceramic Materials, *Nanostr. Mater.*, 9:97–100 (1997)

228. Bykov, Y., Gusev, S., Eremeev, A., Holopstev, V., Malygin, N., Pirarunas, S., Sorokin, A., and Shurov, A., Sintering of Nanophase Oxide Ceramics by Using Millimeter Wave Radiation, *Nanostr. Mater.*, 6:855–858 (1995)

229. Fliflet, A. W., Bruce, R. W., Kinkead, A. K., Fischer, R. P., Lewis, D., III, Rayne, R., Bender, B., Kwihara, L. K., Chow, G. M, and Schoen, P. E., Application of Microwave Heating to Ceramic Processing: Design and Initial Operation of a 2.45-GHz Single-Mode Furnace, *IEEE Trans.Plasma Sci.*, 24:1041–1049 (1996)

230. Freim, J., McKittrick, J., Katz, J., and Sickafus, K., Microwave Sintering of Nanocrystalline γ-Al_2O_3, *Nanostr. Mater.*, 4:371–385 (1994)

231. Link, G., Ivanov, V., Paranin, S., Khrustov, V., Bohme, R., Muller, G., Schumacher, G., Thumm, M., and Weisenburger, A., *A Comparison of mm-Wave Sintering and Fast Conventional Sintering of Nanocrystalline* Al_2O_3, in: *Microwave Processing of Materials V*, (M. F. Iskander, J. O. Kiggans, Jr., and J. C. Bolomey, eds.), 430:157–162, MRS, Pittsburgh, PA (1996)

232. Misra, R. S., Mukherjee, A. M., and Yamazaki, K., Effect of TiO_2 Doping on Rapid Densification of Alumina by Plasma Activated Sintering, *J. Mater. Res.*, 11:1144–1148 (1996)

233. Goodwin, T. J., Yoo, S. H., Matteazzi, P., and Groza, J. R., Cementite-Iron Nanocomposite, *Nanostr. Mater.*, 8:559–566 (1997)

234. Kimura, H., and Kobayashi, S., Synthesis of High Hardness Intermetallic Compound by Pulse Electro-Discharge Consolidation of Amorphous TiAl Powder, *J. Japan Inst. Met.*, 58:201–207 (1994)

235. Nishimura, T., Mitomo, M., Hirotsuru, H., Kawahara, M., Fabrication of Silicon Nitride Nano-Ceramics by Spark Plasma Sintering, *J. Mater. Sci. Lett.*, 14:1046–1047 (1995)

236. Ivanov, V., Kotov, Y. A., Samatov, O. H., Bohme, R., Karow, H. U., and Schumacher, G., Synthesis and Dynamic Compaction of Ceramic Nanopowders by Techniques Based on Electric Pulsed Power, *Nanostr. Mater.*, 6:287–290 (1995)

237. Chelluri, B., Barber, J. P., Gonzales, E. J., and Thadani, N. N., Nano Grain Size Component Fabrication Using Dynamic Magnetic Compaction (DMC) Process, *Adv. Powder Met. Particulate Mater.*, 3:3-65–3-76 (1997)

238. Kondo, K. I., and Sawai, S., Fabricating Nanocrystalline Diamond Ceramics by a Shock Compaction Method, *J. Am. Ceram. Soc.*, 73:1983–1991 (1990)

239. Suryanarayana, C., Korth, G. E., and Froes, F. H., Compaction and Characterization of Mechanically Alloyed Nanocrystalline Titanium Aluminides, *Met. Mater. Trans.*, 28A:293–302 (1997)

240. Glade, S. C., and Thadhani, N. N., Shock Consolidation of Mechanically Alloyed Amorphous Ti-Si Powders, *Met. Mater. Trans.*, 26A:2565–2569(1995)

241. Hirai, H., and Kondo, K., Shock-Compacted Si_3N_4 Nanocrystalline Ceramics: Mechanisms of Consolidation and of Transition from Alpha- to Beta-Form, *J. Am. Ceram. Soc.*, 77:487–492(1994)

242. Dogan, C. P., Rawers, J. C., Govier, R. D., and Korth, G., Mechanical Processing, Compaction, and Thermal Processing of Alpha-Fe Powder, *Nanostr. Mater.*, 4:631–644 (1994)s

5

Electrodeposited Nanocrystalline Materials

Uwe Erb, Karl T. Aust, and Gino Palumbo

1.0 INTRODUCTION

Over the past decade, the synthesis of nanostructured materials by electrodeposition has been advanced from a laboratory scale phenomenon to a practical industrial materials technology. This chapter addresses the synthesis of nanocrystalline materials by electrodeposition methods as well as structure-property relationships for a variety of pure metals and alloys. Comparison with structure-property relationships observed for materials produced by other synthesis methods are given wherever possible. Some emerging industrial applications are also presented.

2.0 SYNTHESIS OF NANOSTRUCTURED MATERIALS BY ELECTRODEPOSITION

From the synthesis point of view, electrodeposition is one of the oldest methods used to produce nanostructured materials for many years, probably inadvertently in most cases. Consequently, there are numerous early reports in the literature describing electrodeposits with ultrafine structures; many examples are given in Ref. 1. However, no systematic studies were published before the late 1980s[2][3] on the synthesis of nanocrystalline materials by electrodeposition in an attempt to optimize certain properties by deliberately controlling the volume fractions of grain boundaries and triple junctions in the material. In fact, the synthesis of nanostructured materials with grain size control during the electrodeposition process can be considered a distinct form of grain boundary engineering in which the grain boundary content (types and quantities of grain boundaries) of a material are controlled during material processing to achieve certain physical, chemical and mechanical properties.[4]–[6] The final result is thus a bulk interfacial material, as originally defined by Gleiter,[7] which does not require any further processing of precursor powder material. In this respect, electrodeposited nanocrystals are quite different from other nanostructures which are based on consolidated particles.

Potentially there are a very large number of pure metals, alloys, composites, and ceramics which can be electrodeposited with grain sizes less than 100 nm. For example, the literature contains numerous examples giving electrochemical processing windows for the synthesis of nanocrystalline pure metals (e.g., Ni,[8]–[10] Co,[11] Pd,[12] and Cu[11]), binary alloys (e.g., Ni-P,[2][3] Ni-Fe,[13][14] Zn-Ni,[15][16] Pd-Fe,[17] and Co-W[18]), and ternary alloys (e.g., Ni-Fe-Cr[19]–[21]). Even multilayered structures or compositionally modulated alloys (e.g., Cu-Pb,[22] Cu-Ni,[23]–[25] Ag-Pd,[26] Ni-P[27]), metal matrix composites (e.g., Ni-Si C[9]), ceramics (e.g., ZrO_2[28]), and ceramic nanocomposites (e.g., $Tl_aPb_bO_c$[29]) have been successfully produced by electrodeposition methods. However, the latter are not considered in this chapter; this review is limited to equiaxed pure metals and alloys with grain sizes less than 100 nm, without considering grain shape modifications.[30]

Crystalline metal electrodeposits exhibit several types of growth forms including layers, blocks, pyramids, ridges, spiral growth forms, dendrites, powders, and whiskers.[31] These morphologies have been studied extensively and various models have been advanced to correlate specific growth forms with electrodeposition parameters and substrate microstructure.[31][32] Electrodeposition parameters are bath composition, pH, temperature, overpotential, bath additives, etc., while important microstructural features of the substrate include grain size, crystallographic texture, dislocation density, internal stress, and the like.[31][32]

Electrocrystallization (Fig. 1) occurs either by the build up of existing crystals or the formation of new ones.[33] These two processes are in competition with each other and are influenced by different factors. The two key mechanisms which have been identified as the major rate-determining steps for nanocrystal formation are charge transfer at the electrode surface and surface diffusion of adions on the crystal surface.[34] Earlier, Fischer presented a classification of microstructures typically observed in electrodeposits.[35] One of the key factors in the microstructural evolution of electrodeposits in terms of grain size and shape is inhibition, for example, resulting from reduced surface diffusion of adions by adsorption of foreign species (such as grain refiners) on the growing surface. With increasing inhibition, the deposit structure changes from basis oriented and reproduction type (BR) to twin transition types (TT), to field oriented type (FT), and finally to unoriented dispersion type (UD).[36] A large number of grain refiners have been described in the literature (see for example, Ref. 37); their effectiveness depends upon surface adsorption characteristics, compatibility with the electrolyte, temperature stability, etc. For example, saccharin,[38] coumarin,[39] thiorea,[39] and HCOOH[40] have all been successfully applied to achieve grain refinement down to the nanocrystalline range for nickel electrodeposits.

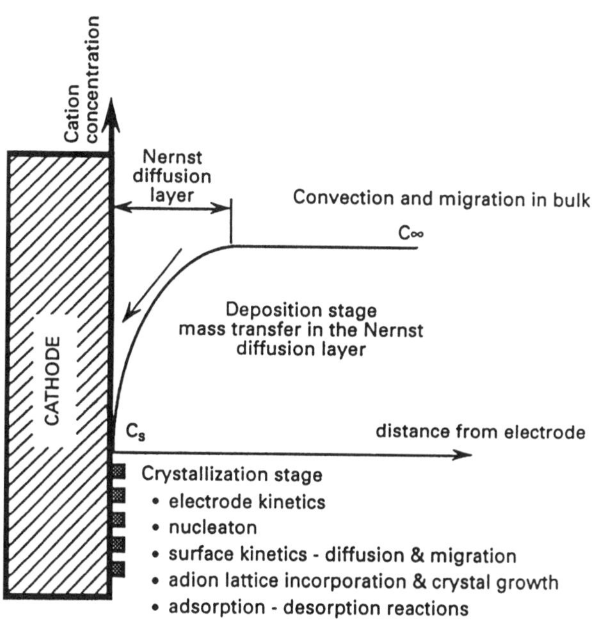

Figure 1. Two stages of electrocrystallization according to Bockris, et al.[34]

The second important factor in nanocrystal formation during electrocrystallization is overpotential.[33][34] Grain growth is favored at low overpotential and high surface diffusion rates. On the other hand, high overpotential and low diffusion rates promote the formation of new nuclei. These conditions can be experimentally achieved when using pulse plating (Fig. 2), where the peak current density can be considerably higher than the limiting current density attained for the same electrolyte during direct current plating.

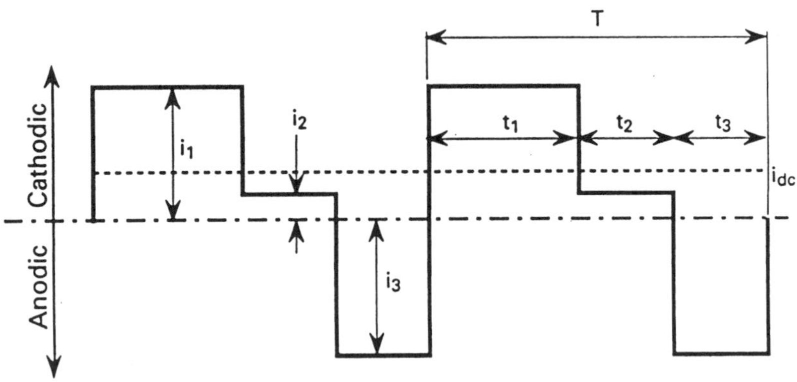

Figure 2. Generalized pulse current waveform. T is the period of the waveform, i_n are current densities and t_n are pulse durations.[33]

While many of the processes associated with the crystallization stage (Fig. 1) are still poorly understood, the previous work has shown that electrodeposition will result in nanostructured materials when electrodeposition variables (e.g., bath composition, pH, temperature, current density, etc.) are chosen such that electrocrystallization results in massive nucleation and reduced grain growth. Under these conditions the effect of the substrate on the resulting bulk electrodeposit often becomes negligible (for example, see Ref. 41).

Electrodeposition of nanocrystalline materials is not limited to applications as coating in-production or in-situ on structures and components. As discussed in more detail in Sec. 5, this method also provides for cost-effective production of freestanding forms such as ultrathin foil, wire, sheet, and plate, as well as complex shapes.

3.0 STRUCTURE OF NANOCRYSTALLINE METAL ELECTRODEPOSITS

This chapter deals mainly with equiaxed nanostructured electrodeposits, although layered or grain-shape modified structures can also be synthesized by electrodeposition.[30] Figure 3 shows bright field, dark field, diffraction pattern, and grain size distribution of a nanocrystalline Ni specimen produced by direct current plating from a modified Watts bath.[14] Electrodeposition of nanocrystals typically operates far from equilibrium conditions. Consequently, the material obtained is a non-equilibrium structure which is primarily manifested in the small grain size and the associated large volume fraction of grain boundaries and triple junctions. In addition, alloys produced by this method can show considerable extensions of the solid solubility range similar to what is observed in materials produced by other non-equilibrium processing routes, such as rapid solidification. For example, the room temperature solid solubility for P in Ni is negligible.[42] On the other hand, electrodeposited Ni-P can form solid solutions containing phosphorus levels of 10 wt% or more.[2][3] Similarly, extended solubility ranges were also observed in other alloys, such as Co-W,[18] Zn-Ni,[16] and Ni-Mo.[43]

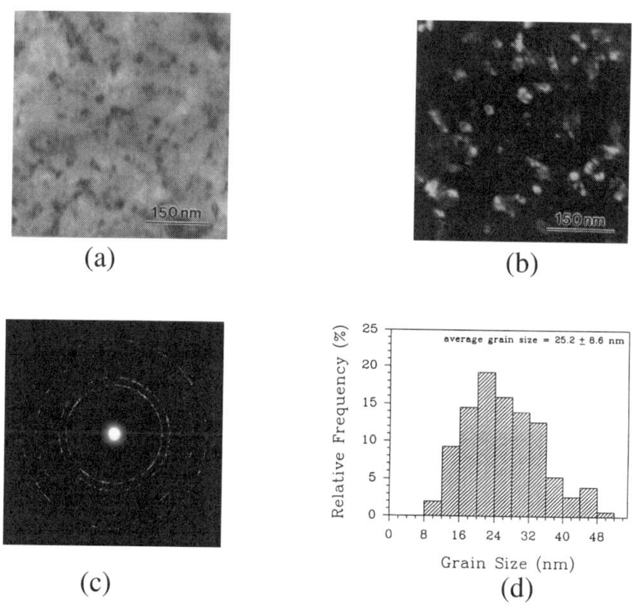

Figure 3. TEM micrographs for electrodeposited nanocrystalline Ni (*a*) bright field, (*b*) dark field, (*c*) electron diffraction pattern, and (*d*) grain size distribution.[14]

Depending on the electrodeposition parameters, the grain size distribution of electrodeposited nanocrystals is relatively narrow as shown, for example, for Ni in Fig. 3. The crystallographic texture depends strongly on the electroplating parameters as demonstrated in Fig. 4 for pulse plated nickel nanocrystals.[38] In this example, the crystallographic texture changes from a strong (200) fiber texture to a (111) (200) double fiber texture with increasing saccharin content in the plating bath. The series of x-ray diffraction scans presented in Fig. 4 also show that the saccharin concentration in the plating bath has a strong effect on the grain size of the material. This is evident from the increasing line broadening with increasing saccharin concentration.

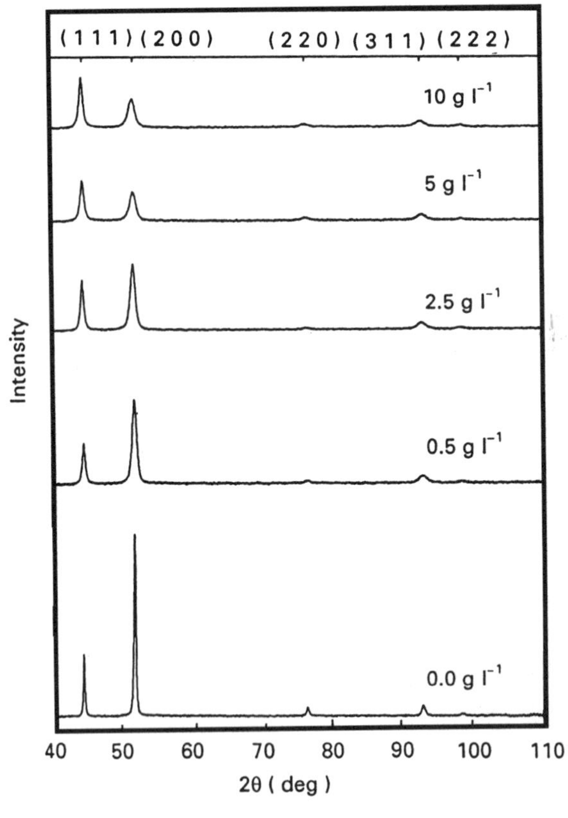

Figure 4. X-ray diffraction patterns showing the influence of saccharin concentration in the electrolyte on the preferred orientation of nickel electrodeposits produced by pulse plating.[38]

High-resolution electron microscopy has revealed that the grain boundary structure in electrodeposited nanocrystals is similar to the structure found in conventional polycrystalline materials.[44] This finding is in agreement with previous results by Thomas, et al.,[45] but in contrast to earlier work by Wunderlich, et al.,[46] who observed extended grain boundary structures in nanocrystalline palladium produced by inert gas condensation.

Using position annihilation spectroscopy, Würschum, et al.,[12] recently reported relatively large free volumes in electrodeposited nanocrystalline Pd which they described as nanopores (4 missing atoms) or nanovoids (10–15 missing atoms) containing light impurity atoms. However, such large free volumes may be a particular microstructural feature of electrodeposited Pd. For other electrodeposited nanocrystals, porosity is usually negligible as recently demonstrated by detailed density measurements[47] and positron annihilation spectroscopy.[48]

Porosity-free electrodeposited nanocrystals are a distinct form of grain boundary engineered materials. If the intercrystalline region of a material is considered to consist of distinct grain boundary and triple junction defects, the influence of these defects on the bulk properties of nanocrystalline materials will depend upon their relative volume fractions. A three-dimensional treatment involving tetrakaidecahedral grains, where grain boundaries are represented by the faces of the polyhedron, and triple junctions by the edges, has been applied,[4] and more recently generalized to any grain shape.[49] Figure 5 shows calculated volume fractions for the grain boundary, triple junction, and total intercrystalline component in the grain size range from 2 nm to 1000 nm, calculated for a boundary thickness of 1 nm. The intercrystalline volume fraction increases from a value of ~0.3% at 1000 nm to ≥50% at grain sizes smaller than 5 nm. In the range of 100 nm to 2 nm, the triple junction volume fraction increases by three orders of magnitude, while the grain boundary volume fraction increases by a little over one order of magnitude. The grain boundary volume fraction also shows a plateau at a grain size of ~3 nm, while the triple junction volume fraction continues to increase and becomes equivalent to the grain boundary volume fraction at a grain size of ~2 nm.

The plateau in grain boundary volume fraction coincides with values of grain size below which a transition to the noncrystalline state is usually observed. Figure 6 schematically illustrates a limiting case for this phenomenon.[50] When the mean grain size becomes very small, individual crystals can be better represented as spherical clusters of atoms. Under these conditions, the grain boundary can be represented by the point of contact between adjacent spherical clusters, while the triple junction region assumes a relatively large volume. Thus, it was postulated[50] that the transition to the noncrystalline state can be defined as the crystal (cluster) size where the ratio of triple junctions to grain boundary volume fraction begins to approach infinite values.

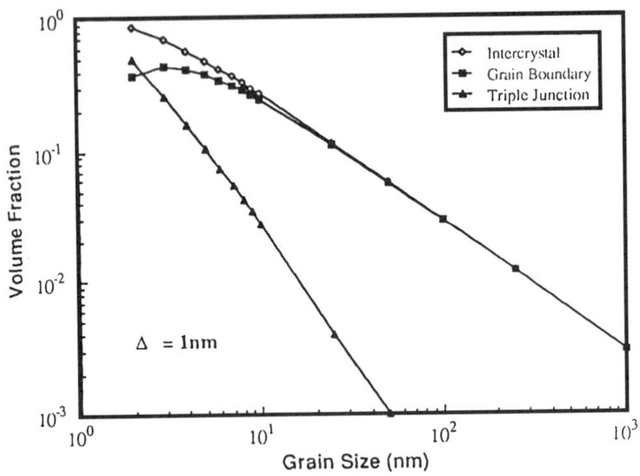

Figure 5. The effect of grain size (d) on calculated volume fractions for intercrystalline regions, grain boundaries, and triple junctions, assuming a grain boundary thickness of 1 nm.[4]

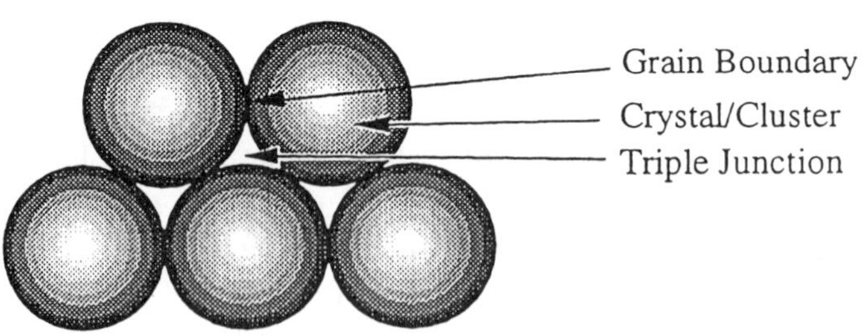

Figure 6. Schematic representation of a postulated limiting grain size for crystallinity where the triple junction to grain boundary volume fraction ratio approaches infinite values.[50]

4.0 PROPERTIES

A critical assessment of the properties measured to date on electrodeposited nanocrystals shows that these can be classified into two basic categories. The first group of properties are strongly dependent on grain size. These include strength, ductility and hardness,[14][21][50]–[61] wear resistance and coefficient of friction,[62] electrical resistivity,[10][11][63] coercivity,[64] solid solubility,[2][3][16][18][43] hydrogen solubility and diffusivity,[65][66] resistance to localized corrosion and intergranular stress corrosion cracking,[59][60][67]–[70] and thermal stability.[44][48][50][56][59][60][71]–[78] On the other hand, the second group of properties including bulk density,[47] thermal expansion,[48][79] Young's modulus,[53][57][60][80]–[83] resistance to salt spray environment,[41] and saturation magnetization[39][54][64][80][84]–[89] are little affected by grain size. In the following sections some of these properties are discussed in more detail and comparisons with properties observed in nanostructured materials produced by other methods are made.

4.1 Mechanical Properties

As expected, the plastic deformation behavior of electrodeposited nanocrystalline materials is strongly dependent on grain size. Much of the early work was concerned with room temperature microhardness measurements on free-standing sheet samples (typical thickness 0.1–0.5 mm) which were initially electrodeposited onto a Ti substrate and then removed from the Ti for hardness measurements. Figure 7a shows the results obtained by Palumbo, et al.,[51] for room temperature Vickers hardness measurements of Ni-P electrodeposits. Also shown are the results by Chokshi, et al.,[91] on nanocrystalline Pd and Cu produced by the inert gas condensation technique. Initial increases, followed by significant decreases in hardness are noted with decreasing grain size (d) in the nanocrystal range, i.e., $d \leq 20$ nm. The observed decreases in hardness are contrary to Hall-Petch behavior and consistent with results reported elsewhere[92][93] for nanocrystalline materials. Others, e.g., Ref. 94, have only reported a reduction in the Hall-Petch slope in the nanometer range. Recently, a study of room temperature tensile strength of nanocrystalline Ni[61] showed a behavior consistent with that of the hardness studies (Fig. 8).

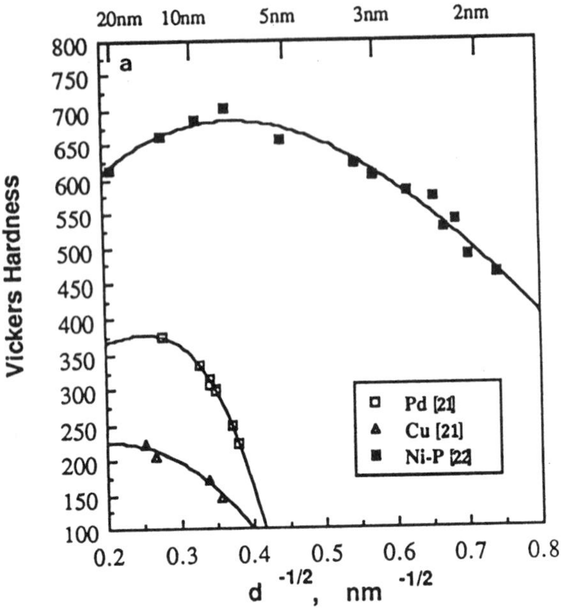

(a)

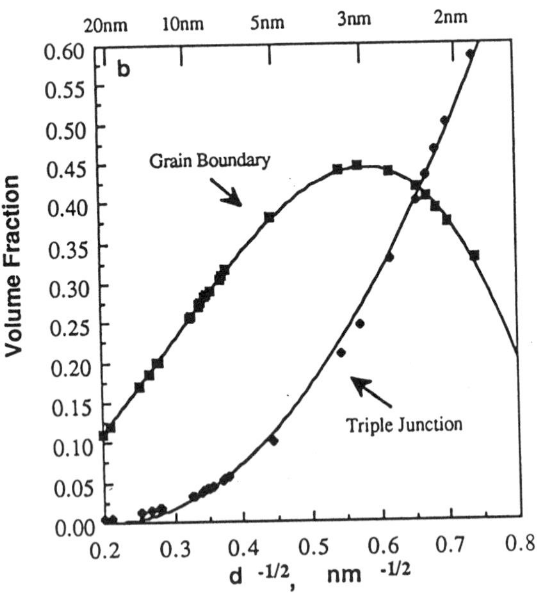

(b)

Figure 7. (*a*) Vickers hardness measurements for nanocrystalline Ni-P,[51] Pd,[91] and Cu.[91] (*b*) Corresponding intercrystalline volume fractions.[4]

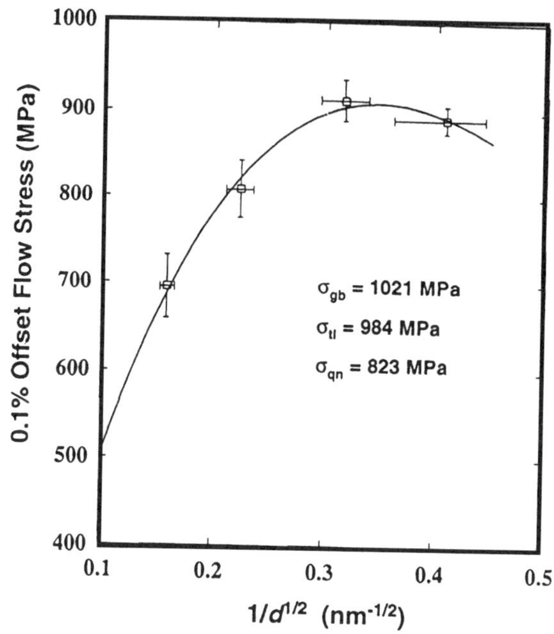

Figure 8. The result of fitting the yielding strength of nanocrystalline nickel electrodeposits to a composite model incorporating strength contributions from grain boundaries (σ_{gb}), triple junctions (σ_{tj}), and quadruple nodes (σ_{qn}).[61]

Chokshi, et al.,[91] interpreted their results in terms of room temperature Coble creep, arising from the disorder associated with large intercrystalline volume fractions. However, in one study[95] it appeared that grain boundary diffusional creep is not an appreciable factor in determining the room temperature mechanical behavior of nanocrystalline Cu and Pd. The onset of decreasing hardness, i.e., deviation from Hall-Petch behavior, in these systems occurs at grain sizes where triple lines begin to comprise a significant fraction of the bulk specimen volume (see Fig. 7b). The observed phenomena are in general agreement with the triple line softening effects first reported by Rabukhin,[96] who investigated the effect of triple junctions on the room temperature tensile properties of conventional polycrystalline wires (Al, Cu, W) having various grain sizes. By electrochemical thinning of the wires to a diameter less than the average grain size, triple junctions could be eliminated from the microstructure. In all cases, an increase in strength and decrease in ductility was noted on such a transition from an equiaxed to bamboo grain structure. The grain size dependence of the proof stress was found to obey the Hall-Petch relationship; however, at constant grain size, lower values were always obtained with the equiaxed geometry. More recently, using a similar experimental approach, Lehockey and co-workers[97] also confirmed triple line softening effects in Ni.

Modified dislocation pile-up theories involving small numbers of dislocations[98][99] can be used to explain the deviation behavior of the Hall-Petch relationship but not the negative slopes shown in Figs. 7 and 8. A significant reduction in the Hall-Petch slope value was predicted by Smith, et al.,[100] to occur for the extreme case of only one dislocation loop being expanded against the grain boundary obstacle stress. Wang, et al.,[55] concluded that the dislocation pile-up mechanism no longer applies to nanocrystalline materials below a critical grain size, e.g., about 10 nm for fcc metals. A composite model based upon geometric considerations in terms of the volume fraction of crystalline and intercrystalline components was proposed by Wang, et al.,[55][61] to evaluate the strength of nanocrystalline materials. It was shown that this model can be used for interpreting the various observations involving deviation from the Hall-Petch relationship and a negative Hall-Petch slope.

In addition to grain boundaries and triple junctions, this analysis also included quadruple nodes where triple lines (usually four) are linked up.[49] The strength contributions for grain boundaries (σ_{gb}), triple junctions (σ_{tl}) and quadruple modes (σ_{qn}) was shown to have the following sequence:[61] $\sigma_{gb} > \sigma_{tl} > \sigma_{qn}$.

Wang, et al.,[61] also derived an analytical expression for assessing the creep rate of nanocrystalline materials by a diffusion mechanism, including triple line diffusion. The overall creep rate is the sum of the creep rate due to lattice diffusion, grain boundary diffusion, and triple line diffusion. It was predicted that the creep rate due to triple line diffusion will exhibit a stronger grain size dependence than that due to grain boundary diffusion. For example, the contribution of triple line diffusion to steady-state creep rate appears to be the inverse of d^4 (d = grain size), which is one order higher than grain boundary diffusion and two orders higher than lattice diffusion in terms of grain size dependence. In addition, the secondary creep rate is still linearly proportional to the applied tensile stress, compared to the dislocation mechanism in which the exponent of the applied stress is usually greater than three.

The upshot of the work by Wang, et al.,[61] is that, at high stress levels, grain boundary sliding is the major room temperature deformation mechanism in nanocrystalline pure Ni electrodeposits. However, the contribution from creep mechanisms through intercrystalline regions can be significant for smaller grain size. A negative Hall-Petch slope was observed when the grain size was below 10 nm. It was suggested that the deviation from the Hall-Petch relationship can be attributed to a dynamic creep process due to diffusion mechanisms.

Recently, a more complete study on mechanical properties of nanocrystalline materials was performed in conjunction with the development of the first large scale industrial application of electrodeposited nanocrystalline materials, the Electrosleeve™ technology (Electrosleeve™ is a registered Trademark of Ontario Hydro, Canada; see Sec. 5 for details). The results of various mechanical properties of nanocrystalline nickel with grain size of 10 nm and 100 nm in comparison to conventional polycrystalline material are shown in Table 1. In addition to the remarkable increases in hardness, yield strength, and ultimate tensile strength with decreasing grain size, it is interesting to note that the work hardening coefficient decreases with decreasing grain size to virtually zero at a grain size of 10 nm. The ductility of the material decreases with decreasing grain size from 50% elongation to failure in tension for conventional material to 15% at 100 nm grain size and about 1% at 10 nm grain size. Generally somewhat greater ductility was observed in bending. A slight recovery in ductility was observed for grain sizes less than 10 nm.[61]

Compared to conventional polycrystalline Ni, nanocrystalline Ni electrodeposits exhibited drastically reduced wear rates and lower coefficients of friction as determined in dry air pin-on-disc tests.[62]

Contrary to earlier measurements on nanocrystalline materials prepared by consolidation of precursor powder particles,[95][101][102] nanocrystalline nickel electrodeposits do not show a significant reduction in Young's modulus. This result provides further support for earlier findings of Krstic, et al.,[81] and Zugic, et al.,[83] which demonstrated that the previously reported reductions in modulus with nanoprocessing were likely the result of high residual porosity.

With respect to the hardness curve for Ni-P shown in Fig. 7a, it should be noted that the grain size for the smallest grain sizes (<3 nm) was derived from x-ray line broadening measurements. However, the x-ray diffraction scans for these particular samples resembled those typically obtained for amorphous structures. These electrodeposits contained increasing P content with decreasing grain size and it has been previously shown[2][3] that there is a smooth, but not fully characterized, transition from the nanocrystalline to the amorphous structure. The smooth decrease in hardness through the nanocrystalline to the amorphous transition in Fig. 7a

Table 1. Mechanical Properties of Conventional and Nanocrystalline Nickel

Property	Conventional [a]	Nano-Ni 100nm	Nano-Ni 10nm
Yield Strength, MPa (25°C)	103	690	>900
Yield Strength, MPa (350°C)	—	620	—
Ultimate Tensile Strength, MPa (25°C)	403	1100	>2000
Ultimate Tensile Strength, MPa (350°C)	—	760	—
Tensile Elongation, % (25°C)	50	>15	1
Elongation in Bending, % (25°C)	—	>40	—
Modulus of Elasticity, GPa (25°C)	207	214	204
Vickers Hardness, kg/mm^2	140	300	650
Work Hardening Coefficient	0.4	0.15	0.0
Fatigue Strength, MPa (10^8 cycles/air/25°C)	241	275	—
Wear Rate (dry air pin on disc), $\mu m^3/\mu m$	1330	—	7.9
Coefficient of Friction (dry air pin on disc)	0.9	—	0.5

a: ASM Metals Handbook, ASM International, Metals Park, OH, 2:437 (1993) indicates that a common structural element may be responsible for ductilization in both the nanocrystalline and amorphous states. This transition coincides with the region following the plateau in the grain boundary volume fraction (discussed in Sec. 3.0) below which the triple junction volume fractions assume relatively large values. It was speculated that the common structural element which could be responsible for the ductilization is the disclination.[51]

4.2 Corrosion Properties

In general, the corrosion resistance of nanocrystalline materials in aqueous solutions is of great importance in assessing a wide range of potential future applications. To date, research in this area is still scarce and relatively few studies have addressed this issue. For the case of the corrosion behavior of nanocrystalline materials produced by crystallization of amorphous precursor materials (e.g., Refs. 103–108), both beneficial and detrimental effects of the nanostructure formation on the corrosion performance were observed. The conflicting results are, to a large extent, due to the poorly characterized microstructures of the crystallized amorphous materials. On the other hand, for nanostructured materials produced by electrodeposition, considerable advances in the understanding of microstructure on the corrosion properties have been made in recent years.[59][60][67]–[70]

In previous studies,[67][68] potentiodynamic and potentiostatic polarizations in de-areated 2N H_2SO_4 (pH = 0) were conducted on bulk (2 cm square coupons, 0.2 mm thick) nanocrystalline pure Ni at grain sizes of 32, 50, and 500 nanometers and compared with polycrystalline pure Ni (grain size of 100 μm). Figure 9 shows the potentiodynamic anodic polarization curves of these specimens. The nanocrystalline specimens exhibit the same active-passive-transpassive behavior typical of conventional Ni. However, differences are evident in the passive current density and the open circuit potential. The nanocrystalline specimens show a higher current density in the passive region resulting in higher corrosion rates. These higher current densities were attributed to the higher grain boundary and triple junction content in the nanocrystalline specimens, which provide sites for electrochemical activity. However, this difference in current density diminishes at higher potentials (1100 mV SCE) at which the overall dissolution rate overwhelms the structure-controlled dissolution rate observed at lower potentials.

Another notable difference in the potentiodynamic response of nanocrystalline and polycrystalline specimens is the open circuit potential. The positive shift of the open circuit potential for the nanocrystalline specimens is thought to be the result of the catalysis of the hydrogen evolution reaction.[67]

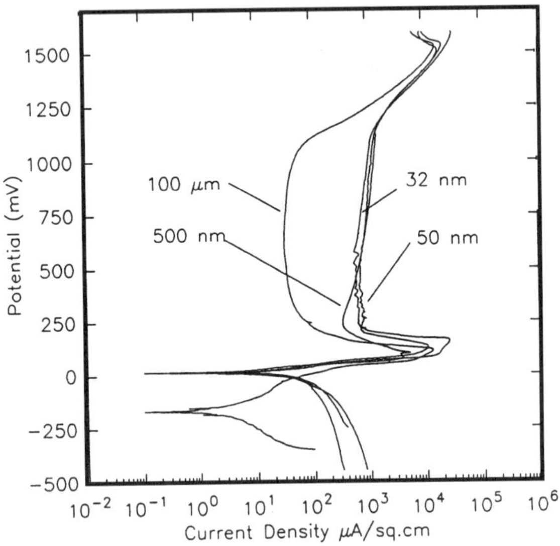

Figure 9. Potentiodynamic polarization curves for nanocrystalline and polycrystalline Ni in 2N H_2SO_4 at ambient temperature.[67]

Figure 10 shows scanning electron micrographs of nickel with *a)* 32 nm and *b)* 100 µm grain size, held potentiostatically at 1200 mV (SCE) in 2N H_2SO_4 for 2000 seconds.[67] Both specimens exhibit extensive corrosion but the nanocrystalline Ni is more uniformly corroded while the specimen with 100 µm grain size shows extensive localized attack along the grain boundaries and triple junctions. X-ray photoelectron spectroscopy of the specimens polarized in the passive region showed that the passive film formed on the nanostructured specimen is more defective than that formed on the polycrystalline specimen, while the thickness of the passive layer was the same on both specimens.[109] This higher defective film on the nanocrystalline specimen allows for a more uniform breakdown of the passive film, which in turn leads to a more uniform corrosion. In contrast, as has been previously shown in Ref. 110, in coarse-grained Ni the breakdown of the passive film occurs first at the grain boundaries and triple junctions rather than the crystal surface, leading to preferential attack at these defects.

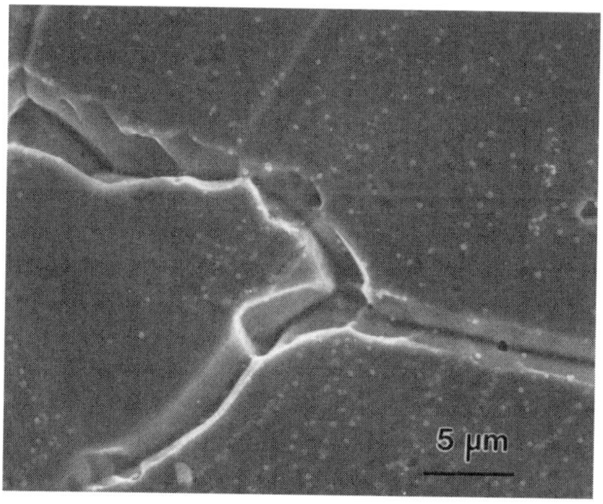

(a)

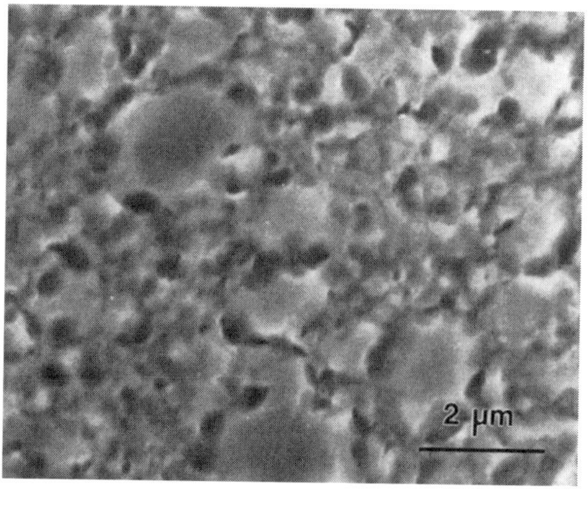

(b)

Figure 10. SEM micrographs of Ni with (*a*) 100 μm and (*b*) 32nm grain size held potentiostatically at 1200 mV (SCE) in 2N H_2SO_4 for 2000 seconds.[67][68]

Similar observations were made for the corrosion behavior of nanocrystalline 304 stainless steel (grain size 25 nm) in HC1 produced by sputtering.[111] The reduced susceptibility to localized corrosion was attributed to the fine-grained microstructure, which allowed for a uniform distribution of $C1^-$ ions.

More recently, the corrosion behavior of nanocrystalline Ni was also studied in 30 wt% KOH solution[70] and pH neutral solution containing 3 wt% sodium chloride.[43] The results were similar to the corrosion behavior observed in sulfuric acid. The general corrosion was somewhat enhanced compared to conventional polycrystalline Ni; however, the nanostructured materials were much more immune to localized attack which often can lead to catastrophic failures.

Using the ASTM B-117 salt spray test, it was found that the microstructure of Ni has little effect on the overall corrosion performance under these electrochemical conditions.[41] Both conventional polycrystalline and nanostructured coatings gave the same corrosion protection to mild steel substrates.

Further corrosion testing was performed on nanocrystalline Ni under conditions required for steam generator alloy application as part of the Electrosleeve™ development program.[59] Tests ASTM G28 (susceptibility to intergranular attack), ASTM G48 (susceptibility to pitting and crevice corrosion), ASTM G35, G36, and G44 (susceptibility to stress corrosion cracking in polythionic acids, magnesium chloride, and alternate immersion in sodium chloride, respectively) were performed. The results showed that electrodeposited nanostructured Ni with a grain size of 100 nm is intrinsically resistant to intergranular processes such as intergranular attack and intergranular stress corrosion cracking. The material was found to be resistant to pitting attack and only slightly susceptible to crevice corrosion.

A second series of tests focused on specific environments that are known to be detrimental to steam generator materials.[59] Environments included alkaline, acidic, and a combination of oxidizing and reducing species. The tests revealed excellent resistance of the nanocrystalline nickel to alkaline environments and reducing acidic environments. The corrosion resistance to oxidizing and acidic environments was found to be limited.

4.3 Hydrogen Transport and Activity

The transport behavior of hydrogen in electrodeposited nanocrystalline Ni foil (average grain size of 17 nm) at 293 K was determined using an electrochemical double cell.[65] Figure 11 shows a typical hydrogen permeation curve, where the anodic exit current (*I*) is plotted as a function of cathodic charging time (*t*). Three distinct breakthrough events are clearly evident, as indicated by the arrows in Fig. 11. On the basis of determined diffusivities, permeation flux values, and area (volume) fraction considerations,[65] these breakthrough events were considered to be due to hydrogen transport through distinct triple junction, grain boundary, and lattice paths, respectively. The triple junction diffusivity was determined to be approximately three times faster than grain boundary diffusivity, and 70 times faster than lattice diffusion. Other studies[112] have also shown that diffusive transport occurs at a considerably faster rate through the triple junctions than along the adjoining grain boundaries. These results provide support for the defect character of triple junctions. Furthermore, the existence of a "measurable" triple junction diffusivity in nanocrystalline Ni indicates the importance of triple junction defects in the bulk properties of nanocrystalline materials.

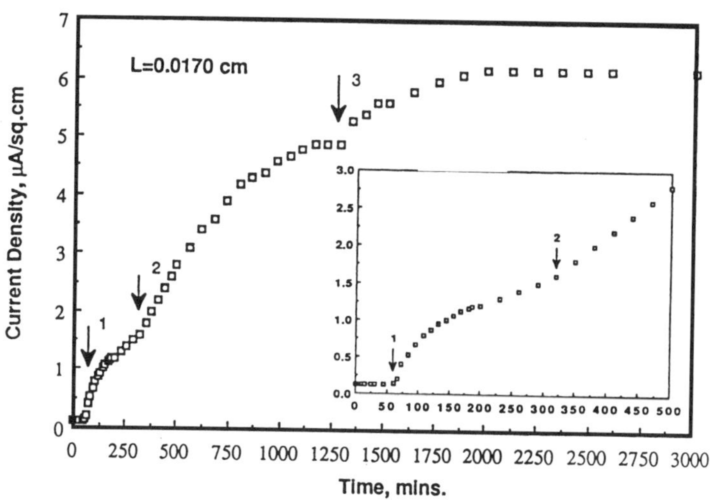

Figure 11. Anodic exit current density (*I*) as a function of cathodic charging time (*t*) at 0.1 mA/cm^2 for nanocrystalline Ni (17 nm grain size) foil of 0.017 cm thickness.[65]

As shown in Fig. 12, nanocrystalline Ni having an average grain size of 20 nm is also observed to display significantly higher electrocatalytic behavior when compared to 1) cold worked, 2) fine-grained, and 3) fully annealed reference structures with regard to the hydrogen evolution reaction (HER) for alkaline water electrolysis at room temperature.[66]

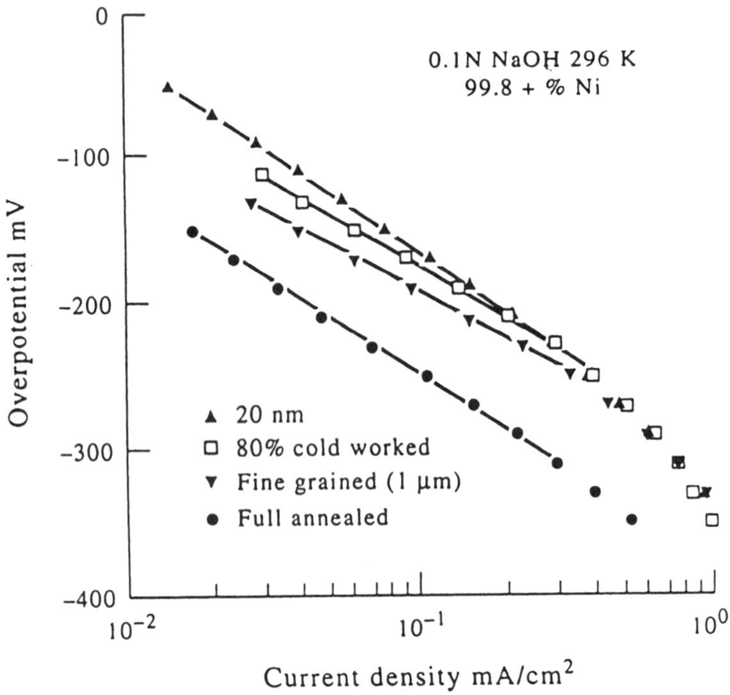

Figure 12. Room temperature Tafel plots of HER for electropolished nanocrystalline, 80% cold worked, fine grained (1 μm) and fully annealed Ni in 0.1N NaOH.[66]

The enhanced HER kinetics observed here are considered to be the direct result of the high area fraction of grain boundaries (and to some extent, triple junctions) intersecting the free surface of the electrode. In a more recent study,[43] it was shown that the HER kinetics can be further enhanced by alloying nanocrystalline Ni with molybdenum.

An additional study[66] into the transport behavior of hydrogen in nickel as determined by an electrolytic charging technique revealed that substantial increases in hydrogen diffusivity and capacity are obtained when Ni is in nanocrystalline form. Figure 13 illustrates three representative permeation transients corresponding to hydrogen transport through nanocrystalline (20 nm), fine grained (1 μm), and single crystalline Ni foils of 140 μm thickness. Detection of permeated hydrogen in the Ni bi-electrodes of identical thickness is observed in the following order: 1) nanocrystalline, 2) fine grained, 3) single crystal structures.

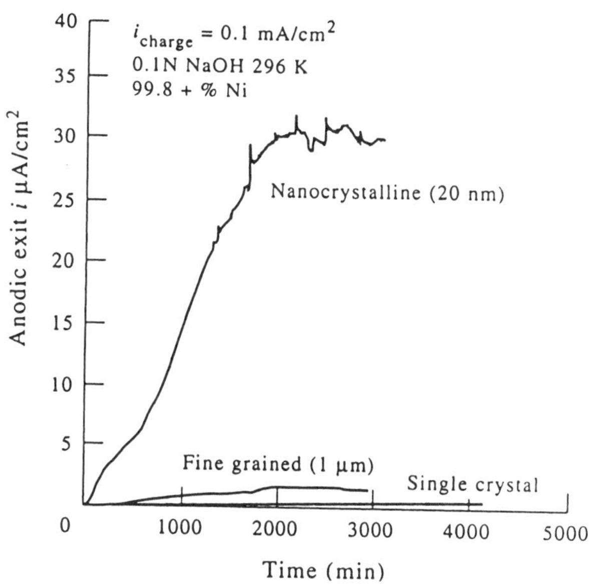

Figure 13. Hydrogen permeation transients showing anodic exit current density (flux) vs. time for nanocrystalline (20 nm), fine grained (1 μm), and single crystal Ni foils.[66]

In addition, the apparent concentration of hydrogen in the 20 nm sample is found to be approximately 60 times greater than that of the single crystal structure with regard to the permeation transients shown in Fig. 13. The increased hydrogen diffusivity and capacity are attributed to high intercrystalline content, which provides 1) a high density of short circuit diffusion paths and 2) large free volumes to which increased segregation of hydrogen can occur.

Recently, permeation experiments were conducted in a double chamber ultra-high vacuum system separated by a test nickel specimen.[113] Hydrogen permeabilities and diffusivities through microcrystalline and nanocrystalline Ni were measured in the temperature range of 30°C to 200°C. Steady-state permeability measurements indicate that nanocrystalline Ni (average grain size of 78 nm) displays enhanced permeability below 50°C (e.g., a factor of six at 30°C), as compared to the microcrystalline Ni (average grain size of 3 μm). Also, diffusivity measurements in combination with hydrogen trapping site density measurements suggest that there are more intercrystalline hydrogen trapping sites in the nanocrystalline Ni.

4.4 Magnetic Properties

Conflicting results have been reported regarding the dependence of certain magnetic properties on the grain size of the material. While the understanding of the magnetic structure of nanostructured materials is still far from complete, a clear picture is now emerging regarding the saturation magnetization, M_s, where the early contradictory results can be explained in terms of the chemical and physical microstructure of the nanocrystalline materials.

Initially many studies reported that, for nanocrystalline materials, there is a large reduction in saturation magnetization with decreasing grain size.[101][114]–[117] Gleiter first reported a 40% decrease in saturation magnetization compared to bulk α-iron for nanocrystalline iron with 6 nm grain size which was produced by consolidating nanocrystalline particles produced by the inert gas condensation technique.[101]

This behavior was attributed to differences in the magnetic microstructure between nanocrystalline and conventional polycrystalline iron. Similarly, strong effects of particle size on saturation magnetization have been observed in the study of uncompacted ultrafine particles produced by the gas evaporation method.[114][115] In the case of ultrafine particles (10–50 nm) of Ni, Co, and Fe, Gong, et al.,[114] observed a rapid decrease in saturation magnetization with decreasing grain size which they attributed to antiferromagnetic oxide layers on the ultrafine metal particles. In another study on ultrafine particles it was found that the normalized magnetization ratio decreases with decreasing particle diameter.[115] The reduction in saturation magnetization was linked to surface effects, which were considered more important in the case of smaller particles. Schaefer, et al.,[116] also noted a decrease in M_s in consolidated nanocrystalline nickel powder produced by gas evaporation which they explained in terms of structural disorder of the interfaces. The magnetic moment of the interfacial atoms was calculated to be nearly half that of the atoms in the bulk material. Furthermore, Yao, et al.,[117] also found that the saturation magnetization of ultrafine Ni particles decreases drastically with decreasing grain size. Krill, et al.,[118] reported that the spontaneous magnetization of nanocrystalline Gd samples produced by gas condensation and subsequent compaction was approximately 75% of the value for polycrystalline Gd. It should be noted that all of the previous materials were produced using the gas condensation method which yields materials with high internal porosity that can provide large surface areas for oxide formation after exposing the samples to air.

In contrast, Aus, et al.,[64] reported, for the first time, that the saturation magnetization of nanocrystalline Ni was not strongly dependent on the grain size. In this study, the grain sizes of Ni varied from 100 μm to 10 nm, and for the Ni samples with the smallest grain size, the observed M_s was only 10% less than for conventional polycrystalline Ni. These results were obtained on bulk nanocrystalline Ni produced by electrodeposition and were explained in terms of the negligible porosity/oxide formation in this material.

The finding by Aus, et al.,[64] agrees well with results of recent calculations which assessed the effect of structural disorder, introduced by grain boundaries, on the magnetic properties of nanocrystalline metals.[86]–[88] In these studies, grain boundary configurations representing various degrees of disorder were generated using molecular dynamics simulations with embedded-atom potentials. They ranged from $\Sigma3$ boundaries with minimum structural disorder through $\Sigma5$ and $\Sigma13$ special grain boundaries of intermediate structural disorder to random amorphous grain boundaries with maximum disorder. Electronic structure calculations were performed using the tight-binding linear muffin-tin orbital atomic-sphere-approximation method. These calculations have shown that the magnetic moment is rather insensitive to the degree of structural disorder associated with grain boundaries. Even when the entire material was amorphous, the average moment was found to be reduced by only 15%. It was concluded that, for the case of nanocrystalline Ni with a grain size of 10 nm at which the grain boundary atoms comprise about 30% of the volume, the overall effect of structural disorder on the average moment is very small, in good agreement with the experimental data reported by Aus, et al.,[64] for electrodeposited nickel.

More recently, there have been other reports confirming the early results by Aus, et al.,[64] for nanocrystalline Ni. For example, Daroczi, et, al.,[119][120] reported for nanocrystalline nickel prepared by ball milling, that there is no observable difference in M_s for materials with 7 nm and 50 μm grain sizes. Bakonyi, et al.,[11] observed the same trend for nanocrystalline Ni, also prepared by electrodeposition. Weissmüller, et al.,[89] confirmed the earlier measurements by Aus, et al.,[64] reporting only small changes in M_s for electrodeposited nanocrystalline Ni with 18 nm grain size. Kisker, et al.,[121] presented new results for gas condensed Ni which, in contrast to their earlier work,[116] now showed the saturation magnetization to be independent of grain size as long as the gas condensed material was not exposed to air. However, after exposure to air, M_s decreased to about 80% of its original value.

Aus, et al.,[122] and Szpunar, et al.,[90] have recently presented further experimental evidence and detailed calculations for nanocrystalline Ni-Fe, Ni-P, Co, and Co-W which further support their earlier findings that structural disorder introduced by grain boundaries and triple junction has an insignificant effect on saturation magnetization compared with chemical disorder introduced by alloying additions.

4.5 Thermal Stability

The thermal stability of nanocrystalline materials is of considerable importance for applications at elevated temperatures. For electrodeposited nanocrystals, the thermal stability has been assessed by in-situ transmission electron microscopy,[44][71][73] conventional annealing followed by TEM analysis,[74]–[76] and differential scanning calorimetry (DSC) experiments from which activation energies for grain growth were determined using the Kissinger[123] analysis.[48] Additional indirect experiments on the thermal stability involved hardness measurements as a function of annealing time.[59][60]

Figure 14 shows the grain growth kinetics for a nanocrystalline Ni-1.2 wt% P alloy as evaluated from in-situ electron microscopy studies.[71] At 473 K, no grain growth was observed and the material was stable as a solid solution. At 673 K, substantial grain growth was observed within the first few minutes of annealing, resulting in a microcrystalline two phase (Ni + Ni_3P) structure. However, at 573 K and 623 K, the grain size initially increased rapidly by a factor of 2–3 and then became essentially independent of annealing time. Similar behavior was observed for a Ni-S alloy at 573 K.[73]

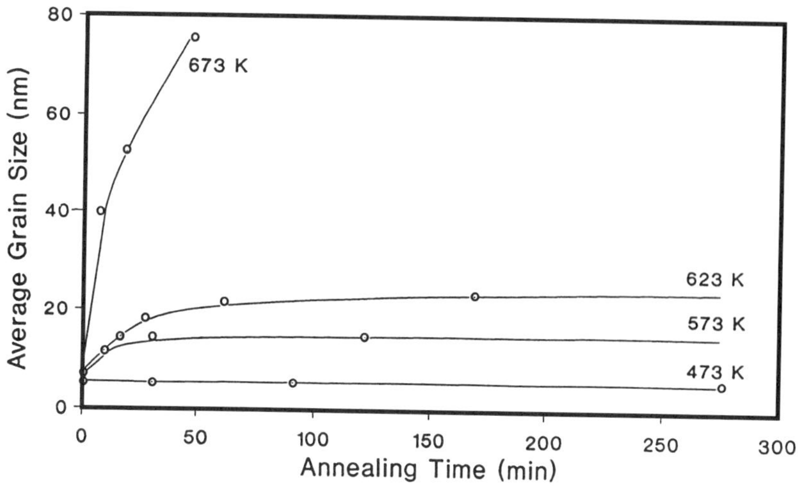

Figure 14. Grain size as a function of annealing time for electrodeposited nanocrystalline Ni-1.2 wt % P.[71]

Grain growth kinetics leading to a constant characteristic grain size is common for systems subjected to large grain boundary dragging forces. The most obvious dragging mechanism for these alloy systems is precipitate-induced Zener type drag.[124] However, considering the extremely large driving forces for grain growth expected in these materials (e.g., about 200 J/cm^3 at a grain size of 20 nm[72]), the observed thermal stabilization may not be attributed solely to such a mechanism. In nanocrystalline materials, an additional dragging force may be due to triple junctions.[71] It has been shown that grain growth in fine-grained polycrystalline materials may be controlled by the intrinsic mobility of triple junctions.[125] A further contribution of triple junctions to the thermal stability of nanostructured materials is the result of preferential solute segregation to these sites.[126] Such solute enrichment at triple junctions in annealed nanostructured Ni-0.12 wt% S was recently observed by scanning transmission microscopy.[75]

Klement, et al.,[75] investigated the thermal stability of 10 nm and 20 nm Ni using DSC (heating rate 10 K/min) and TEM. The temperature at which the material tends to become unstable was found to be as low as 353 K. This instability was attributed to "nucleation" and abnormal grain growth producing a dual-sized microstructure after annealing in the range of about 400 to 550 K. The origin of the abnormal grain growth may be the "clusters" of subgrains observed in the nanocrystalline nickel deposits, as indicated by Moiré patterns in TEM studies[75][76] and high resolution microscopy.[44] In order to nucleate the grains subsequently observed in abnormal grain growth, these nanometer-sized subgrains have only to rotate slightly towards each other to form a larger grain. This mechanism is analogous to the subgrain coalescence model of primary recrystallization.[127][128] In fact, changes of grain orientation caused by rigid body rotations have been observed directly during the annealing of nanocrystalline thin films of gold; the observed rates of grain rotation were consistent with a mechanism based upon diffusion-limited grain boundary sliding in response to the variation of grain boundary energy with misorientation.[129] Gertsman and Birringer[130] have suggested that inhomogeneity of grain boundary structure and non-uniform interface segregation contribute to abnormal grain growth observed at ambient temperature in nanocrystalline copper. The results obtained by Klement, et al.,[75] for abnormal grain growth and S segregation at grain boundaries and triple lines in annealed nanocrystalline Ni provide support for this interpretation.

For nanocrystalline Ni (starting grain sizes ranging from 15–30 nm), the activation energies for grain growth as determined by Kissinger analysis[123] from differential scanning calorimetry studies[48] were in the range of 1.2–1.4 eV which corresponds to the activation energy of grain boundary diffusion in Ni.[131] Considerably higher activation energies were measured for nanocrystalline Ni alloys containing P alloying additions. For example, for nanocrystalline Ni-1.2% P (starting grain size 10 nm), the activation energy was 2.25 eV[44] which was likely due to additional solute, Zener and triple junction drag.

The beneficial effect of microalloying on the thermal stability has recently been further demonstrated for nanocrystalline nickel (approximately 100 nm grain size) developed for the Electrosleeve™ application.[59][60] In this case, the thermal stability was assessed indirectly by measuring the hardness of pure nanocrystalline Ni and nanocrystalline Ni-P (<3000 ppm P) annealed at 616 K as a function of annealing time. For pure nanocrystalline Ni, the hardness decreased rapidly from about 420 VHN to 150 VHN within the first 100 minutes of annealing. However, for nanocrystalline Ni-P the hardness remained unchanged at 420 VHN for annealing times of in excess of 10^6 minutes. In this case, the thermal stability of microalloyed Ni was attributed mainly to solute drag and possible Zener drag by micro-precipitates.

A recent theoretical and experimental study[132] on the effect of grain growth on resultant grain boundary character distributions indicates that the thermal stability of nanocrystalline materials may be further enhanced by the tendency for these ultrafine grained materials to form "special" low-energy grain boundaries during the early stages of grain growth.

4.6 Thermal Expansion and Heat Capacity

Thermal expansion, α_L, and specific heat, C_P, measurements on nanocrystalline electrodeposits[48][79] showed quite different results from those reported earlier on nanostructures produced by inert gas condensation[133] or crystallization of amorphous precursors.[134] Rupp and Birringer[133] found α_L for gas condensed nanocrystalline copper of 8 nm grain size to be nearly twice that of regular polycrystalline copper. However, the nanocrystalline copper was only 90% dense. Heat capacities of nanocrystalline copper and palladium were also found to be increased by 10% and 40%, respectively.[133] Thermodynamic models based on quasiharmonic approximations used to predict theoretically the thermodynamic properties as a function of grain boundary free volume have produced similar increases in α_L (70–85%) and C_P (10–25%) for nanocrystalline palladium with respect to conventional polycrystalline palladium.[135] Fecht's calculations[136] also predict increases by a factor of two in both volumetric thermal expansion and heat capacity. Lu, et al.,[134] observed an increase in α_L of about 60% and 12% in C_P for Ni-P alloys crystallized from amorphous precursors. However, these materials consisted of two phases (fcc Ni and bct Ni_3P) and cannot be compared directly with single-phase materials.

More recent results by Gleiter[137] showed that the differences for α_L between nanocrystalline and conventional polycrystalline copper depend strongly on the applied pressure during consolidation of gas condensed powder. The general trend was that this difference decreased with increasing compaction pressure. Some compacted nanocrystals even showed a slight reduction of α_L from the value of polycrystalline material.

In contrast, for electrodeposited nanocrystals, neither thermal expansion nor specific heat showed significant differences when compared to their polycrystalline counterparts.[48][79] Figure 15 shows the temperature dependence of the linear thermal expansion coefficient, α_L, for both the as-plated nanocrystalline nickel (20 nm grain size) and conventional nickel (100 μm grain size) between 140 and 500 K. Between 140 and 205 K, α_L is slightly higher for the nanocrystalline nickel. However, between 205 and 500 K, α_L of nanocrystalline nickel is reduced from the value observed for polycrystalline nickel; the maximum reduction is 2.6% at 500 K. Overall, the effect of grain size on α_L for pure, fully-dense Ni is rather small. Figure 16 shows the heat capacity, C_P, as a function of temperature for nanocrystalline (20 nm grain size) and normal polycrystalline (100 μm grain size) nickel. Over the entire temperature range, the heat capacity of nanocrystalline nickel is marginally increased above C_P for polycrystalline nickel by 2.5–5%.

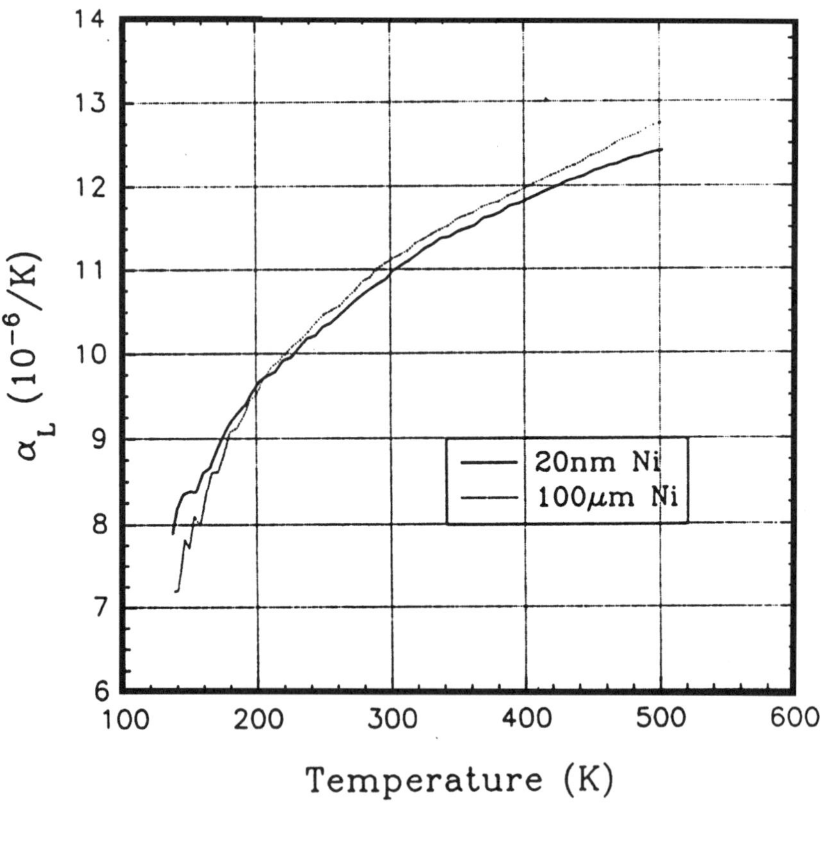

Figure 15. Temperature dependent coefficient of linear thermal expansion of nanocrystalline (20 nm grain size) and conventional polycrystalline (100 μm grain size) nickel.[79]

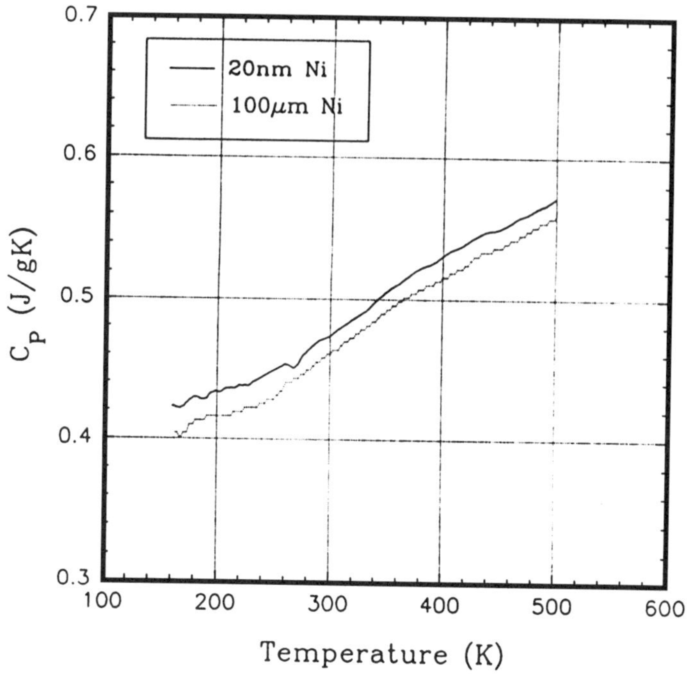

Figure 16. Temperature dependent isobaric heat capacity of nanocrystalline (20 nm grain size) and polycrystalline (100 μm grain size) nickel.[79]

As for the case of Young's modulus and saturation magnetization, these results clearly show that certain properties measured on nanostructured materials are not necessarily the result of increased volume fractions of interfaces in the materials. Rather than attributing major property changes to the presence of large densities of grain boundaries and triple junctions, other microstructural defects such as porosity, impurities, and the like must be considered.

4.7 Electrical Properties

A comparison of results of electrical property measurements performed on nanostructured materials produced by different synthesis routes (e.g., gas condensation,[101] electrodeposition[10][11][63]) show very similar trends. In most cases, the electrical resistivity was observed to increase with decreasing grain size. For example, the room temperature resistivity for Ni was increased from about 6 μΩ cm at 100 μm grain size in fully-annealed material, to about 22 μΩ cm at 11 nm grain size in electrodeposited material.[63] This can be attributed to electron scattering at defects, such as grain boundaries and triple junctions. In fact, a linear relationship between excess resistivity—defined as the total resistivity of the nanocrystalline material minus the resistivity of conventional polycrystalline material (100 μm grain size) with negligible intercrystalline volume fraction—was observed for nanocrystalline Ni of varying grain size.[63]

There is also good agreement as far as the temperature coefficient of resistivity is concerned; both nanocrystalline materials produced by inert gas condensation,[101] electrodeposited Ni,[11][63] and Co[11] show decreasing values with decreasing grain size.

For electrodeposited Cu, Bakonyi, et al.,[11] found no effect of grain size on electrical transport properties which they attributed to the negligible effect of atomic disorder on the density of states around the Fermi level of copper.

5.0 APPLICATIONS

Electrodeposited nanostructures have advanced rapidly to commercial applications as a result of:

1. An established industrial infrastructure (i.e., electroplating and electroforming industries).
2. A relatively low cost of application, whereby nanomaterials can be produced by simple modification of bath chemistries and electrical parameters used in current plating and electroforming operations.
3. The capability in a single-step process of producing metals, alloys, and metal-matrix composites in various forms (i.e., coatings, free-standing complex shapes).
4. The ability to produce fully dense nanostructures free of extraneous porosity.

The importance of the latter cannot be overemphasized with regard to industrial applications since, as has been outlined in previous sections, many of the extraordinary properties initially attributed to nanostructures have since been demonstrated to be an artifact of residual porosity in these materials. From the outset, the fully dense nanomaterials produced by electrodeposition have displayed predictable material properties based upon their increased content of intercrystalline defects. This "predictability" in ultimate material performance has accelerated the adoption of nanomaterials by industry, whereby such extreme grain refinement simply represents another metallurgical tool for microstructural optimization. In this section, an overview of some current and emerging practical applications for electrodeposited nanocrystalline materials are presented and discussed in light of the importance of property-specific grain size "optimization" rather than grain miniaturization for its own sake.

5.1. Structural Applications

As would be expected from Hall-Petch considerations, numerous practical applications for nanocrystalline materials are based upon opportunities for high-strength coatings and free-standing structural components. The superior mechanical properties of these electrodeposited nanostructures have led to one of their first large scale industrial applications—the Electrosleeve™ process for in-situ repair of nuclear steam generator tubing.[60] This proprietary process[138] has been successfully implemented in both Canadian CANDU and U.S. Pressurized Water Reactors, and has been incorporated as a standard procedure for pressure tubing repair.[139] In this application, nanocrystalline Ni (100 nm) is electroformed on the inside surface of steam generator tubes to effect a complete structural repair at sites where the structural integrity of the original tube has been compromised (e.g., corrosion, stress corrosion cracking, etc.). Figure 17 shows a cut-away view of an installed Electrosleeve™. The high strength and good ductility of this 100 nm grain size material permits the use of a thin "sleeve" (0.5–1 mm) which minimizes the impact on fluid flow and heat transfer in the steam generator.

Recent geometric models and experimental findings[140][141] have shown that nanostructured materials can also possess a high resistance to intergranular cracking processes, including creep cracking. Several emerging applications for nanocrystalline materials possessing high intergranular cracking resistance include lead-acid battery (positive) grids, and

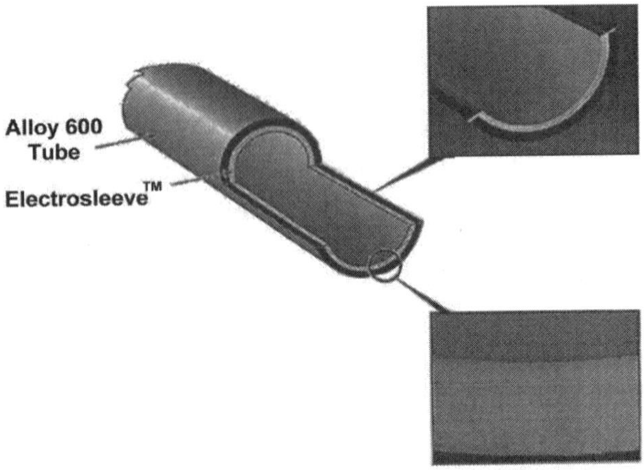

Figure 17. Cut-away view of an installed nanocrystalline Ni Electrosleeve™ on a host Alloy 600 nuclear steam generator tube.

Figure 18. Nanocrystalline-Cu (left), and -Ni (middle) shaped charge liners (81 mm), electroformed on a Ti mandrel (right).

shaped charge liners (Cu, Pb, Ni) for military and industrial applications (e.g., demolition, oil well penetrators, etc.) (see Fig. 18); applications in which durability and performance are frequently compromised by premature intergranular failure.

5.2 Functional Applications

Some of the most promising industrial applications for nanostructured materials are in the area of soft magnets for high-efficiency transformers, motors, etc. Anticipated reductions in magnetocrystalline anisotropy and coercivity, as grain size is reduced below the mean thickness of a magnetic domain wall in conventional materials, have generated considerable development activity in this area. Figure 19 summarizes the potential opportunity for nanocrystalline soft magnets, whereby these electrodeposited nanocrystals can possess a low coercivity without compromise of saturation magnetization.

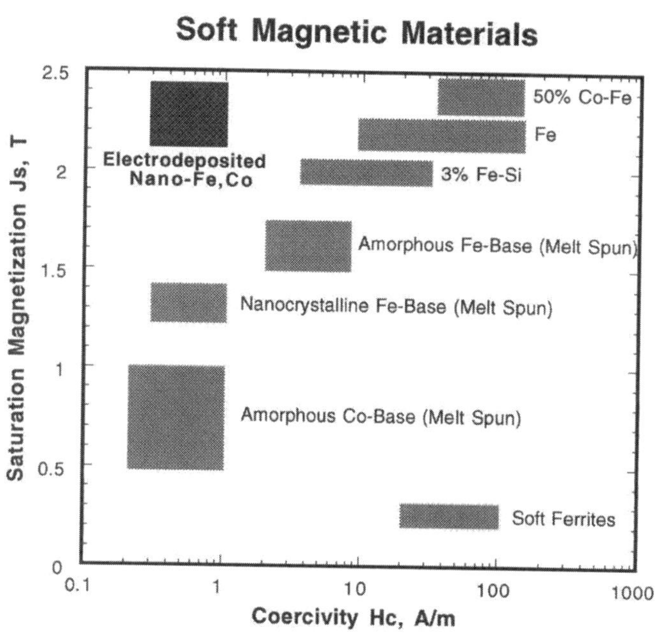

Figure 19. Typical coercivity and saturation magnetization ranges for several ferromagnetic materials.

As depicted in Fig. 20, the industrial use of these high-performance ferromagnetic materials in motor, transformer, and shielding applications have been accelerated by the recent development of a drum plating process for cost-effectively producing large quantities of sheet, foil, and wire in nanocrystalline form.

Figure 20. Prototype drum-plater for producing nanocrystalline sheet, foil and wire products.

Another major application for drum-plated nanocrystalline material (as in Fig. 20) is in the production of copper foil for printed circuit boards, where enhanced etching rates and reduced line spacing/pitch can be achieved by reducing grain size. Figure 21 shows a cross-sectional field emission scanning electron micrograph of nanocrystalline Cu foil produced for this application. Grain size has been optimized on the basis of calculated electrical resistivity for nanocrystalline Cu[143] as summarized in Fig. 22. A 50 nm to 100 nm grain size provides optimum etchability while maintaining good electrical conductivity.

As previously discussed in Ref. 66, the high density of intercrystalline defects present within the bulk, and intersecting the free surface of nanostructured materials, provides considerable opportunity in catalytic and hydrogen storage applications. Several applications are being developed for the use of these materials, either as an electrodeposited coating or electroformed free-standing component in Nickel Metal Hydride battery systems, and as alkaline fuel cell electrodes (see Fig. 23).

Figure 21. Cross-sectional field emission scanning electron micrograph of electrodeposited nanocrystalline Cu foil having an average grain size of 50 nm.

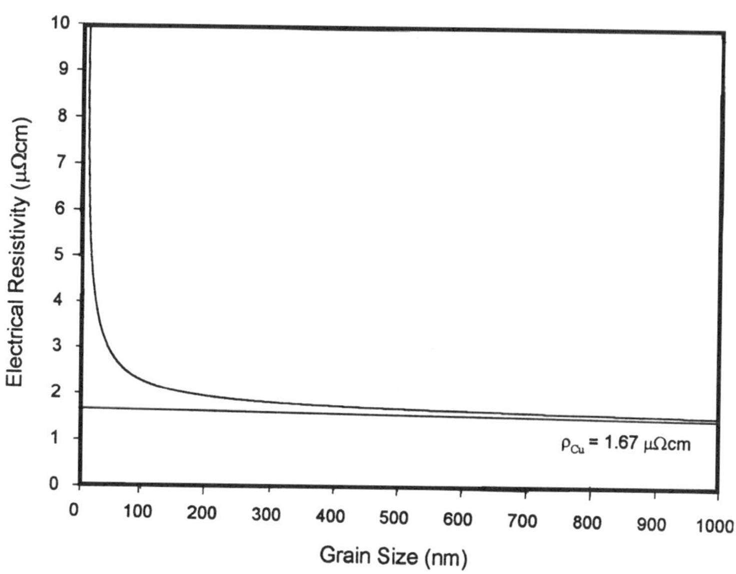

Figure 22. Calculated room temperature electrical resistivity of Cu as a function of grain size.[142]

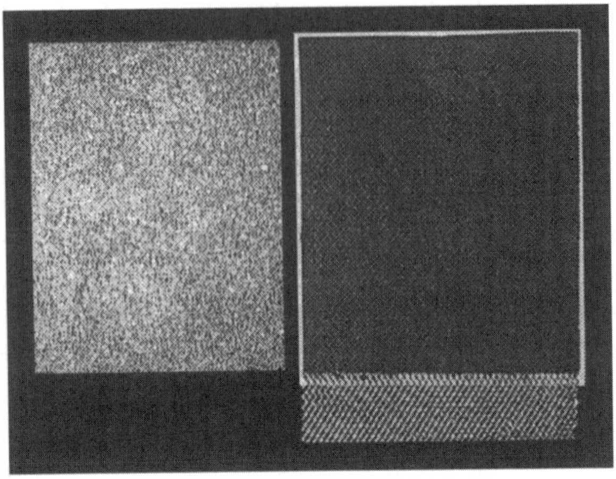

Figure 23. Alkaline fuel cell electrodes coated with a nanocrystalline Ni-Mo alloy.

5.3 Coating Applications

The improved hardness, wear resistance, and corrosion resistance, coupled with undiminished saturation magnetization and predictable thermal expansion, elastic properties, and electrical resistivity, make nanocrystalline coatings ideal candidates for protective and functional coatings (e.g., as used in hard facing on softer, less wear resistant coatings, recording heads, electronic connectors, replacement coatings for chromium and cadmium in automotive and aerospace applications). For applications as thin coatings (on the order of a few nanometers thick), the microstructural evolution of the deposit with increasing coating thickness can be a major concern. Many previous studies on electrodeposited metals, not necessarily in nanocrystalline form, have shown that the grain size usually increases considerably with increasing coating thickness. (See, for example, Ref. 143.) For the nanocrystalline Ni electrodeposits studied by Bakonyi, et al.,[40] it was reported that the deposit initially was amorphous right at the substrate interface. This was followed by the transition to nanocrystalline structure, and then a gradual increase in grain size was observed. In contrast, nanocrystalline Ni electrodeposits produced following the procedures given by Erb, et al.,[8][39] showed that in most cases the nanostructure was fully established right at the interface with the substrate and that the grain size was essentially independent of coating thickness

(Fig. 24). For certain electrochemical and substrate conditions, a thin transition layer of larger grains was observed in which the initial structure was influenced by the larger grains of the substrate, presumably by an epitaxy mechanism. However, even in this case, a constant, thickness-independent structure was fully established within the first 200 nm from the interface.

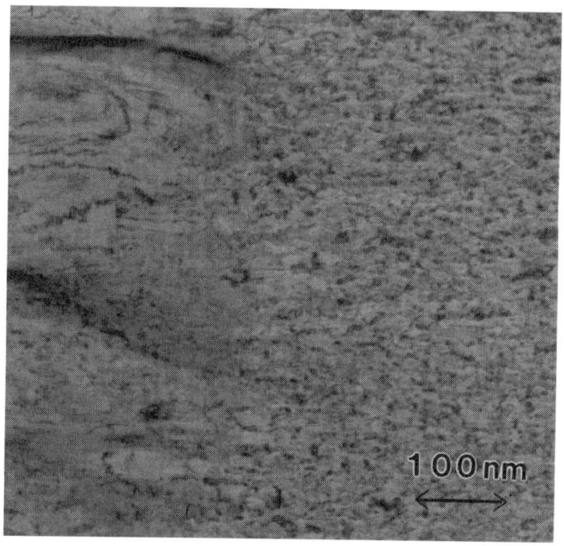

Figure 24. TEM bright field micrograph of cross section showing nanocrystalline nickel coating (right) deposited onto polycrystalline bronze electronic connector substrate (left). Cross section prepared by ultramicrotomy.

REFERENCES

1. Brenner, A., *Electrodeposition of Alloys-Principles and Practice*, Academic Press, NY (1963)
2. McMahon, G., and Erb, U., *Microstr. Sci.*, 17:447 (1989)
3. McMahon, G., and Erb, U., *J. Mat. Sci. Lett.*, 8:865 (1989)
4. Palumbo, G., Thorpe, S. J., and Aust, K. T., *Scripta Metall. Mater.*, 24:1347 (1990)

5. Aust, K. T., Erb, U., and Palumbo, G., in: *Mechanical Properties and Deformation Behaviour of Materials Having Ultrafine Microstructures*, (M. Nastasi, et al., eds.), p. 107, Kluwer Academic Publishers (1993)
6. Erb, U., *Can. Met. Quart*, 34:275 (1995)
7. Gleiter, H., in: *Deformation of Polycrystals: Mechanisms and Microstructures*, Proc. 2nd Risø Int. Symp. on Metallurgy and Materials Science, p. 15, Risø National Laboratory, Roskilde, Denmark (1991)
8. Erb, U., and El-Sherik, A. M., US Patent No. 5,352, 266 (1994)
9. Erb, U., El-Sherik, A. M., Palumbo, G., and Aust, K. T., *Nanostr. Mat.*, 2:383 (1993)
10. Bakonyi, I., Toth-Kadar, E., Tarnoczi, T., Varga, L. K., Cziraki, A., Gerocs, I., and Fogarassy, B., *Nanostr. Mat.*, 3:155 (1993)
11. Bakonyi, I., Toth-Kadar, E., Toth, J., Tarnoczi, T., and Cziraki, A., in: *Processing and Properties of Nanocrystalline Materials*, (C. Suryanarayana, et.al., eds.), p. 465, TMS, Warrendale (1996)
12. Würschum, R., Gruss, S., Gissibl, B., Natter, H., Hempelmann, R., and Schäfer, H. E., *Nanostr. Mat.*, 9:615 (1997)
13. Grimmett, D. L., Ph. D. Thesis, University of California, Los Angeles (1988)
14. Cheung, C., Djuanda, F., Erb, U., and Palumbo, G., *Nanostr. Mat.*, 5:513 (1995)
15. Alfantazi, A. M., El-Sherik, A. M., and Erb, U., *Scripta Metall. Mater.*, 30:1245 (1994)
16. Alfantazi, A. M., and Erb, U., *J. Mat. Sci. Lett.*, 15:1361 (1996)
17. Bryden, K. J., and Ying, J. Y., *Nanostr. Mat.*, 9:485 (1997)
18. Osmola, D., Renaud, E., Erb, U., Wong, L., Palumbo, G., and Aust, K. T., *Mat. Res. Soc. Symp. Proc.,* 286:161 (1993)
19. Cheung, C., Erb, U., and Palumbo, G., *Mat. Sci. Eng. A*, 185:39 (1994)
20. Cheung, C., Nolan, P., and Erb, U., *Mat. Lett.*, 20:135 (1994)
21. Cheung, C., Palumbo, G., and Erb, U., *Scripta Metall. Mater.*, 31:735 (1994)
22. Despic, A. R., and Jovic, V. D., *J. Electrochem., Soc.*, 134:3004 (1987)
23. Lashmore, D. S., and Dariel, M. P., *J. Electrochem. Soc.*, 135:1218 (1988)
24. Tench, D., and White, J., *Metall. Trans.*, 15A:2039 (1994)
25. Yahalom, J., and Zadok, O., *J. Mat. Sci.*, 22:499 (1987)
26. Cohen, U., Koch, F. B., and Sand, R., *J. Electrochem. Soc.*, 130:1987 (1983)

27. Haseeb, A., Blanpain, B., Wouters, G., Celis, J. P., and Roos, J. R., *J. Mat. Sci. Eng. A,* 168:137 (1993)

28. Shirkhanzadeh, M., *Mat. Lett.*, 16:189 (1993)

29. Switzer, J. A., *Nanostr. Mat.*, 1:43 (1992)

30. Cheung, C., Erb, U., and Palumbo, G., *Mat. Sci. Eng.*, A185:39 (1994)

31. Bockris, J. O. M., in: *Modern Aspects of Electrochemistry*, 3:224 (J. O. M. Bockris and B. Conway, eds.), Butterworths, London (1964)

32. Gorbunova, K. M., and Polukarov, Y. M., in: *Advances in Electrochemistry and Electrochemical Engineering*, (W. Tobias , ed.), pp. 249, Interscience, NY (1967)

33. Choo, R. T. C., El-Sherik, A. M., Toguri, J., and Erb, U., *J. Appl. Electrochem.*, 25:384 (1995)

34. Bockris, J. O. M., and Razumney, G. A., *Fundametal Aspects of Electrocrystallization*, p. 27, Plenum Press, NY (1967)

35. Fischer, H., *Angew. Chem. Int. Edit,* 8:108 (1969)

36. Sheppard, K., Smith, D. A., Hentzell, H. T. G., and Ibrahim, A.,in: *Defect Structure, Morphology and Properties of Deposits*, (H. Merchant, ed.), p. 413, TMS, Warrendale, (1995)

37. Dini, J. W., *Electrodeposition*, Noyes Publications, Park Ridge, NJ (1993)

38. El-Sherik, A. M., and Erb, U., *J. Mat. Sci.*, 30:5743 (1995)

39. Erb, U., El-Sherik, A. M., Cheung, C. K. S., and Aus, M., US Patent No. 5,433,797 (1995)

40. Cziraki, A., Fogarassy, B., Gerocs, I., Toth-Kadar, E., and Bakonyi, I., *J. Mat. Sci*, 29:4771 (1994)

41. El-Sherik, A. M., and Erb, U., *Plat. & Surf. Fin.*, p. 85 (1995)

42. Hansen, M., and Anderko, K., *Constitution of Binary Alloys*, p. 1027, McGraw Hill, NY (1958)

43. Wang, S., *Electrochemical Properties of Nanocrystalline Nickel and Nickel-Molybdenum Alloys*, Ph.D. Thesis, Queen's University, Kingston, Ontario, Canada, (1997)

44. Mehta, S. C., Smith, D. A., and Erb, U., *Mat. Sci. Eng.*, A204:227 (1995)

45. Thomas, G. J., Siegel, R. W., and Eastman, J. A., *Scripta Metall. Mater.*, 24:201 (1990)

46. Wunderlich, W., Ishida, Y., and Maurer, R., *Scripta Metall. Mater.*, 24:403 (1990)

47. Haasz, T. R., Aust, K. T., Palumbo, G., El-Sherik, A. M., and Erb, U., *Scripta Metall. Mater.*, 32:423 (1995)

48. Turi, T., *Thermal and Thermodynamic Properties of Fully Dense Nanocrystalline Ni and Ni-Fe Alloys*, Ph.D. Thesis, Queen's University, Kingston, Ontario, Canada (1997)

49. Wang, N., Palumbo, G., Wang, Z., Erb, U., and Aust, K. T., *Scripta Metall. Mater.*, 28:253 (1993)

50. Palumbo, G., Aust, K. T., and Erb, U., *Mat. Sci. Forum*, 225–227:281 (1996)

51. Palumbo, G., Erb, U., and Aust, K. T., *Scripta Metall. Mater.*, 24:2347 (1990)

52. El-Sherik, A. M., Erb, U, Palumbo, G., and Aust, K. T., *Scripta Metall. Mater.*, 27:1185 (1992)

53. Wong, L., Ostrander, D., Erb, U., Palumbo, G., and Aust, K. T., in: *Nanophases and Nanocrystalline Structures*, (R. D. Shull and J. M. Sanchez, eds.) p. 85, TMS, Warrendale (1994)

54. Rofagha, R. and Erb, U., in: *Defect Structure, Morphology and Properties of Deposits*, (H. Merchant, ed.), p. 245, TMS, Warrendale, PA (1995)

55. Wang, N., Wang, Z., Aust, K. T., and Erb, U., *Acta Metall. Mater.*, 43:519 (1995)

56. Alfantazi, A. M., and Erb, U., *Mat. Sci. Eng. A*, 212:123 (1996)

57. Erb, U., Palumbo, G., Zugic, R., and Aust, K. T., in: *Processing and Properties of Nanocrystalline Materials*, (C. Suryanarayana, et al., eds.), p. 93, TMS, Warrendale (1996)

58. Aust, K. T., Erb, U., and Palumbo, G., in: *Processing and Properties of Nanocrystalline Materials* (C. Suryanarayana, et al., eds.), p. 11, TMS, Warrendale (1996)

59. Gonzalez, F., Brennenstuhl, A. M., Palumbo, G., Erb, U., and Lichtenberger, P. C., *Mat. Sci. Forum*, 225–227:831 (1996)

60. Palumbo, G., Gonzalez, F., Brennenstuhl, A. M., Erb, U., Shmayda, W., and Lichtenberger, P. C., *Nanostr. Mat.*, 9:737 (1997)

61. Wang, N., Wang, Z., Aust, K. T., and Erb, U., *Mat. Sci. Eng. A*, 237:150 (1997)

62. El-Sherik, A. M., and Erb, U., *Nickel-Cobalt 97,* in: *Applications and Materials Performance* (F. N. Smith, et al., eds.), IV:257, The Metallurgical Society of CIM, Montreal (1997)

63. Aus, M. J., Szpunar, B., Erb, U., El-Sherik, A. M., Palumbo, G., and Aust, K. T., *J. Appl. Phys.*, 75:3632 (1994)

64. Aus, M. J., Szpunar, B., El-Sherik, A. M., Erb, U., Palumbo, G., and Aust, K. T., *Scripta Metall. Mater.*, 27:1639 (1992)

65. Palumbo, G., Doyle, D. M., El-Sherik, A. M., Erb, U., and Aust, K. T., *Scripta Metall. Mater.*, 25:679 (1991)

66. Doyle, D. M., Palumbo, G., Aust, K. T., El-Sherik, A. M., and Erb, U., *Acta Metall. Mater.*, 43:3027 (1995)

67. Rofagha, R., Langer, R., El-Sherik, A. M., Erb, U., Palumbo, G., and Aust, K. T., *Scripta Metall. Mater.*, 25:2867 (1991)

68. Rofagha, R., Langer, R., El-Sherik, A. M., Erb, U., Palumbo, G., and Aust, K. T., *Mat. Res. Soc. Symp. Proc.*, 238:751 (1992)

69. Rofagha, R., Erb, U., Ostrander, D., Palumbo, G., and Aust, K. T., *Nanostr. Mat.*, 2:1 (1993)

70. Wang, S., Rofagha, R., Roberge, P. R., and Erb, U., *Electrochem. Soc. Proc.*, 95-98:224 (1995)

71. Boylan, K., Ostrander, D., Erb, U., Palumbo, G., and Aust, K. T., *Scripta Metall. Mater.*, 25:2711 (1991)

72. Osmola, D., Nolan, P., Erb, U., Palumbo, G., and Aust, K. T., *Phys. Stat. Sol.*, A131:569 (1992)

73. El-Sherik, A. M., Boylan, K., Erb, U., Palumbo, G., and Aust, K. T., *Mat. Res. Soc. Symp. Proc.*, 238:727 (1992)

74. Klement, U., Erb, U., and Aust, K. T., *Nanostr. Mat.*, 6:581 (1995)

75. Klement, U., Erb, U., El-Sherik, A. M., and Aust, K. T., *Mat. Sci. Eng. A*, 203:177 (1995)

76. Wang, N., Wang, Z., Aust, K. T., and Erb, U., *Acta Mater.*, 45:1655 (1997)

77. Czerwinski, F., Li, H., Megret, M., and Szpunar, J. A., *Scripta Mater.*, 37:1967 (1997)

78. Czerwinski, F., Li, H., Megret, F., Szpunar, J. A., Clark, D. G., and Erb, U., *Mat. Res. Soc. Symp. Proc.*, 451:501 (1997)

79. Turi, T., and Erb, U., *Mat. Sci. Eng. A*, 204:34 (1995)

80. El-Sherik, A. M., Erb, U., Krstic, V., Szpunar, B., Aus, M. J., Palumbo, G., and Aust, K. T., *Mat. Res. Soc. Symp. Proc.,* 286:173 (1993)

81. Krstic, V., Erb, U., and Palumbo, G., *Scripta Metall. Mater.*, 29:1501 (1993)

82. Erb, U., Palumbo, G., Szpunar, B., and Aust, K. T., *Nanostr. Mat.*, 9:261 (1997)

83. Zugic, R., Szpunar, B., Krstic, V., and Erb, U., *Phil. Mag. A*, 75:1041 (1997)

84. Erb, U., El-Sherik, A. M., Palumbo, G., and Aust, K. T., *Nanostr. Mater.*, 2:383 (1993)

85. Aus, M. J., Szpunar, B., Erb, U., Palumbo, G., and Aust, K. T., *Mat. Res. Soc. Symp. Proc.*, 318:39 (1994)

86. Szpunar, B., Erb, U., Aust, K. T., Palumbo, G., and Lewis, L., *Mat. Res. Soc. Symp. Proc.*, 318:477 (1994)

87. Szpunar, B., Zugic, R., Erb, U., and Lewis, L., *Can, Met. Quart.*, 349:281 (1995)
88. Szpunar, B., Erb, U., Palumbo, G , Aust, K. T., and Lewis, L. J., *Phys, Rev. B,* 53:5547 (1996)
89. Weissmüller, J., McMichael, R. D., Barker, J., Brown, H. J., Erb, U., and Shull, R. D., *Mat. Res. Soc. Symp. Proc.*, 457:231 (1997)
90. Szpunar, B., Aus, M. J., Cheung, C., Erb, U., Palumbo, G., and Szpunar, J. A., *J. Magn. Magn. Mat.*, 187:325 (1998)
91. Chokshi, A. H., Rosen, A. H., Karch, J., and Gleiter, H., *Scripta Metall. Mater.*, 23:1679 (1989)
92. Lu, K., Wei, W. D., and Wang, J. T., *Scripta Metall. Mater.*, 24:2319 (1990)
93. Christman, T., and Jain, M., *Scripta Metall. Mater.*, 25:767 (1991)
94. Nieman, G. W., Weertman, J. R., and Siegel, R. W., *Nanostr. Mat.*, 1:185 (1992)
95. Nieman, G. W., Weertman, J. R., and Siegel, R. W., *Scripta Metall. Mater.*, 24:145 (1990)
96. Rabukhin, V. B., *Phys. Met. Metalloved.*, 43:3027 (1995)
97. Lehockey, E. M., Palumbo, G., Aust, K. T., Erb, U., and Lin. P., *Scripta Metall.*, 39:341 (1998)
98. Armstrong, R. W., et al., *Phil Mag.*, A14:943 (1966)
99. Pande, C. S., Masumura, R. A., and Armstrong, R. W., *Nanostr, Mater.*, 2:323(1993)
100. Smith, T. R., et al., *Mat. Res. Soc. Symp. Proc.*, 362:31 (1995)
101. Gleiter, H., *Progr. Mater. Sci.*, 33:224, (1989)
102. Nieman, G. W., Weertman, J. R., and Siegel, R. W., *J. Mat. Res.*, 6:1012 (1991)
103. Diegle, R. B., and Slater, J. E., *Corrosion*, 32:155 (1976)
104. Thorpe, S. J., Ramaswami, B., and Aust, K.T., *J. Electrochem. Soc.*, 135:2162 (1988)
105. Hashimoto, K., Osada, K., Masumoto, T., and Shimodaira, S., *Corrosion Sci.*, 16:71 (1976)
106. Naka, M., Hashimoto, K., and Masumoto, T., *Corrosion*, 36:679 (1980)
107. Turn, J. C., and Latanison, R. M., *Corrosion*, 39:271 (1983)
108. Bragagnolo, P., Waseda, Y., Palumbo, G., and Aust, K. T., *MRS Int. Mtg. Adv. Mat.*, 4:469 (1989)
109. Rofagha, R., Splinter, S. J., Erb, U., and McIntyre, S. N., *Nanostr. Mat.*, 4:69 (1994)

110. Palumbo, G., Intergranular Corrosion in High Purity Nickel, Ph.D. Thesis, University of Toronto, Toronto, Ontario, Canada (1989)
111. Inturi, R. B., and Szklarska-Smialowska, Z., *Corrosion*, 48:398 (1992)
112. Rabukhin, V. B., and Panikarski, A. S., *Phys. Chem. and Mech. of Surfaces*, 5:1304 (1990) 1304
113. Haasz, A. R. A., Aust, K. T., Shmayda, W. T., and Palumbo, G., *Fusion Techn.*, 28:1169 (1995)
114. Gong, W., Li, H., Zhao, Z., and Chen, J., *J. Appl. Phys.*, 69:5119 (1991)
115. Gangopadhyay, S., Hadjipanayis, G. C., Dale, B., Sorensen, C. M., and Klabunde, K. J., *Nanostr. Mat.*, 1:77 (1992)
116. Schaefer, H. E., Kisker, H., Kronmüller, H., and Würschum, R., *Nanostr. Mat.*, 1:523 (1992)
117. Yao, Y. D., Chen, Y. Y., Hsu, C. M., Lin, H. M., Tung, C. Y., Tai, M. F., Wang, D .H., Wu, K. T., and Suo, C. T., *Nanostr. Mat.*, 6:933 (1995)
118. Krill, C. E., Merzoug, F., Krauss, W., and Birringer, R., *Nanostr. Mat.*, 9:455 (1997)
119. Daroczi, L., Beke, D. L., Posgay, G., Zhou, G. F., and Bakker, H., *Nanostr. Mat.*, 2:512 (1993)
120. Daroczi, L., Beke, D. L., Posgay, G., and Kis-Varga, M., *Nanostr. Mat.*, 6:981 (1995)
121. Kisker, H., Gessmann, T., Würschum, R., Kronmüller, H., and Schaefer, H. E., *Nanostr. Mat.*, 6:925(1995)
122. Aus, M. J., Cheung, C., Szpunar, B., Erb, U., and Szpunar, J. A., *J. Mater. Sci. Lett.*, 17:1949 (1998)
123. Kissinger, H. E., *Anal. Chem.*, 29:1702 (1957)
124. Zener, C., private communication to Smith, C. S., *Trans AIME*, 15:175 (1948)
125. Galina, A. V., Fradkov, V. YE., and Shvindlerman, L. V., *Phys. Met. Metalloved.*, 63:1220 (1987)
126. Palumbo, G., and Aust, K. T., *Mater. Sci. Eng.*, A113:139 (1989)
127. Hu, H., in: *Recovery and Recrystallization of Metals*, (H. Himmel, ed.), p. 311, J. Wiley & Sons, NY (1963)
128. Li, J. C. M., *J. Appl. Phys.*, 33:2958 (1962)
129. Harris, K. E., Singh, V. V., and King, A. H., *Acta Mater.*, 46:2623 (1998)
130. Gertsman, V. Y., and Birringer, R., *Scripta Metall. Mater.*, 30:577 (1994)
131. Kaur, I., Gust, W., and Kozma, L., eds., *Handbook of Grain and Interface Boundary Diffusion Data*, 2:1037, Ziegler Press, Stuttgart (1989)

132. Palumbo, G., and Aust, K. T., *Grain Growth in Polycrystalline Materials,* (H. Weiland, B. L. Adams, and A. D. Rollett, eds.), 3:311, TMS, Pittsburgh, PA (1998)

133. Rupp, J., and Birringer, R., *Phys. Rev. B*, 36:7888 (1987)

134. Lu, K., Wang, J. T., and Wei, W. D., *J. Phys. D*, 25:808 (1992)

135. Wagner, M., *Phys. Rev. B*, 45:376 (1992)

136. Fecht, H. J., *Phys. Rev. Lett.*, 65:610 (1990)

137. Gleiter, H., presented at 2nd Int. Conf. on Nanostructured Materials, Stuttgart (Oct. 1994)

138. Palumbo, G., Lichtenberger, P. C., Gonzalez, F., and Brennenstuhl, A. M., US Patents: 5,527,445 (1996); 5,516,415 (1996); 5,538,615 (1996)

139. ASME Code Case 96-189-BC96-206 Case N-569; Section XI, Division 1; Alternative Rules for Repair by Electrochemical Deposition of Class 1 and 2 Steam Generator Tubing (1996)

140. Palumbo, G., King, P. J., Aust, K. T., Erb, U., and Lichtenberger, P. C., *Scripta Metall.*, 25:1775 (1991)

141. Palumbo, G., Lehockey, E. M., Lin, P., Erb, U., and Aust, K. T., *Mat. Res. Soc. Symp.Proc*. 458:273 (1997)

142. McCrea, J., M.A.Sc. Thesis, Department of Metallurgy and Materials Science, University of Toronto (2000)

143. Merchant, H. K., in: *Defect Structure, Morphology and Properties of Deposits*, (H. D. Merchant, ed.), p. 1, TMS, Warrendale (1995)

6

Computer Simulation of Nanomaterials

Philip C. Clapp

1.0 INTRODUCTION

In the last decade or so, the field of computer simulation of nanomaterials has advanced extremely rapidly for several reasons. The first is the tremendous increase in computing power (along with simulation techniques) that has occurred in this time span, making the simulation of nanoscale arrays possible. The second is that atomic scale simulations (such as Molecular Dynamics) are ideally suited to explore nanoscale phenomena at a time when experimental exploration and theoretical understanding of nanomaterials has become of intense interest. In fact, several regularly published journals are now devoted to articles illustrating simulations of new material properties spanning length scales from the sub-atomic to the macroscopic.[1] Because publications in this field are appearing at an exponentially increasing rate, a review such as this can only offer a very blurry snapshot of the state of progress, and can only provide a starting point for the reader to obtain an overview of the subject.

One quite important recent development in the art of computer simulation has been the introduction of several hybrid techniques for extending the spatial scale of simulations from the nanometer to the micron and above. This is highly desirable in situations where some areas of the simulation require precise resolution at the atomic level, while other areas should be treated at micron or even millimeter scale resolution, such as is encountered in problems containing singularities in elastic continuum treatments. One example of this is in the simulation of crack propagation. Very near the crack tip it is necessary to model the atomic behavior very accurately, but at distances of microns or millimeters from the crack tip, one must have the elastic displacement field accurately represented, and knowing the momentary position of individual atoms is not so important. Another example is in modeling one or more dislocations, where again it may be necessary to simulate the structure of the cores with atomic resolution, but the intermediate regions between dislocation cores with a coarser grained focus.

One of the first hybrid methods developed[2] to attack this kind of problem uses a network where the nodes near the elastic singularity are individual atoms whose movements are modeled with a conventional Molecular Dynamics routine, but the nodes far from the singularity are treated as lumped masses representing the behavior of the center of mass of a large region of atoms. An intermediate network is also required to provide for the transition between the near-field and far-field zones, and the best way to model this region remains a thorny problem—for example, it is difficult to adjust conditions so that dislocations can pass smoothly from the atomic region into the far-field region.

Another technique of simulating across different spatial scales has been to extend the finite element method to a quasi-continuous multiple scale analysis by refining the mesh near singularities to near atomic scale, but not finer, by using the intraatomic distance as a lower cut-off.[3] In the coarse grain limit, the mesh behaves as a nonlinear elastic medium, and in the finest grain limit, the calculations become equivalent to lattice statics. An important departure from usual finite element treatments is that, rather than using ad hoc constitutive relations to calculate the energy and forces on a node, these quantities are instead determined by carrying out a full atomic energy calculation using semi-empirical quantum mechanical or ab initio quantum mechanical interactions based on the strain condition and environment of the node in question. Thus far, the method has been applied to quasi-static calculations of the plastic deformation and dislocation emission occurring as a nanoindenter is pressed into a metal single crystal, for example. But the true kinetics of the events are not correctly represented, since they might be in a large-scale fully-atomistic Molecular Dynamics simulation, for instance.

A third approach[4] has developed from the opposite direction by extending atomistic calculations upwards so that each node now represents some small cubical volume of atoms, whose dimensions can be chosen arbitrarily as long as it is some multiple of the unit cell parameter. Again, a quasi-continuum is assumed, and the constitutive relations are determined, in this case, by a Ginzberg-Landau strain-free energy function which may include terms up to the fourth order in the elastic constants, as well as strain-gradient terms. The strain-gradient terms provide the effect of the varying environment on a given cell and, in particular, provide the energies associated with discontinuities such as interfaces or free surfaces. The Ginzberg-Landau strain-free energy function is determined for a chosen material via a lattice statics calculation using atomistic interactions of any desired degree of accuracy.

Including terms up to the fourth order in elastic constants allows the modeling of highly non-linear elastic media and, in particular, permits the modeling of materials that may undergo a phase transformation involving volume and shape changes. A quasi-static calculation, based on some set of imposed boundary conditions, is carried out using a Monte Carlo simulation of the displacement field. Although a changing pattern of displacements evolves in the simulation, there is not a true time scale because the activation energies are not sufficiently accurate, but the final equilibrium state should be reasonably correct. In fact, this approach was developed principally to model transformation toughening processes[5] wherein regions ahead of an advancing crack tip might alter their strain states by undergoing a martensitic transformation, but the method can be applied to many other types of problems, with the caveat that the time scaling is not dependable. Thus, the problem of simulation across both different space scales and time scales remains to be solved.

Several necessary limits have been observed in this survey. First, although there are many excellent "first principles" quantum mechanical calculations and simulations that are becoming increasingly important for understanding nanoscale phenomena, the simulations themselves (due to current computer limitations) still contain less than a hundred atoms typically and, therefore, are not yet of sufficient size to be regarded as "nanoscale." No study of this type has been explicitly included in this review, although many will be found in references within papers that are covered here. Second, the entire area of polymer material nanoscale simulations, although of great interest and considerable fascination, has not been attempted here because it is a subject peripheral to the main material topics of this volume and, in truth, would deserve an entire review in its own right. A third topic, which has not been covered here, is the numerous atomistic simulations of fracture, except as they apply to nanograin materials, because to be accurate, the simulations must include mesoscopic and even macroscopic scale effects. There is currently much progress in the direction of hybrid simulation techniques which can span large ranges of scale, as briefly described above, but this topic is best left for a future review. For much the same reason, simulations of individual or multiple dislocations have not been discussed here.

This review begins with studies of the stability of individual isolated nanoparticles, then moves on to a variety of subjects related to interface properties in nanosystems, and finally covers simulations of fully three-dimensional systems of nanograin materials.

2.0 NANOPARTICLES

2.1 Phase Stability (Liquid, Amorphous, and Crystalline)

The relative stability of different phases (e.g., crystalline, amorphous, and liquid) can change in dramatic and non-intuitive ways as particles are reduced to nanoscale dimensions. For instance, it has been found in computer simulations[6][7] that an isolated crystalline particle of Cu or Au will "melt" when it is slowly heated from room temperature to a temperature twenty or thirty percent below the bulk crystalline melting point, if it is sufficiently small (i.e., a few nanometers in diameter). This phenomenon has also been carefully examined and confirmed experimentally in the case of Pb nanocrystals by means of x-ray powder diffraction in ultrahigh vacuum.[8] This instability can be understood from the classical nucleation theory[9] which gives the radius, R_c, of a critical sized nucleus of crystalline solid, immersed in a super-cooled liquid as $R_c \sim \sigma T_m / L \Delta T$, where these symbols refer to:

σ = surface energy/unit area

T_m = bulk melting temperature

L = latent heat of fusion

ΔL = degrees of supercooling

Since for most metals, $\sigma/L \sim 0.46$ (in units of atomic volume)$^{1/3}$ for a supercooling of twenty percent, the critical radius is of the order of 1 nm.

One only has to imagine that the free surface of the isolated particle is sufficiently irregular to approximate the structure of a very thin surrounding film of liquid to complete the analogy and conclude that the undercooling at which isolated nanoparticles will melt is inversely proportional to their radius. This is equally true if, instead of isolated nanoparticles, one is dealing with nanoparticle agglomerations (as in sintering) or with nanograin solids, which will result in the melting of the very smallest particles (or grains) at quite low temperatures (with subsequent rapid recrystallization generated by the larger neighbors). This phenomenon can have a very significant impact on sintering (and/or grain coarsening) kinetics in nanoscale assemblies and has been the subject of some recent extensive computer simulation work.[10]–[14]

A detailed Molecular Dynamics examination of the relative stability and configuration of the liquid and amorphous states of silicon has been carried out using two- and three-body Stillinger-Weber potentials in a system of 588 atoms using periodic boundary conditions.[15][16] The study found that by using a sufficiently slow process of cooling from the melt, while simultaneously increasing the strength of the three-body interaction, an amorphous silicon state could be achieved which agreed very well with experimental data for the static structure factor and phonon density of states. Upon heating, the amorphous state was seen to melt via a first-order transition about 230 K below the crystalline-liquid melting temperature, also in excellent agreement with experimental values.

2.2 Surface Properties

The surface free energy of spherical crystallites (Ag, Au, Cu and Pt) was studied[17][18] as a function of radius (ranging from 3 a_0 to 14 a_0) and temperature (from 0 K to 1000 K) using a technique called the *free energy simulation method*. It is based on an approximate free-energy functional, which uses a static lattice energy (computed from many body potentials) combined with a local harmonic approximation for the vibrational entropy contribution.[19] It was found that the surface free energy measured in this way, when plotted against the inverse cluster radius, extrapolated to a well-defined value, which could be interpreted as the surface free energy of an indefinitely large equiaxed particle averaged over all crystalline orientations. This could also be translated as the surface free energy for a macroscopic polycrystalline sample without texture and, hence, of considerable practical use. Another interesting result of the study was that this orientationally averaged surface free energy came out to invariably be within a percent of the {110} surface free energy (for the fcc metals studied), which led the authors to suggest that this number, if available either experimentally or from computation, would be a useful substitute.

3.0 NANOCONTACTS

3.1 Adhesion

A good overview[20] of the physics and chemistry of adhesion, substantially based on computer simulation studies, examines the atomic processes operating at the contact interface between unlike as well as like materials. Oxide-metal, polyimide-metal, and metal-metal contacts are discussed. A point the authors[20] make is that long range Coulomb forces, and not just short range "bonding" forces, can dominate in adhesion. This leads to the idea that in metal-ionic material contacts, adhesion could be substantially affected by controlling charged defect populations. Another interesting point is that the force of adhesion is always reduced by the insertion, at the contact zone, of a film of a material different than either of the contacting materials, even when the new interlayer material has stronger bonds than the other two. This is explained as due to an inevitable reduction of the neck radius during pull-off, as demonstrated in a number of simulations, thus reducing the maximum tensile force required for fracture.

A further point that is made in this overview is the importance of statistical roughness factors in the mating surfaces, and the necessity of being able to model the influence of these geometrical properties at different length scales on adhesion. A more recent paper by another group has also emphasized the importance of misfit effects in predicting the degree of adhesion.[21]

A small number of Molecular Dynamics simulations have studied the details of a nanometer size metallic tip approaching, adhering to, and being withdrawn from a substrate of the same or different material at the atom by atom level. In one of these simulations,[22] the plastic deformation of a metallic single crystal (fcc Lennard-Jonesium) atomic force microscope tip, as it was brought into contact with a single crystal substrate of the same material (oriented identically) and subsequently pulled off, was studied. The temperature was maintained at about $0.3T_m$ (30% of the melting point). Elastic shear strains of up to 8% (implying near-theoretical shear strength) were observed prior to the sudden penetration of the tip into the surface with concomitant inelastic flow. The inelastic flow processes were not analyzed but were presumed to be some combination of local diffusion, dislocation motion, and block slip.

A form of sintering was observed, preceded by individual atoms jumping across the gap from the tip to the substrate as the tip got within range of the atomic attractions from the substrate. A localized soft phonon process was suggested for this pre-sintering effect. That sintering actually occurred was shown by the fact that material from the tip remained behind on the substrate when the tip was pulled back. Quite similar results were obtained in Molecular Dynamics studies of analogous configurations by other workers[23] who, however, used more realistic potentials to simulate a nickel tip touching down on a gold substrate.

Another simulation of this general type was performed using the covalent material, silicon, as both the tip and the substrate.[24] A well-tested Stillinger-Weber semi-empirical potential was used for the silicon interactions and, although most of the large plastic deformation effects observed in the metallic systems were not seen here (as might be expected), some defects were found to be injected into the substrate by the stress gradients caused by the proximity of the tip.

3.2 Friction

Understanding friction at the atomic level involves the complex process of separating the role of structure, dynamics, energy factors, and transport mechanisms in terms of their relative importance in the total result. As a consequence, until the advent of computer simulations, most theoretical attempts to understand friction were based on approximating all the parts of the tribological system as continuous media. However, with the advances in the past decade of computer simulation techniques, and especially the development of reliable semi-empirical interatomic potentials, the use of Molecular Dynamics in examining tribological problems has provided considerable insight into the microscopic mechanisms involved in friction.

In some cases, the Molecular Dynamics simulations,[25][26] agree substantially with predictions made from continuum analyses, but in other cases, the atomistic simulations have revealed unsuspected and surprising mechanisms underlying familiar tribological phenomena. In fact, these atomic level pictures have often shown that the physical behavior at tribological interfaces can be remarkably different from the behavior of the same materials in bulk. For instance, phenomena which show up at the atomic level include, but are not limited to, dramatic interfacial wetting effects,[27][28] atomic-scale neck constructions,[29][30] stick-slip friction with atomic periodicities,[31]–[34] and substantial configurational phase changes at the sliding interface.[35]–[39]

An auspicious marriage has recently occurred between the extremely high-resolution experimental methods of studying friction (i.e., atomic force microscopy, scanning tunneling microscopy, and quartz micro-balance techniques) and atomic level simulations. The synergistic interaction between simulations[40] and experiments, both with essentially atomic scale resolution, has provided a number of important new insights. For instance, stick-slip phenomena previously were commonly explained by continuum analysis in terms of a negative slope in the friction-velocity characteristic, but now these same phenomena are seen to be a result of alternating bond shearing and reformation processes[41]–[45] which, in turn, cause the negative friction-velocity characteristic. Another important example is the discovery of configurational phase changes[46] (i.e., crystalline–amorphous, martensitic, etc.), which substantially alter the physical and chemical properties of the tribological interfaces, and yet, because they are reversible, leave no trace of their existence in a macroscopic examination after stress is removed.

Several excellent reviews of many of these advances are available in the literature,[47][48] and make a good introduction to the subject prior to consulting the more specialized papers cited above.

3.3 Electrical Conductance

Sutton[49] and co-workers,[50]–[53] carried out an extensive series of Molecular Dynamics simulations studying the process of an atomically sharp scanning tunneling microscope (STM) tip coming in contact with a conductive substrate. Their papers address the physical origin of jumps in the electronic conductance that have been observed experimentally when an STM tip is brought into contact with a substrate and subsequently pulled off. It was shown, through a combination of Molecular Dynamics simulations and tight binding conductivity calculations, that the jumps were due to mechanical instabilities producing abrupt substantial structural rearrangements within the tip-surface contact, and not due to conductance quantizations as had been previously claimed by other investigators. This is perhaps an excellent illustrative example of the synergistic power of using computer simulations, in conjunction with both experimental and theoretical investigations, to completely analyze a complex phenomenon.

4.0 NANOFILMS

4.1 Formation: General Methods

In a series of three papers,[54]–[56] a general method has been proposed for modeling the synthesis of various thin film materials from molecular precursors. The method uses a tight-binding quantum-mechanical approach which is not too expensive in computational time, but still fairly accurate, especially for small hydrocarbons. Some of the examples used for illustration include Molecular Dynamics simulations of the gas phase polymerization of ethene to polythene, using a butyl radical to start the process off, as well as the simulated formation of diamond-like carbon thin films from organic molecular precursors.

4.2 Formation: Liquid Droplet and Cluster Beam Deposition

Molecular Dynamics simulations of either liquid or solid nanoclusters of atoms landing on a solid crystalline substrate show quite similar behavior, essentially because even a solid nanocluster is seen to "melt" just prior to impact due to the strong interactions with the approaching substrate. In studies of molten Cu droplets impinging on a cold Cu substrate at different angles and different energies, a variety of spreading behaviors were observed.[57] At the start, the molten droplet, equilibrated at 1500 K, was placed on the surface of the substrate at 10 K. The spreading spontaneously occurred under the attractive atomic forces between the molten droplet and substrate atoms. The snapshots of the spreading and solidification processes, as well as the dynamic spreading behavior described by a dimensionless spreading index, are quite analogous to those observed in recently reported experiments. A multiply-twinned crystal structure was formed when the droplet was completely solidified. It was also shown that the contact line atoms are quickly solidified and trapped, but spreading can persist via downward motion of the unsolidified atoms to the outside neighbor positions of the frozen contact line, which suggests that a no-slip boundary condition may apply to the spreading of a molten droplet on a cold substrate because of simultaneous rapid solidification.

A sizable number of simulations have been carried out by different groups to emulate the process of low-energy cluster beam deposition.[58]–[62] Most of this work has concentrated on silicon cluster deposition on silicon single crystal substrates due to its obvious commercial importance in the semiconductor industry. The principle features studied have been the efficiency of spreading and frequency of defect formation as functions of substrate temperature, cluster temperature, cluster size, and cluster impingement velocity in order to try to predict the optimum conditions for perfect film growth.

Although quite interesting and suggestive results have emerged from these studies, there is a concern (often expressed by the authors themselves) that Molecular Dynamics simulations will not adequately measure the important role that surface diffusion and atomic rearrangement are likely to play in a real situation. This is because the time span of any Molecular Dynamics run is still of the order of nanoseconds and, due to this limitation, the bombardment rates studied tend to be unrealistically high and the time for redistribution of atoms between bombardments, unrealistically short.

One quite promising innovation, recently introduced to deal with this problem, utilizes a hybrid approach combining Monte Carlo and Molecular Dynamics techniques, resulting in a very substantial lengthening of the time scale represented by the full simulation.[63] However, so far this technique has been used only for single atom deposition processes, but apparently could be applied to cluster deposition as well.

4.3 Formation: Vapor and Molecular Beam Deposition

A large number of studies have been carried out, mostly by the Molecular Dynamics approach, to simulate the atom-by-atom formation of thin films, e.g., via chemical vapor deposition (CVD), molecular beam epitaxy (MBE), or related techniques.[64][65] Many of these have been specifically addressed to diamond[66]–[70] or silicon[71][72] nanofilm deposition, due partly to the enormous practical interest in finding optimum fabrication conditions, but also due to the convenient availability of accurate semi-empirical short range potentials for these materials.

4.4 Mechanical Instabilities and Defects in Thin Films

A considerable amount of effort has been invested in simulations to understand, in particular, the origin of various types of defects that may occur during the production of thin films, and also to understand the causes of diverse mechanical instabilities, which can also adversely affect the performance of thin film devices. Some simulations[73][74] have concentrated on the mechanisms for accidental void formation, such as shadowing effects during deposition. Others[75][76] have studied the defects that may be generated by the increasing epitaxial stress that develops during film growth, as well as the effect of this stress on grain morphology. The very important question of whether surface instabilities driven by capillary forces will cause the surface morphology of a growing thin film to become chaotically rough has also been given a certain amount of study.[78]–[80]

4.5 Chemical Instabilities and Phase Separation

Several different approaches have been used to simulate the various stages of phase separation of a homogeneous disordered alloy into a two-phase mixture. The evolving microstructures have been calculated by using either continuum mean field approaches or Monte Carlo simulations and, since almost all the work has been limited to two dimensions, they are more correctly regarded as nanofilm studies. In principle, both approaches can be used in fully three-dimensional simulations (and this work is beginning to appear) but the calculations become an order of magnitude more time consuming and are much more difficult to analyze visually.

The continuum mean field model was pioneered by Khachaturyan[81] and has been applied by him and his coworkers to a number of interesting cases.[82][83] In this approach, it is assumed that a free-energy functional exists which can be evaluated for both equilibrium and non-equilibrium states of a system and, furthermore, that the time evolution of the local composition order parameter is linearly proportional to the local thermodynamic driving force (determined from the free-energy functional). What emerges in this modeling is a sequence of quite realistic looking "micrographs" of successive coarsening stages whose morphologies and size distributions can be compared favorably with real micrographs showing experimental coarsening sequences.

However, a number of authors have criticized this approach on the grounds that there is no inherent length or time scale in the model so that the "micrographs" cannot be truly compared with real ones, in addition to the obvious disparity between the dimensions of the simulations and the real samples in the laboratory.

Another criticism is that although the basic assumptions of the mean field model are probably reasonably valid close to equilibrium, most of the exercises carried out with the model start in states quite far from equilibrium. This problem has been addressed, to some degree, by several authors[84]–[87] and the whole issue has been nicely reviewed recently.[88]

A further problem with the mean field model is that the mechanisms of atomic diffusion are not incorporated in any way, nor are the effects of dynamical fluctuations included. A comparative analysis of the importance of these factors has been looked at in some detail by Rautiainen and Sutton[89] by contrasting the results of a stochastic Monte Carlo model and the mean field model. To quote from the abstract to their paper:

> "We have carried out a comparative study of phase separation and subsequent coarsening of the microstructure in a two dimensional atomistic model system using two approaches: a stochastic Monte Carlo model and a deterministic mean field model. The differences between these approaches in the microstructural morphology, coarsening mechanisms, and kinetics are discussed. As a main result, we have found that using a realistic diffusion mechanism of vacancy motion in a Monte Carlo model produces a variety of coarsening mechanisms over a range of temperatures, which is reflected in the corresponding kinetic behavior. At low and intermediate temperatures coarsening proceeds through Brownian motion and coalescence of smaller particles, but at higher temperatures the Lifshitz-Slyozow-Wagner evaporation of smaller particles and growth of larger particles dominates. By contrast, Brownian motion of particles is not observed in a microscopic mean field model, either with or without an environment dependence in the mobility, nor is it observed within a Monte Carlo model in which diffusion is affected by direct atomic exchange (Kawasaki dynamics). The

time scale associated with microscopic mean field models is discussed critically."

Some of these difficulties can be overcome by using various types of Monte Carlo simulations, and quite a bit of work has been carried out along these lines. If the Monte Carlo simulation is performed atomistically, then the ordinary method (starting from some chosen initial pattern) is to try a series of atomic interchanges with the change permitted if the energy of the array decreases. But if the change produces an energy increase it is allowed only with a probability given by a Boltzmann factor expression according to the "temperature" that is assumed for the array. This process, if carried far enough, should lead to the lowest chemical-free energy state, if the accident of the chosen starting state does not lead to some metastable configuration, which is always a concern. However, there is no natural time-scale in this type of simulation, so it is questionable to try to say that the sequences generated by the computer are related in any significant way with the kinetics of the actual phase separation process.

One way that has been proposed to bring a real time measure into this type of modeling is to mimic the actual diffusion process by inserting vacancies into the array and allowing atomic movements only via vacancy jumps.[90] Then a reasonably realistic time measure enters the picture through the average time that is assumed for a vacancy to make a jump, hopefully based on some good experimental data or fundamental calculation. Even so, significant complications remain in that the actual jump time will generally depend on the details of the chemical configuration in the immediate vicinity of the vacancy, as well as the effect of any strain fields that might be present. To varying degrees, these factors have been included in the more advanced Monte Carlo calculations, which can then be said to present a reasonable approximation to the kinetics of a phase separation.

The Monte Carlo approach can be extended to model systems at the mesoscopic scale as well by regarding each "site" in the array as representative of a region of many atoms, rather than just one. Then, a parameter associated with each site will give the measure of, say, average chemical composition in that region if it is phase separation that one desires to model; or average grain orientation if it is grain growth to be modeled.

One of the more interesting examples of this type was carried out to study the effect of two chemically different phases on the evolution of grain structure in a nanofilm.[91] It was imagined that a continuum microstructure was mapped onto a two-dimensional discrete triangular lattice. In this case, the site parameter, S, represented both the phase composition (either pure α or pure β) by its arithmetic sign, and the average grain orientation by a number between 1 and 50. Ad hoc interaction energies were then assigned between "nearest neighbor" and "next nearest neighbor" sites to account for both grain boundary energies and interphase energies in an essentially Ising Model Hamiltonian, which provided the energy of any imaginable configuration. It had been predicted[92] that stable four-grain junctions could exist in a two-phase polycrystalline array, as compared to the single-phase two dimensional situation where it is known that only three-grain junctions are stable, and these simulations were carried out to explore the consequences of this difference on grain evolution in nanofilms. The authors found that structures having four-grain junctions would eventually stop coarsening, as compared with structures having only three-grain junctions which continue to coarsen indefinitely. This, of course, could lead to quite significant practical differences since it is often very desirable to preserve extremely fine grain structure in nanomaterials.

There are whole hosts of other simulations on similar topics, utilizing closely related Monte Carlo models which have been pioneered by Srolovitz, among others, to which the reader is referred.[93]–[97] However, if the primarily requirement is a determination of just the final equilibrium state, then the Free Energy Minimization simulation method[98] produces results[99] for this purpose, at least as good, if not better, than any of the Monte Carlo approaches.

4.6 Free Surfaces

The properties of the free surface of a macroscopic sample, to the extent that long range interactions with the interior are unimportant, can be considered analogous to that of a nanofilm having one side free. This, in fact, is the configuration most often used in computer simulations studying the properties of free surfaces of bulk materials. One of the interesting questions addressed with this technique has been an examination of the elastic field of surface line steps, and the interaction between such steps. Theoretical calculations predicted that a single step would have the elastic displacement field of a surface force dipole, and that the interaction energy would be inversely proportional to the square of the distance between parallel steps.[100][101] However, it was very difficult to calculate analytically what the strength of this dipole was, or even whether it had in-plane components in addition to components perpendicular to the surface plane. Several computer studies[102][103] were made addressing these questions, which helped the analysis considerably. One approach[104] (using EAM potentials) resolved these questions very nicely by setting up arrays of Ni or Au atoms with stepped surfaces of varying periodicities, and then used a conjugate gradient method to determine the ground state energy, displacement fields, and surface stresses in each case. The thickness of the array perpendicular to the surface was increased until there was no significant change in the simulated outcome. By comparing these results with the analytical expressions, the range of validity of the theoretical approximations could be determined and the magnitude of the dipole components estimated, which were found to vary according to the degree of anisotropy. This work is a good example of the very beneficial synergy that can exist between analytical theory and computer simulation.

A second area of interest has been the problem of chemical segregation to the free surface,[105]–[108] which has been generally studied via Monte Carlo modeling with some success. An interesting departure, which has allowed a fairly accurate inclusion of vibrational entropy contributions (usually the bête noir of order-disorder calculations) to the free energy of the system, has been called the "free-energy simulation method,"[109] and has been used to study a variety of surface segregation situations in different materials.[110]–[112] It is based on an approximate free-energy functional, which, besides a local harmonic approximation for the vibrational entropy, uses an averaged composition "mean field" value for the chemical identity of each site, along with the mean field (regular solution) approximation for the configurational entropy. Then, an appropriately chemically averaged many-body Embedded Atom Method potential is used between the atoms.

The method claims to be much more computer-time efficient than comparable Monte Carlo studies, because, in this procedure, only a straightforward minimization of the free-energy functional, with respect to position and composition of each atomic site, is required. Temperature only appears as a parameter to be set in the free-energy functional. Probably the weakest elements in this approach are the mean field approximations, which are well known in Ising model studies to lead to false phase transformations in quasi one-dimensional lattices. Since typically the atomic array used to model surface segregation is made periodic (with a short period) in the two directions parallel to the surface, and the composition within each plane parallel to the surface is uniform, this is potentially a quasi one-dimensional lattice case, and perhaps the results need to be checked with a more accurate expression for the configurational energy and entropy.

5.0 NANOGRAIN MATERIALS

It is virtually impossible to discuss the properties and behavior of nanograin materials without a detailed consideration of the nature of the grain boundaries which hold them together. This is for the obvious reason that the grain boundaries now represent a very substantial volume fraction of the whole and, thus, may be expected to play a dominant role in the overall behavior of the material. Consequently, it is important to begin by reviewing some of the very extensive body of work aimed at simulating the characteristics of grain boundaries. This has been a very fruitful field for simulation because it is extremely difficult to determine the properties of individual grain boundaries experimentally and, thus, the detailed knowledge that atomistic simulations can provide has been invaluable. Following the review of grain boundary simulations in Secs. 5.1 and 5.2, a discussion of the computer modeling that has been done on "bulk" nanograin materials is presented in Secs. 5.3–5.6.

5.1 Grain Boundary Structure and Energy

Generally, the prime objectives of the simulations have been to determine the atomic structure of the grain boundary and the energy that can be associated with it. The energy naturally relates to the ease of fracture of the boundary and has been a subject of intense interest over the last several decades. The studies of grain boundaries (without segregation effects) are reviewed in this section. It has long been known that segregation of some chemical elements to grain boundaries can substantially affect the resistance to fracture, and simulations examining this question are covered in Sec. 5.2. An excellent overview of most of our current knowledge of grain boundaries (as well as other types of interfaces), as derived from simulation, experiment, and theory is available in the book by Sutton and Balluffi.[113]

The usual method employed in modeling grain boundaries is to choose relatively simple misalignments of the crystalline arrays on either side of the grain boundary so that the grain boundary itself will have a two dimensional unit cell with an area of no more than a square nanometer or so. Then, the atomic arrays on either side of the boundary are extended for perhaps several nanometers normal to the interface and ended with either periodic boundary conditions (thus creating a second grain boundary) or with free surfaces. Periodic boundary conditions are used in the two directions parallel to the grain boundary, typically. Thus, arrays of this kind contain only a few thousand atoms, which is a convenient size for performing extensive Molecular Dynamics or Molecular Statics calculations, and a large range of materials and grain boundary orientations have been studied in just this way, a small selection of which may be cited.[114]–[122]

However, the requirement to keep the unit cell area of the grain boundary at a reasonable size in the simulations has meant that high angle or irrational grain boundary orientations (which probably occur most frequently in real materials) have not been explored in this approach, and analytic methods (assuming pair potentials) have been developed by Sutton[123][124] to fill in this important area of knowledge.

In this calculation method, the cleavage energies and expansions of general grain boundaries were analyzed within the framework of a pair potential model of atomic interactions, and Molecular Dynamics simulations were performed to study the effect of localized atomic relaxations near the boundary on the energy. By using a decomposition in terms of layer interaction energies it was found that the effect of periodicity in the boundary plane was to provide additional terms in the boundary energy that are absent when the boundary is incommensurate.

Wolf[125] proposed a rule, suggested by simulations, that the lowest energy boundary (of all boundaries sharing the same Miller indices) is the periodic boundary with the smallest area unit cell. This rule has been tested by others and generally found to be true.[126] However, as pointed out by Sutton,[127] since a large incommensurate energy term can come into the calculation when the boundary plane changes, this rule does not necessarily hold for grain boundaries with differing plane normals.

In addition to the considerable body of work on the structure and energy of grain boundaries, as revealed by simulation, Vitek and co-workers have studied the elastic properties (including phonon characteristics) of grain boundaries and found that there can be a significant population of localized phonons in their close proximity.[128]–[131]

5.2 Grain Boundary Segregation Effects

The fracture resistance of polycrystalline materials often critically depends on the degree of segregation of various solutes to the grain boundaries and, yet, the trace amounts of impurities and their effects on the binding and structure of the boundaries are extremely hard to detect experimentally. As a result, much of the advance in understanding of this phenomenon has come from simulations. A number of excellent studies are in the literature,[132]–[142] and it may be useful to provide the details of several as examples of possible procedures.

As an illustration when the segregant does not change the crystallography of the boundary (e.g., Ag in Cu), Menyhard, Yan, and Vitek combined Molecular Statics followed by Monte Carlo methods.[143] First they constructed the boundary geometrically for pure Cu, using the coincidence site lattice (CSL) method and taking the simplest cases of tilt boundaries (i.e., $\Sigma = 3$ and 5 in CSL notation). This configuration was then relaxed via Molecular Statics (or Molecular Dynamics at ~0 K) with regard to relative rigid body displacements of the adjoining grains, as well as individual atom positions.

The enthalpy of segregation (at 0 K) was then determined by substituting a Ag atom at various sites in the grain boundary zone, again allowing relaxation, and comparing the energy of the array to that when the Ag atom was substituted at a bulk site instead. Once the most favorable site for segregation was found, the next most favorable was sought, keeping a Ag atom in the first, etc. In this way, the enthalpy of segregation was found as a function of the concentration of the segregant at the boundary.

To model the segregation pattern at finite temperatures, the authors used a Monte Carlo method in which the relative number of atoms of each species is allowed to vary, but the total number remains constant. Each Monte Carlo step involved both variations of atomic position and species, and began from the equilibrium configuration obtained in the static (T ~0 K) relaxation. Thus, both configurational and chemical entropy terms could contribute to the final stable arrangement.

One interesting result from their study was that, at high temperatures and/or segregate concentrations, the interactions between segregation sites was quite significant, and that segregation occurred to varying degrees over a zone extending as much as five or more lattice spacings away from the boundary. A substantial range of segregation enthalpies was found, depending on site location, leading to the conclusion that comparison with simple thermodynamic theories of segregation employing only a single enthalpy value was problematical.

As another illustration when the segregant does change the grain boundary structure, Siegl, Yan, and Vitek[144] modeled the case of Cu-Bi alloys using Finnis-Sinclair many-body semi-empirical potentials, and compared the results with ab initio quantum mechanical calculations and high resolution electron microscopy (HREM). The simulations predicted that an ordered Cu-Bi double layer would form at coherent twin boundaries and cause a characteristic faceting if the Bi concentration was high enough. Both the presence and the structure of the predicted double layer were confirmed by HREM. When the question of whether this double layer structure could be grown into a three dimensional intermetallic compound under pressure was posed, both the semi-empirical potential calculations and the ab initio work agreed that it could not and, therefore, was a unique two-dimensional structure requiring the grain boundary environment to provide stability.

5.3 Sintering

Theoretically, nanograined materials offer many advantages over micron grain size solids, and one of the more promising routes to produce such structures is via the sintering of nanoparticles. However, rather little is known about the kinetics of such a process and the question arises as to whether it is safe to extrapolate the classical theories of sintering which are reasonably useful for micron powders to the regime of nanopowders. There have been relatively few simulations of nanoscale sintering,[145]–[151] but the results are quite startling and provocative.

In standard sintering theory,[152] six distinct mechanisms are thought to contribute to the sintering of crystalline particles of conventional sizes:

1. Surface diffusion
2. Lattice diffusion from the surface
3. Vapor transport
4. Grain boundary diffusion
5. Lattice diffusion from the grain boundary
6. Lattice diffusion through dislocations

However, at the level of nanoscale processes, due to the high degree of surface curvatures and the fact that atomic forces are now comparable to the mass of an individual particle, it may well be that sintering mechanisms are radically different.

In one set of studies,[153] Molecular Dynamics techniques were used to simulate Cu and Au nanoparticle arrays at different temperatures to study surface energies, grain boundary mobility, and sintering. The results on multi-particle arrays several hundred degrees below the melting point (T_m) showed unexpectedly strong contributions from plastic deformation, mechanical rotations, amorphization, and ultra-rapid atomic force driven diffusion effects. Evidence was offered that the basic processes and kinetics of nanoscale sintering are fundamentally different from those of normal micron-scale sintering. For instance, only two of the six classical mechanisms for matter transport during the initial stages of sintering were found to make significant contributions for nanoscale sintering. They were

1. Surface diffusion
2. Grain boundary diffusion

Both were considerably accelerated by large atomic forces near interfacial cusps. A corollary of this observation was that standard "zero potential gradient" diffusion coefficients would be very inaccurate (if not useless!) guides for predicting nanoscale sintering rates in any material system.

In addition, three unconventional mechanisms were seen to contribute in a major way to the early stages of nanoscale sintering:

1. Mechanical rotation
2. Plastic deformation via dislocation generation and transmission
3. Amorphization of subcritical grains

Since the simulations showed that dislocations can be generated at strongly curved void surfaces (leading to spontaneous shear of the nanoparticles) without the application of external stress, a previous statement of sintering theory[154]

> "*...the pressure resulting from surface tension is almost insignificant compared to that from the external pressure during the initial stage of sintering, and we shall ignore it.*"

should be re-examined in the case of nanoscale sintering, at least in the case of reasonably ductile metals. A final conclusion from these simulations was that because of the operation of a host of unconventional sintering processes that came into play at the nanoscale, the kinetics of sintering (at least in ductile metals) would be six to nine orders of magnitude more rapid than the rate estimated by standard sintering theory, and could even proceed fairly rapidly at room temperature in many materials.

A very accurate test at the nanoscale of Herring's scaling law[155] for sintering kinetics was performed via Molecular Dynamics for pairs of Cu cylinders below ten nanometers in diameter.[156] The simulated sintering, using accurate EAM Cu potentials, was carried out to virtual completion on a series of two equal sized, but misoriented, copper cylinders with diameters of 2.2 nm, 4.4 nm, and 8.8 nm at two different temperatures, 1100 K and 1200 K, and analyzed in terms of Herring's scaling law. This law asserts that the relative times to reach the same isomorphic sintering configuration should scale as some fixed power of the relative sizes if a single sintering mechanism is dominant. By monitoring the neck-width to center-separation ratio continuously over the entire sintering span, it was found that Herring's law was not obeyed at any stage of sintering, but that the "fixed power" required to fit the data varied from a numerical value of approximately zero at early stages to values approaching four at the latest stages.

Since computer simulations avoid the major difficulties of testing Herring's scaling law experimentally, (e.g., oxides, impurities, nonisomorphous arrays, continuous measurements, etc.), this result is probably the most accurate test to date of this sintering "law." Its failure is probably ascribable to the observation that in simulated sintering at the nanoscale, a series of mechanisms come into play in rapid succession, and each of them are strongly dependent on size, thus violating Herring's fundamental assumption of just one mechanism being dominant at a time. Whether simulations of larger scale arrays will eventually find sizes at which Herring's "law" is reasonably obeyed is a moot question at this point in time.

5.4 Recrystallization

Most of the simulation studies on recrystallization (both static and dynamic) have been performed by Srolovitz and co-workers.[157]–[161] They developed a Monte Carlo method capable of simulating grain growth, as well as different types of recrystallization processes, mapped onto a two-dimensional triangular lattice. As such, the modeling is not of any specific material system nor is there any inherent size scale in the simulations, but they obtain quite interesting qualitative results which broadly mimic the behavior of real systems, and can be expected to most closely match the response of nanofilms with columnar grain structures.

Their basic approach is to imagine an overlay of a triangular lattice (TL) onto the microstructure of some system where the nearest neighbor spacing of the TL is significantly smaller than the average grain diameter of the underlying system. Then each lattice point of the TL is given one number, S_i, representing the crystallographic orientation and a second number, H_i, representing the stored energy (proportional to the dislocation density) of the underlying material grain. If two nearest neighbor sites of the TL are found to have different values of S_i, then an additional energy, J, representing an isotropic grain boundary energy, is included in the energy calculation. The Monte Carlo method proceeds by choosing a TL site at random, changing it to an arbitrarily different S_i value and calculating the energy difference. If the energy decreases, the change is kept, otherwise not, so it is essentially a zero temperature Monte Carlo procedure.

The nucleation of recrystallized grains is performed by randomly picking a triangular cluster of three nearest neighbor sites of the TL, and at a random moment giving these three the same new arbitrary value of S_i along with an initial H_i value of zero. Thus, there is no inherent nucleation barrier (e.g., in terms of minimum stored energy, work hardening rate, etc.) in this procedure. If it is static recrystallization which is being modeled, then this rezeroed value of H_i is not changed further, but if the modeling is of dynamic recrystallization, then all the H_i's of the TL (new or otherwise) will continue to be increased linearly with time during the simulation. They have argued[162] that (under certain assumptions) this linear increase of stored energy with time corresponds to a parabolic stress-strain relation for constant strain rate situations. They assert that other work hardening assumptions can be suitably accommodated by appropriately non-linear stored-energy–time functions.

As an example of what can be gleaned from their simulations, if one considers the dynamic recrystallization work,[163] a plot of total stored energy in the TL as a function of time shows a monotonic increase up to a peak value followed by a series of damped oscillations asymptotically approaching some value below the peak. These oscillations are also seen in the average grain size with time plots, and are found to a varying degree with changes in the control parameters of the simulations, which are:

1. Energy storage rate
2. Nucleation rate
3. Initial grain size

The oscillations in stored energy and grain size were found to have the same period but were out of phase by approximately ninety degrees. By arguing that the stored energy can be related to the flow stress, these plots can be compared with laboratory experiments (e.g., in mild carbon steels) which show a similar damped oscillatory behavior.

5.5 Grain Growth

Above a certain temperature, generally in the range of two-thirds of the melting point, the average grain size in a polycrystalline material will spontaneously increase with time. This occurs by larger grains growing at the expense of smaller grains. Many coarse-grained simulations of this process have been performed on two-dimensional, and occasionally three-dimensional, generic systems, but only one atomistic Molecular Dynamics study for a particular nanoscale three-dimensional material (Au) has been reported to date to this author's knowledge.

An excellent overview of the coarse-grained simulation work has been given by Frost and Thompson[164] who roughly divide the approaches into:

1. Monte Carlo methods (usually based on a Potts model)
2. Front tracking methods

The first approach has been extensively employed by Srolovitz and co-workers,[165]–[181] whereas the latter has been preferred by Frost, Thompson, and co-workers.[182]–[191] Some other techniques related to each of these approaches are also described in the cited overview.

The Monte Carlo methods use a similar technique to that described above in the section on recrystallization, in that a mapping lattice (ML), usually chosen as triangular in two-dimensions, is overlaid onto the microstructure of some system, where the nearest neighbor spacing of the ML is made significantly smaller than the average grain diameter of the underlying system. Then each lattice point of the ML is given a number, S_i, representing the initial crystallographic orientation of the region in which it occurs. Energy is assigned to the configuration by calculating an interaction energy, which is higher if nearest neighbor ML sites have different S values and lower if not. Nearest neighbor sites having different S values are interpreted as being on opposite sides of a grain boundary. Various assumptions may be made about this energy characteristic (e.g., that the nearest neighbor interaction energy monotonically increases with some functionality of the difference in neighboring S values, etc.). A finite temperature Monte Carlo simulation then proceeds by picking a ML site at random, changing its S value by a random amount, and then allowing the change with a probability given by the Boltzmann factor, $\exp\{-\Delta E/kT\}$, where ΔE is the energy change caused by the new configuration and T is the assumed "temperature" of the system. As ML sites at grain boundaries change their S values the location of the intervening grain boundary moves, and the kinetics of grain growth can be observed and measured as a function of the assumed grain boundary energy characteristic and temperature.

The symmetry of the ML (e.g., triangular or square in two-dimensions, simple cubic or face centered cubic in three-dimensions) and the limited number of neighbors used to calculate energies can cause significant variations in the grain boundary energies and mobilities that are simply an artifact of the chosen ML symmetry. The effects of this choice have been examined by Holm, et al.,[192] and recommendations made for minimizing these artifacts.

By contrast, the front-tracking methods focus on the grain boundaries themselves as continuous curving lines (or planes in three-dimensions). Discrete points along each grain boundary are identified and moved with a velocity related to the local curvature and any other driving forces to be included in the model. Points at grain boundary intersections are moved in a manner to maintain a balance of grain boundary tensions.

The relative advantage of the Monte Carlo Potts model is that it can be easily extended to three-dimensions, without much extra programming and without an undue increase in computational time. However, the simulations only approach true accuracy when the spacing of the nodes used in the calculations is very small compared to the average grain diameter, thus requiring quite lengthy computations. Another disadvantage is that because of the discreteness of the mapping process, local curvature of the grain boundaries is difficult to assess, making it correspondingly difficult to relate boundary mobility to local curvature.

On the other hand, in the front tracking methods, the grain boundary segments can be moved according to particular relationships between the boundary migration velocity and the local characteristics of the boundary, such as its curvature, the relative orientation of the neighboring grains, and their relative stress states. Dependence of kinetics on temperature (possibly including temperature gradients) can be introduced via an Arrhenius relation between boundary mobility and local temperature. This method is relatively easy to program and compute for two-dimensional simulations, but is much more difficult and time consuming in three-dimensional cases, although Brakke has published a coding for both types, which is available for general use.[193][194]

In two-dimensions, both these methods show kinetics which generally agree with the prediction of Mullins[195] that the rate of change of a grain's area is proportional to its number of sides less six. It should be noticed, however, that there is no inherent length scale in either of these types of simulation, thus leading to scale-invariant kinetics, which for real micron or larger scale structures is probably correct. But this could have serious drawbacks if used to predict the behavior of actual nanograin structures, since it may be expected that non-diffusive and non-stochastic mechanisms may then play a very important role in grain growth, similar to that found in nanoscale sintering phenomena.[196] In fact, an atomistic Molecular Dynamics study[197] of a thirteen grain columnar structure of gold (average grain diameter ~4 nm) shows extremely rapid non-stochastic grain growth mechanisms (such as amorphization followed by recrystallization, as well as dislocation mediated grain rotations), which have never been considered before in grain growth analysis.

5.6 Strength

A good review of simulation studies addressing the question of strength versus grain size at the nanoscale by Weertman, et al., has recently appeared.[198] In this review article, they point out that although it is rather difficult to get good reliable experimental data on whether the Hall-Petch relation (for strength vs grain size) begins to break down as nanoscale grain sizes are approached, a number of recent Molecular Dynamics simulations[199]–[206] have begun to shed light on this important question. The simulations done on Cu and Ni arrays by Van Swygenhoven and colleagues indicate that once the grain size is below about 10 nm, the Hall-Petch slope becomes negative, i.e., diminishing grain size now results in diminishing strength. Some of the simulations have suggested that at grain sizes below 10 nm, the deformation mechanism can be described as grain boundary sliding controlled by self-diffusion at the boundaries, so that below a critical grain size a transition is made from intragranular to intergranular deformation mechanisms. Furthermore, at nanoscale grain sizes, the increasing grain boundary volume fraction is found to cause diminishing hardness.

In similar work on pure Cu, Schiøtz, et al.,[207] carried out Molecular Dynamics studies of nanoscale grained arrays with grain diameters between 3.3 nm and 6.6 nm, with the overall dimension of the array being about 10 nm, simulating low temperature tensile tests. The simulations showed yield and plastic flow phenomena, but it was also found that contrary to the prediction of the Hall-Petch relation, the strength decreased with diminishing grain size. The underlying mechanism responsible for yield at these small grain sizes was also argued to be multiple independent slip events at the grain boundaries.

This result shows that at nanoscale grain sizes, the Hall-Petch relation begins to break down and progressive softening occurs as grain size is further reduced below some material dependent critical dimension. Yip[208] has argued that this critical dimension (for greatest strength) can be related to the smallest grain size that can sustain a dislocation pile-up, and calculates it to be 19.3 nm for copper and 11.2 nm for palladium. Since the grain sizes used in any of the Cu simulations were below these numbers, and no simulations have yet been carried out on palladium, the results can be considered to be roughly consistent with Yip's theory of the realm in which the Hall-Petch law should fail.

However, it should be stressed that, at this time, no Molecular Dynamics simulations have been carried out that are of sufficient scale to demonstrate normal Hall-Petch behavior so, strictly speaking, the *transition regime* between normal positive slope Hall-Petch hardening and the anomalous negative slope Hall-Petch grain boundary range has not been displayed. The reason that this may be important is that it is not entirely clear that Molecular Dynamics simulations performed with periodic boundary conditions (as all of those cited have been) do not artificially suppress dislocation activity, as would be necessary to allow normal Hall-Petch behavior. Thus, a key link in this demonstration will have to be simulations with grain sizes of sufficient magnitude to reach the positive slope regime and, with the current ability to simulate arrays on the order of 10^8 atoms, this should now be possible, or nearly so.

REFERENCES

1. See, for example, *Mod. Sim. Mat. Sci. Eng.* or *J. Computer-Aided Materials Design*
2. Kohlhoff, S., Gumbsch, P., and Fischmeister, H. F., *Phil. Mag.*, 64A:851 (1991)
3. Tadmor, E. B., Ortiz, M., and Phillips, R., *Phil Mag.*, A73:1529–1563 (1996)
4. Clapp, P. C., Zhao, Y., and Rifkin, J. A., *MRS Symposia Proceedings*, 350:299–311 (1994); Clapp, P. C., in: *Stability of Materials*, (A. Gonis, P. E. A. Turchi, and J. Kudrnovsky, eds.), pp. 137–164, Plenum Press, NY (1996)

5. Guenin, R. A., Clapp, P. C., Rifkin, J. A., and Zhao, Y., *Materials Sci. Eng. B*, 37:193–196 (1996)
6. Zeng, P., Zajac, S., Clapp, P. C., and Rifkin, J. A., Nanoparticle Sintering Simulations," *Mat. Sci. Eng.*, A252:301–306 (1998)
7. Lewis, L. J., Jensen, P., and Barrat, J.-L., Melting, Freezing, and Coalescence of Gold Nanoclusters, *Phys. Rev. B,* 56:2248–2257 (1997)
8. Peters, K. F., Cohen, J. B., and Chung, Y.W., Melting of Pb Nanocrystals, *Phys. Rev. B*, 57:13430–13438 (1998)
9. See, for example, Haasen, P., *Physical Metallurgy,* 3rd ed., p. 58, Cambridge Univ. Press, (1996)
10. Zhu, H., and Averback, R. S., Sintering Processes of Two Nanoparticles: A Study by Molecular Dynamics Simulations, *Phil. Mag. Lett.*, 73:27–33 (1996)
11. Lewis, L. J., Jensen, P., and Barrat, J. L., Melting, Freezing, and Coalescence of Gold Nanoclusters, *Phys. Rev. B,* 56:2248–2257 (1997)
12. Zeng, P., Clapp, P. C., and Rifkin, J. A., *Failure of Herring's Scaling Laws at the Nanoscale,* in: *Computer Aided Design of High Temperature Materials,* (A. Pechenik, R. K. Kalia, and P. Vashishta, eds.), pp. 483–490, Oxford Univ. Press (1998)
13. Zeng, P., Zajac, S., Clapp, P. C., and Rifkin, J. A., Nanoparticle Sintering Simulations, *Mat. Sci. Eng. A*, 252:301–306 (1998)
14. Zeng, P., *Computer Simulation of Nanoparticle Sintering*, Ph. D. Thesis, Univ. of Connecticut, Storrs, CT (1999)
15. Luedtke, W. D., and Landman, U., Preparation and Melting of Amorphous Silicon by Molecular Dynamics Simulations, *Phys. Rev. B*, 37:4656–4663 (1988)
16. Luedtke, W. D., and Landman, U., Preparation, Structure, Dynamics and Energetics of Amorphous Silicon: A Molecular Dynamics Study, *Phys. Rev. B,* 40:1164–1174 (1989)
17. Swaminarayan, S., *Thermodynamics and Segregation Behavior of Nanoscale Solids*, Ph. D. Thesis, Univ. of Michigan, Ann Arbor, MI (1995)
18. Swaminarayan, S., Najafabadi, R., and Srolovitz, D. J., Polycrystalline Surface Properties From Spherical Crystallites: Ag, Au, Cu and Pt, *Surf. Science,* 306:367–380 (1994)
19. Najafabadi, R., Srolovitz, D. J., and LeSar, R., Finite Temperature Structure and Thermodynamics of the Gold Ó5 (001) Twist Boundary, *J. Mater. Res.*, 5:2663–2676 (1990)

20. Stoneham, A. M., Ramos, M. M. D., and Sutton, A. P., Feature Article: How Do They Stick Together? The Statics and Dynamics of Interfaces, *Phil. Mag.* A, 67:797–811 (1993)

21. Schnitker, J., and Srolovitz, D. J., Misfit Effects in Adhesion Calculations, *Modelling and Simulation in Materials Science and Engineering,* 6:153–164 (1998)

22. Sutton, A. P., and Pethica J. B., Inelastic Flow Processes in Nanometre Volumes of Solids, *J. Phys: Condens. Matter,* 2:5317–5326 (1990)

23. Landman, U., Luedtke, W. D., Burnham, N. A., and Colton, R. J., Atomistic Mechanisms and Dynamics of Adhesion, Nanoindentation, and Fracture, *Science,* 248:458–461 (1990)

24. Landman, U., Luedtke, W. D., and Nitzan, A., Dynamics of Tip-Substrate Interactions in Atomic Force Microscopy, *Surf. Sci.,* 210:L177–L184 (1989)

25. Luedtke, W. D., and Landman, U., *Comput. Mater. Sci,.* 1:1–24 (1992)

26. Landman, U, Luedtke, W. D., Ringer, E. M., in: *Fundamentals of Friction: Macroscopic and Microscopic Processes*, (J. L. Singer and H. M. Pollock, eds.), pp. 463–508, Kluwer, Dordrecht (1991)

27. Landman, U., Luedtke, W. D., Burnham, N. A., and Colton, R. J, Atomistic Mechanisms and Dynamics of Adhesion, Nanoindentation, and Fracture, *Science*, 248:458–461 (1990)

28 Landman, U., Luedtke, W. D., Ringer, E. M., in: *Fundamentals of Friction: Macroscopic and Microscopic Processes* (J. L. Singer and H. M. Pollock, eds.), pp. 463–508, Kluwer, Dordrecht (1991)

29. Landman, U., Luedtke, W. D., Burnham, N. A., and Colton, R. J., *Science,* 248:458–461 (1990)

30. Landman, U., and Luedtke, W. D., *J. Vac. Sci. Technol.*, 9:414–423 (1991)

31. Sorensen, M. R., Jacobsen, K. W., and Stoltze P., Simulations of Atomic-Scale Sliding Friction, *Phys. Rev.* B, 53:2101–2113 (1996)

32. Landman, U., Luedtke, W. D., and Ringer, E. M., *Wear*, 153:3–30 (1992)

33. Landman, U., Luedtke, W. D., and Nitzan, A., Dynamics of Tip-Substrate Interactions in Atomic Force Microscopy, *Surf. Sci.*, 210:L177–L184 (1989)

34. Landman, U., Luedtke, W. D., and Ribarsky, M. W., *J. Vac. Sci. Technol.*, A7:2829–2839 (1989)

35. Thompson, P. A., and Robbins, M. O., *Science*, 250:792–794 (1990)

36. Robbins, M. O., and Thompson, P. A., *Science*, 253:916 (1991)

37. Persson, B. N., *J., Phys. Rev. B*, 50:4771–4786 (1994)

38. Thompson, P. A., Robbins, M. O., and Grest, G. S., in: *Computations for the Nanoscale*, (P. E. Bloohl, C. Joachim, and A. J. Fisher, eds.), pp. 127–138, Kluwer, Dordrecht (1993)
39. Gao, J. P., Luedtke, W. D., and Landman, U., Nano-Elastohydrodynamics-Structure, Dynamics and Flow in Nonuniform Lubricated Junctions, *Science*, 270:605–608 (1995)
40. Landman, U, Luedtke, W. D., and Nitzan, A., Dynamics of Tip-Substrate Interactions in Atomic Force Microscopy, *Surf. Sci.*, 210:L177–L184 (1989)
41. Sutton, A. P., Pethica, J. B., Raffi-Tabar, H., and Nieminen, J. A., Mechanical Properties of Metals at the Nanometre Scale, Ch. 7 of *Electron Theory in Alloy Design* (D. G. Pettifor, and A. H. Cottrell, eds.), Institute of Materials, London (1992)
42. Li, B., Molecular Dynamics Simulation of Friction Between Metals, M. Sc. Thesis, Univ. of Connecticut (1999)
43. Nieminen, J. A., Sutton, A. P., and Pethica, J. B., Static Junction Growth During Frictional Siding of Metals, *Acta Metall. Mater*, 40:2503–2509 (1992)
44. Nieminen, J. A., Sutton, A. P., Pethica, J. B., and Kaski, K., Mechanism of Lubrication by a Thin Solid Film on a Solid Surface, *Mod. Sim. Mat. Sci. Eng.*, 1:83–90 (1992)
45. Pethica, J. B., and Sutton, A. P., Nanomechanics: Atomic Resolution and Frictional Energy Dissipation in Atomic Force Microscopy, in: *Forces in Scanning Probe Methods*, (Guntherodt, et al., eds.), pp. 353–366, Kluwer (1995)
46. Li, B., Molecular Dynamics Simulation of Friction Between Metals, M. Sc. Thesis, Univ. of Connecticut (1999)
47. Stoneham, A. M., Ramos, M. M. D., and Sutton, A. P., Feature Article: How Do They Stick Together? The Statics and Dynamics of Interfaces., *Phil. Mag. A*, 67:797–811 (1993)
48. Bhushan B., Israelachvili, J. N., Landman U., Nanotribology: Friction, Wear and Lubrication at the Atomic Scale (review article), *Nature*, 374:607–616 (1995)
49. Sutton, A. P., Deformation Mechanisms, Electronic Conductance and Friction of Metallic Nano-Contacts, *Current Opinion in Solid State and Materials Science*, 1:827–833 (1996)
50. Todorov, T. N., and Sutton, A. P., Jumps in Electronic Conductance Due to Mechanical Instabilities, *Phys. Rev. Letts.*, 70:2138–2141 (1993)

51. Krans, J. M., Muller, C. J., van der Post, N., Postma, F. R., Sutton, A. P., Todorov, T. N., and van Ruitenbeek, J. M., Comment on: Quantized Conductance in an Atom-Sized Point Contact, *Phys. Rev. Lett.*, 74:2146 (1995)

52. Todorov, T. N., and Sutton, A. P., Force and Conductance Jumps in Atomic-Scale Metallic Contacts, *Phys. Rev. B*, 54:14234 (1996)

53. Bratkovsky, A. M., Sutton, A. P., and Todorov, T. N., Conditions for Conductance Quantization in Realistic Models of Atomic Scale Metallic Contacts, *Phys. Rev. B*, 52:5036–5051 (1995)

54. Horsfield, A. P., Godwin, P. D., Pettifor, D. G., and Sutton, A. P., Computational Materials Synthesis I: A Tight Binding Scheme for Hydrocarbons, *Phys. Rev. B*, 54:15773–15775 (1996)

55. Godwin, P. D, Horsfield, A. P., Pettifor, D. G., and Sutton, A. P., Computational Materials Synthesis II: A Study of Polymerization, *Phys. Rev. B*, 54:15776–15784 (1996)

56. Godwin, P. D., Horsfield, A. P., Stoneham, A. M., Bull, S. J., Ford, I. J., Harker, A. H., Pettifor, D. G., and Sutton, A. P., Computational Materials Synthesis III: Synthesis of Hydrogenated Amorphous Carbon from Molecular Precursors, *Phys. Rev. B*, 54:15785–15794 (1996)

57. Ge, R., Clapp, P. C., and Rifkin, J. A., Molecular Dynamics of a Molten Cu Droplet Spreading on a Cold Cu Substrate, *Surface Science Letters*, 426:L413–L419 (1999)

58. Gilmer, G. H., and Roland, C., Applications of Molecular Dynamics Methods to Low Energy Ion Beams and Film Deposition Processes, *Radiation Effects and Defects in Solids*, 130/131:321–343 (1994)

59. Melinon, P., Prevel, B., Dupuis, V., Perez, A., Champagnon, B., Guyot, Y., Boudeulle, M., Pellarin, M., Dugourd, P., and Broyer, M., New Phases of Amorphous Carbon and Silicon Films Obtained by Low Energy Cluster Beam Deposition, *Mat. Sci. Eng. A*, 217/218:69–73 (1996)

60. Perez, A., Melinon, P., Dupuis, V., Jensen, P., Prevel, B., Tuaillon, J., Bardotti, L., Martet, C., Treilleux, M., Broyer, M., Pellarin, M., Vaille, J. L., Palpant, B., and Lerme, J., Cluster Assembled Materials: A Novel Class of Nanostructured Solids with Original Structures and Properties, *J. Phys. D: Appl. Phys.*, 30:709–721 (1997)

61. Kwon, I., Biswas, R., Grest, G. S., and Soukoulis, C. M., Molecular Dynamics Simulation of Amorphous and Epitaxial Si Film Growth on Si (111), *Phys. Rev. B*, 41:3678 (1990)

62. Biswas, R., Grest, G. S., and Soukoulis, C. M., Molecular Dynamics Simulation of Cluster and Atom Deposition on Silicon (111), *Phys. Rev. B*, 38:8154–8162 (1988)

63. Battaile, C. C., Srolovitz, D. J., and Butler, J. E., A Kinetic Monte Carlo Method for the Atomic Scale Simulation of Chemical Vapor Deposition: Application to Diamond, *J. Appl. Phys.*, 82:6293–6300 (1997)

64. Dong, L., Schnitker, J., Smith, R. W., and Srolovitz, D. J., Stress Relaxation and Misfit Dislocation in Heteroepitaxial Film Growth: A Molecular Dynamics Simulation Study, *J. Appl. Phys.*, 83(1):217–227 (1998)

65. Smith, R. W., and Srolovitz, D. J., Void Formation During Film Growth: A Molecular Dynamics Simulation Study, *J. Appl. Phys.*, 79(3):1448–1457 (1996)

66. Battaile, C. C., Srolovitz, D. J., and Butler, J. E , A Kinetic Monte Carlo Method for the Atomic Scale Simulation of Chemical Vapor Deposition: Application to Diamond, *J. Appl. Phys.*, 82:6293–6300 (1997)

67. Battaile, C. C., Srolovitz, D. J., and Butler, J. E., Morphologies of Diamond Films from Atomistic Simulations of Chemical Vapor Deposition, *Diamond and Related Materials*, 6(9):1198–1206 (1997)

68. Battaile, C. C., Srolovitz, D. J., and Butler, J. E., Molecular View of Diamond CVD Growth, *J. Elect. Mater.*, 26(9):960–965 (1997)

69. Srolovitz, D. J., Dandy, D. S., Butler, J. E., Battaile, C. C., and Paritosh, Integrated Multi-Scale Modeling of Diamond Chemical Vapor Deposition, *J. Met.*, S49(9):42–47 (1997)

70. Battaile, C. C., Srolovitz, D. J., and Butler, J. E., Atomic-Scale Simulations of Chemical Vapor Deposition on Flat and Vicinal Diamond Substrates, *J. Cryst. Growth*, 194:353–368 (1998)

71. Ethier, S., and Lewis, L. J., Epitaxial Growth of $Si_{1-x}Ge_x$ on Si(100)2x1: A Molecular Dynamics Study, *J. Mater. Res.*, 7:2817–2827 (1992)

72. Srivastava, D., Garrison, B. J., and Brenner, D. W., Anisotropic Spread of Surface Dimer Openings in the Initial Stages of the Epitaxial Growth of Si on Si{100}, *Phys. Rev. Letts.*, 63:302 (1989)

73. Smith, R. W., and Srolovitz, D. J., Void Formation During Film Growth: A Molecular Dynamics Simulation Study, *J. Appl. Phys.*, 79(3):1448–1457 (1996)

74. Dong, L., Smith, R. W., and Srolovitz, D. J., A Two Dimensional Molecular Dynamics Simulation of Thin Film Growth by Oblique Deposition, *J. Appl. Phys.*, 80(10):5682–5690 (1996)

75. Sridhar, N., Rickman, J. M., and Srolovitz, D. J., Twinning in Thin Films: I. Elastic Analysis, *Acta Materialia* 44(10)4085–4096 (1996); Twinning in Thin Films: II. Equilibrium Microstructures, *Acta Materialia*, 44(10):4097–4113 (1996)

76. Dong, L., Schnitker, J., Smith, R. W., and Srolovitz, D. J., Stress Relaxation and Misfit Dislocation in Heteroepitaxial Film Growth: A Molecular Dynamics Simulation Study, *J. Appl. Phys.*, 83(1):217–227 (1998)
77. Carel, R., Thompson, C. V., and Frost, H. J., Computer Simulation of Strain Energy Effects vs. Surface and Interface Energy Effects on Grain Growth in Thin Films, *Acta Materialia*, 44:2479–2494 (1996)
78. Srolovitz, D. J., and Safran, S. A., Capillary Instabilities in Thin Films: I. Energetics, *J. Appl. Phys.*, 60:247–254 (1986); Capillary Instabilities in Thin Films: II. Kinetics, *J. Appl. Phys.*, 60:255–260 (1986)
79. Srolovitz, D. J., and Thompson, C. V., Beading Instabilities in Thin Film Lines With Bamboo Microstructures, *Thin Solid Films*, 139:133–141 (1986)
80. Sridhar, N., Rickman, J. M., and Srolovitz, D. J., Multilayer Film Stability, *J. Appl. Phys.*, 82(10):4852–4859 (1997)
81. Khachaturyan, A. G., *Theory of Structural Transformations in Solids*, Wiley, New York (1983)
82. Wang, Y., Chen, L. Q., and Khachaturyan, A. G., in: *Computer Simulation in Materials Science*, (H. O. Kirchner, et al., eds.), pp. 325–371, Kluwer, Netherlands (1996)
83. Chen, L. Q., and Khachaturyan, A. G., Computer Simulation of Structural Transformations During Precipitation of an Ordered Intermetallic Phase, *Acta Metall. Mater.*, 39:2533–2551 (1991)
84. Vaks, V. G., Beiden, S. V., and Dobretsov, V. Yu, *JETP Lett.*, 61:68–74 (1995)
85. Dobretsov, V. Yu, Vaks, V. G., and Martin, G., Kinetic Features of Phase Separation Under Alloy Ordering, *Phys. Rev. B*, 54:3227–3239 (1996)
86. Chen, L. Q., Thermodynamics and Kinetics of Order-Disorder Processes Derived from the Cluster-Activation Method and Microscopic Diffusion-Theory, *Phys. Rev. B*, 49:3791–3799 (1994)
87. Chen, L. Q., and Simmons, J. A., Microscopic Master Equation Approach To Diffusion Transformation in Inhomogeneous Systems Single-Site Approximation and Direct Exchange Mechanism, *Acta Metall. Mater.*, 42:2943–2954 (1994)
88. Rautiainen, T. T., and Sutton, A. P., The Influence of the Atomic Diffusion Mechanism on Morphologies, Kinetics, and the Mechanisms of Coarsening During Phase Separation., *Phys. Rev. B*, 59:13681–13692 (1999)
89. Rautiainen, T. T., and Sutton, A. P., The Influence of the Atomic Diffusion Mechanism on Morphologies, Kinetics, and the Mechanisms of Coarsening During Phase Separation., *Phys. Rev. B*, 59:13681–13692 (1999)

90. Rautiainen, T. T., and Sutton, A. P., The Influence of the Atomic Diffusion Mechanism on Morphologies, Kinetics, and the Mechanisms of Coarsening During Phase Separation., *Phys. Rev. B*, 59:13681–13692 (1999)

91. Holm, E. A., Srolovitz, D. J., and Cahn, J. W., Microstructural Evolution in Two-Dimensional Two-Phase Polycrystals, *Acta Metall. Mater.*, 41:1119–1136 (1993)

92. Cahn, J. W., Stability, Microstructural Evolution, Grain Growth, and Coarsening in a Two-Dimensional Two-Phase Microstructure, *Acta Metall. Mater.*, 39:2189–2199 (1991)

93. Adams, C. D., Srolovitz, D. J., and Atzmon, M., Monte Carlo Simulation of Phase Separation During Thin Film Co-Deposition. *J. Appl. Phys.*, 74:1707–1715 (1993)

94. Swaminarayan, S., Najafabadi, R., and Srolovitz, D. J., Finite Size Effects on the Thermodynamics of Cu-Ni Alloys: {100} and {110} Thin Films, *Acta Materialia*, 45(4):1715–1724 (1997)

95. Atzmon, M., Kessler, D. A., and Srolovitz, D. J., Phase Separation During Film Growth, *J. Appl. Phys.*, 72:442–446 (1992)

96. Adams, C. D., Atzmon, M., Cheng, Y. T., and Srolovitz, D. J., Phase Separation During Co-Deposition of Al-Ge Thin Films, *J. Mater. Res.*, 7:653–666 (1992)

97. Srolovitz, D. J., and Scott, J., Clock-Model Descriptions of Incommensurate Ferro-Electric Films and Nematic Liquid Crystals, *Phys. Rev. B*, 34:1815–1819 (1986)

98. LeSar, R., Najafabadi, R., and Srolovitz, D. J., Finite Temperature Defect Properties from Free-Energy Minimization, *Phys. Rev. Letts.*, 63:624–627 (1989)

99. Swaminarayan, S., and Srolovitz, D. J., Surface Segregation in Thin Films, *Acta Metall. Mater.*, 44(5):2067–2072 (1996)

100. Marchenko, V. I., and Parshin, A. Ya, *Sov. Phys. JETP* 52:129–132 (1980)

101. Srolovitz, D. J., and Hirth, J. P., Elastic Analysis of the Energy and Relaxation of Stepped Surfaces, *Surface Science* 255:111–119 (1991)

102. Wolf, D., and Jaszczak, J. A., On the Interaction Between Steps in Vicinal fcc Surfaces. I. Steps Along <001>, *Surf. Sci.*, 277:301–322 (1992)

103. Najafabadi, R., and Srolovitz, D. J., Elastic Step Interactions on Vicinal Surfaces of fcc Metals, *Surf. Sci.*, 317:221–234 (1994)

104. Shilkrot, L. E., and Srolovitz, D. J., The Elastic Field of a Surface Step: Atomistic Simulations and Anisotropic Elastic Theory, *Phys. Rev. B*, 53(16):11120–11127 (1996)

105. Foiles, S. M.,Calculation of the Surface Segregation of Nickel-Copper Alloys with the Use of the Embedded-Atom Method, *Phys. Rev. B*, 32:7685–7693 (1985)

106. Eykholt, R., and Srolovitz, D. J., Surface Segregation During Deposition, *J. Appl. Phys.* 60:1793–1796 (1986)

107. Foiles, S. M., Calculation of Grain-Boundary Segregation in Nickel-Copper Alloys, *Phys. Rev. B*, 40:11502–11506 (1989)

108. Eymery, J., and Joud, J. C.,Surface Segregation in Binary Copper-Nickel and Platinum-Nickel Alloys Using Monte Carlo Simulation, *Surf. Sci.*, 231:419–426 (1990)

109. Najafabadi, R., Wang, H. Y., Srolovitz, D. J., and LeSar, R., A New Method for the Simulation of Alloys: Application to Interfacial Segregation, *Acta Metall. Mater.*, 39:3071–3082 (1991)

110. Wang, H. Y., Najafabadi, R., Srolovitz, D. J., and LeSar, R., (100) Surface Segregation in Cu-Ni Alloys, *Phys. Rev. B*, 45:12028–12042 (1992)

111. Battaile, C. C., Najafabadi, R., and Srolovitz, D. J., Simulation of Segregation to Free Surfaces in Cubic Oxides, *J. Amer. Cer. Soc.*, 78(12):3195–3200 (1995)

112. Wang, H. Y., Najafabadi, R., Srolovitz, D. J., and LeSar, R., Interfacial Segregation in Ag-Au, Au-Pd, and Cu-Ni Alloys: I. (001) Surfaces, *Interface Science* 1:7–30 (1993)

113. Sutton, A. P., and Balluffi, R. W., *Interfaces in Crystalline Materials,* Oxford University Press, Oxford (1995)

114. Dahmen, U., Hetherington, C. J. D., O'Keefe, M. A., Westmacott, K. H., Mills, M. J., Daw, M. S., and Vitek, V., Atomic Structure of a S99 Grain Boundary in Aluminum, *Phil. Mag. Lett.*, 62:327 (1990)

115. Alber, I., Bassani, J. L., Khantha, M., Vitek, V., and Wang, G. J., Grain Boundaries as Heterogeneous Systems: Atomic and Continuum Elastic Properties, *Phil. Trans.*, 339:555, Roy. Soc. London (1992)

116. Vitek, V., Gutekunst, G., Mayer, J., and Rühle, M., Atomic Structure of Misfit Dislocations in Metal-Ceramic Interfaces, *Philos. Mag. A*, 71:1219 (1995)

117. Schmidt, C., Ernst, F., Finnis, M. W., and Vitek, V., Prediction and Observation of the bcc Structure in Pure Copper at a S3 Grain Boundary, *Phys. Rev. Lett.*, 75:2160 (1995)

118. Marinopoulos, A. G., Vitek, V., and Carlsson, A. E., Significance of Non-Central Forces in Atomistic Studies of Grain Boundaries in bcc Transition Metals, *Philos. Mag. A*, 72:1311 (1995)

119. Vitek, V., *Atomistic Studies of the Structure of Grain Boundaries and Dislocations, in Stability of Materials, NATO Advanced Science Institute* (A. Gonis, P. E. A. Turchi, and J. Kudrnovsky, eds.), p. 53, Plenum Press, New York (1996)

120. Vellinga, W. P., DeHosson, J. Th. M., and Vitek, V., Misfit Dislocations: An Atomistic and Elastic Continuum Approach, *Acta Mater.*, 45:1525 (1997)

121. Gutekunst, G., Mayer, J., Vitek, V., and Rühle, M., Atomic Structure of Epitaxial Nb/Al_2O_3 Interfaces: Misfit Dislocations, *Phil. Mag. A*, 75:1357 (1997)

122. Schmidt, C., Finnis, M. W., Ernst, F., and Vitek, V., Theoretical and Experimental Investigations of Structures and Energies of S = 3/[112] Tilt Grain Boundaries in Copper, *Philos. Mag. A*, 77:1161 (1998)

123. Sutton, A. P., An Analytic Model for Grain Boundary Expansions and Cleavage Energies, *Phil. Mag. A,* 63:793–818 (1991)

124. Sutton, A. P., Irrational Interfaces, *Prog. Mat. Sci.*, 36:167–202 (1992)

125. Wolf, D., *J. Phys. Paris,* 46:C4–C197 (1985)

126. Sutton, A. P., and Balluffi, R. W., *Acta Metall.*, 35:2177 (1987)

127. Sutton, A. P., An Analytic Model for Grain Boundary Expansions and Cleavage Energies, *Phil. Mag. A*, 63:793–818 (1991)

128. Vitek, V., Wang, G. J., Alber, E. S., and Bassani, J. L., Relationship Between Modeling of the Atomic Structure of Grain Boundaries and Studies of Mechanical Properties, *J. Phys. Chem. Sol.*, 55:1147 (1994)

129. Bassani, J. L., Vitek, V., and Alber, I., Atomic-Level Elastic Properties of Interfaces and Their Relation to Continua, *Acta Metall.*, 40:S-307 (1992)

130. Alber, E. S., Bassani, J. L., Vitek, V., and Wang, G. J., *Stoneley Waves and Grain Boundary Phonons, Modeling, and Simulation in Materials Science and Engineering*, 2:455 (1994)

131. Alber, E. S., Bassani, J. L., Vitek, V., and Wang, G. J., Elastic Interfacial Waves in Discrete and Continuous Media, *Phys. Rev. B*, 53:8398 (1996)

132. Chen, S. P., Voter, A. F., and Srolovitz, D. J., Computer Simulation of Surfaces and [001] Symmetric Tilt Grain Boundaries in Ni, Al and Ni_3Al, *J. Mater. Res.*, 4:62–77 (1989)

133. Vitek, V., Chen, S. P., Voter, A. F., Kruisman, J. J., and DeHosson, J. Th. M., Grain Boundary Structure and Intergranular Fracture in L12 Ordered Alloys, *Mater. Sci. Forum*, 46:237 (1989)

134. Yan, M., Vitek, V., and Ackland, G. J., Atomic Structure and Chemical Composition of Grain Boundaries in L12 Intermetallic Compounds: Relation to Intergranular Brittleness, *Proc. NATO Symposium on Ordered Intermetallics-Physical Metallurgy and Mechanical Behaviour* (C. T. Liu, R. W. Cahn, and G. Sauthoff, eds.), p. 335, Kluwer, Dodrecht (1992)

135. Yan, M., Sob, M., Luzzi, D. E., Vitek, V., Ackland, G. J., Methfessel, M., and Rodriguez, C. O., Interatomic Forces and Atomic Structure of Grain Boundaries in Copper-Bismuth Alloys, *Phys. Rev. B*, 47:5571 (1993)

136. Menyhard, M., Yan, M., and Vitek, V., Atomistic Versus Phenomenological Approaches to Grain Boundary Segregation: Computer Modeling of Cu-Ag Alloys, *Acta Met. Mater.*, 42:2783–2796 (1994)

137. Yan, M., and Vitek, V., Atomistic Studies of the Structure and Composition of Boundaries in Cu_3Au and Ni_3Al, *Interface Science*, 3:17 (1995)

138. Vitek, V., and Yan, M., Grain Boundary Structure and Chemistry in Intermetallic Compounds, *Physical Metallurgy and Processing of Intermetallic Compounds*, (N. S. Stoloff and V. K. Sikka, eds.) p. 28, Chapman & Hall, New York (1995)

139. Yan, M., Vitek, V., and Chen, S. P., Many-Body Central Force Potentials and Properties of Grain Boundaries in NiAl, *Acta Mater.*, 44:4351 (1996)

140. Siegl, R., Vitek, V., Inui, H., Kishida, K., and Yamaguchi, M., Directional Bonding and Asymmetry of Interfacial Structure in Intermetallic TiAl: Combined Theoretical and Electron Microscope Study, *Philos. Mag. A*, 75:1447 (1997)

141. Siegl, R., Vitek, V., Luzzi, D. E., and Yan, M., Phase Stability and Grain Boundary Structure in Cu-Bi System, *J. Phase Equilibria*, 18:562 (1997)

142. Siegl, R., Yan, M., and Vitek, V., Atomic Structures of Grain Boundaries in Copper Bismuth Alloys: Ab Initio and Empirical Modeling, *Modelling and Simulation in Materials Science and Eng.*, 5:105 (1997)

143. Menyhard, M., Yan, M., and Vitek, V., Atomistic Versus Phenomenological Approaches to Grain Boundary Segregation: Computer Modeling of Cu-Ag Alloys, *Acta Met. Mater.*, 42:2783–2796 (1994)

144. Siegl, R., Yan, M., and Vitek, V., Atomic Structures of Grain Boundaries in Copper Bismuth Alloys: Ab Initio and Empirical Modeling, *Modelling and Simulation in Materials Science and Eng.*, 5:105 (1997)

145. Zhu, H., and Averback, R. S., Sintering Processes of Two Nanoparticles: A Study By Molecular Dynamics Simulations, *Phil. Mag. Lett.*, 73:27–33 (1996)

146. Lewis, L. J., Jensen, P., and Barrat, J. L., Melting, Freezing, and Coalescence of Gold Nanoclusters, *Phys. Rev. B*, 56:2248–2257 (1997)

147. Lewis, L. J., Deltour, P., Jensen, P., and Barrat, J. L., Melting, Freezing, Diffusion and Coalescence of Gold Nanoclusters, in: *Microscopic Simulation of Interfacial Phenomena in Solids and Liquids*, (S. R. Phillpot, P. D. Bristowe, D. G. Stroud, and J. R. Smith, eds.), *MRS Symp. Proc.*, 492:293–298 (1998)

148. Zeng, P., Zajac, S., Clapp, P. C., and Rifkin, J. A., Nanoparticle Sintering Simulations, *Mat. Sci. Eng. A*, 252:301–306 (1998)

149. Wang, S. H., Clapp, P. C., and Rifkin, J. A., *Molecular Dynamics Simulations of the Sintering Process of β-SiC Nanoparticles,* in: *Computer Aided Design of High Temperature Materials* (A. Pechenik, R. K. Kalia, and P. Vashishta, eds.) pp. 67–80, Oxford Univ. Press (1998)

150. Zeng, P. Computer Simulation of Nanoparticle Sintering, Ph. D. Thesis, Univ. of Connecticut, Storrs, CT (1999)

151. Lewis, L. J., Jensen, P., Combe, N., and Barrat, J. L., Melting, Freezing, and Coalescence of Gold Nanoclusters,, *Phys. Rev. B*, 56:2248–2257 (1997)

152. For example, see: Kingery, W. D., Bowen, H. K., and Uhlmann, D. R., *Introduction to Ceramics*, 2nd ed., p. 449, Wiley-Interscience, NY; or Helle, A. S., Easterling, K. E., and Ashby, M. F., *Acta Metall.*, 33:2163–2174 (1985)

153. Zeng, P., Zajac, S., Clapp, P. C., and Rifkin, J. A., Nanoparticle Sintering Simulations, *Mater. Sci. Eng. A*, 252:301–306 (1998); Zeng, P., Computer Simulation of Nanoparticle Sintering, Ph. D. Thesis, Univ. of Connecticut, Storrs, CT (1999)

154. Helle, A. S., Easterling, K. E., and Ashby, M. F., *Acta Metall.*, 33:2163–2174 (1985)

155. Herring, C., *J. Appl. Phys.*, 21:301–303 (1950)

156. Zeng, P., Clapp, P. C., and Rifkin, J. A., *Failure of Herring's Scaling Laws at the Nanoscale,* in: *Computer Aided Design Of High Temperature Materials* (A. Pechenik, R. K. Kalia, and P. Vashishta, eds.), pp. 483–490, Oxford Univ. Press (1998)

157. Srolovitz, D. J., Grest, G. S., and Anderson, M. P., Computer Simulation of Recrystallization: I. Homogeneous Nucleation and Growth, *Acta Metall.*, 34:1833–1845 (1986)

158. Srolovitz, D. J., Grest, G. S., Anderson, M. P., and Rollett, A. D., Computer Simulation of Recrystallization: II. Heterogeneous Nucleation and Growth, *Acta Metall.*, 36:2115–2128 (1988)

159. Rollett, A. D., Srolovitz, D. J., Doherty, R. D., and Anderson, M. P., Computer Simulation of Recrystallization in Non-Uniformly Deformed Metals, *Acta Metall.*, 37:627–639 (1989)

160. Rollett, R. D., Luton, M. J., and Srolovitz, D. J., Microstructural Simulation of Dynamic Recrystallization, *Acta Metall. Mater.*, 40:43–55 (1992)

161. Rollett, A. D., Srolovitz, D. J., Doherty, R. D., and Anderson, M. P., Computer Simulation of Recrystallization: III. Influence of a Particle Dispersion of Fine Particles, *Acta Metall. Mater.*, 40:3475–3495 (1992)

162. Rollett, R. D., Luton, M. J., and Srolovitz, D. J., Microstructural Simulation of Dynamic Recrystallization, *Acta Metall. Mater.*, 40:43–55 (1992)

163. Rollett, R. D., Luton, M. J., and Srolovitz, D. J., Microstructural Simulation of Dynamic Recrystallization, *Acta Metall. Mater.*, 40:43–55 (1992)

164. Frost, H. J., and Thompson, C. V., Computer Simulation of Grain Growth, *Current Opinion in Solid State and Materials Science*, 1:361–368 (1996)

165. Anderson, M. P., Srolovitz, D. J., Grest, G. S., and Sahni, P. S., Computer Simulation of Grain Growth: I. Kinetics, *Acta Metall.*, 32:783–791 (1984)

166. Srolovitz, D. J., Anderson, M. P., Sahni, P. S., and Grest, G. S., Computer Simulation of Grain Growth: II. Grain Size Distribution, Topology and Local Dynamics, *Acta Metall.*, 32:793–802 (1984)

167. Srolovitz, D. J., Anderson, M. P., Grest, G. S., and Sahni, P. S., Computer Simulation of Grain Growth: III. The Presence of a Particle Dispersion, *Acta Metall.*, 32:1429–1438 (1984)

168. Grest, G. S., and Srolovitz, D. J., Vortex Effects on Domain Growth: The Clock Model, *Phys. Rev. B,* 30:6535–6539 (1984)

169. Grest, G. S., Srolovitz, D. J., and Anderson, M. P., Computer Simulation of Grain Growth: IV. Anisotropic Grain Boundary Energies, *Acta Metall.*, 33:509–520 (1985)

170. Srolovitz, D. J., and Grest, G. S., Impurity Effects on Domain Growth Kinetics: I. Ising Model, *Phys. Rev. B*, 32:3014–3020 (1985)

171. Srolovitz, D. J., and Grest, G. S., Impurity Effects on Domain Growth Kinetics. II. Potts Model, *Phys. Rev. B*, 32:3021–3025 (1985)

172. Srolovitz, D. J., Grest, G. S., and Anderson, M. P., Computer Simulation of Grain Growth: V. Abnormal Grain Growth, *Acta Metall.*, 33:2233–2247 (1985)

173. Srolovitz, D. J., Grain Growth Phenomena in Films, *J. Vac. Sci. Technol. A*, 4:2925–2931 (1986)

174. Srolovitz, D. J., and Hassold, G. N., Effects of Diffusing Impurities on Domain Growth in the Ising Model, *Phys. Rev. B*, 35:6902–6910 (1987)

175. Srolovitz, D. J., Mazor, A., and Bukiet, B. G., Analytical and Numerical Modeling of Columnar Evolution in Thin Films, *J. Vac. Sci. Technol. A*, 6:2371–2380 (1988)

176. Grest, G. S., Anderson, M. P., and Srolovitz, D. J., Computer Simulation of Normal Grain Growth in Three Dimensions, *Philos. Mag. B*, 59:293–329 (1988)

177. Grest, G. S., Anderson, M. P., and Srolovitz, D. J., Domain Growth Kinetics for the Q-State Potts Model in Two and Three Dimensions, *Phys. Rev. B*, 38:4752–4760 (1988)

178. Rollett, A. D., Srolovitz, D. J., and Anderson, M. P., Simulation and Theory of Abnormal Grain Growth–Anisotropic Grain Boundary Energies and Mobilities, *Acta Metall.*, 37:1227–1240 (1989)

179. Mazor, A., Bukiet, B. G., and Srolovitz, D. J., The Effect of Vapor Incidence Angle Upon Thin Film Columnar Growth, *J. Vac. Sci. Technol. A*, 7:1386–1391 (1989)

180. Holm, E. A., Zacharopoulos, N., and Srolovitz, D. J., Nonuniform and Directional Grain Growth Caused by Grain Boundary Mobility Variations, *Acta Mater.*, 46(3):953–964 (1998)

181. Upmanyu, M., Smith, R. W., and Srolovitz, D. J., Atomistic Simulation of Curvature Driven Grain Boundary Migration, *Interface Science*, 6:41–58 (1998)

182. Thompson, C. V., Frost, H. J., and Spaepen, F., The Relative Rates of Secondary and Normal Grain Growth, *Acta Metall.*, 35:887–890 (1987)

183. Frost, H. J., Thompson, C. V., Howe, C. L., and Whang, J., A Two-Dimensional Computer Simulation of Capillarity-Driven Grain Growth: Preliminary Results, *Scripta Metall.*, 22:65–70 (1988)

184. Frost, H. J., and Thompson, C. V., Computer Simulation of Microstructural Evolution in Thin Films, *J. Elect. Mater.*, 17:447–458 (1988)

185. Frost, H. J., Thompson, C. V., and Walton, D. T., Simulation of Thin Film Grain Structures: I. Grain Growth Stagnation, *Acta Metall. Mater.*, 38:1455–1462 (1990)

186. Walton, D. T., Frost, H. J., and Thompson, C. V., The Development of Near-Bamboo and Bamboo Microstructures in Thin Film Strips, *Appl. Phys. Lett.*, 61:40–42 (1992)

187. Frost, H. J., Thompson, C. V., and Walton, D. T., Simulation of Thin Film Grain Structures: II. Abnormal Grain Growth, *Acta Metall. Mater.*, 40:779–793 (1992)

188. Frost, H. J., Microstructural Evolution in Thin Films, *Materials Characterization*, 32:257–273 (1994)

189. Carel, R., Thompson, C. V., and Frost, H. J., Computer Simulation of Strain Energy Effects vs. Surface and Interface Energy Effects on Grain Growth in Thin Films, *Acta Mater.*, 44:2479–2494 (1996)

190. Fayad, W., Thompson, C. V., and Frost, H. J., Steady-State Grain-Size Distributions Resulting from Grain Growth in Two-Dimensions, Submitted to *Scripta Mater.* (June 1998)

191. Riege, S. P., Thompson, C. V., and Frost, H. J., Simulation of the Influence of Particles on Grain Structure Evolution in Thin Films, submitted to *Acta Mater.*, (Nov. 1998)

192. Holm, E. A., Glazier, J. A., Srolovitz, D. J., and Grest, G. S., The Effects of Lattice Anisotropy and Temperature on Domain Growth in the Two Dimensional Potts Model, *Phys. Rev. A*, 43:2662–2668 (1991)

193. Brakke, K. A., The Surface Evolver, *Exp. Math*, 1:144–165 (1992)

194. Brakke, K. A., Grain Growth with the Surface Evolver, in: *Video Proceedings of the Workshop on Computational Crystal Growing*, (J. E. Taylor, ed.), American Mathematical Society, Providence, RI (1992)

195. Mullins, W. W., Two Dimensional Motion of Idealized Grain Boundaries, *J. Appl. Phys*, 27:900–904 (1956)

196. Zeng, P., Zajac, S., Clapp, P. C., and Rifkin, J. A., Nanoparticle Sintering Simulations, *Mat. Sci. Eng. A*, 252:301–306 (1998); Zeng, P., Computer Simulation of Nanoparticle Sintering, Ph. D. Thesis, Univ. of Connecticut, Storrs, CT (1999)

197. Clapp, P. C., Zajac, S., Zeng, P., and Rifkin, J. A., Nanograin Growth in Gold, in preparation.

198. Weertman, J. R., Farkas, D., Hemker, K., Kung, H., Mayo, M., Mitra, R., and Van Swygenhoven, H., Structure and Mechanical Behavior of Bulk Nanocrystalline Materials, *MRS Bulletin*, 24(2):44–50 (1999)

199. Schiøtz, J., Di Tolla, F. D., and Jacobsen, K. W., *Nature*, 391:561–563 (1998)

200. van Swygenhoven, H., Spaczér, M., and Caro, A., Characterisation of the Microstructure of Nanophase Ni: A Molecular Dynamics Simulation Study *Nanostruc. Mater.*, 12:629–632 (1999)

201. van Swygenhoven, H., and Caro, A., Plastic Behavior of Nanophase Metals Studied by Molecular Dynamics, *Phys. Rev. B*, 58:11246–11251 (1998)

202. van Swygenhoven, H., and Caro, A., Plastic Behavior of Nanophase Ni: A Molecular Dynamics Computer Simulation, *Appl. Phys. Lett.*, 71(12):652–654 (1997)

203. Spaczér, M., van Swygenhoven, H., and Caro, A., Microscopic Description of Plasticity in Computer Generated Metallic Nanophase Samples, *Mater. Res. Soc. Symp. Proc.*, 492:29–33 (1998)

204. van Swygenhoven, H., Spaczér, M., Farkas, D., and Caro, A., The Role of Grain Size and the Presence of Low and High Grain Boundaries in the Deformation Mechanism of Nanophase Ni: A Molecular Dynamics Computer Simulation, *Nanostruct. Mater.*, 12:323–326 (1999)

205. van Swygenhoven, H., Spaczér, M., Caro, A., and Farkas, D., Competing Plastic Deformation Mechanisms in Nanophase Metals, *Phys. Rev. B*, 60:22–25 (1999)

206. van Swygenhoven, H., Spaczér, M., and Caro, A Microscopic Description of Plasticity in Computer Generated Metallic Nanophase Samples: A Comparison Between Cu and Ni, *Acta Mater*, 47:3117–3126 (1999)

207. Schiøtz, J., Di Tolla, F. D., and Jacobsen, K. W., *Nature*, 391:561–563 (1998)

208. Yip, S., The Strongest Size, *Nature*, 391:532–533 (1998)

Part II

Properties

7

Diffusion in Nanocrystalline Materials

Roland Würschum, Ulrich Brossmann, and Hans-Eckhardt Schaefer

1.0 INTRODUCTION

Nanocrystalline (n-) materials are polycrystals with an ultra-fine grain size (diameter 5–100 nm) and a high volume fraction of interfaces. Since the pioneering work of Gleiter and coworkers in the beginning of the eighties,[1] the investigation of nanocrystalline materials has gained considerable interest with prospects of attractive application potentials which may arise from improved mechanical and magnetic properties compared to their coarse-grained counterparts.[2] With respect to both the structure and the physical properties of nanocrystalline solids, atomic transport represents a key issue.

In general, atomic transport in nanocrystalline materials differs substantially from that in coarse-grained or single-crystalline materials. This is due to the fact that, in nanocrystalline solids, the crystallite interfaces provide paths of high diffusivity, whereas in more coarse-grained crystals, volume self-diffusion or substitutional atom diffusion is substantial generally only at temperatures greater than approximately half the melting temperature.

Interface diffusion, in combination with a high fraction of interfaces, gives rise to modified physical properties of nanocrystalline solids. For instance, enhanced ductility of nanocrystalline ceramics[3][4] and intermetallic compounds[5] have recently been analyzed in the framework of models of mesoscopic sliding[3] or *grain switching* according to Ashby-Verrall,[5] both of which are controlled by diffusion in interfaces. With respect to nanocrystalline magnetic materials, atomic diffusion enables, for example, a controlled stress-induced adjustment of magnetic anisotropies in soft-magnetic alloys[6] or a texturing of hard-magnetic $Nd_2Fe_{14}B$-nanocomposites.[7] Furthermore, diffusion processes may control the formation of nanocrystalline materials, for example, by means of crystallization of amorphous precursors,[8] as well as the stability of nanocrystalline materials (relaxation, crystallite growth), their reactivity, corrosion behavior, or interaction with gases.

The relevance of diffusion-controlled processes, as outlined above, demands a comprehensive understanding of atomic diffusion in nanocrystalline materials. Since the first publication in this field some ten years ago[9] there is the key question, whether, and to what extent, the interface diffusion in nanocrystalline materials differs from that in conventional grain boundaries. Due to the correlation between the interface structure and diffusion (see below), the answer to this question allows conclusions on the structure of interfaces in nanocrystalline materials.

The present article is an up-to-date review of atomic diffusion in nanocrystalline materials.[10] Since the earlier reviews on this topic,[11]–[13] diffusion data on highly-dense nanocrystalline metals,[14][15] and oxides[16] after gas-phase condensation, as well as on nanocrystalline materials prepared by other routes, e.g., crystallization[17] and severe plastic deformation,[18] became available. Most importantly, diffusion studies of conventional grain boundaries have recently been extended to low temperatures,[19]–[21] thus allowing a conclusive comparison with the diffusion characteristics of nanocrystalline solids studied in a similar temperature regime.

This chapter is organized as follows. After a short summary of the models of interface diffusion (Sec. 2.0) and of the diffusion characteristics in conventional grain boundaries (Sec. 3.0), results of diffusion studies of nanocrystalline metals are presented and discussed (Sec. 4.1). The correlation between the diffusion and rapid grain growth in highly-dense nanocrystalline metals and attempts to stabilize the nanocrystalline structure are considered in Sec. 4.2. As a particular example of diffusion in a structurally stable nanocrystalline system, recent results obtained on the prototype alloy

Finemet are presented in Sec. 5.0 with a subsequent brief summary of the diffusion of hydrogen in nanocrystalline materials (Sec. 6.0) which is used as a probe for free volumes in the interfaces. In the last section (7.0), recent ^{18}O-diffusion studies on n-ZrO_2 and a comparison with diffusion studies on other nanocrystalline ceramics are presented.

2.0 MODELING OF INTERFACE DIFFUSION

Diffusion coefficients are usually determined from the penetration profile of tracer atoms measured in a diffusion experiment. In the general case of interface diffusion in polycrystalline materials, two simultaneous diffusion processes have to be taken into account, i.e., rapid diffusion in the crystallite interfaces (diffusion coefficient D_B) accompanied by diffusion from the interfaces and specimen surface into the volume of the crystallites (diffusion coefficient D_V). According to Harrison,[22] and Levine and MacCallum,[23] three different diffusion regimes, denoted by A, B, and C, occur in polycrystalline materials for which the evaluation of the diffusion profiles is relatively simple.[24] These regimes are characterized by appropriate ratios of the diffusion length in the crystallites ($\propto \sqrt{D_V t}$, diffusion time, t) and the crystallite diameter, d, or the interface thickness, δ.

Mishin and Herzig[25] recently proposed an extension of this scheme for fine-grained metals distinguishing between different ratios of the diffusion length in the interfaces and the crystallite diameter. Klinger, et al.,[26] on the other hand, took into account diffusion at tripleline-intersections of interfaces in addition to diffusion in the interfaces and the crystallites.

In practice however, at the low temperatures at which most diffusion experiments on nanocrystalline materials are performed, the diffusivity in the crystallites is negligibly small (see Sec. 4.1). In this so-called "type-C" diffusion regime, the diffusion coefficient D_B in the interfaces can be directly determined from the Gaussian concentration profile[24]

Eq. (1) $$C \propto \frac{1}{\sqrt{D_B t}} \exp\left(-\frac{x^2}{4D_B t}\right)$$

which is derived for a thin layer of radiotracer atoms on the specimen surface as the starting condition. In Eq. (1), x and t denote the depth and time of diffusion, respectively.

Of particular importance, considering nanocrystalline metals, are the cases where diffusion occurs in the presence of porosity or is accompanied by interface migration. In the latter case, which occurs due to crystallite growth (see Sec. 4.2), the tracer atoms are extracted from fast interfacial diffusion paths and immobilized in the crystallites where the volume diffusivity is negligible. Due to this process, an evaluation of the diffusion profile according to type-C kinetics (Eq. 1) would yield an apparent diffusivity lower than the actual atomic diffusivity in interfaces. A consideration of interface migration in the framework of a simple model yields the relationship[27]

Eq. (2)
$$C \propto \exp\left(-\sqrt{\frac{v}{D_B \delta}}\,x\right)$$

for the depth profile of the tracer concentration assuming a constant velocity, v, of interfaces.

Porosity, on the other hand, provides fast diffusion paths from which diffusion into agglomerates of crystallites may occur via interfaces between crystallites, and via volume diffusion into the crystallites. This has been taken into account by Bokstein, et al., in their *cluster* model[28] for analyzing the diffusion in porous nanocrystalline metals. Between the fast diffusivity on the agglomerate surfaces and the diffusivity in the agglomerate interfaces, relationships analogous to those for the type-B kinetics of grain boundary diffusion are valid.[28] (See Sec. 4.1.)

3.0 DIFFUSION IN GRAIN BOUNDARIES OF METALS

Grain boundary diffusion in conventional polycrystalline materials or in bicrystals represents a subject of comprehensive research, particularly due to its technical relevance for a variety of processes, for example, solid-state transformation, creep, or corrosion. (See recent reviews by Kaur, et al., 1995,[24] and Sutton and Balluffi, 1995.[29]) A first quantitative description of grain boundary diffusion dates back to Langmuir in 1934.[30]

Strong indications of a vacancy-mediated self-diffusion process in grain boundaries of metals are derived from diffusion studies under hydrostatic pressure (Ag[31]), studies of the isotope effect (Ag[32]), and computer simulations (Fe,[33] Ag,[34] Cu[35]).

Recent extensions of diffusion studies of conventional grain boundaries towards relatively low temperatures[19]–[21] are important with respect to an assessment of the diffusion characteristics of nanocrystalline metals. According to these studies, the diffusivities in conventional grain boundaries at low temperatures are higher and the diffusion enthalpies, H^D, are lower than expected on the basis of an extrapolation from high temperatures (Fig. 2[21]). Atomistic simulations indicated that this variation of H^D from high to low temperatures is due to an interstitial-type diffusion mechanism[34] that adds, at low temperatures, to the vacancy mechanism which is believed to prevail in conventional grain boundaries of metals (see above).

Correlations between the grain boundary structure and diffusion are derived from diffusion studies on [001]-tilt boundaries in noble metals.[21][36][37] These studies show that, with increasing tilt angle, the diffusivity parallel to the tilt axis increases up to an order of magnitude[21][36] with local minima near-coincidence orientations of the adjacent crystals ($\Sigma 5$[37]). Since a similar correlation exists between the tilt- or twist-angle and the specific energy of grain boundaries[38] and, on the other hand, the specific energy increases with the excess volume of grain boundaries,[38][39] this result may indicate an increase of the grain boundary diffusivity with the excess volume.

4.0 DIFFUSION IN NANOCRYSTALLINE METALS

4.1 Results and Discussion

A compilation of the available literature data on the diffusion of metals and metalloids in nanocrystalline metals and alloys is given in Table 1 in comparison with diffusion data obtained on crystals and conventional grain boundaries. So far, diffusion studies on nanocrystalline metals have focused primarily on the cubic face-centered metals Pd, Cu, and Ni, the results of which are summarized in Fig. 2.

Table 1. Diffusion coefficients D of metals and metalloids in nanocrystalline metals and alloys at 20% and 25% of the melting temperature T_M according to a linear interpolation of the measuring data (see Fig. 2, diffusor elements are quoted in brackets). D-values obtained from linear extrapolation are put in brackets. T_a: annealing temperature (Rt: room temperature); T_d: temperature interval of diffusion; ρ/ρ_0, d: relative mass density and crystallite size prior to diffusion, respectively. Preparation route: severe plastic deformation (smc-Pd), crystallization of an amorphous precursor (Finemet), ball milling (n-$Al_{91.9}Ti_{7.8}Fe_{0.3}$), crystallite condensation and compaction (all others). Measuring techniques: Radiotracer with sputter sectioning (Tracer), electron-beam microanalysis (EBMA), Auger electron spectroscopy (AES) or secondary ion mass spectroscopy (SIMS) with depth profiling, Rutherford backscattering (RBS), nuclear magnetic resonance (NMR).

	T_a [K]	T_d [K]	$D(0.2\ T_M)$ [m^2s^{-1}]	$D(0.25\ T_M)$ [m^2s^{-1}]	ρ/ρ_0	d [nm]	method
n-Pd (Fe)[n)]	373 §	423–523	(3.2×10^{-21})	2.8×10^{-20}	> 0.97	50	Tracer
smc-Pd (Fe)[o)]	293–673	371–623	1.1×10^{-21}	3×10^{-20}	0.99	80	Tracer
n-Pd (Pt)[n)]	373 §	623		1.3×10^{-18} (0.29 T_M)	0.97	20	RBS
Finemet (Fe)[p)]	810–818	628–773		(4×10^{-29}) £		13	Tracer

(Cont'd.)

Table 1. *(Cont'd.)*

Literature values for nanocrystallline metals

	T_a [K]	T_d [K]	$D(0.2\,T_M)$ [m^2s^{-1}]	$D(0.25\,T_M)$ [m^2s^{-1}]	ρ/ρ_0	d [nm]	method
n-Cu (Cu)[a)]	Rt	293–393	(4.6×10^{-21})	9.2×10^{-19}	0.91	–	Tracer
n-Cu[l)]	Rt	293–420			0.80	11	NMR
n-Cu (Ag)[b)]	Rt	303–373	(6.3×10^{-20})	1.6×10^{-18}	0.90	8	EBMA
	373	(identical D-values for $T_d < 0.25\,T_M$)					
n-Cu (Au)[c,d)]	Rt	373		4.8×10^{-22} ($0.27\,T_M$)	> 0.9	10	AES
n-Cu (Bi)[c,d)]	Rt	293–413	(5×10^{-23})	3×10^{-20}	> 0.9	10	RBS
	353	($D_{intrinsic}$ reduced by 50%, $0.26\,T_M$)					RBS
n-Cu (Sb)[e)] †	Rt	353–373		3.1×10^{-21}	0.75	50	RBS
				1.3×10^{-22}			
n-Ni (Ni)[f)] †	773 §	293–473	2.4×10^{-19}	5.9×10^{-18}	0.92	50	Tracer
				1.6×10^{-19} ($0.27\,T_M$)			
n-Pd (Ag)[c)]	Rt	293–453	3.5×10^{-17}	1.2×10^{-15}	0.90	8	SIMS
n-Pd (Au)[c)]	Rt	293–373	8.3×10^{-18} ‡		0.90	8	SIMS
n-Pd (Cu)[c)]	Rt	373	4.1×10^{-18}		0.90	10	SIMS
	373 §	373	1.0×10^{-18}		0.95	11	SIMS
	373 §	373	6×10^{-20} (p_{max} = 2 GPa) ‡‡			10	SIMS
n-Pd (B)[c)]	Rt	453		3.0×10^{-24}	0.90	10	SIMS
n-Fe (B)[c,g)]	Rt	293–383	9.8×10^{-20}	(4.5×10^{-18})	> 0.9	7	SIMS
n-$Al_{91.9}Ti_{7.8}Fe_{0.3}$ (Cu)[h)]	673 §	371–571		2.9×10^{-17} ††	0.98	22	SIMS

(Cont'd.)

Table 1. *(Cont'd.)*

Values for comparison

	T_d [K]	$D(0.2\ T_M)$ [m^2s^{-1}]	$D(0.25\ T_M)$ [m^2s^{-1}]
crystal:			
c-Pd (Pd)[i)]	1323–1723	(1.0×10^{-42})	(7.5×10^{-36})
grain boundary:			
fcc metals[j)]		(1.8×10^{-25})	(1.6×10^{-21})§§
Σ5-tilt grain-boundary:			
Ag/Au[k)]	382–483	(7.3×10^{-22})	(2.7×10^{-19})§§
Ag[m)]	353–394		(1.3×10^{-18})

§ Temperature of crystallite compaction.

£ Related to $T_M = 1538$ K (solidus temperature) of $Fe_{80}Si_{20}$[41] ($0.25\ T_M = 385$ K).

† Analysis according to *cluster* diffusion model[28] (value quoted in 1. or 2. line refers to diffusion on cluster surfaces or in interfaces of nanocrystalline clusters, respectively).

‡ According to Fig. 5.5 of Ref. 40.

‡‡ Diffusion in dependence of pressure.

†† Related to $T_M = 1493$ K (solidus temperature) of $Al_{92}Ti_8$[42] ($0.25\ T_M = 373$ K).

§§ Related to grain boundary thickness $\delta = 1$ nm.

(Cont'd.)

Table 1. *(Cont'd.)*

References

a)	Horváth, et al., 1987[9]	b)	Schumacher, et al., 1989[43]
c)	Höfler, 1991[40]	d)	Höfler, et al., 1993[44]
e)	Balandin, et al., 1996[45]	f)	Bokstein, et al., 1995[28]
g)	Höfler, et al., 1993[46]	h)	Minamino, etal., 1996[47]
i)	Peterson, 1964[48]	j)	Gust, et al., 1985[49]
k)	Qing Ma and Balluffi, 1993[21]	l)	Dickenscheid, et al., 1991[50]
m)	Sommer and Herzig, 1992[51]	n)	Würschum, et al., 1997[14]
o)	Würschum, et al., 1996[18]	p)	Würschum, et al., 1997[17]

The following discussion of the diffusion data starts with a summary of the Fe-tracer diffusion studies on nanocrystalline Pd performed in the authors' group.[10][14][15][18] Based on earlier results, these studies were aimed at diffusion measurements on highly-dense nanocrystalline metals where the influence of porosity is excluded or negligible. By means of crystallite condensation and compaction, a nearly theoretical mass density of nanocrystalline Pd could be achieved applying a high compaction pressure (4 GPa), at slightly elevated temperatures (380 K), under ultrahigh vacuum conditions.[10][15] In addition, the diffusion studies were extended to porosity-free submicrocrystalline (smc) Pd prepared by severe plastic deformation.[10][18] The diffusion measurements were performed by means of radiotracer techniques, where the activity profiles were determined by ion-beam sectioning.[54]

In both nanocrystalline Pd prepared by cluster condensation and compaction (Fig. 1) and in submicrocrystalline Pd prepared by severe plastic deformation,[18] the fast diffusion of Fe can be detected at temperatures slightly higher than ambient temperature. This is considered as evidence that the atomic transport occurs in the crystallite interfaces since the Fe-diffusion in crystalline Pd, which is similar to Pd self-diffusivity, can be neglected at these temperatures (Fig. 2).

The strong curvature of the ^{59}Fe diffusion profiles in an x^2 representation (Fig. 1) is attributed to crystallite growth during diffusion. Crystallite growth gives rise to a decrease of the fraction of interfaces and, as a result of growth-induced interface migration, to a slowing-down of tracer diffusion, since the tracer atoms are immobilized by incorporation on lattice sites in the crystallites. Since different types of interfaces migrate with different time-dependent velocities, v, in the course of crystal growth, both the migration and the elimination of interfaces lead to a curvature of the interface-diffusion controlled penetration profiles unlike the simple case described by Eq. (2). The diffusivities derived from fitting Gaussians (Eq. 1 and Fig. 2) to the diffusion tails, according to type-C kinetics,[24] are considered as the actual values of D_B since the tails of the diffusion profiles are governed by those crystallites which grow at the lowest rate. This is particularly true for n-Pd prepared by crystallite condensation and compaction where inhomogeneous crystallite growth occurs.[10][15]

The ^{59}Fe diffusivities in n-Pd are considered to be characteristic of self-diffusion in interfaces since initial studies with ^{103}Pd indicate a rather similar diffusion behavior. Moreover, the diffusion in cluster-compacted n-Pd is probably not affected by porosity as indicated by the high relative mass density (Table 1), the rapid grain growth (see Sec. 4.2), and the similar ^{59}Fe-diffusivities compared with deformation-prepared smc-Pd (Fig. 2).

The interfacial diffusivity in nanocrystalline Pd appears to be similar to conventional grain boundaries which can be concluded from a comparison with the diffusion studies of bicrystals or coarse-grained polycrystals. (See Fig. 2.) According to the correlation between the structure and diffusion (see Sec. 3.0), the similar diffusion behavior suggests an equilibrium-like structure of the interfaces in nanocrystalline metals. For obvious reasons this may be related to interface migration. During this process, interfaces with an initially non-equilibrium structure due to the preparation can easily relax within a short migration distance. Strong evidence for an interfacial relaxation in cluster-compacted nanocrystalline metals, at slightly elevated temperatures, is obtained from a decrease of the excess free volume[55] and of internal strains.[56]

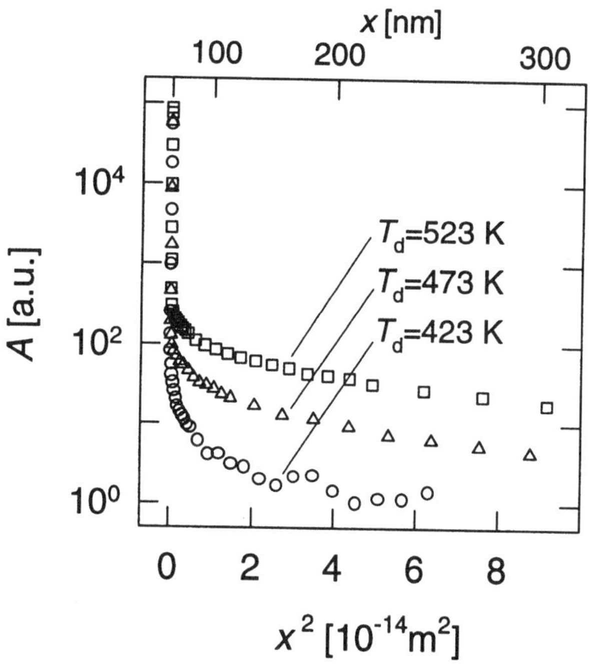

Figure 1. ^{59}Fe diffusion profiles in cluster-compacted n-Pd measured after diffusion annealing at the temperatures T_d = 523 K (annealing time t_d = 35 h), T_d = 473 K (t_d = 72 h), and T_d = 423 K (t_d = 121 h) without pre-annealing.

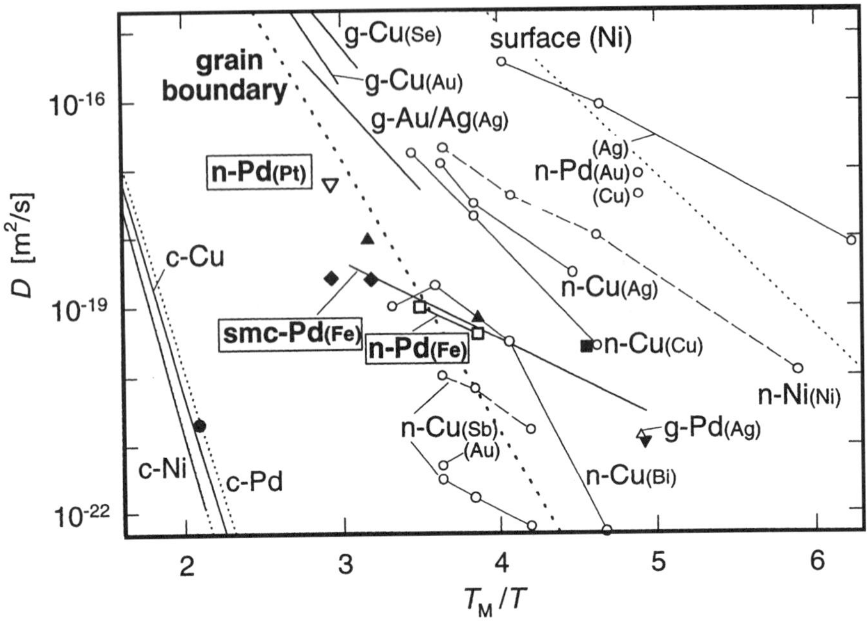

Figure 2. Arrhenius plots of tracer diffusivities D in face-centered cubic nanocrystalline metals (the diffusion temperatures are scaled to the melting temperatures T_M; the tracer atoms are quoted in brackets): Diffusion of Fe (□)[14] and Pt (∇)[14] in cluster-compacted n-Pd without pre-annealing; diffusion of Fe in submicrocrystalline (smc) Pd prepared by severe plastic deformation [18] without pre-annealing (▼) and after pre-annealing at T_a = 453 K (■), 553 K (▲) and 673 K (◆) for 2400 s. The data for the ^{59}Fe diffusion in smc- and n-Pd were derived from the deep-penetration tails of the diffusion profiles assuming Gaussian-type solutions (type-C kinetics[24]). Literature data for nanocrystalline metals (o): Diffusion of Cu,[9] Ag,[43] Au,[40][44] Bi,[40][44] and Sb[45] in n-Cu; diffusion of Ag, Cu, and Au in n-Pd,[40] Ni diffusion in n-Ni[28]. Analysis of n-Ni(Ni) and n-Cu(Sb) according to a model of atomic diffusion in a porous solid (see Ref. 28): diffusion on cluster surfaces (- - -), diffusion in interfaces of crystallite clusters [—, n-Cu(Sb)]. Data of coarse-grained metals (dotted lines denote extrapolated values): self-diffusion on (110)-Ni surfaces parallel to <110>,[52] self-diffusion in crystalline (c-) Ni,[53] Cu,[54] and Pd,[48] Fe diffusion in Pd after annealing of deformation-prepared Pd at 977 K (•);[18] grain-boundary diffusivity in fcc metals[49] extrapolated from high temperatures (assuming a grain-boundary thickness δ = 1 nm). Grain-boundary diffusion (g) in the type-C regime: Au[19][20] and Se[20] in grain boundaries of Cu; Ag in Σ5(310)-tilt grain boundaries of Au/Ag[21] (δ = 1 nm), Ag in evaporated Pd layers (Δ).[40]

Diffusivities in cluster-condensed n-Pd similar to conventional grain boundaries can also be estimated from a preliminary study of the Pt diffusion by means of Rutherford backscattering (Fig. 2[15]), taking into account interface migration during the diffusion annealing. In combination with the high fraction of interfaces, the interface diffusion, much faster than lattice self-diffusion (Fig. 2), strongly favors diffusion-controlled processes in nanocrystalline solids.

Literature data, according to which diffusivities in cluster-compacted nanocrystalline metals are substantially enhanced compared to grain boundaries (Fig. 2), might have been affected by fast diffusion paths due to residual porosity as signified by the reduced mass density in the earlier experiments (Table 1). Taking into account porosity in the framework of the aforementioned cluster model (Sec. 3.0), Bokstein, et al.,[28] estimated a self-diffusivity similar to conventional grain boundaries for the crystallite interfaces in cluster-compacted n-Ni.

Diffusion coefficients, for example, that of Sb in n-Cu, which are, on the other hand, considerably lower than self-diffusivities in conventional grain boundaries (Fig. 2) presumably arise from the effect of chemical diffusion and the formation of intermetallic compounds due to a high concentration of diffusing atoms used in SIMS experiments.[45] In addition, the diffusion may be hampered by oxides which are formed in the interfaces due to a penetration of oxygen in the case of insufficient powder densification.[45]

Diffusion studies on nanocrystalline cluster-condensed metals have been recently extended to body-centered cubic Fe.[56a] In that case, direct evidence for the influence of a structural relaxation on the interface diffusion could be derived from the isothermal time-dependence of the self-diffusion coefficient. In the relaxed state of n-Fe, interface diffusivities characteristic of conventional grain boundaries occur similarly to nanocrystalline fcc metals.

4.2 Correlation Between Diffusion and Crystallite Growth

The interface diffusion in highly-dense n-metals after cluster synthesis or severe plastic deformation is accompanied by strong crystallite growth (Fig. 3). A quantitative analysis shows that the rapid crystallite growth can be attributed to the high driving forces due to the small crystallite size, taking into account the dynamics of equilibrium-like grain boundaries (Fig. 4).[14] The analysis is based on the reasonable assumption that the activation enthalpy H^G of crystallite growth is given by the activation enthalpy H^M of grain boundary migration which—according to Sutton and Balluffi[29]—is lower than the grain boundary diffusion enthalpy H^D. The reduced value of H^M, compared to H^D, is considered one indication that the migration of large-angle boundaries occurs by a local shuffling of atoms across the interface of the adjacent crystals rather than by a transport of atoms in the grain boundary plane.[29][61]

In conclusion, both the mobility of interfaces (Fig. 4) and the interfacial diffusivities (Fig. 2) in highly-dense nanocrystalline metals appear to be characteristic of conventional grain boundaries.

A retardation of the normal crystallite growth, which is frequently observed in nanocrystalline metals (Fig. 3) (for a review see Ref. 62) and which is combined with abnormal crystallite growth, presumably arises from interface pinning due to solute impurities, impurity-related second phases, or residual porosity. (See Ref. 63 for a review on drag mechanisms.)

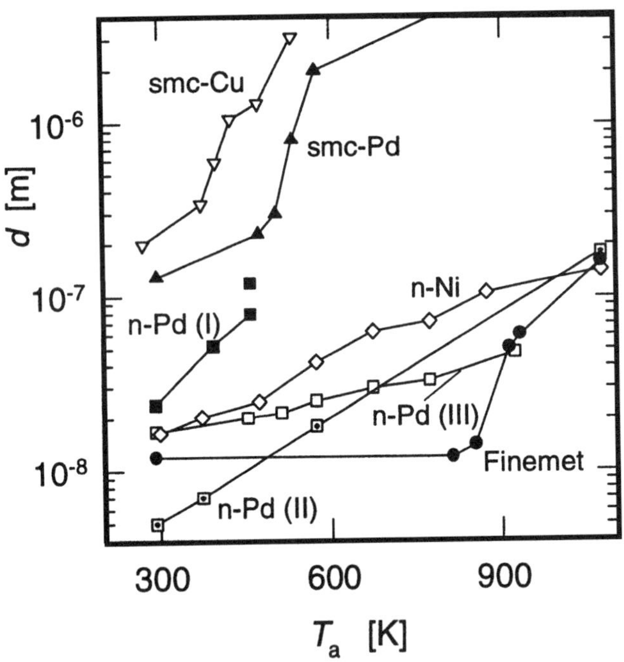

Figure 3. Variation of the crystallite size d with the annealing temperature T_a in submicrocrystalline (smc) Cu[57] and Pd[18] prepared by severe plastic deformation, in n-Pd (I, $\rho/\rho_0 > 97\%$;[15] II, $\rho/\rho_0 = 86\%$[58]) and n-Ni ($\rho/\rho_0 = 79\%$[10]) prepared by cluster condensation and compaction, in n-Pd prepared by electrochemical deposition (III),[59] and in crystallized $Fe_{73.5}Si_{13.5}B_9Nb_3Cu_1$ (Finemet[60]). ρ/ρ_0 denotes the relative mass density of the cluster-compacted metals after preparation.

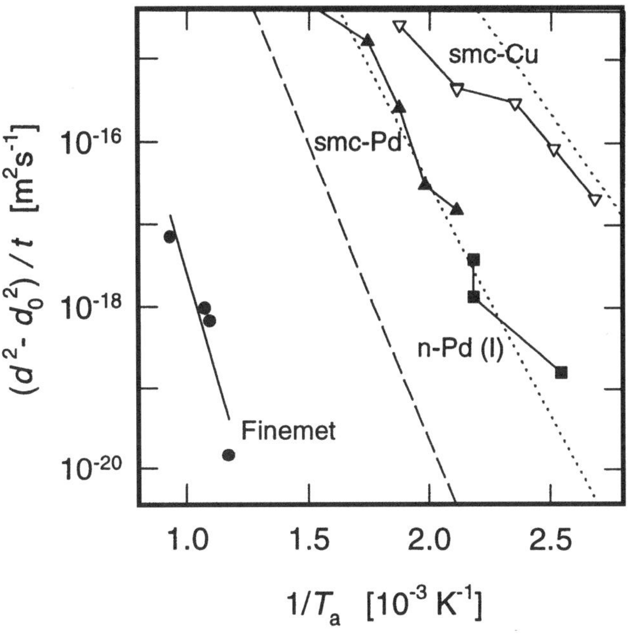

Figure 4. Arrhenius-representation of crystallite sizes d in submicrocrystalline and nanocrystalline metals and in Finemet (cf. Fig. 3) according to a parabolic growth behavior d_0^2 - ∝ exp($-H^G/k_BT$) (for details see Ref. 14, d_0: initial crystallite size). The dotted or dashed lines refer to a parabolic growth behavior with the activation enthalpy of crystallite growth H^G as given by the activation enthalpy H^M of the grain boundary mobility [H^M = 0.83 eV (Cu), 1.12 eV (Pd), see Ref. 14] or the grain boundary self-diffusion [H^D = 1.43 eV (Pd)],[49] respectively.

Porosity-retarded crystallite growth[64] is discussed for nanocrystalline ceramics.[65] It may also prevail in nanocrystalline metals[59] with relative mass densities, ρ/ρ_0, below the theoretical limit. This is supported, for example, by the correlation found between the mass density and (abnormal) grain growth at ambient temperature in cluster-compacted n-Cu.[66] Similarly, recent combined studies of crystallite growth (Fig. 3) and positron lifetime in n-Pd prepared by pulsed electrodeposition suggest a correlation between the thermal stability and the presence of nanopores.[59]

A promising strategy for the controlled stabilization of nanocrystalline metals, in addition, for example, to interface pinning by oxide or carbide nanoparticles,[67][68] consists of decreasing of the driving force for crystallite growth by means of a doping-induced decrease of the specific grain boundary energy.[69] The diffusion characteristics of Bi in nanocrystalline Cu, as reported by Höfler, et al.,[44] is of particular interest in this respect. Since Bi is extremely insoluble in crystalline Cu, it exhibits a strong tendency for segregation at interfaces. Similar to that observed for conventional grain boundaries, e.g., in Fe-Sn alloys,[70] the segregation gives rise to a decrease of the Bi diffusivity in the interfaces of n-Cu with increasing Bi concentration [44] which is attributed to a segregation-induced decrease of the specific grain boundary energy. Probably as a consequence of that, a retardation of the crystallite growth occurs in Bi-doped n-Cu in comparison to undoped n-Cu.[71] An even stronger stabilization reported for Zr-doped Pd is considered to arise from a decrease of the grain boundary energy, too.[69]

5.0 DIFFUSION IN THE NANOCRYSTALLINE ALLOY FINEMET

A way of stabilizing the nanocrystalline structure, similar to that described at the end of the preceding section (Sec. 4.2), is opened up by the use of intergranular amorphous phases. This is accomplished in the case of the soft-magnetic nanocomposite $Fe_{73.5}Si_{13.5}B_9Nb_3Cu_1$ (Finemet[72]) which consists of ordered intermetallic nanocrystallites ($D0_3$-type Fe_3Si, crystallite size d = 12 nm) embedded in a Nb- and B-enriched amorphous matrix.[72][73] This nanocomposite is structurally stable until well above the temperature (T_a = 810 K) where the nanocrystalline structure is formed by crystallization of the initially amorphous alloy (Fig. 3). The thermal stability makes this alloy highly attractive for technical applications as well as for diffusion studies which are summarized in the following.[17]

High-temperature studies of positron lifetime (Fig. 7) and ^{59}Fe-tracer diffusion of the Finemet-alloy (Fig. 6) show thermal-vacancy formation and rapid self-diffusion in the $D0_3$-$Fe_{80}Si_{20}$-nanocrystallites similar to that observed in single-crystals of this intermetallic compound.[75][76] This is concluded from the increase of the diffusivity upon crystallization (Figs. 5 and 6) and from the high-temperature increase of the positron lifetime which can be analyzed in terms of competitive positron trapping and annihilation at thermal vacancies in the nanocrystallites and at free volumes in the crystallite interfaces (Fig. 7[17]).

The rapid Fe-diffusion in the nanocrystallites can be taken into account in a diffusion modelin which the crystallites and the interfaces with the amorphous intergranular phase are treated as alternating layers perpendicular to the direction of diffusion (see Ref. 17). This model leads to the conclusion that the diffusivity, D_{IF}, in the amorphous interfaces is similar to that measured in the relaxed amorphous state prior to crystallization (Fig. 2) which is characteristic of Fe-B-Si amorphous alloys.[77]

The interfacial diffusivity, which is substantially lower than in grain boundaries of coarse-grained or nanocrystalline pure metals (Fig. 6), reflects a high packing density of both the amorphous intergranular phase *and* the interfaces between these interlayers and the $D0_3$-$Fe_{80}Si_{20}$ nanocrystallites of the Finemet alloy. Metallic amorphous phases exhibit densities higher than those of large angle boundaries.[78] A dense interface structure in the Finemet alloy is also concluded from positron lifetime measurements at ambient temperature which show the existence of the same type of free volumes as in the amorphous state, i.e., free volumes of smaller size than lattice vacancies.[17] These direct indications of a densely packed structure of the crystallite-amorphous interfaces are in accordance with the low specific energy of this type of interface.[79] Interface diffusivities lower than those of n-metals have been recently found in other n-alloys prepared by crystallization of melt-spun amorphous ribbons.[79a]

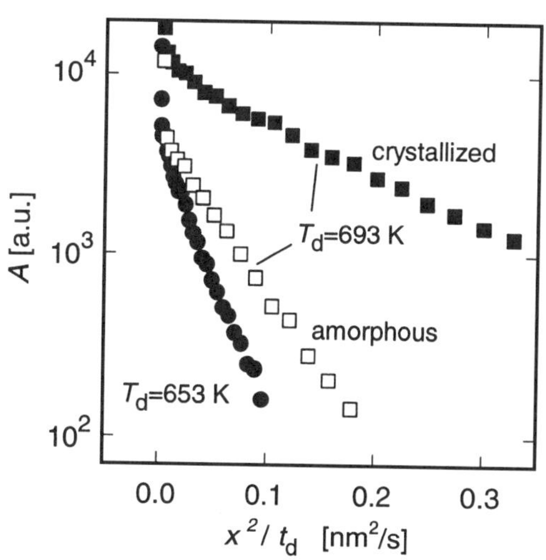

Figure 5. ^{59}Fe diffusion profiles in amorphous (□) and nanocrystalline (■,●) $Fe_{73.5}Si_{13.5}B_9Nb_3Cu_1$ (Finemet) crystallized at $T_a = 810$ K for $t_a = 1$h. T_d and t_d denote the diffusion temperature and time, respectively.

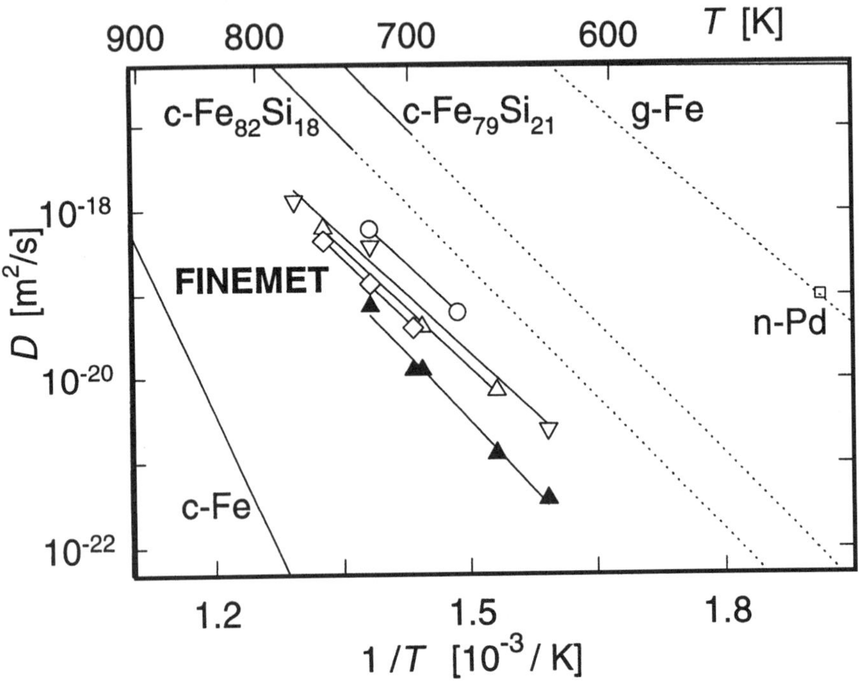

Figure 6. Arrhenius plots of ^{59}Fe-tracer diffusivities in nanocrystalline $Fe_{73.5}Si_{13.5}B_9Nb_3Cu_1$ (Finemet) crystallized at T_{crys} = 793 K (◊), 810 K (Δ), 813 K (∇), 818 K (o) and in the relaxed amorphous state (▲, pre-annealed at 723 K) prior to crystallization.[17] Literature data are shown for comparison (extrapolations are dotted): Fe diffusion in cluster-compacted nanocrystalline Pd (□,[14]), in the ferromagnetic phase of crystalline α-Fe (c-Fe,[74]), in grain boundaries of Fe (*g*-Fe, [70]) as well as in the intermetallic compounds $D0_3$-$Fe_{79}Si_{21}$ and $D0_3$-$Fe_{82}Si_{18}$.[75]

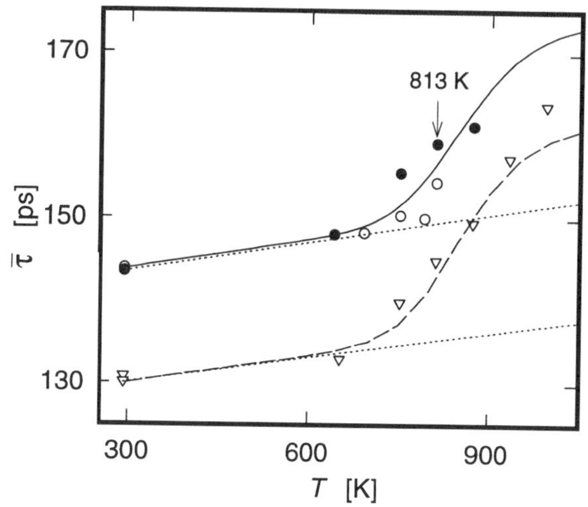

Figure 7. Thermal-vacancy formation in $Fe_{73.5}Si_{13.5}B_9Nb_3Cu_1$ (Finemet).[17] Mean positron lifetime in the nanocrystalline state after annealing at 813 K for 1 h (o) or 21 h (•) and in the microcrystalline state after annealing at 993 K for 31 h (▽). The solid curve is a numerical fit taking into account competitive positron trapping and annihilation at thermal vacancies in the nanocrystallites and at free volumes in the crystallite interfaces (for details see Ref. 17). The dotted straight lines show the mean positron lifetime in the absence of thermal vacancy formation assuming a linear temperature variation of like in the free state of pure metals.[17]

The combination of a dense interface structure, due to an intergranular amorphous phase and a high thermal-vacancy concentration in the nanocrystallites of the Finemet-nanocomposite, gives rise to the unique behavior where self-diffusion in the interfacial regions is slower than in the crystallites. So far, a similar situation has only been taken into consideration for ZrO_2-based ionic conductors where the diffusivity in the crystalline state is very high, due to a high concentration of dopant-induced oxygen vacancies.[80]

The variations of the Fe diffusivity (Fig. 6) and the thermal-vacancy formation (Fig. 7) with temperature and time of crystallization indicate atomic ordering processes in the nanocrystallites.[17] This is concluded from the fact that, in the coarse-grained intermetallic compound Fe_3Si, the Fe diffusivity is observed to increase with increasing Si-content or degree of order[75] due to an increase of the thermal vacancy concentration C_V.[76] The time scale and temperature range suggest that these atomic ordering processes in Finemet are controlled by the slow diffusivity on the

Si sublattice of the nanocrystallites.[17] It is worthwhile to note that the annealing behavior of the stress-induced magnetic anisotropy, which is of interest with respect to tailoring the soft-magnetic properties of Finemet,[6] can be quantitatively understood on the basis of the slow ordering processes mentioned above.

The crystallite growth during crystallization of Finemet, on the other hand, is controlled by the diffusion of large atoms, i.e., Nb, in the amorphous matrix.[8] This can be concluded from the high values of the activation enthalpy and pre-exponential factor of crystallization, a correlation which is similar to that existing between the activation enthalpy and the pre-exponential factor of tracer diffusion in amorphous alloys (see Ref. 8).

The studies of n-$Fe_{73.5}Si_{13.5}B_9Nb_3Cu_1$[17] described above have led, for the first time, to the detection of vacancy-mediated rapid self-diffusion in nanocrystallites. These combined high-temperature studies of positron lifetime and tracer diffusion may be extended to structurally stabilized nanocrystalline metals[67][69] in order to assess the relation between thermal-vacancy formation and diffusion in grain boundaries of metals[33] (cf. Sec. 3.0). Furthermore, diffusion studies on stabilized nanocrystalline metals without concomitant crystallite growth might enable a distinction of the various diffusion paths available in nanocrystalline metals, e.g., interfaces and triple lines between interfaces.

6.0 DIFFUSION OF HYDROGEN IN NANOCRYSTALLINE METALS AND ALLOYS

The diffusion of hydrogen in nanocrystalline metals and alloys is being intensively studied to explore the application potentials in the field of hydrogen storage. The diffusion or local atomic jumps of hydrogen in dependence of the hydrogen concentration C_H are used as a local probe technique that yields information on the spectra of available trapping sites in the crystallite interfaces. This information is deduced from measurements of the chemical potential (n-Pd[81]), hydrogen permeation (n-Ni[82]), thermal desorption (n-$Fe_{90}Zr_{10}$[83][84]), magnetic after-effect (e.g., n-$Fe_{90}Zr_{10}$,[84] n-Pd_3Fe[85]), internal friction (n-$CoZr_2$,[86] n-Pd[87]), or inelastic neutron scattering (n-Pd[88][89]).

Studies of hydrogen spectroscopy on n-$Fe_{90}Zr_{10}$ by Hirscher, et al.,[84][90] support the results of positron lifetime spectroscopy and Fe-tracer diffusion according to which the interfaces in nanocrystalline alloys crystallized from amorphous precursors (see Sec. 5.0) are densely packed similar to that in the initial amorphous state. A decrease of the mean activation energy, $\bar{Q}$, of reorientation jumps with increasing hydrogen concentration is observed by measurements of the magnetic after-effect.[84][90] Following a model on the H diffusion in amorphous alloys,[81] this decrease of $\bar{Q}$ is ascribed to a successive filling of energetically favorable sites (trapping centers) in the interfaces with hydrogen. The trapping centers are characterized by a continuous spectrum of binding energies witha width similar to that in the initial amorphous state.[90] The shift of the $\bar{Q}$-C_H relation of n-$Fe_{90}Zr_{10}$ to lower hydrogen concentrations in comparison to amorphous $Fe_{90}Zr_{10}$[90] is due to the fact that hydrogen in the nanocrystalline state is primarily located in the interfaces, the volume fraction of which is reduced compared to the initial amorphous phase.[10]

In $CoZr_2$, a slight increase (~ 6%) of the mean activation energy of hydrogen jumps is observed upon the amorphous-to-nanocrystalline transition by means of internal friction.[86]

In contrast to crystallization-prepared nanocrystalline alloys, the spectrum of hydrogen sites in the interfaces of cluster-compacted metals (n-Pd[81]) appears to differ from that in amorphous alloys, e.g., a-$Pd_{83}Si_{17}$.[81] On the one hand, the energy distribution of the H sites as deduced from the variation of the H diffusivity with concentration is enhanced by ca. 30% compared to a-$Pd_{83}Si_{17}$.[81] On the other hand, the H diffusivity in n-Pd at low H concentrations with respect to that in the crystalline state is reduced more strongly than in a-$Pd_{83}Si_{17}$[81] which indicates traps in the interfaces of cluster-compacted n-Pd with higher binding energies than in amorphous alloys. This fits the different interface structures which are found for cluster-compacted n-metals and crystallized alloys by means of positron lifetime spectroscopy[91][92] and Fe tracer diffusion (see Sec. 5.0). In the case of n-Pd prepared by pulsed electrodeposition, initial studies by means of quasielastic neutron scattering indicate deep traps in the interfaces, too.[89]

7.0 DIFFUSION IN NANOCRYSTALLINE CERAMICS

In the case of nanocrystalline ceramics, interphase diffusion is particularly important since it gives rise to enhanced sintering rates and improved deformation characteristics in comparison to coarse-grained ceramics.[4][93] So far, investigations of the diffusion in nanocrystalline ceramics have focused on the transition metal oxides ZrO_2[16] and TiO_2.[40][94]

Ceramics with stoichiometric compositions, in general, exhibit high activation energies for the bulk self-diffusion due to the strong binding of the constituents.[95] Deviations from stoichiometry may provide mechanisms of fast diffusion with a reduced activation energy and, therefore, exert a strong influence on the diffusion behavior. This particularly applies to zirconium dioxide.[96] In this interesting model system both the oxygen diffusivity and the crystallite structure[97] are determined by the presence of cationic dopants with a lower valency which induce extrinsic vacancies on the oxygen sublattice. The diffusion studies on nanocrystalline ZrO_2 are aimed at assessing the role of crystallite interfaces on the diffusion behavior.

The oxygen diffusion in undoped n-ZrO_2 was studied in collaboration with the group of U. Södervall (Chalmers University of Technology, Gothenburg, Sweden) using ^{18}O as a tracer and secondary ion mass spectroscopy (SIMS) for measuring the diffusion profiles.[16] The nanocrystalline ZrO_2 samples were prepared by inert-gas condensation, post-oxidation and in-situ consolidation at ambient temperature under a pressure of 1.8 GPa. [98] A relative mass density of about 97% and an average grain size, d, of 80 nm was obtained by subsequent pressureless sintering at T_{sinter} = 950–970°C for 2–3 hours.[16] Further sintering at 1050°C for 12 hours led to nearly full density and an increase of the grain size to 300 nm.

The shape of the ^{18}O diffusion profiles (Fig. 8) indicates two concomitant diffusion processes, i.e., volume diffusion in the crystallites (D_V) and interface diffusion (D_B). Analyses of the diffusion profiles according to type-B diffusion kinetics (see Sec. 2.0) reveal interface diffusivities, D_B, 3–4 orders of magnitude higher than the volume diffusivities, D_V, within the entire temperature range of 450 to 950°C studied[16] (Fig. 9).

The volume diffusion coefficients, $D_V = (2.5 \pm 1.5) \times 10^{-7} \exp[-(2.29 \pm 0.1)\ \mathrm{eV}/k_BT]\ \mathrm{m^2/s}$ were determined from the n-ZrO_2 samples with the larger grain size (T_{sinter} = 1050°C) where the surface-near parts of the ^{18}O-diffusion profiles up to depths of several 100 nm are dominated by the diffusion from the surface into the volume of the crystallites. The interface diffusion coefficients, $D_B = (3.3 \pm 1.5) \times 10^{-5} \exp[-(1.95 \pm 0.05)\ \mathrm{eV}/k_BT]\ \mathrm{m^2/s}$, determined from the slopes at deeper penetration are almost identical for both sintering temperatures (Fig. 9) and independent of grain size.

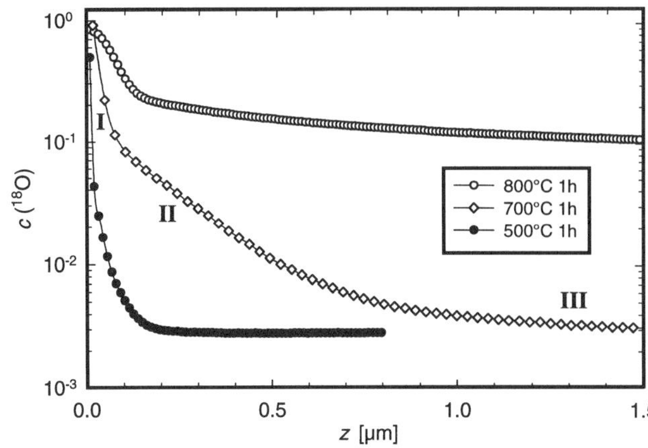

Figure 8. ^{18}O diffusion profiles in nanocrystalline undoped ZrO_2 sintered at 950–970 °C.[16] Diffusion anneals were performed for 1 h in an atmosphere of 100 hPa 97 at%-enriched ^{18}O at the temperatures quoted in the inset. The shape of the diffusion profiles is typical for diffusion from a constant source in the type B regime. The ranges of prevailing diffusion from the surface in to the volume (I), via interfaces (II) and into the bulk of the sample (III) due to residual pores are indicated.

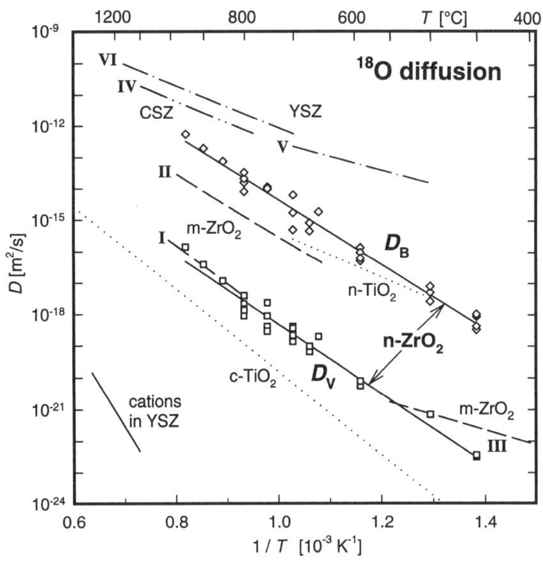

Figure 9. Arrhenius plot of the ^{18}O tracer diffusion coefficients D_V and D_B measured in undoped, monoclinic n-ZrO_2 with a grain size of 80 to 300 nm.[16] Literature data: Oxygen diffusion in undoped ZrO_2 (– – –) obtained by gas exchange techniques (I: Madeyski, et al.,[99] II: Keneshea, et al.,[100]) or thermogravimetry (III: Ikuma, et al.,[101]), in Ca- (CSZ, – · · –) or Y-stabilized zirconia (– · –) as measured by SIMS profiling (IV: Simpson, et al.,[102]), V: Tannhauser, et al.,[103]) or spectroscopy of the exchanged gas (VI: Kim, et al.,[104]) and in n-TiO_2 and bulk (c-)TiO_2 (· · ·, Hoefler, et al., [40][94]). Values for the cation self-diffusion in YSZ as deduced from the shrinkage of dislocation loops (Heuer, et al.[105]) are shown for comparison.

At larger penetration depths a constant ^{18}O background appears (Fig. 8) which increases with temperature and duration of the diffusion anneals. This can be ascribed to ^{18}O penetration through a few residual pores and cracks with subsequent diffusion into the dense agglomerates of crystallites by interface and volume diffusion.[16] Taking into account the crystallite size and the ^{18}O diffusivities D_V and D_B determined from the ^{18}O profiles, the size of these agglomerates can be determined to a few tens of micrometers. Structures of the same size are confirmed by optical microscopy.

The oxygen diffusivities directly determined from diffusion profiles in undoped n-ZrO_2 support earlier results where diffusivities in coarse-grained undoped monoclinic ZrO_2 were indirectly obtained using gas exchange techniques[99][100] (Fig. 9). The interface diffusion coefficients in n-ZrO_2 are lower than the oxygen diffusivities in Ca- (CSZ) or Y-stabilized zirconia (YSZ)[102]–[104] where high volume diffusion coefficients with a low activation energy, Q_V = 1.2–1.3 eV, occur due to a high concentration of structural oxygen vacancies (Fig. 9). Nevertheless, on the basis of the results on undoped n-ZrO_2 it may be anticipated that the oxygen diffusivity in YSZ and CSZ may be further enhanced by the introduction of a high number of crystallite interfaces.

A comparison of the results for ZrO_2 with the chemically related TiO_2 is of particular interest. The diffusion of both oxygen and hafnium was studied in rutile n-TiO_2[40][94] with relative densities above 95% and crystallite sizes $d \cong 30$ nm after sinter forging at 570°C (pressure of 1.5 GPa). In n-TiO_2 the oxygen diffusivity is much faster ($\cong 10^5 \times$) and, accordingly, the diffusion activation energy, $Q_B \approx 1.5$ eV,[40][94] lower than in bulk-TiO_2 ($Q_V = 2.58$ eV,[107] Fig. 9). Although the differences between the bulk and interface O-diffusion are similar in TiO_2 and ZrO_2, the interface O-diffusion in n-ZrO_2 appears to be much faster than in TiO_2 when these diffusivities are related to the melting temperatures (Fig. 9).

Regarding the cation self-diffusion in nanocrystalline ceramics, some information is available from studies of Hf diffusion in n-TiO_2,[40][94] since Hf as a substitutional cation is considered to reflect the Ti self-diffusion. The Hf diffusivities in n-TiO_2 are similar to extrapolated values for the Ti volume diffusion in TiO_2 based on data of Hoshino, et al.,[108] The cationic self-diffusion in rutile is considered to occur via an interstitial mechanism.[95][108] This view is based on the tendency that TiO_2 exhibits an excess of the metallic constituent[95] and on the observation that the Ti self-diffusivity in rutile increases under reducing conditions due to the formation of Ti interstitials.[108] The results on the Hf diffusion in n-TiO_2 indicate that self-diffusion via self-interstitials in interfaces is not enhanced in comparison to volume diffusion in crystals. This situation in TiO_2 appears to differ from metals where diffusion via self-interstitials may contribute to the fast grain boundary diffusion.[34] (See Sec. 3.0.)

Compared to TiO_2, the cationic self-diffusion in ZrO_2 is characterized by slower bulk diffusivities with a higher activation energy (4–5 eV) as deduced from radio tracer experiments (CSZ[106]) and the shrinking of dislocation rings (YSZ[105]). Some indirect information on the cation self-diffusion may also be deduced from sintering processes where both the cation and the anion diffusion are involved. The onset temperatures of full densification and massive grain growth are substantially higher for n-ZrO_2[16] than for n-TiO_2.[40][94] This again indicates that cation diffusivities in ZrO_2 are lower than in n-TiO_2. On the other hand, the fact that n-ZrO_2 can be sintered to full density at temperatures below 0.45 T_m has to be regarded as evidence that in ZrO_2, in contrast to n-TiO_2, the cation diffusivity in the interfaces is enhanced compared to that in the crystalline state. This item could be further clarified by studies of the Zr self-diffusion in fully sintered n-ZrO_2 where both the grain boundary and volume diffusion should be accessible.

ACKNOWLEDGMENT

The authors are indebted to W. Frank, P. Scharwaechter, U. Södervall, and the coworkers of their group, particularly K. Reimann, P. Farber, and R. Dittmar for fruitful collaboration. The work is financially supported by Deutsche Forschungsgemeinschaft (SFB 277, Project B9).

REFERENCES

1. Birringer, R., Gleiter, H., Klein, H. P., and Marquardt, P., Nanocrystalline Materials: An Approach to a Novel Solid Structure With Gas-Like Disorder?, *Phys. Lett. A*, 102:365–369 (1984)
2. Edelsteinand, A. S., and Cammarata, R. C., (eds.), *Nanomaterials: Synthesis, Properties, and Applications*, Institute of Physics, Bristol (1996)
3. Padmanabhan, K. A., Mondal, P., and Hahn, H., Plastic Deformation in Nanocrystalline Materials, *Nanostruct. Mater.*, 9:603–606 (1997)

4. Mayo, M. J., Processing of Nanocrystalline Ceramics From Ultrafine Particles, *Internat. Mater. Rev.*, 41:85–115 (1996)
5. Altstetter, C., Superplasticity of Nanophase Intermetallics?, in: *Mechanical Properties and Deformation Behavior of Materials Having Ultrafine Microstructure*, (M. A. Nastasi, D. M. Parker, and H. Gleiter, eds.) E233:381–399, NATO ASI Series, Kluwer Academic Press, Dordrecht (1993)
6. Hofmann, B., and Kronmüller, H., Stress-Induced Magnetic Anisotropy in Nanocrystalline FeCuNbSi Alloy, *J. Magnetism Magnetic Mater.*, Academic Press, 152:91–98 (1996)
7. Grünberger, W., Bürger, G., Hinz, D., Müller, K. H., and Schultz, L., Warmumformung Nanokristalliner NdFeB-Legierungen, *Verhandl. DPG (VI)* 31:AMV (1996); Cao, L., Müller, K. H., Handstein, A., Grünberger, W., Neu, V., and Schultz, L., High Performance $Sm_2Fe_{15}Ga_2C_2$ Permanent Magnets Made by Mechanical Alloying and Hot Pressing, *J. Phys. D: Appl. Phys.*, 29:271–272 (1996)
8. Damson, B., and Würschum, R., Correlation Between the Kinetics of the Amorphous-to-Nanocrystalline Transformation and the Diffusion in Alloys, *J. Appl. Phys.*, 80:747–751 (1996)
9. Horvath, J., Birringer, R., and Gleiter, H., Diffusion in Nanocrystalline Material, *Sol. Stat. Comm.*, 62:319–322 (1987)
10. Würschum, R., Struktur und Diffusionseigenschaften Nanokristalliner Metalle und Legierungen, Habilitation Thesis, Universität Stuttgart (1997)
11. Gleiter, H., Diffusion in Nanostructured Metals, *Phys. Stat. Sol. B*, 172:41–51 (1992)
12. Schaefer, H. E., Nanostructured Solids—Diffusional Properties and Nanocrystalline Metal Oxides, in: *Defects in Insulating Materials*, (O. Kanert and J. M. Spaeth, eds.), pp. 122–136, World Scientific Publ., Singapore (1993)
13. Larikov, L. N., Diffusion in Nanocrystalline Materials, *Met. Phys. Adv. Techn.*, 15:1–37 (1995)
14. Würschum, R., Reimann, K., and Farber, P., Correlation Between Interfacial Structure, Tracer Diffusion, and Crystal Growth in Nanocrystalline Metals and Alloys, *Defect and Diffusion Forum*, 143–147:1463–1468 (1997)
15. Würschum, R., Reimann, K., Gruß, S., Kübler, A., Scharwaechter, P., Frank, W., Kruse, O., Carstanjen, H. D., and Schaefer, H. E., Structure and Diffusional Properties of Nanocrystalline Pd, *Phil. Mag. B*, 76:401–417 (1997)
16. Brossmann, U., Untersuchung der Sauerstoffdiffusion in Nanokristallinem Monoklinem Zirkondioxid, Dissertation, Universität Stuttgart (1998); Ossmann, U., Würschum, R., Södervall, U., and Schaefer, H. E., Oxygen Diffusion in Ultra-Fine Grained Monoclinic ZrO_2, *J. Appl. Phys.*, 85:7646–7654 (1999)

17. Würschum, R., Farber, P., Dittmar, R., Scharwaechter, P., Frank, W., and Schaefer, H. E., Thermal-Vacancy Formation and Self-Diffusion in Intermetallic Fe_3Si-Crystallites of Nanocomposite Alloys, *Phys. Rev. Lett.*, 79:4918–4921 (1997)

18. Würschum, R., Kübler, A., Gruß, S., Scharwaechter, P., Frank, W., Valiev, R. Z., Mulyukov, R. R., and Schaefer, H. E., Tracer Diffusion and Crystallite Growth in Ultra-Fine Grained Pd Prepared by Severe Plastic Deformation, *Annales de Chimie — Science des Matériaux*, 21:471–482 (1996)

19. Surholt, T., Mishin, Y. M., and Herzig, C., Grain Boundary Diffusion and Segregation of Gold in Copper: Investigation in the Type-B and Type-C Kinetic Regimes, *Phys. Rev. B*, 50:3577–3587 (1994)

20. Surholt, T., and Herzig, C., Grain Boundary Self- and Solute-Diffusion and Segregation in General Large Angle Grain Boundaries in Copper, *Defect Diff. Forum*, 143–147:1391–1396 (1997)

21. Qing Ma, and Balluffi, R. W., Diffusion Along [001] Tilt Boundaries in the Au/Ag System, I. Experimental Results, *Acta Metall. Mater.*, 41:133-141 (1993)

22. Harrison, L. G., Influence of Dislocations on Diffusion Kinetics in Solids With Particular Reference to the Alkali Halides, *Trans. Faraday Soc.*, 57:1191–1199 (1961)

23. Levine, H. S., and MacCallum, C. J., Grain Boundary and Lattice Diffusion in Polycrystalline Bodies, *J. Appl. Phys.*, 31:595–599 (1960)

24. Kaur, I., Mishin, Y., and Gust, W., *Fundamentals of Grain and Interphase Boundary Diffusion*, John Wiley, Chichester (1995)

25. Mishin, Y., and Herzig, C., Diffusion in Fine Grained Materials: Theoretical Aspects and Experimental Possibilities, *Nanostruct. Mater.* 6:859–862 (1995)

26. Klinger, L., Levin, L., and Petelin, A., A Model of Accelerated Diffusion Along Triple Junction of Grain Boundaries, *Mater. Sci. Forum*, 207–209:501–504 (1996)

27. Glaeser, A. M., and Evans, J. W., Effect of Grain Boundary Migration on Apparent Boundary Diffusion Coefficients, *Acta Metall.*, 34:1545–1552 (1986); Güthoff, F., Mishin, Y., and Herzig, C., Self-diffusion Along Stationary and Moving Grain Boundaries in α-Hf, *Z. Metallkd.*, 84:584–591 (1993)

28. Bokstein, B. S., Bröse, H. D., Trusov, L. I., and Khvostantseva, T. P., Diffusion in Nanocrystalline Ni, *Nanostruct. Mater.*, 6:873–876 (1995)

29. Sutton, A. P., and Balluffi, R. W., *Interfaces in Crystalline Materials*, Monographs on the Physics and Chemistry of Materials, Clarendon Press, Oxford (1995)

30. Langmuir, I., Thoriated Tungsten Filaments, *J. Franklin Inst.*, 217:543–570 (1934)

31. Martin, G., Blackburn, D. A., and Adda, Y., Autodiffusion Au Joint de Grains de Bicristaux d'argent Soumis à une Pression Hydrostatique, *Phys. Stat. Sol.*, 23:223–228 (1967)

32. Robinson, J. T., and Peterson, N. L., Correlation Effects in Grain Boundary Diffusion, *Surf. Sci.* 31:586–616 (1972); Robinson, J. T., and Peterson, N. L., *Acta Metall.*, 21:1181–1186 (1973)

33. Kwok, T., Ho, P. S., Yip, S., Balluffi, R. W., Bristowe, P. D., and Brokman, A., Evidence for Vacancy Mechanism in Grain Boundary Diffusion in bcc Iron: A Molecular-Dynamics Study, *Phys. Rev. Lett.*, 47:1148–1151 (1981); Balluffi, R. W., Kwok, T., Bristowe, P. D., Brokman, A., Ho, P. S., and Yip, S., Determination of Vacancy Mechanism for Grain Boundary Self-Diffusion by Computer Simulation, *Scripta Metall.*, 15:951–956 (1981)

34. Qing Ma, Liu, C. L., Adams, J. B., and Balluffi, R. W., Diffusion Along [001] Tilt Boundaries in the Au/Ag System-II. Atomistic Modelling and Interpretation, *Acta Metall. Mater.*, 41:143–151 (1993)

35. Nomura, M., and Adams, J. B., Self-Diffusion Along Twist Boundaries in Cu, *J. Mater. Res.*, 7:3202–3212 (1992)

36. Sommer, J., Herzig, C., Mayer, S., and Gust, W., Grain Boundary Self-Diffusion in Silver Bicrystals, *Defect Diff. Forum*, 66–69:843–848 (1989)

37. Budke, E., Orientierungsabhängigkeit der Radiotracer-Diffusion von ^{195}Au, ^{64}Cu und ^{75}Se in Symmetrischen [001]-Kupferkippkorngrenzen, Dissertation, Universität Münster (1994)

38. Wolf, D., and Merkle, K. L., Correlation Between the Structure and Energy of Grain Boundaries in Metals, in: *Material Interfaces: Atomic-Level Structure and Properties,* (D. Wolf, and S. Yip, eds.), pp. 87–150, Chapman and Hall, London (1992)

39. Seeger, A., and Schottky, G., Die Energie und der elektrische Widerstand von Großwinkelkorngrenzen in Metallen, *Acta Metall.*, 7:495–503 (1959)

40. Höfler, H. J., Diffusion und Festkörperreaktionen in Nanokristallinen Materialien, Dissertation, Universität des Saarlandes (1991) (See also Ref. 11.)

41. Massalski, T. B., Murray, J. L., Bennet, L. H., and Baker, H., *Binary Alloy Phase Diagrams*, American Soc. Metals, ASM International, Ohio (1986)

42. Murray, J. L., *Phase Diagrams of Binary Titanium Alloys,* American Soc. Metals, ASM International, Ohio (1987)

43. Schumacher, S., Birringer, R., Strauß, R., and Gleiter, H., Diffusion of Silver in Nanocrystalline Copper Between 303 K and 373 K, *Acta Metall.*, 37:2485–2488 (1989)

44. Höfler, H. J., Averback, R. S., Hahn, H., and Gleiter, H., Diffusion of Bismuth and Gold in Nanocrystalline Copper, *J. Appl. Phys.*, 74:3832–3939 (1993)

45. Balandin, I. L., Bokstein, B. S., Egorov, V. K., and Kurkin, P. V., Antimony Diffusion in Fine Scale Copper, *Defect Diff. Forum*, 143–147:1475–1480 (1997)

46. Höfler, H. J., Averback, R. S., and Gleiter, H., Diffusion of Boron in Nanocrystalline Iron–A New Type of Diffusion Kinetics: Type C, *Phil. Mag. Lett.*, 68:99–105 (1993)

47. Minamino, Y., Saji, S., Hirao, K., Ogawa, K., Araki, H., Miyamoto, Y., and Yamane, T., Diffusion of Copper in Nanocrystalline Al-7.8 at% Ti-0.3 at% Fe Alloy Prepared by Mechanical Alloying, *Mater. Trans., Jpn. Inst. Metals (JIM)*, 37:130–137 (1996)

48. Peterson, N. L., Isotope Effect in Self-Diffusion in Palladium, *Phys. Rev.*, 136:568–574 (1964)

49. Gust, W., Mayer, S., Bögel, A., and Predel, B., Generalized Representation of Grain Boundary Self-Diffusion Data, *J. de Physique*, 46 Colloq., C4:537–544 (1985)

50. Dickenscheid, W., Birringer, R., Gleiter, H., Kanert, O., Michel, B., and Günther, B., Investigation of Self-Diffusion in Nanocrystalline Copper by NMR, *Solid State Commun.*, 79:683–686 (1991)

51. Sommer, J., and Herzig, C., Direct Determination of Grain Boundary and Dislocation Self-Diffusion Coefficients in Silver From Experiments in Type-C Kinetics, *J. Appl. Phys.*, 72:2758–2766 (1992)

52. Bonzel, H. P., and Latta, E. E., Surface Self-Diffusion on Ni(110): Temperature Dependence and Dislocation Anisotropy, *Surface Science*, 76:275–295 (1978)

53. Maier, K., Mehrer, H., Lessmann, E., and Schüle, W., Self-Diffusion in Nickel at Low Temperatures, *Phys. Stat. Sol. B*, 78:689–698 (1976)

54. Maier, K., Self-Diffusion in Copper at "Low" Temperatures, *Phys. Stat. Sol. A*, 44:567–576 (1977)

55. Birringer, R., Krill, C. E., and Klingel, M., Orientation-Phase-Space-Averaged Properties of Grain Boundaries, *Phil. Mag. Lett.*, 72:71–77 (1995)

56. Reimann, K., and Würschum, R., Distribution of Internal Strains in Nanocrystalline Pd Studied By X-ray Diffraction, *J. Appl. Phys.*, 81:7186–7192 (1997)

56a. Tanimoto, H., Farber, P., Würschum, R., Valiev, R. Z., and Schaefer, H. E., Self-Diffusion in High-Density Nanocrystalline Fe, *Nanostruct. Mater.*, 12:681–684 (1999)

57. Akhamadeev, N. A., Kobelev, N. P., Mulyukov, R. R., Soifer, Y. M., and Valiev, R. Z., The Effect of Heat Treatment on the Elastic and Dissipative Properties of Copper with the Submicrocrystalline Structure, *Acta Metall. Mater.* 41:1041–1046 (1993); Jianshe Lian, Valiev, R. Z., and Baudelet, B., On the Enhanced Grain Growth in Ultrafine Grained Metals, *Acta Metall. Mater.*, 43:4165–4170 (1995)

58. Sanders, P. G., Weertman, J. R., Barker, J. G., and Siegel, R. W., Small Angle Neutron Scattering From Nanocrystalline Palladium as a Function of Annealing, *Scripta Metall. Mater.*, 29:91–96 (1993)

59. Würschum, R., Gruß, S., Gissibl, B., Natter, H., Hempelmann, R., and Schaefer, H.-E., Free Volumes and Thermal Stability of Electro-Deposited Nanocrystalline Pd, *Nanostruct. Mater.*, 9:615–618 (1997)

60. Hofmann, B., Reininger, T., and Kronmüller, H., Influence of the Microstructure on the Magnetization Processes in Nanocrystalline $Fe_{73.5}Si_{13.5}B_9Nb_3Cu_1$, *Phys. Stat. Sol. A*, 134:247–261 (1992)

61. Babcock, S. E., and Balluffi, R. W., Grain Boundary Kinetics-II: In-situ Observation of the Role of Grain Boundary Dislocations in High-Angle Grain Boundary Migration, *Acta Metall.*, 37:2367–2376 (1989)

62. Malow, T. R., and Koch, C. C., Grain Growth of Nanocrystalline Materials -A Review, in: *Synthesis and Processing of Nanocrystalline Powder*, (D. L Bourrell, ed.), p. 33, TMS, Warrendale, USA (1996)

63. Humphreys, F. J., and Hatherly, M., *Recrystallization and Related Annealing Phenomena,* Pergamon, Elsevier, Oxford (1995)

64. Liu, Y., and Patterson, R. R., Grain Growth Inhibition by Porosity, *Acta Metall. Mater.*, 41:2651–2656 (1993)

65. Mayo, M. J., and Hague, D. C., Porosity-Grain Growth Relationships in the Sintering of Nanocrystalline Ceramics, *Nanostruct. Mater.*, 3:43–52 (1993)

66. Gertsman, V. Y., and Birringer, R., On the Room-Temperature Grain Growth in Nanocrystalline Copper, *Scripta Metall. Mater.*, 30:577–581 (1994)

67. Lebedev, A. B., Pulnev, S. A., Kopylov, V. I., Burenkov, Y. A., Vetrov, V. V., and Vylegzhanin, O. V., Thermal Stability of Submicrocrystalline Copper and Cu:ZrO_2 Composite, *Scripta Mater.*, 35:1077–1080 (1996)

68. Morris, D. G., and Morris, M. A., Microstructure and Strength of Nanocrystalline Copper Alloy Prepared by Mechanical Alloying, *Acta Metall. Mater.*, 39:1763–1770 (1991)

69. Krill, C. E., Klein, R., Janes, S., and Birringer, R., Thermodynamic Stabilization of Grain Boundaries in Nanocrystalline Alloys, *Mater. Sci. Forum*, 179–181:443–448 (1995)

70. Bernardini, J., Gas, P., Hondros, E. D., and Seah, M. P., The Role of Solute Segregation in Grain Boundary Diffusion, *Proc. Roy. Soc. Lond. A*, 379:159–178 (1982)

71. Konrad, H., Haubold, T., Birringer, R., and Gleiter, H., Nanostructured Cu-Bi Alloys Prepared by Co-evaporation in a Continuous Gas Flow, *Nanostruct. Mater.*, 7:605–610 (1996)

72. Yoshizawa, Y., Oguma, S., and Yamauchi, K., New Fe-Based Soft Magnetic Alloys Composed of Ultrafine Grain Structure, *J. Appl. Phys.* 64:6044–6046 (1988); Herzer, G., Grain Structure and Magnetism of Nanocrystalline Ferromagnets, *IEEE Transactions on Magnetics*, 25:3327–3329 (1989)

73. Hampel, G., Pundt, A., and Hesse, J., Crystallization of $Fe_{73.5}Si_{13.5}B_9Nb_3Cu_1$: Structure and Kinetics Examined by X-ray Diffraction and Mössbauer Spectroscopy, *J. Phys.; Condens. Matter* 4:3195–3214 (1992)

74. Lübbehusen, M., and Mehrer, H., Self-Diffusion in α-Iron: The Influence of Dislocations and the Effect of the Magnetic Phase Transition, *Acta Metall. Mater.*, 38:283–292 (1990)

75. Gude, A., and Mehrer, H., Diffusion in the $D0_3$ Type Intermetallic Phase Fe_3Si, *Phil. Mag. A*, 76:1–30 (1997)

76. Kümmerle, E. A., Badura, K., Sepiol, B., Mehrer, H., and Schaefer, H. E., Thermal Formation of Vacancies in Fe_3Si, *Phys. Rev. B*, 52:R6947–R6950 (1995)

77. Ulfert, W., Horváth, J., and Frank, W., Self-Diffusion of ^{59}Fe Tracer Atoms in Amorphous $Fe_{78}Si_9B_{13}$ and $Fe_{40}Ni_{40}B_{20}$, *Cryst. Lattice Def. Amorph. Mater.*, 18:519–531 (1989)

78. Wang, J., Wolf, D., Phillpot, S. R., and Gleiter, H., Computer Simulation of the Structure and Thermo-Elastic Properties of a Model Nanocrystalline Material, *Phil. Mag. A*, 73:517–555 (1996)

79. Nishi, Y., Solid-Undercooled Liquid Interfacial Energy of Pd-Cu-Si Alloy Glass, in: *Rapidly Quenched Metals,* (S. Steeb and H. Warlimont, eds.), pp. 231–234, Elsevier, Amsterdam (1985)

79a. Herth, S., Michel, T., Tanimoto, H., Eggersmann, M., Dittmar, R., Schaefer, H.-E., Frank, W., and Würschum, R., Self-Diffusion in Nanocrystalline Fe and Fe-Rich Alloys, *Defect and Diffusion Forum*, Proc. of the Int. Conf. on Diffusion in Materials, 194–199:1199–1204 (2001)

80. Filal, M., Petot, C., Mokchah, M., Chateau, C., and Carpentier, J. L., Ionic Conductivity of Yttrium-Doped Zirconia and the "Composite Effect," *Solid State Ionics*, 80:27–35 (1995)

81. Kirchheim, R., Mütschele, T., Kieninger, W., Gleiter, H., Birringer, R., and Koblé, T. D., Hydrogen in Amorphous and Nanocrystalline Metals, *Mater. Sci. Eng.*, 99:457–462 (1988)

82. Palumbo, G., Doyle, D. M., El-Sherik, A. M., Erb, U., and Aust, K. T., Intercrystalline Hydrogen Transport in Nanocrystalline Nickel, *Scripta Metall. Mater.*, 25:679–684 (1991)

83. Maier, C. U., and Kronmüller, H., Diffusion and Relaxation Processes of Hydrogen in Heterogenous Microcrystalline Fe-Zr Alloys, *Z. Metallkd.*, 84:410–418 (1993)

84. Hirscher, M., Zimmer, S., and Kronmüller, H., Diffusion of Hydrogen in Nanocrystalline $Fe_{90}Zr_{10}$, *Z. Phys. Chem.*, 183:51–58 (1994)

85. Mössinger, J., Hirscher, M., and Kronmüller, H., Diffusion of Hydrogen in Nanocrystalline Pd_3Fe, *Phil. Mag.B*, 73:503–510 (1996)

86. Sinning, H.-R., Mechanical Spectroscopy With Hydrogen in Intermetallic Phases, *J. Alloys Comp.*, 211/212:216–221 (1994)

87. Kappesser, B., Stuhr, U., Wipf, H., Weissmüller, J., Clos, C., and Gleiter, H., Hydrogen-Induced Internal Friction in Nanocrystalline Palladium, *J. Alloys Compounds*, 231:337–342 (1995)

88. Stuhr, U., Wipf, H., Udovic, T. J., Weissmüller, J., and Gleiter, H., Inelastic Neutron Scattering of Hydrogen in Nanocrystalline Pd, *Nanostruct. Mater.*, 6:555–558 (1995)

89. Janßen, S., Natter, H., Hempelmann, R., Striffler, T., Stuhr, U., Wipf, H., Hahn, H., and Cook, J. C., Hydrogen Diffusion in Nanocrystalline Pd by Means of Quasielastic Neutron Scattering, *Nanostruct. Mater.*, 9:579–582 (1997)

90. Hirscher, M., Diffusion of Hydrogen in Amorphous and Nanocrystalline Alloys, in: *Interstitial Intermetallic Alloys*, (F. Grandjean, et al., eds.), pp. 333–347, Kluwer Academic Press, Dordrecht (1995)

91. Würschum, R., Greiner, W., Valiev, R. Z., Rapp, M., Sigle, W., Schneeweiss, O., and Schaefer, H.-E., Interfacial Free Volumes in Ultra-Fine Grained Metals Prepared by Severe Plastic Deformation, by Spark-Erosion, or by Crystallization of Amorphous Alloys, *Scripta Metall. Mater.*, 25:2451–2456 (1991)

92. Würschum, R., and Schaefer, H.-E., Interfacial Free Volumes and Atomic Diffusion in Nanostructured Solids, in Ref. 2, pp. 277–301.

93. Chen, I. W., and Xue, L. A., Development of Superplastic Structural Ceramics, *J. Amer. Ceram. Soc.*, 73:2585–2609 (1990)

94. Höfler, H. J., Hahn, H., and Averback, R. S., Diffusion in Nanocrystalline Materials, *Def. Diff. Forum*, 75:195–210 (1991)

95. Atkinson, A., Diffusion in Ceramics, in: *Materials Science and Technology,* Vol. 11, *Structure and Properties of Ceramics,* (M. V. Swain, ed.), pp. 299–338, VCH Weinheim (1994)
96. Subbarao, E. C., Zirconia–An Overview, in: *Advances in Ceramics, Science and Technology of Zirkonia,* (A. H. Heuer and L. W. Hobbs, eds.), 3:1–13, American Ceramic Soc., Columbus, Ohio (1981)
97. Scott, H. G., Phase Relationships in the Zirconia-Yttria System, *J. Mater. Sci.*, 10:1527–1535 (1975)
98. Würschum, R., Soyez, G., and Schaefer, H.-E., Preparation and Sintering Studies of Nanometer-Sized Polycrystalline ZrO_2, in: *Structure and Properties of Interfaces in Materials*, (W. A. T. Clark, C. L. Briant, and U. Dahmen, eds.) 238:733–738, *Mat. Res. Soc. Symp. Proc.,* Pittsburgh (1992)
99. Madeyski, A., and Smeltzer, W. W., Oxygen Diffusion in Monoclinic Zirconia, *Mat. Res. Bull.*, 3:369–375 (1968)
100. Keneshea, F. J., and Douglas, D. L., The Diffusion of Oxygen in Zirconia as a Function of Oxygen Pressure, *Oxidation of Metals,* 3:1–14 (1971)
101. Ikuma, Y., Komatsu, K., and Komatsu, W., in: *Advances in Ceramics,* Vol. 24: *Science and Technology of Zirkonia III,* (S. Somiya, N. Yamamoto, and H. Yanagida, eds.) pp. 749–758, American Ceramic Soc., Columbus, Ohio (1988)
102. Simpson, L. A., and Carter, R. E., Oxygen Exchange and Diffusion in Calcia-Stabilized Zirconia, *J. Amer. Ceram. Soc.*, 49:139–144 (1966)
103. Tannhauser, D. S., Kilchner, J. A., and Steele, B. C. H., The Determination of the Oxygen Self-Diffusion and Gas-Solid Exchange Coefficients for Stabilized Zirconia by SIMS, *Nucl. Instr. Meth.*, 281:504–508 (1983)
104. Kim, B. K., Park, S. J., and Hamaguchi, H., Determination of the Oxygen Self-Diffusion Coefficients in Y_2O_3-Containing Tetragonal Zirconia Polycrystals by Raman Spectrometric Monitoring of the ^{16}O-^{18}O Exchange Reaction, *J. Amer. Ceram. Soc.*, 76:2119--2122 (1993)
105. Chien, F., and Heuer, A., Lattice Diffusion Kinetics in Y_2O_3-Stabilized ZrO_2 Single Crystals: A Dislocation Loop Annealing Study, *Phil. Mag A*, 73:681–697 (1996)
106. Rhodes, W. H., and Carter, R. E., Cationic Self-Diffusion in Calcia-Stabilized Zirconia, *J. Amer. Ceram. Soc.*, 49:244–249 (1966)
107. Kingery, W., Bowen, H. K., and Uhlmann, D. R., *Introduction to Ceramics*, J. Wiley & Sons, New York (1976)
108. Hoshino, K., Peterson, N. L., and Wiley, C. L., Diffusion and Point Defects in TiO_2, *J. Phys. Chem. Solids*, 46:1397–1411 (1985)

8

Nanostructured Materials for Gas Reactive Applications

Michel L. Trudeau

1.0 INTRODUCTION

The last ten years have revealed the unique properties of nanostructured materials as reflected by the various chapters in this book, and many technological areas clearly stand to benefit from the development of materials on the nanoscale. However, there is one field already based on nanometer crystals that could probably benefit most by such developments: materials used in gas reactive applications. Interestingly, this topic is not the first that comes to mind when discussing nanostructured materials, even if their potential for technological applications is tremendous. It is not clear whether the reason for this stems from the fact that the majority of researchers who first explored the field of nanocrystalline materials belonged to science and physics groups, or the fact that chemists and chemical engineers had already been working in this field for so many decades that they did not notice the headline news about nanosize crystals. Examples of the technological importance of nanostructured materials in this field abound. For instance, metallic nanoparticles highly dispersed heterogeneous catalysts are important to 20% of the GNP of the United States[1] while in Japan in 1986, the number of gas sensors used for home gas leak alarms, based on

SnO_2, was about five million units,[2] most of which were based on nanostructured grains.

The purpose of this chapter is to review recent developments in the field of nanostructured materials in gas reactive applications. Three different research areas are reviewed: catalysis and electrocatalysis, an area which has been working with nanostructured materials for more than fifty years; semiconductor gas sensors, materials which represent a perfect example of nanostructure-related properties; and hydrogen storage, an area where, in latter years, nanostructured materials and designs have succeeded in considerably improving the hydrogen absorption and desorption cycle. The main objective is to briefly describe some of the basic principles in order to show the advantages of nanostructured materials. For each section, recent developments in synthesizing nanostructured materials are presented, with the emphasis less on wet chemistry and more on nontraditional synthesis techniques. This work does not go into detail about the different chemical processes and reactions but deals more with nanostructured material development and the large potential of these fields as well as the already diverse research studies that have emerged. It is hoped that the examples presented here will encourage researchers from different branches of materials science to combine their efforts to produce new materials and to "nanodesign" new structures that will show improved or unique properties.

2.0 CATALYSIS AND ELECTROCATALYSIS

Fundamentally, catalysis involves a cyclic process in which a site on a catalyst forms a complex with reactants from which products are desorbed, thereby restoring the original site and continuing the cycle.[3] In these materials, the active site of a reaction may be a group or cluster of neighboring atoms on the catalyst surface, or a species adsorbed onto the catalyst. These sites are associated with surface structures, lattice defects, and edges or corners of crystallites. Using a number of different synthesis processes, it is now possible to derive nanocrystals with a specific average crystallite size, and with some control over aspects such as the lattice parameters and the amount or type of internal defects with a metastable structure quite different from what is expected with a normal equilibrium process. This will result, for instance, in an extension of the concentration range of a structural phase or of the solubility limit of an element. Furthermore, as the crystallite size decreases below 10 nm and even below 5 nm, surface atoms will start to dominate while, at the same time, other structures, such as triple junctions, will be more and more present. The surface design of materials, as well as the increased presence of these structures, could have a large impact on the properties of catalytic materials.

Nanostructured materials are not something new in catalysis or electrocatalysis. Metallic nanoparticles with an average grain size between 1 to 20 nm, such as Pt or Rh, dispersed on various substrates like SiO_2 or Al_2O_3 have been used for the past fifty years in heterogeneous catalysis.[1][4] These materials are used extensively in many industries such as petrochemical production, automobile emission control, and fine-chemicals synthesis. The exciting news generated by the recent works on nanostructured materials is that it is now possible to consciously "nanodesign" the structure of a material in order to improve its catalytic response or to better understand a particular catalytic process.

Because nanostructured materials have played a major role in catalysis for the past fifty years, this section is not an exhaustive survey of all the relevant studies that show the impact of a crystalline structure on the nanometer scale. Its main purpose is to look at recent studies based on new materials synthesis techniques that have been recently developed and which show the importance of nanoscale materials for the development of catalytic research and the related industries.

2.1 Impact of Structure on Catalysis and Electrocatalysis Processes

Before looking at the different methods for producing nanoscale particles, it is important to consider briefly how the structure can be related to a catalytic reaction. To arrive at new catalysts with nanostructure-level design, different phenomena need to be considered:[1][5] the cluster size and distribution, the sensitivity of the reaction to the structure of the catalyst (crystalline structure, interatomic spacing, number of defects, etc.), the thermodynamic properties (surface energy, heat of adsorption), the electronic state of the active site, the interaction with the support (if one is used), and the spillover of one reaction species to another component of the catalyst or to the catalytic support.

It is well-known that many catalytic reactions depend on the structure of the catalyst such as its particle size, the amount and types of internal defects, the surface structure,[6] etc. Since it is now well known that all these morphological parameters are influenced by the size of the crystals,[7] it is clear that catalytic activity will also be different for large crystalline materials (> 20 nm) and nanocrystalline compounds (< 20 nm), especially at very low crystallite sizes (≈ 5 nm). Zhdanov and Kasemo[8] modeled the effect of catalyst size on the reaction kinetics and have shown that the reaction on small nanometer particles could be very different than on larger surfaces. Such behavior was, in fact, observed by Peuckert, et al.,[8] who found an optimum size of about 3.5 nm for the oxygen reduction kinetics on carbon-supported Pt particles in acid electrolytes. For their part, Zhao, et al.,[10] and Feng, et al.,[11] have also shown that the catalytic activity of a 3-nm iron oxide catalyst used for direct coal liquefaction was greatly dependent on its surface chemistry and, in particular, on the number of water molecules adsorbed at the surface of coordinately unsaturated sites. Another good example of the importance of the crystallite size was presented by Wang, et al.,[12] for the photocatalytic decomposition of $CHCl_3$ by nanocrystalline TiO_2 ranging in size from 6 to 21 nm. They found a blue shift of the absorption edge with decreasing particle size in optical absorption, confirming that quantum-sized particles were produced, and determined that particles with an average size of 11 nm had the highest photoactivity. Also, as catalytic materials are designed, the structure sensitivity of some reactions has to be considered. Davis and Klabunde[13] have shown that some reactions depend on the presence of a single active surface but, in other cases, on several adjacent active sites. Another example is the enhanced adsorption at low pressure of CO_2 and SO_2 of nanoscale MgO (4 nm) produced by an autoclave compared to larger (150 nm) crystalline materials, owing to the abundance of edge and corner sites and crystal planes.[14]

In heterogeneous catalysis, the active catalytic component may be a metal or a metal oxide, which is usually present in small quantities. It relies on a relatively inert support component to disperse it and to stabilize it as a reactive phase. The importance of the catalytic support as well as the amount of catalytic material has been reported in many studies.[15]–[18] Recently, Lalande, et al.,[19] have shown that, for nanocrystalline Co particles supported on high-surface-area carbon used for the electrocatalytic oxygen reduction reaction in fuel cells, there is an optimum Co concentration. These nanoparticles were produced by the pyrolysis of cobalt phthalocyanine

(CoPc). The maximum activity, achieved at 3.5 wt% of Co, corresponded to full monolayer coverage of the CoPc on the support, indicating that the activity of the nano-Co was directly affected by its interaction with the carbon support. In another example, Yao, et al.,[20] found that, during the thermal processing of a Pt catalyst supported on TiO_2 (Pt/TiO_2), there was a catalyst-support interaction which led to the formation of a Ti suboxide coating at the Pt particle surface, resulting in a decrease in the catalyst activity. No such interaction was found by Zafiris and Gorte[21] in the case of Pt/amorphous CeO_2. Yao, et al.,[20] also noted that the structure of the Pt particles was dependent on the nature of the support used. For example, in the case of TiO_2 or CeO_2 support, the small Pt particles (diameter of about 4 nm) were well-faceted cuboctahedral, whereas Pt particles on Al_2O_3 were found to be more spherical. Fukushima, et al.,[22] found that Au 1 nm clusters prepared by an ionized cluster beam had a catalytic activity for the oxidation of Co that was dependent on both the size of the clusters and the nature of the support. Catalytic activity at 150°C was higher than for platinum catalyst at 300°C for Au islands on TiO_2 anatase. If the substrate used was TiO_2 rutile, the activity was found to be reduced by half. In addition, the pore structure of the support can contribute to the selectivity of a catalytic reaction by allowing only molecules that are smaller than the pore opening to gain access to the catalytic sites or to be produced.

Detailed calculations of different bimetallic clusters have shown the possibility of variation in the surface site composition as a function of the bimetal composition.[23][24] The interaction between a catalytic material and its support has also been modeled in some cases.[25] One interesting result is the size dependence of this interaction. For alloy particles, surface segregation was shown to be affected by the particle size. For instance, in systems where segregation of one component is expected in normal coarse-grained particles, a reduction in segregation will result from grain size decrease. This behavior was also demonstrated to be dependent on the bond strength between the different alloy components and the support. For example, an increase in segregation with increased crystallite size was observed if both elements equally strongly bonded to the support. A reverse of this trend would be observed if the segregating component had a stronger bond with the support than the non-segregating one. Control of the particle size and its relation to the designed surface composition can, thus, be seen as an important factor for the development of new catalysts.

2.2 Nanostructure Design

The development of gas phase condensation has led to fundamental advances in the synthesis of new catalytic materials.[26] Some of the first results, presented by Beck and Siegel, studied the activity for the dissociative adsorption of H_2S in a H_2 environment on nanocrystalline TiO_2, produced by the gas-phase evaporation of Ti, followed by post-oxidation.[27] The nano-TiO_2 had a rutile structure with an average diameter of 12 nm and a specific surface area of 76 m^2/g. Better initial activity was found for the nanocrystalline materials than for the commercially available materials. This was partially attributed to the high surface area of the gas-phase-condensed rutile and, also, to its high oxygen deficiency, suggesting a relatively high concentration of anion vacancies or defects in the structure. These anion vacancies were shown to be partly responsible for the dissociation of the H_2S at the surface, which was verified by oxygen annealing at 800°C. After a 10 minute treatment, the surface area of the nano-TiO_2 did not decrease but the stoichiometry showed a measurable increase, which corresponded to a reduction in the overall activity of this catalyst.

Sarkas, et al.,[28] looked at the catalytic activity of nanocrystalline 5 wt% Li-MgO produced by co-evaporation of Li and Mg metals followed by post-oxidation at room temperature for 12 hours in 20% wt% O_2/He. The resulting crystallites have an average diameter of about 5.2 nm with a surface area of 750 m^2/g. The study of this nanocomposite catalyst for the oxidative conversion of methane to higher hydrocarbons revealed excellent results. The nanocomposite provided catalytic activity at least 200°C below the temperature requirements of conventional Li-impregnated MgO catalysts. At higher temperatures (above 300°C), even though some sintering occurred, the average activity of the nanostructured material was about 3.3 times greater than that of the conventional sample.

In another example, non-stoichiometric and highly surface-reactive nanocrystalline cerium oxides have been tailored with a view to catalyzing redox reactions[15][29] such as selective SO_2 reduction by CO:

Eq. (1) $$SO_2 + 2CO \rightarrow S + 2CO_2$$

For this reaction, the surface of an oxide catalyst first needed to be reduced by oxidation of the adsorbed CO species. The oxygen vacancy produced in the catalyst could then enable an adsorbed SO_2 species to be reduced to SO. If there were another oxygen vacancy in the vicinity, then the SO species could be further reduced to elemental sulfur. Cerium oxide was the catalytic material of choice because of its high oxygen mobility. The rare earth Ce may exist in +3 and +4 oxidation states. Reduction of CeO_2, however, typically requires high-temperature treatment, which will eliminate the surface area of the catalyst. Through magnetron sputtering of Ce in Ar followed by controlled post-oxidation, Tschöpe and Ying were able to derive unique CeO_{2-x} nanocrystals with a high surface area and oxygen vacancy concentration.[29][30] X-ray photoelectron spectroscopy indicated that the as-prepared sample consisted of a 22% Ce^{+3} component and a 78% Ce^{+4} component, and required an oxidizing heat treatment at ≥ 500°C to be fully oxidized to CeO_2.[31][32] The nanocrystalline material was found to be an excellent catalyst, which selectively reduced SO_2 to elemental sulfur with 100% conversion at 500°C[33] compared to the 600°C needed by ultrafine stoichiometric CeO_2 powder derived by chemical precipitation.[34] If Cu dopants were used, the SO_2 reaction temperature could be further reduced to 420°C.[35] This nanocomposite system was obtained by sputtering from a mixed Cu-Ce target. From controlled post-oxidation, Cu was forced to segregate to the surface of the CeO_{2-x} nanocrystals, since the two components have low miscibility in their oxide phases. The resulting ultrahigh Cu dispersion on CeO_{2-x} nanocrystals[36][37] created many semiconductor Schottky junctions that facilitated the formation of oxygen vacancy. It also improved the CO adsorption of the catalyst through spillover effects such that CO oxidation could occur at the remarkably low temperature of 80°C over the Cu/CeO_{2-x} catalyst.[38] This model nanocomposite catalyst also has an excellent poisoning resistance against CO_2 and is promising as an effective catalyst for SO_2 removal from the industrial flue gas streams of fossil fuel combustion.

Similarly, nanocrystalline Cu with an average size ≤ 8 nm was prepared by quenching ultrafine particles or clusters from a high-temperature furnace into condensed inert substrates at cryogenic temperature.[39] Using substrates that can be dissolved allows the formation of heterogeneous catalysts which are active for the oxidation of CO to CO_2, several orders of magnitude higher than that of similar catalysts prepared by conventional methods.

Another synthesis technique offering great potential for the development of new active materials is high-energy mechanical alloying or high-energy milling. The main advantages of mechanical milling over most other techniques are its simplicity, the ease of scale-up and the cost, especially when compared to processes such as gas-phase condensation. A major disadvantage, however, is the specific surface of the milled powders, which normally varies between 1 and 10 m^2/g, too small to be used directly for most catalytic applications. Another drawback of this process, as mentioned by Fecht in Ch.3 of this book, is the volume of impurities unavoidably incorporated in the powders during milling. However, in catalysis, these impurities are not necessarily detrimental to the material's properties and may even be beneficial in some cases, although their presence often complicates the phenomenon under study and makes the experimental data more difficult to interpret.

Because the specific surface is not such an important issue in electrocatalysis compared to direct catalysis, some of the first studies on the properties of milled materials were performed on specific electrocatalytic reactions. Trudeau, et al.,[40] and Huot, et al.,[41] examined the electrocatalytic activity for the hydrogen evolution reaction (HER) in an alkaline solution of nanocomposite Fe-based powders produced by the mechanical crystallization of Fe-(Co,Ni)-Si-B amorphous alloy. High mechanical deformation made these amorphous alloys crystallize, producing a two-phase material: α-Fe-rich crystallites surrounded by an amorphous Fe-B-based region. Compared to amorphous ribbons, these nanocomposite powders showed an improved activity for the HER. Also, before use, amorphous ribbons of similar composition first had to be activated by means of an oxidation/reduction treatment, which resulted in the formation of Fe crystals at the surface. No such activation was necessary for the nanocomposite powders because α-Fe nanocrystals were already present. Moreover, the Ni-containing amorphous ribbons were found not to be active even after the oxidation/reduction treatment, since, in this case, a passive Ni-oxide layer was produced in the process. On the other hand, a milled powder of similar composition revealed an activity comparable to the non-Ni-containing nanocomposites, even if its structure remained x-ray amorphous. This improved activity for the ball-milled sample is still not well understood, but could be due to the presence of a large number of structural defects and a rough surface morphology, or to the presence of very small iron nanoclusters in the powder not detectable by x-ray.

Benameur, et al.,[42] for their part, have looked at the HER activity of metastable nickel produced by milling 65 at% of Al and 35 at% of Ni for 150 hours followed by leaching in alkaline solution (20 wt% KOH). In agreement with Ivanov, et al.,[43][44] they found that mechanical alloying gave rise to the formation of a CsCl-type NiAl structure. After leaching, nearly all the aluminum could be removed, leading to the formation of a porous bcc Raney nickel. Interestingly, even though the remaining material is structurally different from the conventional fcc Raney nickel, the first HER study revealed no major difference in activity between these two materials.

Another system offering considerable interest for HER is Ni-Mo, which was first investigated more than nineteen years ago for its high activity. At that time, some Ni-Mo coated electrodes were produced by a complex oxidation-reduction process at elevated temperatures.[45][46] Based on these studies, substantial work was devoted to investigating the structural transformation produced by mechanically alloying Ni and Mo powders, and the impact of such transformation on the electrocatalytic activity for HER.[47][48] The results showed that Ni nanocrystals, with up to 27 at% of Mo in solution, could be produced by mechanical alloying. The pressed-powder electrodes produced an HER activity superior to that of any known materials in alkaline solution at 70°C. These studies suggested that the increased Mo dissolution in the Ni lattice resulting from a reduction in the crystallite size to the nanometer range during milling was the main reason for the improved properties of this material, as presented in Fig. 1a. However, further investigation revealed (Fig. 1b) that oxygen impurities, which were unintentionally introduced to the level of 2 at% in these first experiments, were also key to the unique performance of these electrodes.[49] Figure 1*b* indicates that, to be active, at least 1 wt% of oxygen had to be incorporated into the milled powder. Surface analysis indicated that a Mo suboxide is present at the surface of the active nanostructured grains. As mentioned previously, non-stoichiometric oxides are often good catalytic materials and these Ni-Mo-O powders are probably another example of such systems. At present, the detailed microstructure is still not well understood. In the presence of oxygen, reactive milling could have incorporated oxygen into the Ni structure or caused some Mo atoms to segregate to the surface of the nanocrystals to form an oxide or suboxide shell structure affecting the surface potential, leading to the observed high activity. One result that tends to support this second hypothesis is the decrease in the amount of Mo in solution with further milling in oxygen, which is followed by the formation of MoO_2. Similarly, new non-stoichiometric Ti-Fe-Ru-O nanostructured high-performance cathodes for the electrochemical synthesis of sodium chlorate were produced by milling pure metals, oxide powders, and mixtures of the two.[50][51]

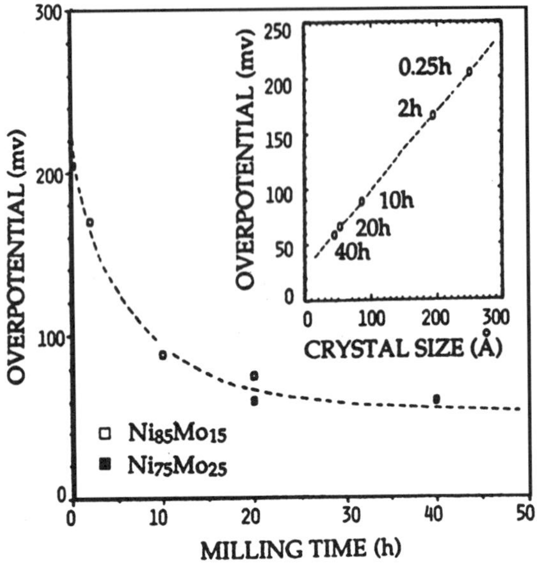

(a)

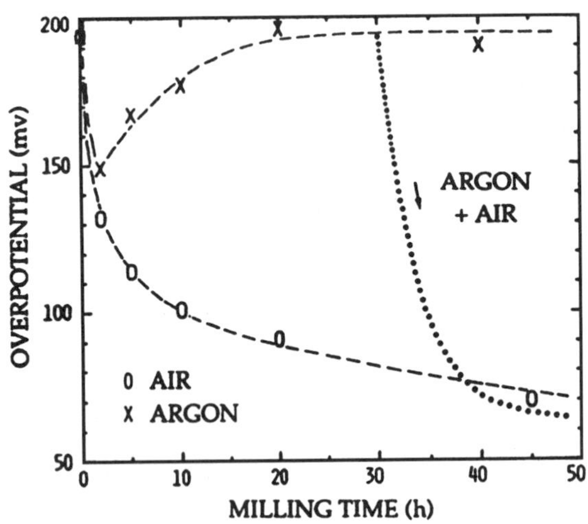

(b)

Figure 1. Hydrogen overpotential at 250 $mAcm^{-2}$ (*a*) as a function of milling time and, in the inset, in relation with the crystallite size and (*b*) as a function of milling time for a $Ni_{75}Mo_{25}$ powder mixture milled under air and argon.[49]

Partially nanocrystalline $Fe_{78}Si_{12}B_{10}$ materials, produced by milling crystalline foils, were also found to be much more active for the catalytic hydrogenation of CO than cast or thermally crystallized materials.[52][53] The structure of the low-temperature annealed sample (500°C) was found to be mainly amorphous with very small α-Fe crystallites of < 10 nm. The activity of this material was found to be three times higher than the fully crystallized sample (crystallite size ≈ 22 nm) and six times higher than the crystalline cast sample (crystallite size > 30 nm).

Trovarelli, et al.,[54] and Miani, et al.,[55] studied the catalytic properties of iron carbides produced by mechanical alloying for the hydrogenation of CO_2. Using WC vial and balls, iron powder (75 at%) and graphite (25 at%) were milled together for 7.5 hours. The result was the production of a composite powder composed of Fe, orthorhombic Fe_3C, hexagonal ε-Fe_2C and monoclinic χ-Fe_5C_2 phase, with an average crystallite size of about 13 nm. The catalytic activity of this powder for the hydrogenation of CO_2 was found to be comparable to noble metal catalysts dispersed on high-surface-area supports.

As mentioned previously in this section, the use of high-energy milled powders in catalysis is restricted due to the low surface area of the powders. One way to circumvent this difficulty is to synthesize a heterogeneous nanocomposite catalyst using high-energy mechanical milling. For example, this technique was used to mill α-Al_2O_3 nano-powders together with nanocrystalline Ni-Ru powders, resulting in a high dispersion of metallic particles on the alumina support with a high specific surface area.[56] TPD experiments indicated that the specific surface of the metallic catalyst increased considerably upon milling with the 100 m^2/g nanoscale α-alumina. In this case, the α-Al_2O_3 was also obtained by high-energy milling from the transformation of γ-Al_2O_3 to the α–phase.[57] High-energy mechanochemistry was also used to prepare metallic nanocrystals (10 nm) supported catalyst directly by high-energy milling metallic oxide M_xO_y (such as Fe_2O_3) and aluminum powder following the reaction:

Eq. (2) $$\frac{3}{y}M_xO_y + 2Al = Al_2O_3 + 3\frac{x}{y}M$$

Surface activation can also be used to improve the surface area of high-energy-produced powders. For example, Sun, et al.,[58] looked at the catalytic activity of mechanically milled $Cu_{30}Al_{70}$ activated in NaOH nanocrystalline for the hydrogenation of xylose. After activation, the powder had a specific surface of about 24 m^2/g and an activity six times higher than that of vacuum-melted samples with a high surface area.

Lastly, even if high-energy milling mostly yields low-surface-area powders, high-surface-area active materials have been synthesized using high-surface-area precursors. De Leitenburg, et al.,[59] produced Ce-Zr-O solid solution by milling CeO_2 (surface area of 55 m^2/g) with ZrO_2 (surface area of 15 m^2/g) and obtained compounds with an average size between 9 and 13 nm and a surface area around 25 m^2/g for Ce-rich compounds, which show a high oxygen mobility and a large volume of surface defects introduced by the milling, which should enhance its catalytic activity.

Another synthesis technique that shows good potential is the sonochemical decomposition of volatile organometallic precursors in a high-boiling solvent. Found to be a very efficient technique for producing highly porous nanostructured catalysts,[60][61] this process is based on the formation, growth, and collapse of bubbles in a liquid which can generate hot spots of ~ 5000 K and pressures of ~ 1800 atm with a cooling rate that exceeds 10^{10} K/s. Fe, Co, Fe-Co, and Mo_2C metallic clusters prepared using this technique, with sizes ranging from 2 to 20 nm, were used to form colloids and nanostructured supported metal catalysts and showed better catalytic properties than normal catalysts for the dehydrogenation and hydrogenolysis of cyclohexane. The dehydrogenation catalytic activity of molybdenum carbide was found to be comparable to commercial ultrafine platinum powder.

Chemical precipitation, followed in some cases by calcination, has been used extensively in the past to prepare catalytic materials. New refinements in this processing have emerged to enable a better control of the structure and the average particle size of the end product. For instance, a variety of pure nanoscale metallic or even intermetallic compounds, with an average grain size between 2 and 15 nm and with little or no surface oxidation, were prepared by a reduction process in alkalides or electrides.[62] Wang, et al.,[63] used reductive synthesis of colloidal particles in solution at room temperature and was able to prepare not only Pt particles with an average size between 5 to 15 nm but also cubic, tetrahedral, and octahedral particles with a particular atomic structure such as steps, ledges, and kinks, which could be critical for catalytic activity and selectivity. The catalyst for the CO oxidation, Au/Fe_2O_3, was produced by co-precipitation from hydrogen tetrachloroaurate trihydrate, ferric nitrate, and sodium carbonate precursors dissolved in deionized water, followed by heat treatment between 473 and 773 K.[64] The gold crystallite size was found to vary from 2–6 nm for the lower annealing temperature to 8–11 nm after calcination at 773 K.[64] In addition to the difference in crystallite size, the structures of the nano-Au crystals and the iron oxide support were also found to vary with the temperature of the heat treatment. A study of the catalytic CO oxidation in an oxygen-rich environment revealed a strong sensitivity to these morphological differences, the only stable catalyst being the one annealed at the lowest temperature. The formation of solid solution can also be used to prevent grain growth during high-temperature annealing. For instance, complex Ce-Zr-O solid solution was found to be more thermally stable than pure CeO_2, with an average crystallite size of about 10 nm after annealing at 1000°C (compared to 100 nm for pure ceria).[65] Moreover, the complex oxides were found to have a higher effective surface ($\approx$ 10 m^2/g) and a much better catalytic activity for the removal of CO, NO_x, and HCs.

Using an electrochemical process, Reetz and Helbig[66] have been able to produce transition metal nanoclusters with a control on the particle size. In their system, an anode of the desired metal is decomposed in a tetraalkylammonium salt electrolyte, which also serves as a stabilizer for the nanoparticles. The average particle diameter was found to be a function of the current density during the electrolysis, varying from 1.4 to 4.8 nm in the case of a Pd electrode. The concentration of metal was about 71 at% for this electrode, the remainder being elements such as C, H, N, and Br (in the case of a bromide salt). Using a slurry of carbon black, in addition to the salt, allowed them to produce dispersed nanoparticles on a carbon support (in this case 5.5% Pd/C). For their part, Kost, et al.,[67] used electrodeposition of metal microparticles of Pd, Ir, Ru, and Pt in poly(4-vinylpyridine) to prepare films for electrocatalysis. The crystallite size varied between 10 and 100 nm and showed good properties for hydrogen evolution in 0.5 M H_2SO_4.

Thermal decomposition of an organometallic complex has been used successfully to produce nanoscale metallic particles embedded in a support matrix. One such system is the nano-Co/C produced by the pyrolysis of carbon supported cobalt phthalocyanine at temperatures between 600 and 1000°C. This material is intended to replace platinum as an electrocatalyst in solid polymer electrolyte fuel cells. Thermal treatment of the organometallic-carbon complex at a temperature around 900°C produced a material with a structure illustrated in Fig. 2. The TEM micrograph reveals a bcc-Co nanocrystal embedded in an amorphous carbon matrix. A layered structure typical of graphite can be observed around the crystallite, which seems to be the basis for the long-term stability of the electrode.[19][68][69]

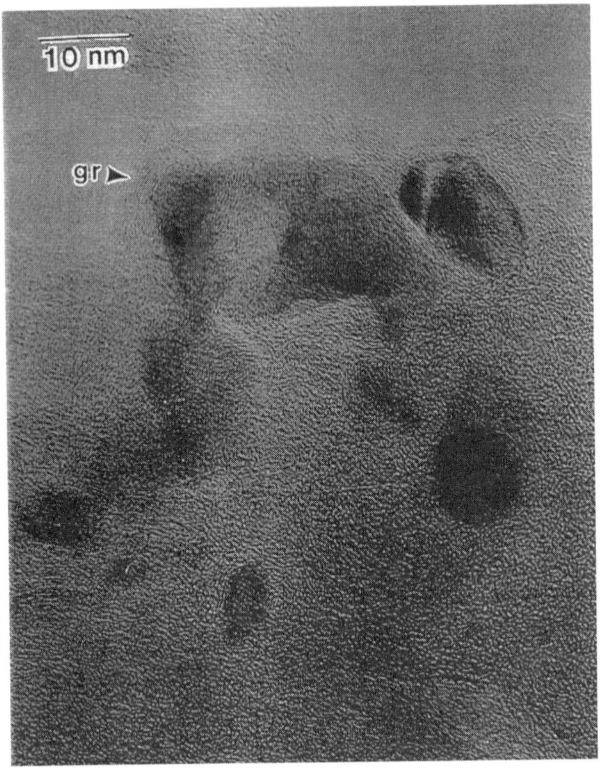

Figure 2. High-resolution TEM micrograph of a cobalt nanocrystal embedded in an amorphous carbon layer. Note the graphitic layers that surround the particle.[3]

Miquel and Katz[70] used a flame burner reactor and subsequent thermal annealing to produce VPO powders, the most widely used catalysts for the selective oxidation of butene and n-butane to maleic anhydride. The design of the catalyst structure in this case is crucial since this catalytic oxidation has been shown to be structure-sensitive (i.e., the selectivity of a given product is related to the crystallographic plane exposed at the surface of the catalyst). Liquid $VOCl_3$ and PCl_3 were used as precursors, and the morphology and the crystalline structure of the final powder could be changed through the variation of flame temperature. At high temperatures (maximum flame temperature of 2800 K), the powder collected is transformed to γ-$VOPO_4$ upon subsequent reheating, while at low temperatures (maximum flame temperature of 2300 K), the final crystalline phases produced are β-$VOPO_4$ and $V(PO_3)_3$.

Nanostructured catalysts have also been prepared by laser vaporization. Samy El-Shall has synthesized a number of metal oxide nanoparticles or nanocomposites using laser vaporization of metal or mixed-metal targets followed by gas-phase chemical reactions and by controlled condensation from the vapor phase. Nanoparticles with an average size between 10–20 nm can be obtained by this process as well as nanostructured $(Ce_xZr_{1-x})O_{2-y}$ catalysts for automobile exhaust treatment. In another study, Rousset, et al.,[71] investigated the formation of PdPt bimetallic nanocrystals with an average size of about 2.5 nm and found that Pd segregation was present at the surface of these crystals.

The thermal crystallization of an amorphous precursor was also used to produce nanostructured reactive materials. Yamashita, et al.,[72] studied the hydrogenation of buta-1,3-diene over $Fe_{90}Zr_{10}$ alloys after different annealing temperatures. They found that the maximum activity was attained when the alloy was in a precrystallized state, with numerous fine particles of α-Fe (5–40 Å) present at the surface, and a distorted crystal lattice from the dissolution of Zr. Peuckert and Baiker[73] have examined the hydrogenation of CO over $Fe_{81}B1_{3.5}Si_{3.5}C_2$ ribbons with different average α-Fe particle sizes. They found that the surface chemistry of their materials was modified upon annealing, as the average iron crystallite size became larger. These results seemed to indicate that maximum activity was obtained when pre-crystallite nuclei were present at the surface of the materials, and that the activity would decrease rapidly with grain growth.

Lastly, it should be mentioned that surface lithography now makes it possible to prepare nanometer-scale model catalysts. Jacobs, et al.,[74] recently presented the catalytic activity of a platinum array formed by 50-nm-diameter disks about 200 nm apart that were prepared on an oxidized silicon wafer as shown in Fig. 3. Ethylene hydrogenation at high temperature revealed that the properties of this catalyst array agree with results on platinum foils and dispersed or supported porous platinum catalysts. Such high-precision nanodesign of catalysts offers a unique possibility to understand the relation between nanostructures and the reaction activities, and especially when coupled with analytical tools such as scanning tunneling microscopy (STM) which can monitor surface changes on the atomic scale during catalytic reactions.[75] Furthermore, this understanding can only improve as smaller structures are developed.

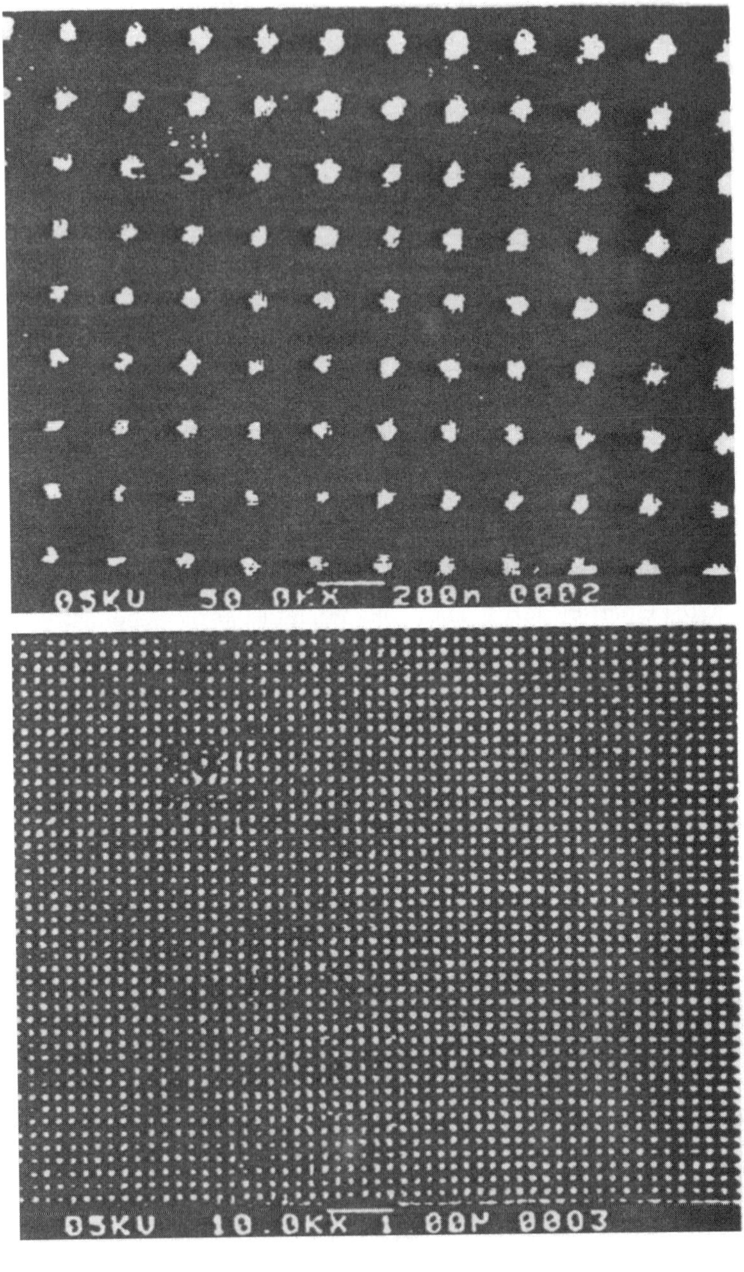

Figure 3. SEM micrographs of the platinum cluster array prepared by electron beam lithography.[75]

3.0 GAS SENSORS

Another area where nanostructured materials already play a prominent role is gas detection. More and more, technological developments and environmental concerns are creating the need to rapidly detect various kinds of toxic gas, such as CO_2, CH_4, fluorocarbon, N_2O, O_3 (greenhouse effects), fluorocarbon, halocarbon (ozone layer destruction), SO_x, HCl (acid rain),[76] with great sensitivity and selectivity. The level of detection for such pollutants would be in the ppm range and even down to the ppb level for some particular cases. A number of different physical principles are presently used for detection purposes:[77]–[79]

- Changes in mass
- Changes in transport properties
- Heat of reaction measurements
- Work function measurements
- Capacitance measurements
- Electrochemical detection
- Optical absorption and reflection

A typical example of the sensitivity of gas sensors is approximately 1 ppm for CO and 0.1 ppm for NO_2.[80] The different chapters of this volume show that a number of material properties are modified as the crystallite size is lowered to the nanosize, and it can be easily concluded that the sensing properties of gas detector materials will also be affected by a reduction in the crystallite size down to a few nanometers.

Because of the cost of the materials and the simplicity of the measurement principles, the method based on changes in the material transport properties is the one chosen for many industrial applications, even if it is not the most sensitive or most accurate at the present time. For this reason, and since this field is one where nanostructured materials are already playing a significant role, the present section will be devoted to this type of sensing technologies and related materials. Again it should be mentioned that, in this technological area, the development of nanostructured materials should not be seen as a recent phenomenon. As early as 1981, in fact, Ogawa, et al.,[81]–[83] were using gas-phase evaporation of tin in an oxygen atmosphere to produce nanoscale SnO_2 crystals with an average size of 6 nm. The novelty in the current work relates more to our understanding of the importance of the structure on the nanoscale and the application of new processing methods which offer numerous possibilities for easier design of materials on this scale.

3.1 Impact of Nanostructure on the Physical Principles of Semiconductor Sensors

The possibility of using the interaction of semiconductors with reactive gas was applied more than 35 years ago.[84] Albeit not the ultimate in accuracy or stability, one of the most common gas sensors is tin oxide, due mainly to its cost and sensitivity.[85]–[87] SnO_2 is an n-type semiconductor with a direct band gap of 3.6 eV as shown in Fig. 4, with the donor levels formed by oxygen vacancies in the lattice. The model most frequently mentioned in the literature is based on the fact that oxygen atoms absorbed or adsorbed at the surface form negative ions resulting in a decrease in the surface conductivity. Three different types of ionic species have been observed on the surface in addition to adsorbed O_2 molecules: O_2^-, O^-, and O^{2-}, which are desorbed at characteristic temperatures.[88] The presence of a reactive gas (oxidant or reducer) in the atmosphere, its adsorption at the surface and its reaction with the oxygen atoms will result in a decrease or an increase in surface electrons and, hence, in the conductivity. Typically, the value measured is the resistance of the semiconductors, R_s,

Eq. (3) $$R_s = KC_s^{-\alpha}$$

where C_s is the concentration of gas in air, K the sensitivity coefficient, and α is the dimensional power exponent between 0 and 1.[85][89] In most cases however, the value given is the sensitivity value, S_G, or its logarithm from:

Eq. (4) $$S_G = \frac{R_a}{R_s}$$

Eq. (5) $$\log S_G = \log\left(\frac{R_a}{R_0}\right) - \log\left(\frac{R_s}{R_0}\right)$$

where R_a is defined as the resistance of the sensor in clean air and R_0 is the value of the resistance at a specific gas concentration, a specific operating temperature or the relative-humidity value.[85] However, because this value cannot fall below 1, some researchers express the sensitivity as a variation of the conductance, G:

Eq. (6.) $$S_G = \frac{G_s - G_a}{G_a}$$

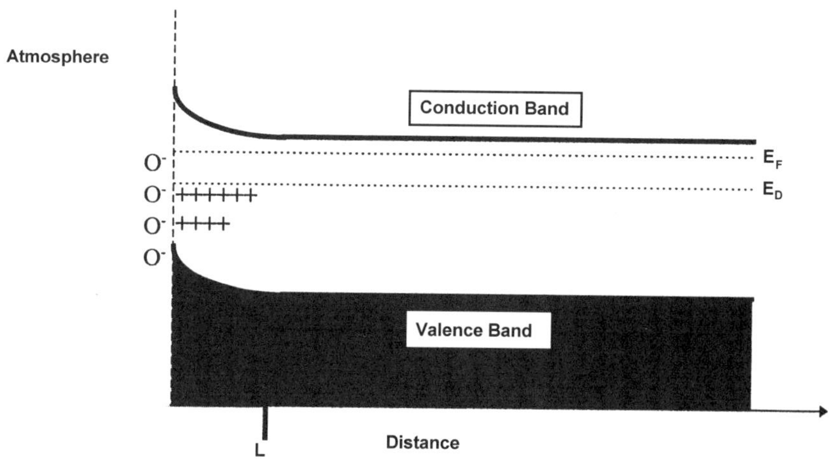

Figure 4. Schematic representation of the band structure with the presence of absorbed oxygen: E_F Fermi level and E_D donor level.

With hydrogen, three different reactions have been identified, depending on the surface species considered.[90][91] At low temperatures (below 150°C), the negative-charged adsorption of hydroxyl species takes place:

Eq. (7) $$H_2 + O_2 + 2e \rightarrow 2OH^-$$ (below 100°C)

Eq. (8) $H_2 + O_2^- + e \rightarrow 2OH^-$ (below 150°C)

both of which result in a decrease in conductivity. At temperatures above 200°C, the desorption of O^- will release one electron in the conduction band:

Eq. (9) $H_2 + O^- \rightarrow H_2O + e$

and, finally, positive-charged adsorption of hydrogen to the surface oxygen occurs at temperatures above 200°C:

Eq. (10) $H_2 + 2O^- \rightarrow 2OH^- + 2e$

A different point of view was presented recently by Kanamori, et al.,[92] who, from ESR (electron spin resonance) on thin films composed of 10 nm SnO_2, concluded that the source of carrier electrons was related more to the amount of unpaired electrons localized in oxygen vacancies.

However, taking into account the amount of adsorbed oxygen, it can be calculated that the observed conductivity variations cannot be explained solely by the above reaction. For SnO_2 the amount of adsorbed species was estimated to be between 10^{12} and $10^{14}/cm^2$, which is negligible compared to the carrier concentration of 10^{18} to $10^{20}/cm^3$.[88][90] A good demonstration of the variation in carrier concentration was made by Ogawa, et al.,[81] who measured it using the Hall effect in ultrafine SnO_2 particle films for the detection of C_2H_5OH.

To explain the high sensitivity of some particular semiconductors for gas detection based on conductivity changes, it is necessary to consider the structure of the semiconductor itself. In fact, early researchers found that the grain size is critical for the response to gas contaminants.[93] Studies of monocrystalline thin films have shown only small electrical resistance variations in the presence of oxidant or reducing gas.[94] Polycrystalline materials, for their part, show a strong response. This effect of polycrystallinity was related to the formation of space charge layers with their potential barriers between grains, the grain boundaries behaving as double Schottky junctions.[95] The height of these barriers was found to depend on the equilibrium concentration of adsorbed oxygen. Charge transport across grain boundaries is thus a major process governing the material response.[94]

Figure 5 shows two different examples of polycrystalline sensing materials: one where the particle size (diameter for spherical grains) is larger than twice the value of the depletion region L, and the other one where D is smaller than $2L$, as well as a schematic description of the gas sensing process for H_2 detection with a typical variation in the material conductivity with the concentration of reducing gas. The value of L corresponds to the Debye length of the materials and the strength of the oxygen chemisorption following the relation.[81][96]

Eq. (11) $$L = \left(\frac{2\varepsilon E_s}{e_0^2 N} \right)^{1/2}$$

where ε is the static dielectric constant and N is the concentration of completely ionized donors or acceptors. It can be seen that if $D \leq 2L$, then all the crystals will be depleted, causing the gas sensitivity of the element to change with D in the presence of a reducing gas.[97] For SnO_2 with the value of $\varepsilon = 1.2 \times 10^{-10}$ F/m, $E_s = kT$ and with a carrier concentration of about 3.6×10^{18} cm^{-3}, L at 250°C can be estimated to be around 3 nm.[81] Nanostructured materials with their very small size and their large number of grain boundaries are, thus, fundamental to this field. If the crystallite diameter is more than twice the depth of the space charge layer, then the material resistance is directly related to the potential barriers, E_s. On the other hand, if the crystallite diameter is less than twice the depth of the space charge, each crystallite contributes to what is effectively the channel of a field-effect transistor, wherein the negative surface charge (due to the adsorbed oxygen ions) acts like a gate voltage.[2][85] The existence of this critical size and, also, morphological considerations could be responsible for the different models proposed by different researchers. For instance, the impact of grain boundaries will be less if a small continuous channel exists between large grains ($D \geq 2L$), the depletion region mainly affecting the neck between the grains.[98] The impact of size can also be schematized by an equivalent circuit as presented in Fig. 6 for different materials: large grains (monocrystals), grain boundary control materials, and nanostructured compounds. The choice of crystallite size is, thus, critical for the properties of a particular compound, especially its sensitivity to a particular gas.[85]

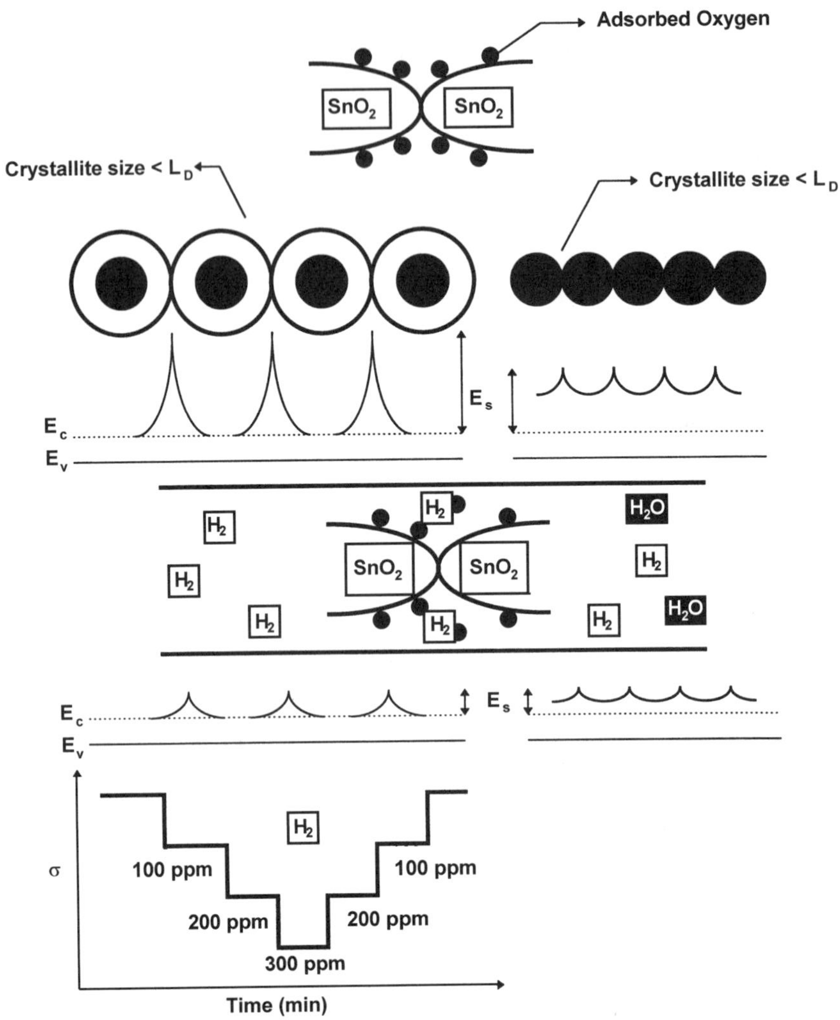

Figure 5. Schematic diagram of the gas sensing process for SnO_2 for two different materials with an average crystallite size larger or smaller than the Debye length (L_D) and with or without the presence of a reducing gas such as H_2. The lower curve represents a typical conductivity variation as a function of the concentration of reducing gas in the surrounding sensor atmosphere.

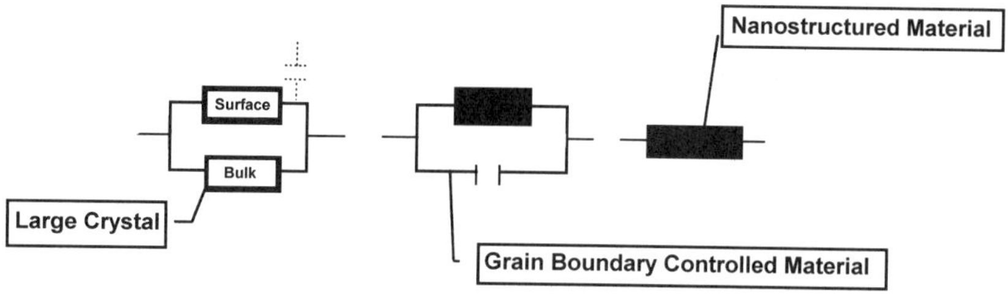

Figure 6. Equivalent circuits in relation to the size and structure of the sensing materials.[87]

It was also shown that a relation exists between the average concentration of the conduction electron $<n_c>$ and the reciprocal of the semiconductor temperature:[99]

Eq. (12) $$\ln\langle n_c\rangle \approx B + \frac{E_a}{kT}$$

where E_a is the activation energy equal to (E_s - V_s), E_s being the surface energy level and V_s the maximum Schottky-barrier height due to chemisorption. Assuming that the electron mobility is independent of the temperature, then

Eq. (13) $$\ln\sigma \approx c + \frac{E_a}{kT}$$

Other factors found to have an impact on the sensing properties include the presence of surface defects or impurities.[95] In fact, the possibility of adding dopants[100][101] to the semiconductor to improve its detection properties was also found as early as the sixties for the case of Pd and Pt[102] doped tungsten oxides (WO_3). Catalytic dopants have the effect of modifying the selectivity mainly by changing the rate of the redox reactions. Chang[103] found that by depositing a 3.5 nm Pd-Au film on a SnO_2 substrate, the sensing properties for H_2 and propylene (C_3H_6) were greatly enhanced compared to CO or methane. Cricenti, et al.,[104] found that Pt doping decreases the sensitivity threshold for CO detection by a factor of almost two while at the same time reducing the peak sensitivity temperature by about 100°C. Recently, Vlachos, et al.,[105] proposed that the enhanced sensitivity, due to metallic additives, was related to an electronic interaction between the metal and the semiconductor. In their model, the presence of metallic additives has a direct impact on the depletion depth in the semiconductor, the depth increasing with the metal work function following:

Eq. (14)
$$W = \sqrt{\frac{2\varepsilon_s}{q^2 n_0}\left[\varphi_m - X - (E_c - E_F) - \frac{\delta}{\varepsilon_i} q^2 D_s y_0\right]}$$

where φ_m is the metal work function, X is the electron affinity of the semiconductor, δ is the thickness of the interfacial layer between the metal and the semiconductor, D_s is the density of surface states per unit area and eV, y_0 is the level below which the surface states are occupied for charge neutrality, and ε_i is the dielectric constant of the interfacial layer. The net effect of adding metallic nanoclusters would be to decrease the electron availability of semiconducting grains, which results in an active size of the semiconductor particles smaller than the geometrical one in relation to the metal work function. A similar result found by Zhang, et al.,[106] was that the addition of dispersed nanoclusters (< 10 nm) of Pt on α- and β-$CdSnO_3$ improved the sensitivity toward ammonia detection at temperatures below 240°C. Cao, et al.,[107] found, by XPS analysis, that the addition of Pd to SnO_2, and the subsequent formation of PdO and PdO_2 during oxidation treatments of thin films produced by sol-gel, decreased the Fermi level of SnO_2 by 0.2 eV. Recently, Matsushima, et al.,[108] correlated the sensing enhancement toward C_2H_5OH detection of La_2O_3-doped SnO_2 to the basic nature of the additives, which favors the reaction selectivity in the oxidation of C_2H_5OH.

Meanwhile, Rastomjee, et al.,[109] used ion implantation to modify the surface structure on the nanoscale and the sensing properties of SnO_2. They found that selectivity toward CO was enhanced by Bi implantation. Barbi, et al.,[110] for their part, used Pt implanted nanostructured SnO_2 for CO detection and found a reduced sensitivity, as well as a decrease in the response rate, which could be caused in part by strong alterations of the lattice structure due to bombardment with heavy platinum shells.

One disadvantage of SnO_2 as a gas sensor is that it is simultaneously sensitive to most reducing gases and often two or more gases will be present at the same time. For example, the detection of a major pollutant such as H_2S

around gas extraction and processing plants will be complicated by the presence of other accompanied gases, such as H_2 and C_nH_m.[111] To improve selectivity, two different approaches can be chosen. First, by modifying the operating temperature, it is possible to promote one oxidation reaction more than another. For example, CO will oxidize more rapidly than CH_4 at high temperature, resulting in a smaller length of interaction for the CO and thus a smaller resistive change and, hence, a lower sensitivity. At lower temperatures, a contrary phenomenon takes place: the CO oxidizing at a smaller rate will interact more with the material.[85] Modifications in the response of solid-state sensors have also been obtained through treatment involving special gases. Zubknas, et al.,[112] changed the sensing response for NO_x (from an increase to a decrease in conductivity) of a field-effect transistor covered by a 10-nm Pt film, by prior exposure to ammonia at 150°C.

Yamazoe, et al.,[88] in their study of the reactivity of SnO_2 with O_2, H_2, and H_2O, found the formation of a number of oxygen species on SnO_2 as a function of the temperature and revealed that the process of conductivity changes as a function of gas present and the detector temperature is quite complex. For example, they found that H_2 will react first with the adsorbed oxygen with an irreversible or reductive adsorption of the H_2 to form surface hydroxyls, which would lead to the adsorption or desorption of the H_2O present on the surface. At higher temperatures (> 500°C) the adsorption of H_2O and H_2 is no longer significant and the change is mainly from the consumption of adsorbed (and lattice) oxygen with H_2.[88]

Finally, mention should be made of one of the important problems of solid-state gas sensors and, in particular, tin oxide, namely the adsorption of water vapor which results in a large increase in the conductivity,[88] due to the presence of hydroxyl groups at the surface which contributes positively to the conductivity. This effect is one of the reasons for the instability of this material and its variation in relative humidity, as well as its long-term conductivity.

3.2 Nanostructured Design

A number of techniques exist for preparing gas-sensing materials of various nanostructures, and most of the different synthesis methods described in this volume can be used to prepare useful materials. An important fact is that compositionally identical materials have a different response to various gases depending on their method of synthesis,[113] their thickness, and the nature of the substrate. For example, Ansari, et al.,[114] produced a number of SnO_2 films using chemical vapor deposition, spray pyrolysis, and physical vapor deposition, and found a very different response to H_2 for these three materials.

Apart from gas-phase evaporation,[81][83] two of the main techniques used to prepare nanostructured films are sputtering and, more often, reactive sputtering. Rickerby, et al.,[80] studied the structure of SnO_2 nanocrystals produced by rf reactive sputtering. The morphology and nanostructure of their films are presented in Fig. 7. The SEM micrograph of Fig. 7a clearly shows the granular nature of the films while the low- and high-resolution TEM micrographs of Figs. 7b and 7c present grains with values ranging from 2 to 20 nm. For their materials, they found that porous films have a higher sensitivity up to a thickness of about 200 nm (at constant grain size) which increase with diminishing grain size, while thick and compact columnar films generally decrease in sensitivity, with increasing film thickness in the range of 0.1 to 1.0 μm. This improved sensitivity of porous films was attributed to the columnar structure, which allows gas molecules to permeate along vertical fissure, which in thin films will also increase the surface area exposed to the gas, with increasing thickness. In a similar work, Vlachos, et al.,[115] produced different SnO_2 films with average crystallite sizes of about 11 nm and a high surface area (50 m^2/g) after deposition at around 460°C. They show that one advantage of reactive sputtering revolves around the potential control of the O:Sn ratio. By controlling the flow of oxygen during deposition it is possible to vary the nature of the films from metallic to semiconductor (nearly stoichiometric SnO_2). Such a control allows the synthesis materials with a large amount of oxygen vacancies which show a higher sensitivity than compounds with a high O:Sn ratio, because the chemisorption of oxygen atoms is directly related to the free electron concentration (or the oxygen vacancy concentration). In an interesting study, Serrini, et al.,[116] studied the effect of the O concentration during rf sputtering, grain sizes and the amount of adsorbed oxygen on the sensing properties for CO and NO_2 of SnO_2 grains. Figure 8, which presents the interrelation between some of these properties, reveals that the sensitivity increases for both gases with decreasing grain sizes and that, as mentioned earlier, the critical size for SnO_2 seems to be indeed around 6 nm if we consider a depletion region of about 3 nm. More interestingly, it can also be seen that the grain size is not the only factor influencing the sensitivity since the response is greatly affected by the amount of adsorbed oxygen. Two samples (4 and 6), which have an average grain size below the threshold for grain-controlled conductivity (as compared to grain boundaries for larger grains), showed a marked difference for NO_2 detection, with the sensitivity nearly doubling when the amount of adsorbed O in the first layers of the materials increases from 17% to 25%.

(a)

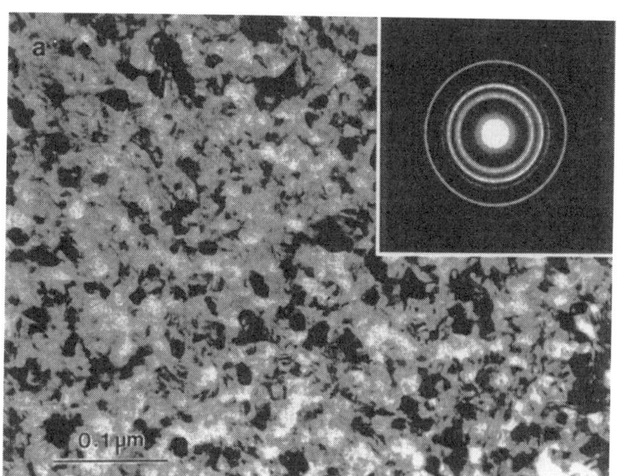

(b)

Figure 7. (*a*) SEM, (*b*) low-, and (*c*) high-resolution micrographs of nanostructured SnO_2 gas sensing materials prepared by reactive sputtering.[80]

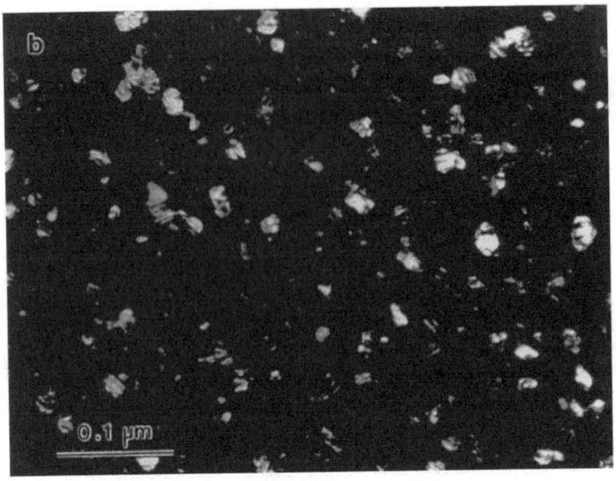

(c)

Figure 7. *(Cont'd.)*

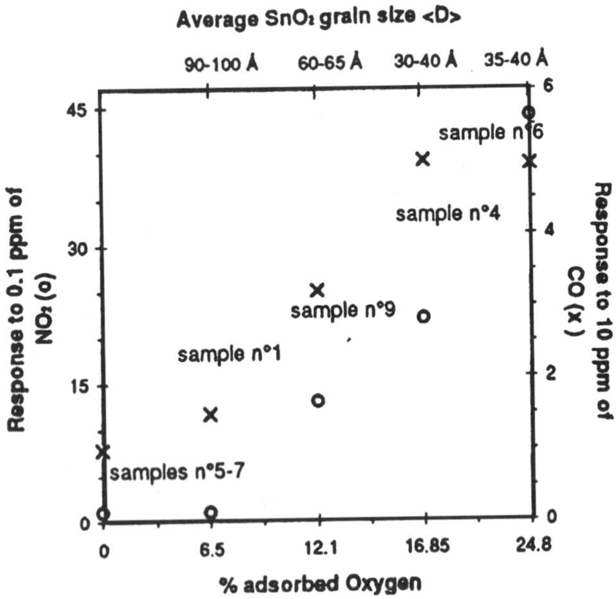

Figure 8. Response to NO_2 (150–200°C) and CO (210 ppm at about 300–350°C) in air as a function of the adsorbed oxygen in the first two layers of SnO_2 films and average grain size.[116]

Hu, et al.,[117] for their part, used laser ablation on SnO_2 and Sn targets to produce nanostructured materials. The lowest average grain sizes were obtained from amorphous films deposited at 100°C followed by crystallization and oxidation, in an oxygen atmosphere at 400°C for 4 hours. The resulting materials had grain sizes between 4.0 and 5.2 nm and showed an improvement in sensitivity by a factor of approximately 2 to 5 for C_2H_5OH detection when compared to films having larger grain sizes (30 to 46 nm).

Dieguez, et al.,[97] found that the nature of the substrate and the nature of the oxidizing atmosphere had an influence on the structure of the materials. They revealed, for nanocrystalline SnO_2 prepared by PVD, that the crystallite size was different if the annealing was done in synthetic air or in oxygen. They also observed a difference in the crystallite structure between materials deposited on sapphire as compared to SnO_2 deposited on SiO_2. In the latter case, diffusion of Sn into the silicon substrate was occurring. Also, they observed that nanocrystalline powders produced by precipitation had a large number of defects after annealing below 450°C. Other work has also shown that the active area can be considerably increased by depositing the film on a rough surface, the surface texture being replicated by the film.[80]

High-energy mechanical alloying was used recently to prepare different nanostructured sensor materials. Jiang, et al.,[113] prepared $(\alpha\text{-}Fe_2O_3)_x\text{-}(SnO_2)_{1-x}$ powders by milling haematite (α-Fe_2O_3) and cassiterite (τ-SnO_2) in air using tungsten carbide balls and vials. They found that their materials, especially one with a 85 mol% of Fe_2O_3, had a much higher sensitivity to alcohol compared to the unmixed powders at temperatures around 250°C. At the same time, the sensitivity to other gases such as CO and CH_4 was minimal and did not change with milling. For this system, high-energy milling produced the formation of an α-Fe_2O_3-SnO_2 solid solution with a haematite-like structure with an average crystallite size which decreased to a stable value of the order of 8 nm after about 25 hours. An amorphous phase and tungsten carbide contamination was also present after a long milling time. Even if the grain size of the haematite phase was stable after 25 hours of milling, the sensitivity to alcohol continued to increase with milling time. This improvement could be related to the fact that, even if the crystallite size remains stable, the lattice parameter of the Fe_2O_3-type crystals showed a continuous increase with milling due to the continuous dissolution of SnO_2. This improvement could also be due to the presence of the amorphous phase or of WC-Co impurities, the possible presence of WO_3 or catalytic effects related to Co.

Liu, et al.,[118] used plasma-enhanced chemical vapor deposition followed by thermal annealing to produce a mixture of α-, β-, and γ-Fe_2O_3 powders with an average size of about 6 nm (which is about 2*L*) after annealing below 600°C, which has a good alcohol-sensing capability at around 400°C while at the same time being less sensitive to methane.

Bhowmik, et al.,[119] studied the electrical characteristics of nanocrystalline TiO_2 prepared by inert-gas-phase condensation and found that the conductivity is dependent on the porosity, grain size, and grain boundary structure. They also observed that materials annealed at low temperature (400°C) had an enhanced sensitivity to changes in the ambient atmosphere, especially the humidity level, which could make them useful for oxygen detection.

Sol-gel was also used to prepare nanostructured or nanocomposite gas detection materials. Rella, et al.,[120] for instance, produced both pure and Pd-doped SnO_2 films with a mean grain diameter which, after annealing at 600°C, was smaller for Pd films compared to the undoped sample (10.6 nm as opposed to 7 nm) and also a higher specific area for the doped material (58 m^2/g compared to 35 m^2/g). These results indicate that the presence of Pd and its subsequent oxidation to PdO can serve as a grain-growth inhibitor by pinning the grain boundaries in some way. As for gas detection, they found that their Pd-doped films show a higher sensitivity to CO with a lower operating temperature. Xu, et al.,[121][122] produced a number of composite materials by impregnating hydrous SnO_2 with different aqueous solutions in which a salt and/or polyoxy acid of different metal or nonmetal elements was dissolved in order to reduce grain growth during high-temperature sintering. They found that large-surface-area films (around 40 m^2/g) could be obtained when tungsten, niobium, or phosphor were used as dopants even after sintering at 900°C, which is about eight times larger than for undoped SnO_2.

Figure 9 presents variations in sensitivity for H_2 and CO as a function of the crystallite size for a number of films doped with 5 at% additives SnO_2 after thermal treatments at 300 and 400°C, showing, at the same time, the impact of various additives on the average crystallite size and the large increases in sensitivity with decreasing grain size, especially below 10 nm. Sun, et al.,[89] studied the sensing properties of sol-gel derived thin films produced by the dissolution of iron ethoxide $Fe(OC_2H_5)_3$ in benzene. After calcination at about 400°C the film consisted of α-Fe_2O_3 grains which varied in size from 20–100 nm and showed good sensing properties for NO_2 and CO below 300°C that depended on the nature and structure of the electrode materials.

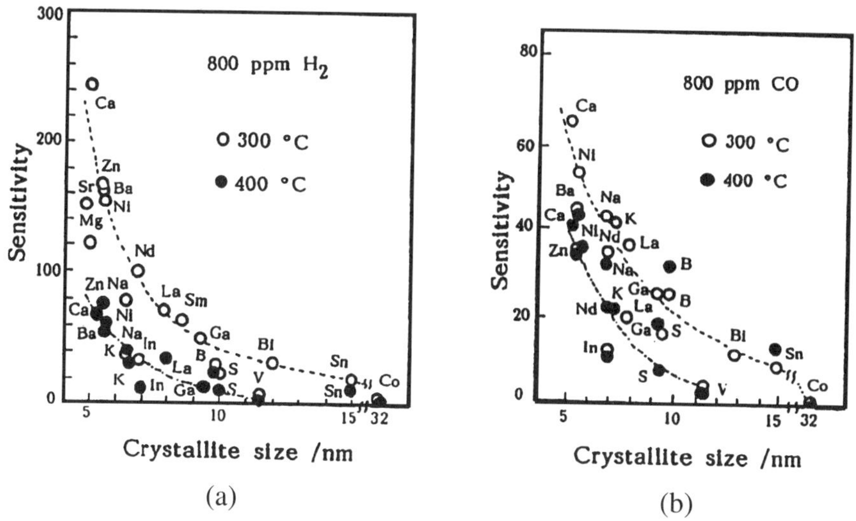

Figure 9. Variation of the gas sensitivity toward (*a*) H_2 and (*b*) CO with crystallite size for doped with 5 at% additives SnO_2.[122]

Recently Li and Kawi[123] prepared very high-surface-area SnO_2 by surfactant-templating. They obtained mesoporous materials with a surface area as high as 157 m^2/g after calcination at 450°C and with a pore diameter centered at about 4 nm. They found a linear increase in sensitivity toward 500 ppm H_2 in air at 300°C with increased surface area.

The effect of adding Cu to SnO_2 5- to 7-nm grains, prepared by aerosol pyrolysis and the subsequent formation of CuO during annealing treatments, were found to greatly increase the material resistance.[111][124] The conductance changed from a value in the range $(0.2–5.0) \times 10^{-4}\ \Omega^{-1}$ at 100°C to a value between $10^{-8}–10^{-9}\ \Omega^{-1}$ for samples containing 1.2–1.5 at% Cu. This increased resistivity was found to have a major impact on the sensitivity to hydrogen sulfide, which was much greater than that of the pure SnO_2 samples. At the same time, as seen in Fig. 10, compared to the sensing properties of pure SnO_2 (labeled "1") and of nanocomposites SnO_2(CuO) (labeled "2") films at 150°C, the sensitivity to other gases such as CO or ethyl alcohol was significantly reduced. It is speculated that this remarkable improvement in sensitivity was due to the reaction between CuO and H_2S which transforms the highly resistive p-CuO segregated at the grain boundaries to highly conductive CuS, a hypothesis that seems to be corroborated by surface analysis which reveals the presence of sulfur only on the surface of the SnO_2(CuO) nanocomposites after exposure to H_2S. A similar sulfur reaction was also found for the case of Ag-doped SnO_2 films.

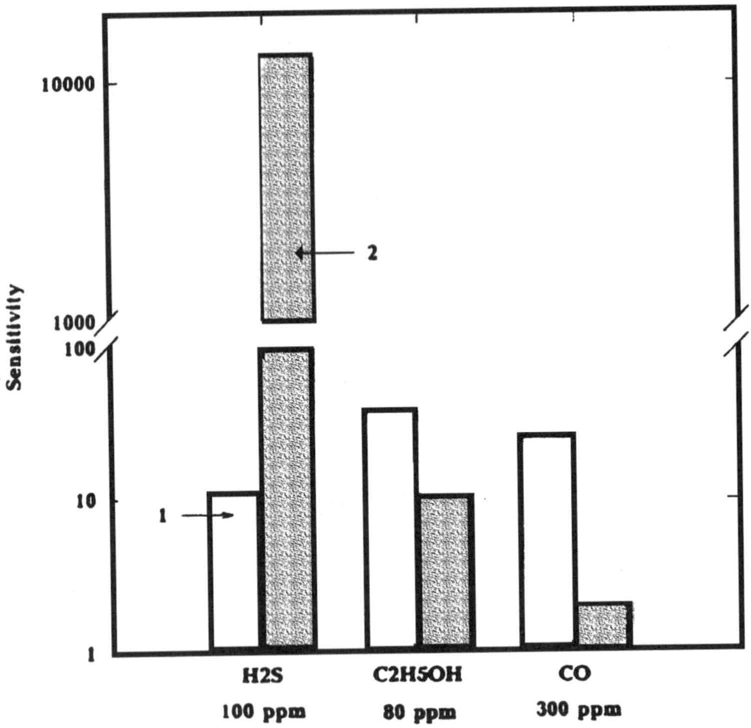

Figure 10. Comparative sensitivity of pure SnO_2 (labeled "1") and nanocomposites $SnO_2(CuO)$ (labeled "2") films to different gases at 150°C.[111]

Neubecker, et al.,[126] used ozone-enhanced evaporation (OEE) to prepare oxygen-deficient NiO sensor materials through molecular beam deposition with an average grain size between 5 and 10 nm and high oxygen deficiency. They found that the sample with the highest oxygen deficiency had better sensing properties toward a number of toxic gases but also observed a tendency toward further oxidation at elevated temperatures which could result in a strong drift of the sensor.

Another question that has not yet been answered is the nature of the defects and their impact on the sensing properties. It is understood that oxygen vacancies play a major role but what role does the surface defect structure of nanostructured materials play exactly?

4.0 HYDROGEN STORAGE

Recent years have also shown that the significance of metallic hydrides relates primarily to the design of safe, more efficient, methods for hydrogen storage in alternative vehicles. A number of storage methods are presently available: hydrogen can be transported as compressed gas or in liquid form, both of which call for very special care to take into account all the safety concerns. Moreover, the liquefaction process consumes almost 30% of the hydrogen energy to compress and cool down the hydrogen to 20 K, making any use of liquid H very difficult. Adsorption on activated carbon with a very high surface area (1500 to 2000 m^3/g), especially at low temperature (77 K) and high pressure (~ 50 atm), can also be used but again the problem of refrigeration increases the difficulties of this approach. Other new forms of carbon such as fullerenes, carbon nanotubes, or different organometallic complexes are presently under investigation and could offer potential in the near future. The same can be said for new mesoporous materials whose composition, pore size, distribution, and connectivity could be tailored for this particular application. Another possibility is to use pure iron and its transformation to oxide by water vapor (rust) to produce H, in situ. One of the drawbacks of this process, apart from the weight aspect, is the need to recycle the oxides. A solution would be to use a high-tech process such as municipal waste to reduce the iron oxide.

Finally, another possibility is metallic hydrides, materials that absorb hydrogen at moderate pressure and are operated at temperatures ranging from ambient to about 200–300°C. They can store hydrogen with a higher volume density than in liquid form. Three types of hydride exist: ionic (such as magnesium), covalent, and metallic (hydrides of transition metals). A major drawback of hydrides is the increase in weight needed for storage. On the other hand, they are much safer, since the hydrogen must be released from the hydride before it can oxidize or burn rapidly. For nearly two decades, car manufacturers have been testing metal hydride storage in hydrogen propulsion test vehicles.[127] A number of alloys have been investigated and it can now be shown that structural design on the nanometer scale will possibly play a significant role in the synthesis of new and more efficient hydrogen storage materials. It should also be mentioned that metallic hydrides are finding applications in the area of rechargeable batteries, with Ni-metal hydride batteries already on the market. Their main advantages are their high energy density, high dischargeability, long charge-discharge cycle life, and environmental cleanliness. Also in this field, the design of new nanostructured materials is expected to prove critical for future technological development.

4.1 Properties of Hydrogen Storage Compounds

Hydrogen gas storage compounds require many unique properties, which is probably the main reason why they are so few in number. In order for hydride materials to be of commercial value, it is essential that the following parameters be optimized:[127]

- ⇒ weight
- ⇒ capacity
- ⇒ kinetics of H_2-exchange
- ⇒ sensitivity to impurity gases
- ⇒ multi-cycle stability
- ⇒ activation procedure
- ⇒ large-scale production possibility at low cost

Except for the last parameter, which is obvious, it is worth discussing these different parameters in order to understand the possibilities offered by the nanostructural design of hydride materials.

Weight: The question of weight, as mentioned above, is essential. To be viable as a hydrogen storage compound the material used must be as light as possible, which is why Mg-based alloys have attracted attention, even if their oxidation characteristics make them more difficult to handle than other intermetallic compounds such as FeTi.

Capacity: This is, in part, related to the weight aspect in that the aim is to be able to store as many hydrogen atoms per metallic atoms as possible.

Kinetics of the H_2-exchange: Typical hydrogen absorption desorption curves are presented in Fig. 11 for a crystalline FeTi. First-order kinetics describe the reaction rate following:

Eq. (15) $$c(t) = c_0(1 - e^{-Kt})$$

where $c(t)$ is the hydrogen concentration in metal hydride, c_0 is the saturation concentration and $K = K(p,T)$ the reaction constant.[130] The important point is that an appropriate material should have an absorption-desorption cycle below 300°C, with rapid rates. In order to identify the rate-limiting step, the hydride formation has been divided into five intermediate processes by Martin, et al.[131]

i) Physisorption of hydrogen molecules.

ii) Dissociation of hydrogen molecules and chemisorption.

iii) Surface penetration of hydrogen atoms.

iv) Diffusion of hydrogen atoms through the hydride layer, either by an interstitial or a vacancy mechanism.

v) Hydride formation at the metal/hydride interface.

A similar five-step process was also developed for hydrogen desorption.

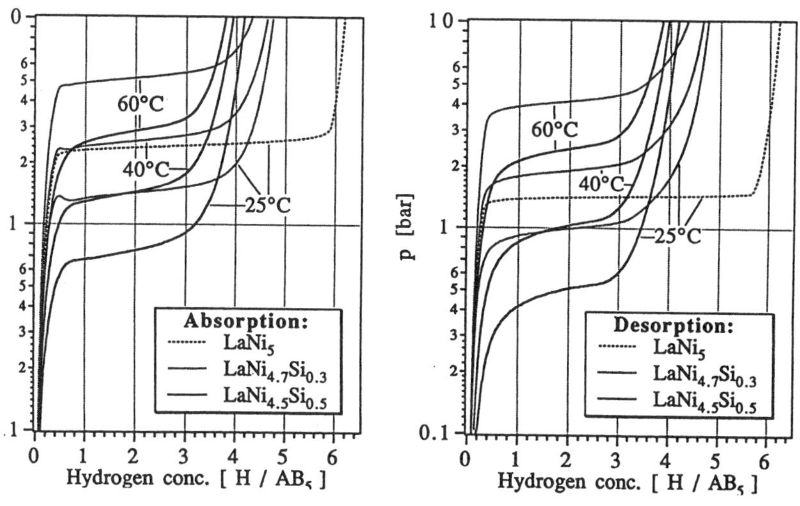

Figure 11. Typical H_2 absorption-desorption curves.[132]

Sensitivity to impurity gases: The presence of gaseous impurities was shown to have an impact on the cycling stability and kinetics of materials. The effect of these gases depends on their nature. For example, N_2 and CH_4 were found to reduce only the kinetics of the hydrogen absorption while even small amounts of impurities such as O_2, CO, CO_2, and H_2O reduce the capacity and cycle stability.[127] In most cases, a reactivation process in pure hydrogen is necessary to regain the initial capacity of the material.

Multi-cycle stability: The material must be able to undergo a number of absorption/desorption cycles without a significant loss in capacity. Moreover, one of the major problems of hydride materials relates to their decripitation with the charging and discharging cycles. A strong variation in the crystalline volume is caused by the absorption-desorption of hydrogen. For example, the unit cell volume in the case of $Ti_{45}Zr_{38}Ni_{17}$ C14 hcp causes a phase increase of about 24% upon hydrogenation.[133] This variation in volume has a direct effect on the generation of crystalline defects, resulting in fracture of the crystals after a number of cycles. One main advantage of nanocrystalline materials can already be seen: high-energy milling studies have revealed that defect-induced breakdown of crystals (such as the presence of dislocations) is very limited at small sizes.[134]

Activation procedure: Most metallic hydrides do not readily absorb hydrogen. Before they can do so, some kind of activation process which normally comprises a number of cycles from room temperature to high temperature (e.g., 400°C), under high hydrogen pressure, must be performed. Two possible explanations have been proposed for this behavior. The first relates the main effect of this process to removal of the surface oxide layer or other impurities present on the surface of the grains. The second is that the increase in surface area caused by the decreased particle size (due to decripitation) provides a larger surface area upon which the hydrogen can react.[135]

4.2 Nanostructured Design

Some of the earlier work on nanostructured hydrogen storage materials was done on FeTi alloys produced by high-energy mechanical milling. As presented in Table 1, FeTi is a well known intermetallic hydrogen storage compound. Trudeau, et al.,[136] and Zaluski, et al.,[137] have shown that it is possible to obtain FeTi powder with an average crystallite size between 7 to 10 nm by milling together the elemental components or by mechanically grinding microcrystalline powders. It was also demonstrated that the presence of a low concentration of oxygen (3.6 at%) was sufficient to produce an amorphous material. Figure 12 presents a comparison of H absorption for microcrystalline (Fig. 12c), nanocrystalline (Fig. 12b) and amorphous FeTi (Fig. 12a) $Fe_{50}Ti_{50}$.[138][139] This data shows that the nanocrystalline sample had absorption properties between normal coarse-grained material and amorphous alloys. Nanocrystalline $Fe_{50}Ti_{50}$ was found to have a lower pressure plateau than the microcrystalline sample, albeit with a reduced capacity of one hydrogen atom per metallic element compared to about 1.2 for its microcrystalline counterpart (at low pressure).[137] Further studies on a number of alloys in the FeTi system revealed the presence of an amorphous layer (from 10 to 30% of the materials) at the surface or at the boundaries of FeTi crystals.[138][139] It is assumed that, below the absorption plateau, the hydrogen absorption in this amorphous layer exerts a negative pressure on the FeTi crystal while above the plateau the expansion of the crystal is restricted by the amorphous layer and that the stress increases with further concentration of the hydrogen in the crystal.[138] Modeling based on this layer has successfully explained the decrease in the absorption plateau as well as its narrowing.

Table 1. Hydrogen Capacity for a Number of H Storage Alloys[127][133]

Materials	wt%
Mg	7.7
FeTi	1.6
$LaNi_5$	1.5
$MgNi_2$	3.6
$Ti_{.98}Zr_{.02}V_{.43}Fe_{.09}Cr_{.05}Mn_{1.5}$	1.8
$Ti_{45}Zr_{38}Ni_{17}$	2.5

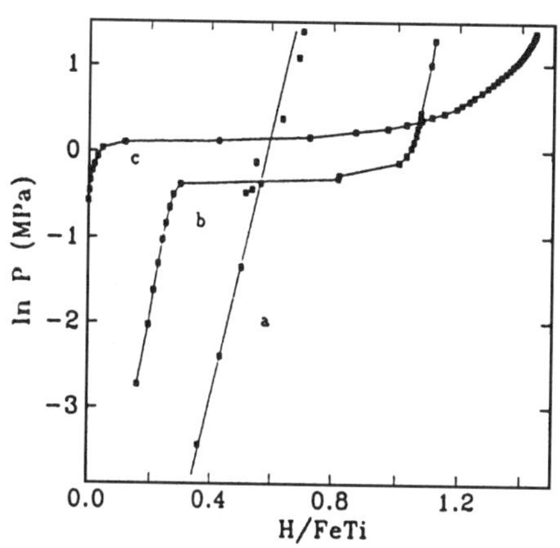

Figure 12. Pressure-composition isotherms for (*a*) amorphous, (*b*) nanocrystalline (5 nm), and (*c*) microcrystalline $Fe_{50}Ti_{50}$ alloy.[137]

Hydrogen absorption also revealed that activation was much easier for the nanocrystalline materials compared to the normal microcrystalline alloy. For the coarse-grained sample, activation consists of several cycles of heating up to the range of 400–450°C in vacuum and annealing in H_2 under 7 bars, followed by cooling down to room temperature and admission of hydrogen at a pressure between 35–65 bars. Apart from the decripitation hypothesis, two models have been proposed to explain the need for this activation. The first proposes that activation transforms the surface iron oxide to metallic iron with the formation of TiO_2 and that the presence of metallic Fe domains would be responsible for dissociating molecular hydrogen. The second model proposes that the formation of an iron-titanium mixed oxide would produce the hydrogen sorption behavior.[140] On the other hand, activation of $Fe_{50}Ti_{50}$ nanocrystals requires only a single annealing treatment at 400°C for 0.5 hours under vacuum. In order to investigate this difference, the oxidation, as well as the surface chemistry of nanocrystalline $Fe_{50}Ti_{50}$, was studied using XPS and compared to the one of the microcrystalline alloys. This study shows that $Fe_{50}Ti_{50}$ oxidizes more easily than its microcrystalline counterpart and that, at temperatures below 500°C, Fe_2TiO_5 and TiO_2 are formed mainly, resulting in dissociation of the intermetallic compound and the formation of metallic iron. Table 2 presents the surface elemental atomic concentration and the Fe:Ti metallic ratio of various $Fe_{50}Ti_{50}$ samples of different grain size. It has been proposed that the simpler activation needed for the nanocrystalline samples is related in part to Fe enrichment of the surface (the surface of nanocrystalline samples having twice the amount of Fe on the arc-melted coarse-grained sample) and to its ability to accomplish the reduction/dissociation transformation. This is further confirmed by the results of two samples analyzed after a hydrogenation cycle which shows a decrease in surface Fe and an increase in Ti and O simultaneously with a drastic hike in the Fe:Ti metallic component ratio, even if the sample was exposed to air prior to the XPS analysis. This tends to indicate that in this nanostructure configuration the Fe metallic surface is stable, since it does not completely oxidize in air. These results support the model that metallic iron acts as the catalytic surface for dissociating the hydrogen molecule in the absorption process in agreement with other results.[141]

In a subsequent work, Zaluski, et al.,[142] studied the storage properties of nanocomposite Pd-doped FeTi prepared by milling FeTi with Pd powder. The result was the formation of a nanocomposite with Pd grains (30 nm) cold-welded to a number of FeTi nanocrystals. The nanocomposite had hydrogenation properties superior to the pure nanocrystalline material, requiring no activation annealing, for instance, even if stored in air. Moreover, absorption was found to be much easier with a hydrogen uptake of about 0.6 H atom/FeTi at pressures lower than 1 bar compared to about 0.2 for the undoped nanocrystalline powders.

Table 2. Variation of Surface Species Obtained by XPS for Different FeTi Samples[140]

Sample Nature	Fe at%	Ti at%	O at%	C at%	Surface Fe/Ti	Fe:Ti metallic
Arc-melted	11.7	7.7	45.6	35.0	1.5	1
Powder - 39 nm	20.1	7.6	46.1	26.3	2.7	0.48
Powder - 24 nm	36.2	6.9	48.4	18.6	3.8	0.43
Powder - 21 nm	23.6	6.6	45.7	24.1	3.6	0.53
Powder - 13 nm	26.9	7.0	45.5	20.6	3.8	0.49
Powder - 10 nm	20.1	5.4	43.1	31.3	3.7	0.47
Powder H - 24 nm	14.7	12.4	52.6	20.3	1.2	> 4.8
Powder H - 10 nm	11.7	13.7	52.4	22.2	0.9	> 3.6

Wasz and Schwarz[143] studied the hydrogen storage properties of different $LaNi_5$-based intermetallic alloys prepared by high-energy mechanical alloying, and especially $LaNi_{5-y}Sn_y$. The presence of Sn was found to reduce the plateau pressures for hydrogen absorption and desorption, to decrease the hysteresis between the pressure for hydride formation and decomposition and, unfortunately, to decrease the hydrogen storage capacity. The latter behavior was explained by changes in the $LaNi_5$ electronic 3d band structure with Sn alloying, resulting in a reduction in the number of holes that can be occupied by electrons from the hydrogen. As in other studies, they found that an activation cycle was not necessary for the powders to absorb hydrogen, which they attributed to the presence of Ni inclusions or, more probably, to the absence (or reduced presence) of La oxides in their powders. In their work, they also observed that nanocrystalline powders were less affected by the volume change variation due to hydrogen charging. In

$LaNi_5S_{0.25}$ the surface-area increase between the first and second hydride cycles was only 1×10^{-3} m^2/g compared to 8×10^{-2} m^2/g in arc-cast powder.

Because of their light weight and their high hydrogen capacity, it is nanostructured Mg-based alloys that have probably drawn more attention in recent years in the field of nanostructured hydrogen storage materials. Zaluski, et al.,[144] found that nanocrystalline Mg_2Ni formed by mechanical alloying, with average grain sizes between 20–30 nm, absorb hydrogen readily without the need for activation at temperatures lower than 250°C. They also observed that these nanocrystalline materials absorb hydrogen at lower temperatures (~ 200°C), i.e., below the temperature of structural transformation of the Mg_2NiH_4 hydride. Moreover, as found for the FeTi and also for $LaNi_5$, the addition of Pd was found to enhance the hydrogen absorption kinetics at 200°C, and absorption at room temperature was observed even without the need for an activation cycle.[145] Cu addition, on the other hand, was found to increase the plateau pressure. Li, et al.,[146] showed that mechanical alloying could be used to prepare Mg_2Ni hydrogen storage compounds with a very fine structure and a surface Mg/Ni atomic ratio of 1:3 after sintering at 400°C for 5 hours. This Ni concentration at the surface, and its aid in the diffusion and decomposition of the hydrogen, could be responsible for the good properties of these materials, especially the greatly reduced need for an activation process.

Imamura and Sakasai[147] studied the hydrogen absorption characteristics of mechanically milled Mg with and without 5 wt% Pd-supporting graphite in the presence of tetrahydrofuran (THF) with an average crystallite size between 17 and 34 nm. The complex nanocomposite materials $(Mg/Pd/G)_{THF}$ were found to have a high surface area and very good absorption properties that were dependent, not so much on the crystallite size as on the amount of THF used. Upon milling, THF was found to decompose the graphite and to form an intimate mixture of Mg (or Mg/Pd) and aromatic carbon molecules, resulting in a charge transfer interaction which could be the reason for the observed enhanced activity.[148]

In a recent work, Holtz and Imam[149] studied the hydrogen storage capacity of nanostructured magnesium-alloys prepared using three different techniques: inert gas condensation of sputtered nanocrystalline powder, co-sputtering of amorphous thin films, and high-energy milling. From these three synthesis methods, the best materials were alloys with 5 and 10 at% Ni which were high-energy milled for 50 hours in mineral oil followed by consolidation which achieved about 6 wt% hydrogen loading without requiring an activation cycle. The other synthesis methods were found to produce materials with low kinetics and very prone to grain growth and oxidation. Based on the previous work of Imamura and Sakasai,[147] it could be interesting to investigate the possible catalytic effect of mineral oil and its decomposition to C group on the hydrogen absorption properties of the mechanically alloyed powder.

Orimo and Fujii[150] studied the properties of the Mg_2Ni-H system synthesized by milling the metallic powders in a hydrogen atmosphere. They found that the hydrogen content in the powder reaches a stable value of about 1.6 wt% after 1 hour of milling, which is 1.5 times higher than the value for H absorption in Mg_2Ni at ambient temperature. Also, the dehydriding reaction was found to happen at 440 K, which is much lower than for the normal low-temperature phase of Mg_2NiH_4 (520 K).

In order to improve the hydrogen absorption-desorption kinetics of Mg-based alloys, Gross, et al.,[151][152] investigated the hydriding properties of composites obtained by mechanically alloying La_2Mg_{17} with various wt% of $LaNi_5$. The composite materials, La_2Mg_{17} + 40 wt% $LaNi_5$, were found to have a reduced absorption capability with a hydrogen capacity of 3.7 wt% but an absorption-desorption rate 50 times that of pure La_2Mg_{17}.

In order to improve the surface area and the absorption-desorption kinetics, Huot, et al.,[153] recently used a lixiviation technique to produce a highly porous hydride material. They produced $Mg_{72}Li_{28}$ by high-energy mechanical alloying with an average size of about 46 nm. Lixiviation was done by immersing the powder in distilled water, with the lithium dissolving with gaseous evolution, resulting in a 10-fold increase in the surface area (from 1 to about 10 m^2/g). Although the lixiviated sample was found to have better absorption and desorption kinetics, its hydrogen capacity was lower than that of the non-lixiviated sample due to the presence of magnesium hydroxides produced by lixiviation.

5.0 CONCLUSION

In this chapter, three different gas reactive applications of nanostructured materials were presented. They bear witness to the great potential of nanostructured materials for large-scale technological applications. In fact, the examples used indicate that large-scale applications of nanostructured materials already exist and, in some circumstances, have been in place for more than fifty years, without the label "nanostructured" ever being voiced.

This review is not an exhaustive description of the possibilities of nanostructured materials as gas reactive applications. Other applications are readily envisaged. For example, because of their surface reactivity, there is an immense potential for the application of nanostructured materials as getters.[154] Moreover, because of their increased diffusivity, nanostructured materials could also be ideal materials for the development of new separating membranes.[155]

It should be remembered, lastly, that the definition of nanostructured materials is somewhat ambiguous. However, it is the feeling of the present author that the large impact of the surface atoms, or of the high proportion of surface atoms, will make itself felt for materials with an average crystallite size well below 10 nm. At 10 nm the number of atoms at grain boundaries is about 20% while this value increases to nearly 50% for an average value of 5 nm. In a number of the studies described here, the average value of the crystals was 10 nm or varied between 10 and 20 nm. The fundamental question to be answered in the near future is what improvement to expect when this average crystallite size reaches the order of 5 nm. For gas-sensing materials, for instance, it was mentioned that the value of the electron-surface depleted layer is of the order of 3 nm in SnO_2, which means that only grains below 6 nm will be completely depleted. Also, point defects and the related oxygen vacancies play a dominant role in the properties of materials. The stronger presence of these structures will have a direct impact on the concentration of vacancies while at the same time improving the surface diffusivity, both effects should enhance the sensitivity as well as the rapidity of detection.

It can now be foreseen that the main difficulty to be circumvented, once materials of such average small size are easily synthesized, will be control of the crystallite size during operation, which could probably be done by means of growth inhibitors or by lowering of the operating temperature.

Much more work in material synthesis and characterization, as well as in the development of theoretical models, awaits researchers for all these domains to bloom. However, the technological potential in all of them is dauntingly enormous. In order that rapid evolution in materials, properties, and technological concepts be achieved, it is essential that the findings from all fields be pooled by common assent.

ACKNOWLEDGEMENTS

The author would like to thank Prof. Jackie Y. Ying and Dr. Virgil Provenzano for their friendship and collaboration. He also acknowledges the contribution of colleagues who have worked at, or in association with, Hydro-Québec's Research Institute. Lastly, thanks go to Lesley Kelley-Régnier for greatly improving the quality of the text.

REFERENCES

1. Klabunde, K. J., Li, Y. X., and Khaleel, A., Catalysis and Surface Chemistry: Metal Clusters/Nanoscale Particles, in: *Nanophase Materials Synthesis-Properties-Applications*, (G. C. Hadjipanayis and R. W. Siegel, eds.), 260:757–769, NATO ASI Series, Kluwer Academic Publ., The Netherlands (1994)
2. Ippommatsu, M., Ohnishi, H., Sasaki, H., and Matsumoto, T., Study of the Sensing Mechanism of Tin Oxide Flammable Gas Sensors using the Hall Effect, *J. Appl. Phys.*, 69:8368–8374 (1991)
3. Trudeau, M., and Ying, J. J., Nanocrystalline Materials in Catalysis and Electrocatalysis: Structure Tailoring and Surface Reactivity, *NanoStruct. Mater.*, 7:245–258 (1996)
4. Davis, S. C., and Klabunde, K. J., Unsupported Small Metal Particles: Preparation, Reactivity, and Characterization, *Chem. Rev.*, 82:153–208 (1982)
5. Poirier, J. A., and Stoner, G. E., Microstructural Effects on Electrocatalytic Oxygen Reduction Activity of Nano-Grained Thin-Film Platinum in Acid Media, *J. Electrochem. Soc.*, 141:425–430 (1994)
6. Rupprechter, G., Hayek, K., and Hofmeister, H., Characterization of Oxide Supported Rhodium Nanocrystals: Habit, Surface Structure and Catalytic Activity, *NanoStruct. Mater.*, 9:311–314 (1997)
7. Gamarnik, M. Y., The Physical Nature of Changes of Lattice Parameters in Small Particles, *Phys. Stat. Sol. B*, 178:59–69 (1993)
8. Zhdanov, V. P., and Kasemo, B., Kinetics of Rapid Reaction on Nanometer Catalyst Particles, *Phys. Rev. B*, 55:4105–4108 (1997)
9. Peuckert, M., Yoneda, T., Dalla Betta, R. A., and Boudart, M., Oxygen Reduction on Small Supported Platinum Particles, *J. Electrochem. Soc.*, 133:944–947 (1986)

10. Zhao, J., Huggins, F. E., Feng, Z., Lu, F., Shah, N., and Huffman, G. P., Structure of a Nanophase Iron Oxide Catalyst, *J. Catal.*, 143:499–509 (1993)
11. Feng, Z., Zhao, J., Huggins, F. E., and Huffman, G. P., Agglomeration and Phase Transition of a Nanophase Iron Oxide Catalyst, *J. Catal.*, 143:510–519 (1993)
12. Wang, C. C., Zhang, Z., and Ying, J. Y., Photocatalytic Decomposition of Halogenated Organics Over Nanocrystalline Titania, *NanoStruct. Mater.*, 9:583–586 (1997)
13. Davis, S. C., and Klabunde, K. J., Unsupported Small Metal Particles: Preparation, Reactivity, and Characterization, *Chem. Rev.*, 82:153–208 (1982)
14. Stark, J. V., Park, D. G., Lagadic, I., and Klabunde, K. J., Nanoscale Metal Oxide Particles/Clusters as Chemical Reagents. Unique Surface Chemistry on Magnesium Oxide as Shown by Enhanced Adsorption of Acid Gases (Sulfur Dioxide and Carbon Dioxide) and Pressure Dependence, *Chem. Mater.*, Vol. 8 (1996)
15. Ying, J. Y., Surface Interaction and Interfacial Chemistry in Multicomponent Nanocrystalline Materials, *Ceramic Trans.*, 44:67–76 (1995)
16. Chernavskii, P. A., and Lunin, V. V., Oxide-Oxide Interaction in Ni, Co and Fe Supported Catalyst, *Kinetics Catal.*, 34:470–477 (1993)
17. Prairie, M. R., Renken, A., Highfield, J. G., Thampi, K. R., and Grätzel, M., A Fourier Transform Infrared Spectroscopic Study of CO_2 Methanation on Supported Ruthenium, *J. Catal.*, 129:130–144 (1991)
18. Trovarelli, A., Mustazza, C., Dolcetti, G., Kaspar, J., and Graziani, M., Carbon Dioxide Hydrogenation on Rhodium Supported on Transition Metal Oxides, *Appl. Catal.*, 65:129–142 (1990)
19. Lalande, G., Tamizhmani, G., Côté, R., Dignard-Bailey, L., Trudeau, M. L., Schulz, R., Guay, D., and Dodelet, J. P., Influence of Loading on the Activity and Stability of Heat-Treated Carbon-Supported Cobalt Phthalocyanine Electrocatalyst in Solid Polymer Electrolyte Fuel Cells, *J. Electrochem. Soc.*, 142:1162–1168 (1995)
20. Yao, M. H., Smith, D. J., Kalakkad, D., and Datye, A. K., HREM Study of Structure of Supported Pt Catalyst, *Z. Phys. D*, 26:S79–S81 (1993)
21. Zafiris, G. S., and Gorte, R. J., A Study of CO, No, and H_2 Adsorption on Model Pt/CeO_2 Catalysts, *Surf. Sci.*, 276:86–94 (1992)
22. Fukushima, K., Takaoka, G. H., Matsuo, J., and Yamada, I., Effects on CO Oxidation Activity of Nano-Scale Au Islands and TiO_2 Support Prepared by the Ionized Cluster Beam Method, *Jpn. J. Appl. Phys.*, 36:813–818 (1997)

23. Yang, L., Raeker, T. J., and DePristo, A. E., *Surf. Sci.*, 290:195– (1993)
24. Schoeb, A. M., Raeker, T. J., Yang, L., Wu, X., King, T. S., and DePristo, A. E., *Surf. Sci.*, 278:L125–129 (1992)
25. Gijzeman, O. L. J., Surface Segregation in Small Supported Particles, *Appl. Surf. Sci.*, 64:9–20 (1993)
26. Ying, J. Y., Synthesis of Nanostructured Catalyst & Catalytic Supports, in: *Nanophase Materials Synthesis-Properties-Applications*, (G. C. Hadjipanayis and R. W. Siegel, eds.), 260:37–44, NATO ASI Series, Kluwer Academic Publ., The Netherlands (1994)
27. Beck, D. D., and Siegel, R. W., The Dissociative Adsorption of Hydrogen Sulfide Over Nanophase Titanium Dioxide, *J. Mater. Res.*, 7:2840–2845 (1992)
28. Sarkas, H. W., Arnold, S. T., Hendricks, J. H., Kidder, L. H., Jones, C. A., and Bowen, K. H., An Investigation of Catalytic Activity in Mixed Oxide Nanophase Materials, *Z. Phys. D*, 26:46–50 (1993)
29. Tschöpe, A. S., and Ying, J. Y., Synthesis of Nanostructured Catalytic Materials Using a Modified Magnetron Sputtering Technique, *Nano Struct. Mater.*, 4:617–623 (1994)
30. Ying, J. Y., Surface Structure of Nanocrystalline Oxides, in: *Nanophase Materials Synthesis-Properties-Applications*, (G. C. Hadjipanayis and R. W. Siegel, eds.), 260:197–204, NATO ASI Series, Kluwer Academic Publ., The Netherlands (1994)
31. Tschöpe, A. S., Ying, J. Y., Amonlirdviman, K., and Trudeau, M. L., Mater. Res. Soc. Symp. Proc., 351:251–256 (1994)
32. Trudeau, M. L., Tschöpe, A. S., and Ying, J. Y., XPS Investigation of Surface Oxidation and Reduction in Nanocrystalline $Ce_xLa_{1-x}O_{2-y}$, *Surf. Interface Anal.,* 23:219–226 (1995)
33. Tschöpe, A. S., Ying, J. Y., Liu, W., and Flytzani-Stephanopoulos, M., *Mater. Res. Soc. Symp. Proc.,* 344:133–138 (1994)
34. Liu, W., Sarofim, A. F., and Flytzani-Stephanopoulos, M., *Appl. Catal. B*, 4:167 (1994)
35. Tschöpe, A. S., Liu, W., Flytzani-Stephanopoulos, M., and Ying, J. Y., *J. Catal.*, 157:42–50 (1995)
36. Tschöpe, A. S., Miani, F., Matteazzi, P., Dolcetti, G., Lutman A.,and Trovarelli, A., *Mater. Sci. Eng. A*, 168:153 (1993)
37. Ying, J. Y., Tschöpe, A. S., and Levin, D., *Nanostruct. Mater.*, 6:237–246 (1995)
38. Tschope, A. S., Schaadt, D., Birringer, R., and Ying, J. Y., Catalytic Properties of Nanostructured Metal Oxides Synthesized by Inert Gas Condensation, *NanoStruct. Mater.*, 9:423–432 (1997)

39. Mile, B., Howard, J. A., Tomietto, M., Joly, H. A., and Sayari, A., Preparation of Small Copper Particles of High Catalytic Activity Using a Rotating Cryostat, *J. Mater. Sci.*, 31:3073–3080 (1996)

40. Trudeau, M. L., Huot, J. Y., Schulz, R., Dussault, D., Van Neste, A., and L'Espérance, G., Nanocrystalline Fe-(Co,Ni)-Si-B: The Mechanical Crystallization of Amorphous Alloys and the Effects on Electrocatalytic Reactions, *Phys. Rev. B,* 45:4626–4636 (1992)

41. Huot, J. Y., Trudeau, M. L., and Schulz, R., unpublished.

42. Benameur, T., Yavari, R., and Durand, R., Electrocatalytic Behavior of Non-Equilibrium bcc Nickel, *Mater. Sci. Eng.*, A181/A182:1145–1149 (1994)

43. Ivanov, E. Y., Golybkova, G. V., and Grigorieva, T. F., The CsCl Structure of the Raney Nickel Catalyst Obtained from Mechanically Alloyed Nickel Aluminides, *Reactivity Sol.*, 8:73–76 (1990)

44. Ivanov, E. Y., Makhlouf, S. A., Sumiyama, K., Yamauchi, H., Suzuki, K., and Golubkuva, G., *J. Alloys Compounds,* 185:25–? (1992)

45. Brown, D. E., Mahmood, M. N., Turner, A. K., Hall, S. M., and Fogarty, P. O., *Int. J. Hydrogen Energy*, 7:405–? (1982)

46. Brown, D. E., Mahmood, M. N., Man, M. C. M., and Turner, A. K., *Electrochimica Acta*, 29:1551–? (1984)

47. Trudeau, M. L., Huot, J. Y., and Schulz, R., Mechanically Alloyed Nanocrystalline Ni-Mo Powders: a New Technique for Producing Active Electrodes for Catalyst, *Appl. Phys. Lett.*, 58:2764-2766 (1991)

48. Huot, J. Y., Trudeau, M. L., and Schulz, R., Low Hydrogen Overpotential Nanocrystalline Ni-Mo Cathodes for Alkaline Water Electrolysis, *J. Electrochem. Soc.*, 138:1316–1321 (1991)

49. Schulz, R., Huot, J. Y., Trudeau, M. L., Dignard-Bailey, L., Yan, Z. H., Jin, S., Lamarre, A., Ghali, E., and Van Neste, A., Nanocrystalline Ni-Mo Alloys and their Applications in Electrocatalysis, *J. Mater. Res.*, 9:2998–3008 (1994)

50. Van Neste, A., Yip, S. H., Jin, S., Boily, S., Ghali, E., Guay, D., and Schulz, R., Low Overpotential Nanocrystalline Ti-Fe-Ru-O Cathodes for the Production of Sodium Chlorate, *Mater. Sci. Forum*, 225–227:795–800 (1996)

51. Blouin, M., Guay, D., Boily, S., Van Neste, A., and Schulz, R., Electrocatalytic Properties of Nanocrystalline Alloys: Effect of the Oxygen Concentration in Ti_2RuFeO_x Alloy on the Structural and Electrochemical Properties, *Mater. Sci. Forum*, 225–227:801–806 (1996)

52. Chui, Z. J., Wang, L., Wang, K. Y., Sun, L., Qiao, G. W., and Wang, J. T., The Surface Structure and Catalytic Properties of Partially Crystallized Fe-Si-B Alloys, *J. Non-Cryst. Sol.*, 150:440–443 (1992)

53. Chui, Zh.J., Wang, L., Wang, K. Y., Song, Q. H., Wang, J. T., and Lu, K., The Catalytic Properties of $Fe_{78}Si_{12}B_{10}$ Alloys Prepared by Ball Milling, *Mater. Sci. Eng. A*, 134:1037–1040 (1991)

54. Trovarelli, A., Matteazzi, P., Dolcetti, G., Lutman, A., and Miani, F., Nanophase Iron Carbides as Catalyst for Carbon Dioxide Hydrogenation, *Appl. Catal.*, 95:L9–L13 (1993)

55. Miani, F., Matteazzi, P., Dolcetti, G., Lutman, A., and Trovarelli, A., Catalytic Properties of Mechanosynthesized Nanophase Iron Carbides, *Mater. Sci. Eng. A*, 168:153–155 (1993)

56. Van Neste, A., Zielinski, P. A., Kaliaguine, S., Trudeau, M. L., and Schulz, R., *Metal Dispersion Enhancement by Ball-Milling with Alumina*, in: *The Proceeding of the Powder Metallurgy Conference*, Paris, France (1994)

57. Zielinski, P. A., Schulz, R., Kaliaguine, S., and Van Neste, A., Structural Transformation of Alumina by High Energy Ball Milling, *J. Mater. Res.*, 8:2985–2992 (1993)

58. Sun, L., Liu, H., and Dou, S., The Effect of Ball Milling and Rapid Quenching on the Catalytic Activity of the $Cu_{30}Al_{70}$ Alloy, *Mater. Lett.*, 33:79–83 (1997)

59. De Leitenburg, C., Trovarelli, A., Zamar, F., Maschio, S., Dolcetti, G., and Liorca, J., A Novel Simple Route to Catalyst with a High Oxygen Storage Capacity: The Direct Room-Temperature Synthesis of CeO_2-ZrO_2 Solid Solutions, *J. Cem. Soc., Chem. Commun.*, 1995:2181 (1995)

60. Suslick, K. S., Hyeon, T., Fang, M., and Cichowlas, A. A., Sonochemical Synthesis of Nanostructured Catalyst, *Mater. Sci. Eng. A*, 204:186–192 (1995)

61. Suslick, K. S., Hyeon, T., and Fang, M., Nanostructured Materials Generated by High-Intensity Ultrasound: Sonochemical Synthesis and Catalytic Studies, *Chem. Mater.*, 8:2172–2179 (1996)

62. Tsai, K. L., and Dye, J. L., Nanoscale Metal Particles by Homogeneous Reduction with Alkalides or Electrides, *J. Amer. Chem. Soc.*, 113:1650–1652 (1991)

63. Wang, Z. L., Ahmad, T. S., and El-Sayed, M. A., Steps, Ledges, and Kinks on the Surfaces of Platinum Nanoparticles of Different Shapes, *Surf. Sci.*, 380:302–310 (1997)

64. Sze, C., Gulari, E., and Demczyk, B. G., Surface Structure—Catalytic Function in Nanophase Gold Catalyst, *Mat. Res. Soc. Symp. Proc.*, 286:143–148 (1993)

65. Ozawa, M., Kimura, M., and Isogai, A., The Application of Ce-Zr Oxide Solid Solution to Oxygen Storage Promoters in Automotive Catalyst, *J. Alloys Comp.*, 193:73–75 (1993)

66. Reetz, M. T., and Helbig, W., Size-Selective Synthesis of Nanostructured Transition Metal Clusters, *J. Amer. Chem. Soc.*, 116:7401–7402 (1994)

67. Kost, K. M., Bartak, D. E., Kazee, B., and Kuwana, T., Electrodeposition of Palladium, Iridium, Ruthenium, and Platinum in Poly(4-vinylpyridine) Films for Electrocatalysis, *Anal. Chem.*, 62:151–157 (1990)

68. Ladouceur, M., Lalande, G., Guay, D., Dodelet, J. P., Dignard-Bailey, L., Trudeau, M. L., and Schulz, R., Pyrolyzed Cobalt Phthalocyanine as Electrocatalyst for Oxygen Reduction, *J. Electrochem. Soc.*, 140:1974–1981 (1993)

69. Dignard-Bailey, L., Trudeau, M. L., Joly, A., Schulz, R., Lalande, G., Guay, D., and Dodelet, J. P., Graphitization and Particle Size Analysis of Pyrolyzed Cobalt Phthalocyanine/Carbon Catalyst for Oxygen Reduction in Fuel Cells., *J. Mater. Res.*, 9:3203–3209 (1994)

70. Miquel, P. F., and Katz, J. L., Formation and Characterization of Nanostructured V-P-O Particles in Flames: A New Route for the Formation of Catalysts, *J. Mater. Res.*, 9:746–754 (1994)

71. Rousset, J. L., Khanra, B. C., Cadrot, A. M., Cadete Santos Aires, F. J., Renouprez, A. J., and Pellarin, M., Investigations on Supported Bimetallic PdPt Nanostructures, *Surf. Sci.*, 352–354:583–587 (1996)

72. Yamashita, H., Sakai, K., Funabiki, T., Yoshida, S., and Isozumi, Y., Catalysis by Amorphous Metal Alloys, *J. Chem. Soc. Faraday Trans.*, I83:2895–2900 (1987)

73. Peuckert, M., and Baiker, A., Hydrogenation of Carbon Monoxide on Amorphous $Fe_{81}B_{13.5}Si_{3.5}$°C, *J. Chem. Soc. Faraday Trans.*, I81:2797–2803 (1985)

74. Jacobs, P. W., Wind, S. J., Ribeiro, F. H., and Somorjai, G. A., Nanometer Size Platinum Particle Arrays: Catalytic and Surface Chemical Properties, *Surf. Sci.*, 372:L249–L253 (1997)

75. Somorjai, G. A., New Model Catalysts (Platinum Nanoparticles) and New Techniques (SFG and STM) for Studies of Reaction Intermediates and Surface Restructuring at High Pressure During Catalytic Reactions, *Appl. Surf. Sci.*, 121/122:1–19 (1997)

76. Yamazoe, N., and Miura, N., Environmental Gas Sensing, *Sensors and Actuators B*, 20:95–102 (1994)

77. Tolles, W. M., *Nanotechnology*, 7:59 (1996)

78. Korolkoff, N. O., Survey of Toxic Gas Sensors and Monitoring Systems, *Solid State Techn.*, pp. 49–64 (Dec. 1989)

79. Sbverveglieri, G., (ed.), *Gas Sensors Principles, Operation and Developments*, Kluwer Acad. Publ., The Netherlands (1992)

80. Rickerby, D. G., Horrillo, M. C., Santos, J. P., and Serrini, P., Microstructural Characterization of Nanograin Tin Oxide Gas Sensors, *NanoStruct. Mater.*, 9:43–52 (1997)

81. Ogawa, H., Abe, A., Nishikawa, M., and Hayakawa, S., Preparation of Tin Oxide Films from Ultrafine Particles, *J. Electrochem. Soc.*, 128:685–689 (1981)

82. Ogawa, H., Abe, A., Nishikawa, M., and Hayakawa, S., Electrical Properties of Tin Oxide Ultrafine Particles Films, *J. Electrochem. Soc.*, 128:2020–2025 (1981)

83. Ogawa, H., Nishikawa, M., and Abe, A., Hall Measurement Studies and an Electrical Conduction Model of Tin Oxide Ultrafine Particle Films, *J. Appl. Phys.*, 53:4448–4455

84. Seiyama, T., Kato, A., Fujiishi, K., and Nagatani, M., A New Detector for Gaseous Components Using Semiconductive Thin Films, *Anal. Chem.*, 34:1502–1053 (1962)

85. Kohl, D., Surface Processes in the Detection of Reducing Gases with SnO_2-Based Device, *Sensors and Actuators*, 18:71–113 (1989)

86. Watson, J., Ihokura, K., and Coles, G. S. V., The Tin Dioxide Gas Sensor, *Meas. Sci. Technol.*, 4:711–719 (1993)

87. Gopel, W., and Schierbaum, K. D., SnO_2 Sensors: Current Status and Future Prospects, *Sensors and Actuators B*, 26–27:1–12 (1995)

88. Yamazoe, N., Fuchigami, J., Kishikawa, M., and Seiyama, T., Interactions of Tin Oxide Surface with O_2, H_2O, and H_2, *Surf. Sci.*, 86:335–344 (1979)

89. Sun, H.-T., Cantalini, C., Faccio, M., and Pelino, M., NO_2 Gas Sensitivity of Sol-Gel-Derived α-Fe_2O_3 Thin Films, *Thin Solid Films*, 269:97–101 (1995)

90. Yea, B., Konishi, R., Osaki, T., Abe, S., Tanioka, H., and Sugahara, K., *Appl. Surf. Sci.*, 100/101:365 (1996)

91. Chang, S. C., Oxygen Chemisorption on Tin Oxide: Correlation Between Electrical Conductivity and EPR Measurements, *J. Vac. Sci. Technol.*, 17:366–369 (1980)

92. Kanamori, M., Suzuki, K., Ohya, Y., and Takahashi Y., Analysis of the Changes in the Carrier Concentration of SnO_2 Thin Film Gas Sensor, *Jpn. J. Appl. Phys.*, 33:6680–6683 (1994)

93. Pink, H., Treitinger, L., and Vite, L., Preparation of Fast Detecting SnO_2 Gas Sensors, *Jpn. J. Appl. Phys.*, 19:513–517 (1980)

94. Sberveglieri, G., Classical and Novel Techniques for the Preparation of SnO_2 Thin-Film Gas Sensor, *Sensors and Actuators B*, 6:239 (1992)

95. Butta, N., Mell, M., and Pizzini, S., Influence of Surface Parameters and Doping on the Sensitivity and on the Response Times of Tin Oxide Resistive Sensors, *Sensors and Actuators B*, 2:151–161 (1990)

96. Windischmann, H., and Mark, P., A Model for the Operation of a Thin-Film SnO_x Conductance-Modulation Carbon Monoxide Sensor, *J. Electrochem. Soc.*, 126:627–633 (1979)

97. Dieguez, A., Romano-Rodriguez, A., Morante, J. R., Weimar, U., Schweizer-Berberich, M., and Gopel, W., Morphological Analysis of Nanocrystalline SnO_2 for Gas Sensor Applications, *Sensors and Actuators B*, 31:1–8 (1996)

98. Xu, C., Tamaki, J., Miura, N., and Yamazoe, N., Grain Size Effects on Gas Sensitivity of Porous SnO_2-Based Elements, *Sensors and Actuators B*, 3:147–155 (1991)

99. Chang, S. C., Thin-Film Semiconductor NO_x Sensor, *IEEE Trans. Elec. Dev.*, ED-26:1875–1880 (1979)

100. Kohl, D., The Role of Noble Metals in the Chemistry of Solid-State Gas Sensor, *Sensors and Actuators B*, 1:158–165 (1990)

101. Morrison, S. R., Selectivity in Semiconductor Gas Sensors, *Sensors and Actuators*, 12:425–440 (1987)

102. Shaver, P. J., Activated Tungsten Oxide Gas Detectors, *Appl. Phys. Lett.*, 11:255–257 (1967)

103. Chang, S. C., The Effects of Palladium-Gold on the Chemisorption Activities of Thin Film Tin Oxide, *J. Vac. Sci. Technol. A*, 1:296–300 (1983)

104. Cricenti, A., Generosi, R., Scarselli, M. A., Perfetti, P., Siciliano, P., Serra, A., Tepore, A., Coluzza, C., Almeida, J., and Margaritondo, G., Morphological, Chemical and Electrical Characterization of Pt-SnO_2 Thin Film Grown on Rough and Mechanically Polished Al_2O_3 Substrates, *J. Phys. D*, 29:2235–2239 (1996)

105. Vlachos, D. S., Papadopoulos, C. A., and Avaritsiotis, J. N., On the Electronic Interaction Between Additives and Semiconducting Oxide Gas Sensors, *Appl. Phys. Lett.*, 69:650–652 (1996)

106. Zhang, T., Shen, Y., Zhang, R., and Liu, X., Ammonia-Sensing Characteristics of Pt-Doped CdSnO3 Semiconducting Ceramic Sensor, *Mater. Lett.,* 27:161–164 (1996)

107. Cao, X., Cao, L., Yao, W., and Ye, X., Structural Characterization of Pd-Doped SnO_2 Thin Films Using XPS, *Surf. & Interf. Anal.*, 24:662–666 (1996)

108. Matsushima, S., Meakawa, T., Tamaki, J., Miura, N., and Yamazoe N., Role of Additives on Alcohol Sensing by Semiconductor Gas Sensor, *Chem. Lett.*, pp. 845–848 (1989)

109. Rastomjee, C. S., Dale, R. S., Schaffer, R. J., Jones, F. H., Egdell, R. G., Georgiadis, G. C., Lee, M. J., Tate, T. J., and Cao, L. L., An Investigation of Doping SnO_2 by Ion Implantation and Application of Ion-Implanted Films as Gas Sensors, *Thin Solid Films*, 279:98–105 (1996)

110. Barbi, G. B., Santos, J. P., Serrini, P., Gibson, P. N., Horrillo, M. C., and Manes, L., Ultrafine Grain-Size Tin-Oxide Films for Carbon Monoxide Monitoring in Urban Environments, *Sensors and Actuators B*, 24–25:559–563 (1995)

111. Rumyantseva, M. N., Labeau, M., Senateur, J. P., Delabouglise, G., Boulova, M. N., and Gaskov, A. M., Influence of Copper on Sensor Properties of Tin Dioxide Films in H_2S, *Mater. Sci. Eng. B*, 41:228–234 (1996)

112. Zubkans, J., Lloyd Spetz, A., Sundgren, H., Winquist, F., Kleperis, J., Lusis A., and Lundstrom, I., In-Situ Modification of the NO_x Sensitivity of Thin Discontinuous Platinum Films as Gates of Chemical Sensors, *Thin Solid Films*, 268:140–143 (1995)

113. Jiang, J. Z., Lin, R., Lin, W., Nielsen, K., Morup, S., Dam-Johansen K., and Clasen, R., Gas-Sensitive Properties and Structure of Nanostructured $(\alpha\text{-}Fe_2O_3)_x\text{-}(SnO_2)_{1\text{-}x}$ Materials Prepared by Mechanical Alloying, *J. Phys. D*, 30:1459–1467 (1997)

114. Ansari, S. G., Gosavi, S. W., Gangal, S. A., Karekar, R. N., and Aiyer, R. C., Characterization of SnO_2-Based H_2 Gas Sensors Fabricated by Different Deposition Techniques, *J. Mater. Sci.*, 8:23–27 (1997)

115. Vlachos, D. S., Papadopoulos, C. A., and Avaritsiotis, J. N., Dependence of Sensitivity of SnO_x Thin-Film Gas Sensors on Vacancy Defects, *J. Appl. Phys.*, 80:6050–6054 (1996)

116. Serrini, P., Briois, V., Horrillo, M. C., Traverse, A., and Manes, L., Chemical Composition and Crystalline Structure of SnO_2 Thin Films Used as Gas Sensors, *Thin Solid Films,* 304:113–122 (1997)

117. Hu, W. S., Liu, Z. G., Wu, Z. C., and Feng, D., Comparative Study of Laser Ablation Techniques for Fabricating Nanocrystalline SnO_2 Thin Films for Sensors, *Mater. Lett.*, 28:369–372 (1996)

118. Liu, Y., Zhu, W., Tan, O. K., and Shen, Y., Structural and Gas Sensing Properties of Ultrafine Fe_2O_3 Prepared by Plasma Enhanced Chemical Vapor Deposition, *Mater. Sci. Eng. B*, 47:171–176 (1997)

119. Bhowmik, S., Constant, K. P., Parker, J. C., and Ali, M., Electrical Characterization of Nanocrystalline Titania, I: Impedance Spectroscopy Studies between 300 K and 473 K, *Mater. Sci. Eng. A*, 204:258–266 (1995)

120. Rella, R., Serra, A., Siciliano, P., Vasanelli, L., De, G., and Licciulli, A., CO Sensing Properties of SnO_2 Thin Films Prepared by the Sol-Gel Process, *Thin Solid Films*, 304:339–343 (1997)

121. Xu, C., Tamaki, J., Miura, N., and Yamazoe, N., Promoting Effects of Additives on Thermal Stability of Thin Oxide (IV) Fine Particles, *J. Mater. Sci. Lett.*, 8:1092–1094 (1989)

122. Xu, C., Tamaki, J., Miura, N., and Yamazoe, N., Correlation between Gas Sensitivity and Crystallite Size in Porous SnO_2-Based Sensors, *Chem. Lett.*, 441–444 (1990)

123. Li, G. J., and Kawi, S., High-Surface-Area SnO_2: A Novel Semiconductor-Oxide Gas Sensor, *Mater. Lett.*, 34:99–102 (1998)

124. Akimov, B. A., Albul, A. V., Gas'kov, A. M., Il'in, V. Yu, Rumyantseva, M. N., and Labeau, M., *Semiconductors*, 31:335 (1997)

125. Liu, C. H., Zhang, L., and He, Y. J., Properties and Mechanism Study of Ag Doped SnO_2 Thin Films as H_2S Sensors, *Thin Solid Films*, 304:13–15 (1997)

126. Neubecker, A., Pompl, T., Doll, T., Hansch, W., and Eisele, I., Ozone-Enhanced Molecular Beam Deposition of Nickel Oxide (NiO) for Sensor Applications, *Thin Solid Films*, 310:19–23 (1997)

127. Topler, J., and Feucht, K., Results of a Test Fleet with Metal Hydride Motor Cars, *Z. fur Phy. Chem. Neue Folge*, 164:1451–1461 (1989)

128. Zaluski, L., Zaluska, A., Tessier, P., Strom-Olsen, J. O., and Schulz, R., Nanocrystalline Hydrogen Absorbing Alloys, *Mater. Sci. Forum*, 225–227:853–858 (1996)

129. Kuriyama, N., Sakai, T., Miyamura, H., Uehara, I., and Ishikawa, H., Characterization of Metal Hydride Electrodes by Means of Electrochemical Impedance Spectroscopy, *J. Alloys Comp.*, 192:161–163 (1993)

130. Bernauer, O., Topler, J., Noreus, D., Hempelmann, R., and Richter, D., Fundamentals and Properties of Some Ti/Mn Based Laves Phase Hydrides, *Int. J. Hydrogen Energy*, 14:187–200 (1989)

131. Martin, M., Gommel, C., Borkhart, C., and Fromm, E., Absorption and Desorption Kinetics of Hydrogen Storage Alloys, *J. Alloys Comp.*, 238:193–201 (1996)

132. Meli, F., Zuettel, A., and Schlapbach, L., Surface and Bulk Properties of $LaNi_{5-x}Si_x$ Alloys from the Viewpoint of Battery Applications, *J. Alloys Comp.*, 190:17–24 (1992)

133. Kim, J. Y., Gibbons, P. C., and Kelton, K. F., Hydrogenation of Pd-Coated Samples of the Ti-Zr-Based Icosahedral Phase and Related Crystalline Phases, *J. Alloys Comp.*, 266:311–317 (1998)

134. Koch, C. C., Synthesis of Nanostructured Materials by Mechanical Milling: Problems and Opportunities, *NanoStruct. Mater.*, 9:13–22 (1997)

135. Murray, J., Miller, H., Bird, P., and Goudy, A. J., The Effect of Particle Size and Surface Composition on the Reaction Rates of Some Hydrogen Storage Alloys, *J. Alloys Comp.*, 231:841–845 (1995)

136. Trudeau, M. L., Schulz, R., Zaluski, L., Hosatte, S., Ryan, D. H., Doner, C. B., Tessier, P., Strom-Olsen, J. O., and Van Neste, A., Nanocrystalline Iron-Titanium Alloys Prepared by High-Energy Mechanical Deformation, *Mater. Sci. Forum*, 88–90:537–544 (1992)

137. Zaluski, L., Tessier, P., Ryan, D. H., Doner, C. B., Zaluska, A., Strom-Olsen, J. O., Trudeau, M. L., and Schulz, R., Amorphous and Nanocrystalline Fe-Ti Prepared by Ball Milling, *J. Meter. Res.*, 8:3059–3068 (1993)

138. Tessier, P., Zaluski, L., Zaluska, A., Strom-Olsen, J. O., and Schulz, R., Effect of Compositional Variations on Hydrogen Storage in Ball-Milled Fe-Ti, *Mater. Sci. Forum*, 225–227:869–874 (1996)

139. Zaluski, L., Zaluska, A., Tessier, P., Strom-Olsen, J. O., and Schulz, R., Investigation of Structural Relaxation by Hydrogen Absorption in Ball-Milled Alloys, *Mater. Sci. Forum*, 225–227:875–880 (1996)

140. Trudeau, M. L., Dignard-Bailey, L., Schulz, R., Tessier, P., Zaluski, L., Ryan, D. H., and Strom-Olsen, J. O., The Oxidation of Nanocrystalline FeTi Hydrogen Storage Compounds, *NanoStruct. Mater.*, 1:457–464 (1992)

141. Selvam, P., Viswanathan B., and Srinivasan, V., XPS and XAES Studies on Hydrogen Storage Magnesium-Based Alloys, *Int. J. Hydrogen Ener.*, 14:899–902 (1989)

142. Zaluski, L., Zaluska, A., Tessier, P., Strom-Olsen, J. O., and Schulz, R., Hydrogen Absorption by Nanocrystalline and Amorphous Fe-Ti with Palladium Catalyst, Produced by Ball Milling, *J. Mater. Sci.*, 31:695–698 (1996)

143. Wasz, M. L., and Schwarz, R. B., Structure and Properties of Metal Hydrides Prepared by Mechanical Alloying, *Mater. Sci. Forum*, 225–227:859–868 (1996)

144. Zaluski, L., Zaluska, A., and Strom-Olsen, J. O., Hydrogen Absorption in Nanocrystalline Mg_2Ni Formed by Mechanical Alloying, *J. Alloys Comp.*, 217:245–249 (1995)

145. Zaluski, L., Zaluska, A., Tessier, P., Strom-Olsen, J. O., and Schulz, R., Catalytic Effect of Pd on Hydrogen Absorption in Mechanically Alloyed Mg_2Ni, $LaNi_5$, and FeTi, *J. Alloys Comp.*, 217:295–300 (1997)

146. Li, J., Ji, X., Wu, F., and Wang. G., A New Method for the Production of Mg-Ni Hydrogen Storage Materials, *Adv. Mater.*, 5:554–555 (1993)

147. Imamura, H., and Sakasai, N., Hydriding Characteristics of Mg-Based Composites Prepared Using a Ball Mill, *J. Alloys Comp.* 231:810–814 (1995)

148. Imamura, H., Sakasai, N., and Kajii, Y., Hydrogen Absorption of Mg-Based Composites Prepared by Mechanical Milling: Factors Affecting its Characteristics, *J. Alloys Comp.*, 232:218–223 (1996)

149. Holtz, R. L., and Imam, M. A., Hydrogen Storage Capacity of Submicron Magnesium-Nickel Alloys, *J. Mater. Sci.*, 32:2267–2274 (1997)

150. Orimo, S., and Fujii, H., Hydriding Properties of the Mg_2Ni-H System Synthesized by Reactive Mechanical Grinding, *J. Alloys Comp.*, 232:L16–L19 (1996)

151. Gross, K. L., Spatz, P., Zuttel, A., and Schlapbach, L., Mechanically Milled Mg Composites for Hydrogen Storage. The Transition to a Steady State Composition, *J. Alloys Comp.*, 240:206–213 (1996)

152. Gross, K. L., Spatz, P., Zuttel, A., and Schlapbach, L., Mg Composites for Hydrogen Storage. The Dependence of Hydriding Properties on Composition, *J. Alloys Comp.*, 261:276–280 (1997)

153. Huot, J., Bouaricha, S., Boily, S., Dodelet, J.-P., Guay, D., and Schulz., R., Increase of Specific Surface Area of Metal Hydrides by Lixiviation, *J. Alloys Comp.*, 266:307–310 (1998)

154. Holtz, R. L., Provenzano, V., and Imam, M. A., Overview of Nanophase Metals and Alloys for Gas Sensors, Getters, and Hydrogen Storage, *NanoStruct. Mater.*, 7:259–264 (1996)

155. Bryden, K. J., and Ying, J. Y., Electrodeposition Synthesis and Hydrogen Absorption Properties of Nanostructured Palladium-Iron Alloys, *NanoStruct. Mater.*, 9:485–488 (1997)

9

Magnetic Properties of Nanocrystalline Materials

Akihisa Inoue and Akihiro Makino

1.0 INTRODUCTION

It was reported in 1976 that an amorphous phase in Pd-Si, Fe-P-C, and (Fe,Co,Ni)-Si-B systems is useful as a precursor to prepare a nanocrystalline structure upon crystallization.[1] For the last two decades, a great effort has been devoted to the development of a new type of high-strength or high-functional material by utilizing the formation of crystallization-induced nanostructures. In a magnetic material field, good hard magnetic properties have been obtained in the crystallized structure of $Nd_2Fe_{14}B$, and amorphous phases obtained from amorphous Fe-Nd-B alloys.[2] In addition to the hard magnetic alloys, it has been found that the crystallization of Fe-Si-B amorphous alloys containing Nb and Cu causes the formation of a nanoscale bcc structure and the bcc alloys exhibit good soft magnetic properties of 1.2 to 1.4 T for saturation magnetization (B_s) and 10×10^4 for effective permeability (μ_e) at 1 kHz.[3] These results indicate that the crystallization-induced nanostructure is useful for the appearance of hard or soft magnetic properties. Although good soft magnetic properties are obtained for the nanoscale bcc $Fe_{73.5}Si_{13.5}B_9Nb_3Cu_1$, the relatively low Fe concentration leads to the limitation of B_s to less than 1.4 T and, hence, the development of a new soft magnetic alloy with high B_s above 1.5 T and high

μ_e above 10^5 at 1 kHz has strongly been desired because the simultaneous achievement of both properties enables the extension of application fields to various kinds of power transformers. It has subsequently been reported[4] that Fe-Ta-C thin films consisting of nanocrystalline bcc and TaC phases obtained by crystallization of a sputtered amorphous phase exhibit good soft magnetic properties (i.e., B_s of 1.54 T and μ_e of 5300 at 1 MHz), though the nanostructure cannot be synthesized by the melt spinning technique. We previously reported that Fe-rich amorphous alloys of 90 at% Fe are formed in Fe-Zr-B and Fe-Hf-B systems by rapid solidification.[5] If a nanostructure is formed in the Fe-rich alloys subjected to crystallization treatment, the nanostructured alloys are expected to exhibit high B_s exceeding 1.5 T. Based on this concept, we examined[6][7] the possibility of synthesizing a nanocrystalline structure in the Fe-M-B (M = Zr, Hf, or Nb) alloys and of obtaining good soft magnetic properties and succeeded[8]–[11] in developing nanocrystalline Fe-Zr-Nb-B-Cu alloys with excellent soft magnetic properties of 1.57 T for B_s and 16×10^4 for μ_e at 1 kHz which had not been hitherto obtained for any soft magnetic materials. It has subsequently been found[12][13] that the dissolution of a large amount of oxygen into the remaining amorphous phase is effective for a drastic increase in electrical resistivity of sputtered Fe-Hf-O and Fe-Zr-O films, leading to the achievement of good high-frequency permeability in the high frequency range of 1 to 100 MHz. This paper reviews our recent results on the formation of a nanogranular bcc and amorphous structure, the development of excellent soft magnetic material by nanocrystallization in the Fe-Zr-Nb-B-Cu and Fe-Hf-O systems and their engineering applications.

2.0 Fe-M-B (M = Zr, Hf, or Nb) AMORPHOUS ALLOYS AND THEIR CRYSTALLIZATION-INDUCED NANOSTRUCTURE

Fe-based amorphous alloys have reportedly been formed in a number of alloy systems such as Fe-(B,C,Si,P,Ge),[14] Fe-(Zr,Hf),[15] Fe-Re (Re = rare earth metals)[16] and Fe-(Zr,Hf,Nb,Ta)-B.[6][7] If attention is paid to a maximum Fe concentration for formation of an amorphous phase by the melt spinning method, the previous data show that the Fe concentration increases in the order Fe-(Zr,Hf) > Fe-Re > Fe-(Zr,Hf,Nb,Ta)-B > Fe-(B,C,Si,P,Ge). This indicates the possibility that the B_s value of the resulting nanocrystalline phase also increases in the same order. Consequently, the relation between the formation tendency of nanocrystalline structure and soft magnetic properties was systematically examined for the Fe-(Zr,Hf), Fe-Re, and Fe-(Zr,Hf,Nb,Ta)-B amorphous ribbons prepared by melt spinning. The crystallized structure of the $Fe_{90}Zr_{10}$, $Fe_{90}Hf_{10}$, and $Fe_{90}Nd_{10}$ amorphous alloys consists of bcc and amorphous phases at the first-stage and α-Fe and compound phases at the second stage, but the grain sizes of the bcc and α-Fe phases are above 40 nm and 80 nm, respectively, which are too large to obtain good soft magnetic properties.[17] Therefore, it is concluded that the binary Fe-based alloys cannot be regarded as an appropriate system leading to the nanocrystalline structure, though high B_s above 1.5 T is obtained.

Figure 1 summarizes the composition ranges in which an amorphous phase is formed in melt-spun Fe-Zr-B, Fe-Hf-B, and Fe-Nb-B alloys containing more than 75 at% Fe and the subsequent crystallized structure consists of a nanoscale bcc-Fe and amorphous structure.[18] The maximum Fe concentration for formation of an amorphous phase is about 92 at% for the Fe-Zr-B and Fe-Hf-B alloys and about 86 at% Fe for the Fe-Nb-B alloy. Besides, it is seen that the nanogranular bcc-Fe phase surrounded by the residual amorphous phase is formed in the Fe-rich concentration range above 86 at% Fe for Fe-Zr-B and Fe-Hf-B alloys. Here, it is important to describe the criteria for formation of the nanoscale bcc-Fe and amorphous mixed structure in the limited Fe-rich composition ranges. Figure 2 shows differential thermal analysis (DTA) curves of the $Fe_{90}Zr_7B_3$ and $Fe_{89}Hf_7B_4$ amorphous alloys. Two exothermic peaks are seen on the DTA curves, indicating that the crystallization takes place through two stages. From x-ray diffraction analysis and transmission electron microscopic observation, the first-stage exothermic reaction is due to the precipitation of bcc-Fe phase and the second exothermic peak results from the precipitation of α-Fe and Fe_2Zr or Fe_2Hf phases from the bcc and remaining amorphous phases. Notice that the temperature interval between the first and second-exothermic peaks is as large as 150 K, indicating that the bcc-Fe and amorphous phases have a high metastability. Figure 3 shows a brightfield electron micrograph and selected-area electron diffraction pattern of the $Fe_{90}Zr_7B_3$ alloy annealed for 3.6 ks at 923 K which is located between the first and second exothermic peaks. The bcc phase has spherical equiaxed grains with a size of about 15 nm and each grain has a random orientation. Furthermore, the diffraction pattern reveals the residual existence of an amorphous phase and the absence of the second crystalline phase. Further heating to a temperature above the second exothermic peak caused the precipitation of α-Fe and Fe_2Zr phases, accompanying the complete disappearance of the remaining amorphous phase and the significant grain growth of the α-Fe phase. Borides of any kind are not observed and hence the B element is thought to be preferentially dissolved into the Fe_2Zr phase because of the similarity of the crystal structure and chemical formula between Fe_2Zr and Fe_2B. The reason for the high thermal stability of the metastable bcc-Fe and amorphous phases is discussed in Sec. 3.0 on the basis of high-resolution TEM image, nanobeam electron diffraction and atom-probe field ion microscopic data.

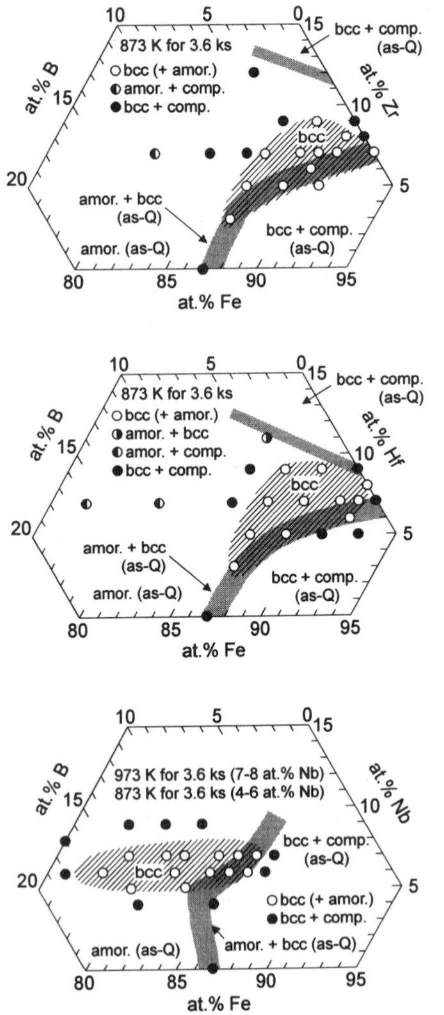

Figure 1. Compositional dependence of structure for Fe-Zr-B, Fe-Hf-B and Fe-Nb-B alloys at as-quenched and annealed states.

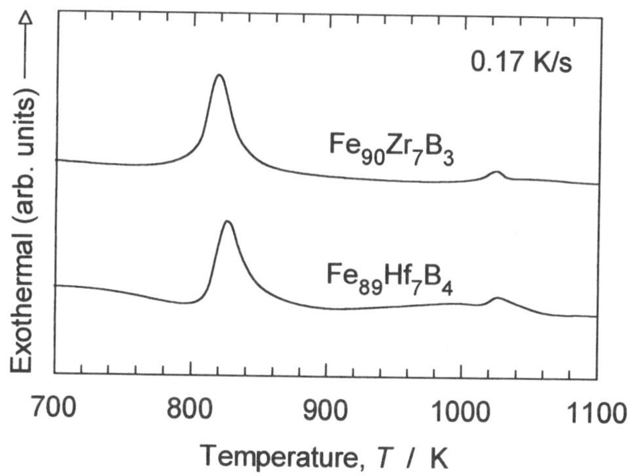

Figure 2. DTA curves of amorphous $Fe_{90}Zr_7B_3$ and $Fe_{89}Hf_7B_4$ alloys.

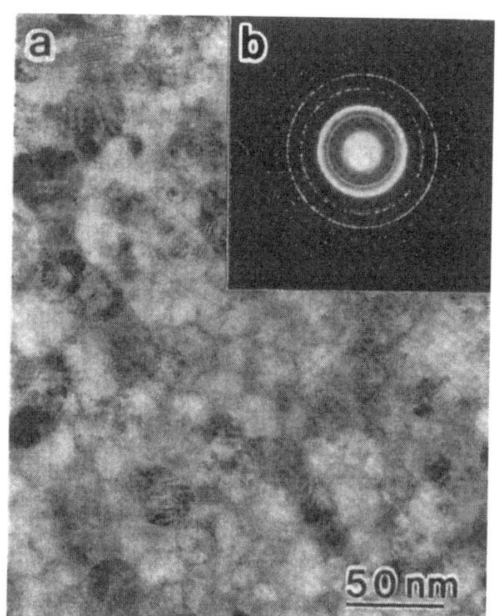

Figure 3. (*a*) Brightfield TEM images and (*b*) selected-area electron diffraction pattern of $Fe_{90}Zr_7B_3$ alloy annealed at 923 K for 3.6 ks.

3.0 SOFT MAGNETIC PROPERTIES AND STRUCTURAL ANALYSES OF Fe-M-B (M = Zr, Hf, or Nb) NANOCRYSTALLINE TERNARY ALLOYS

Figure 4 shows the composition dependence of as-quenched phase B_s and μ_e at 1 kHz for the melt-spun Fe-Zr-B and Fe-Nb-B alloy ribbons subjected to optimum annealing treatments for 3.6 ks at 873 and 923 K. The amorphous phase is formed in the wide composition range up to 92 at% Fe for the Fe-Zr-B alloys and 84 at% Fe for the Fe-Nb-B alloys. The annealed alloys exhibit good soft magnetic properties of high B_s above 1.6 T and high μ_e, exceeding 10^4 around the composition of 90 at% Fe in the former system and 84 at% Fe in the latter system. As seen in Fig. 4, the best soft magnetic properties are obtained in the upper limit range of Fe concentration where the amorphous single phase is obtained in the melt-spun state. The good correspondence is because the nanoscale bcc structure is obtained only from the amorphous single phase in the Fe-rich composition range. Besides, one can notice that the composition range in which the best soft magnetic properties are obtained for the Fe-Nb-B alloys shifts to the higher B concentration in comparison with that for the Fe-Zr-B alloys. The deviation is due to the extension of the formation range of the nanoscale bcc-Fe structure resulting from the decrease in the precipitation tendency of compounds, reflecting the weaker bonding nature of Nb-Fe and Nb-B pairs as compared with Zr-Fe and Zr-B pairs. Similar data as those for the Fe-Zr-B nanocrystalline alloys have been obtained in Fe-Hf-B system.[9][18] Figure 5 shows the changes with annealing temperature (T_a) in the structure, B_s, μ_e, mean grain size of the bcc phase (D), and λ_s for the melt-spun $Fe_{90}Zr_7B_3$, $Fe_{89}Hf_7B_4$, and $Fe_{84}Nb_7B_9$ alloys. As T_a increases, the amorphous phase changes to a mostly single bcc phase in the T_a range of 750 to 930 K and α-Fe plus compound in the T_a range above 930 K. The D value is as small as 10 to 17 nm and increases rapidly upon the phase transition to α-Fe plus compound. The B_s and μ_e keep nearly zero values in the amorphous single phase state because of the Invar effect. The precipitation of bcc phase caused by a further increase in T_a gives rise to significant increases in B_s and μ_e, followed by maximum values of 1.4 to 1.7 T and 2×10^4 to 3×10^4, respectively, in the T_a range of 873 to 923 K just before the phase decomposition of the bcc phase. The λ_s has positive values in the amorphous single-phase state for the three alloys and changes to negative values for the

Zr- and Hf-containing bcc alloys, and to slightly positive or nearly zero value for the Nb-containing bcc alloy. From the correspondence between the structure and magnetic properties, the best soft magnetic properties are obtained in the partially crystallized structure consisting of nanoscale bcc and amorphous phases. Therefore, it is concluded that the residual existence of the amorphous phase plays an important role in the achievement of good soft magnetic properties by the formation of the nanoscale mixed structure.

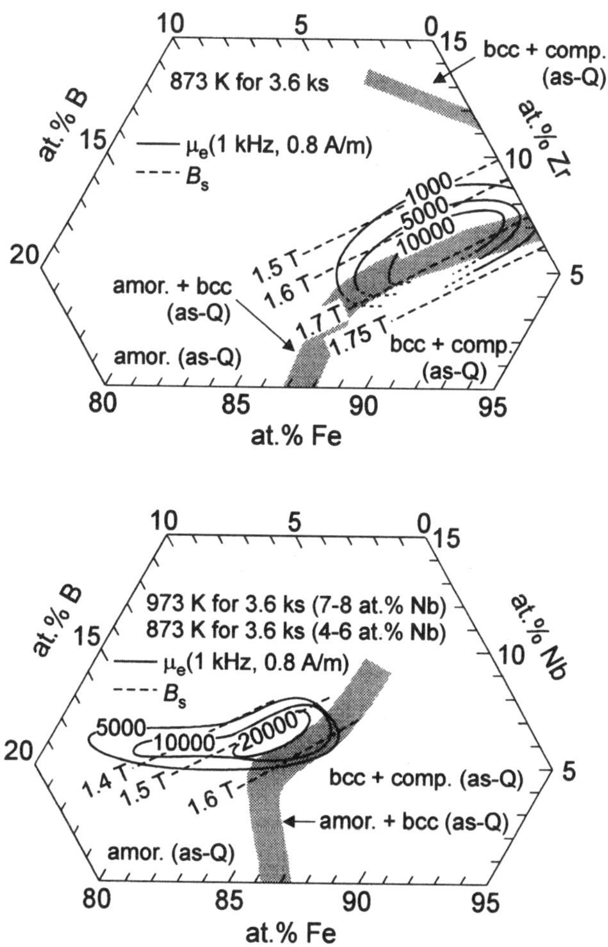

Figure 4. Composition dependence of B_s and μ_e at 1 kHz for Fe-M-B (M = Zr or Nb) amorphous alloys annealed for 3.6 ks at 873 and 923 K. The data of as-quenched phase are also shown for reference.

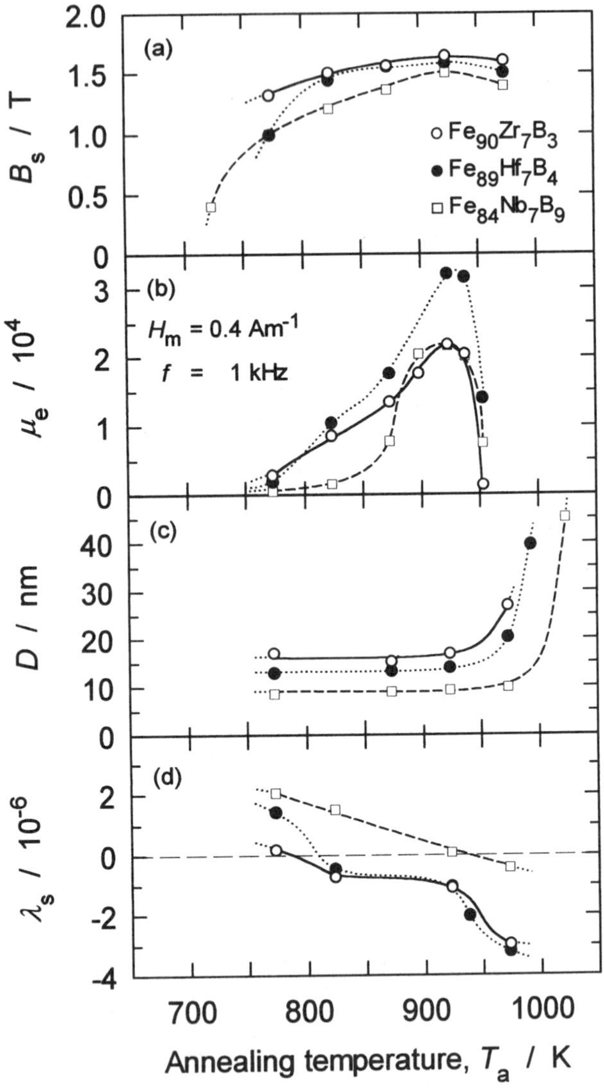

Figure 5. Changes in B_s, μ_e, D, *and* λ_s with T_a for amorphous $Fe_{90}Zr_7B_3$, $Fe_{89}Hf_7B_4$ and $Fe_{84}Nb_7B_9$ alloys.

Here, it is important for understanding the formation mechanism of the nanostructure to confirm the residual existence of the amorphous phase and to examine the solute concentrations of the bcc and amorphous phases. Figure 6 shows a high-resolution TEM image of the $Fe_{88}Hf_{10}B_2$ alloy annealed for 3.6 ks at 873 K, together with the data of nanobeam diffraction patterns and energy-dispersive spectroscopy (EDX) spectra taken from the regions 1 and 2. From the fringe contrast in the TEM image and the diffraction patterns, the regions 1 and 2 are identified to be bcc and amorphous phases, respectively. Thus, the nanoscale bcc phase with a grain size of about 10 nm is surrounded by the remaining amorphous phase. The EDX spectra also indicate that the Hf content is enriched to the remaining amorphous phase. Figure 7 shows a high-resolution TEM image and EDX and electron loss spectroscopy (EELS) profiles for the $Fe_{84}Nb_7B_9$ amorphous alloy annealed for 3.6 ks at 923 K. The structural feature (that the nanogranular bcc-Fe particles with a size of about 12 nm are surrounded by the amorphous phase) is the same as that for the Fe-Hf-B alloy shown in Fig. 6. Similarly, the Nb is enriched to the remaining amorphous phase and no appreciable Nb is detected in the bcc particle. Furthermore, the EELS data indicate that the B is also enriched to the amorphous phase. These results allow us to conclude that the structure consists of nanogranular bcc-Fe particles surrounded by the remaining amorphous phase and the solute elements are significantly enriched to the amorphous phase. The enrichment is presumed to cause an increase in the thermal stability of the residual amorphous phase, leading to the maintenance of the nanogranular bcc-Fe mixed structure, even in the high T_a range. The redistribution of the Nb and B elements is due to an extremely low solid solubility limit of these elements in the bcc-Fe phase.

With the aim of clarifying the redistribution of the solute elements in more detail, the nanostructure was examined by using the atom-probe field ion microscopy technique.[19] Figure 8(*a*) shows the change in the numbers of detected ions across the interface between bcc-Fe and amorphous phases for the $Fe_{90}Zr_7B_3$ alloy annealed for 3.6 ks at 723 K. Although the change near the interface is slight for Fe, the change in the numbers of detected Zr and B ions is significant, indicating that the concentrations of Zr and B differ significantly near the interface between the bcc-Fe and amorphous phases. Figure 8(*b*) shows an illustration of the concentration profiles of Fe and Zr elements near the interface on the basis of the results shown in Fig. 8(*a*). It is seen that the bcc phase has higher Fe and lower Zr concentrations as compared with their nominal concentrations, while the remaining amorphous phase has lower Fe and higher Zr concentrations. Furthermore, notice that the Zr is significantly enriched in the amorphous phase just near the interface and has a steep concentration gradient. By the significant enrichment of Zr in the remaining amorphous phase near the interface, the increase in thermal stability becomes most significant in the amorphous phase region near the interface. As a result, the preferential heterogeneous nucleation at the interface is suppressed, leading to the achievement of the nanoscale bcc structure. Thus, the segregation of the element leading to the increase in thermal stability of the remaining amorphous phase is essential for the maintenance of the nanocrystalline structure in the high T_a range. It has been pointed out that the elements leading to the segregation satisfy the following three criteria:[19]

(1) Large atomic size and high melting temperature with lower diffusivity.

(2) Low solid solubility limit into main constituent metal.

(3) Large negative heat of mixing against the other constituent elements.

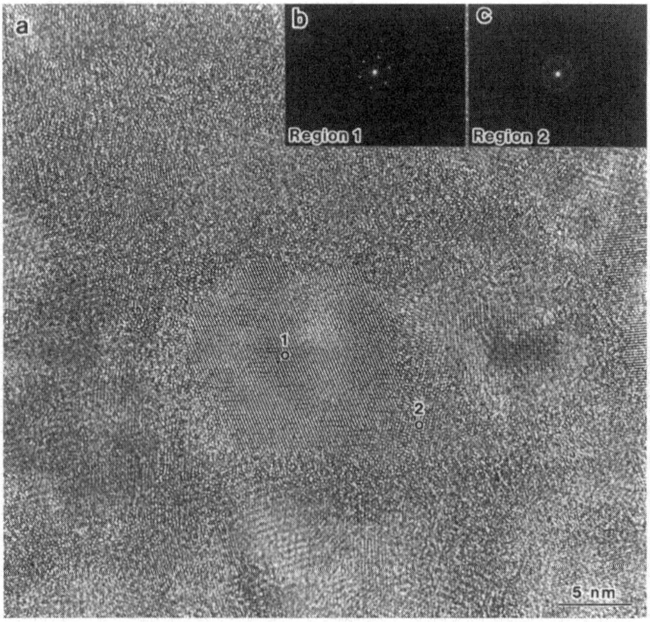

Figure 6. (*a*) High-resolution TEM image, (*b*) and (*c*) nanobeam diffraction patterns and (*d*) and (*e*) EDX spectra taken from the small regions (1) and (2), respectively, with a diameter of 0.6 nm for the amorphous $Fe_{88}Hf_{10}B_2$ alloy annealed for 3.6 ks at 873 K.

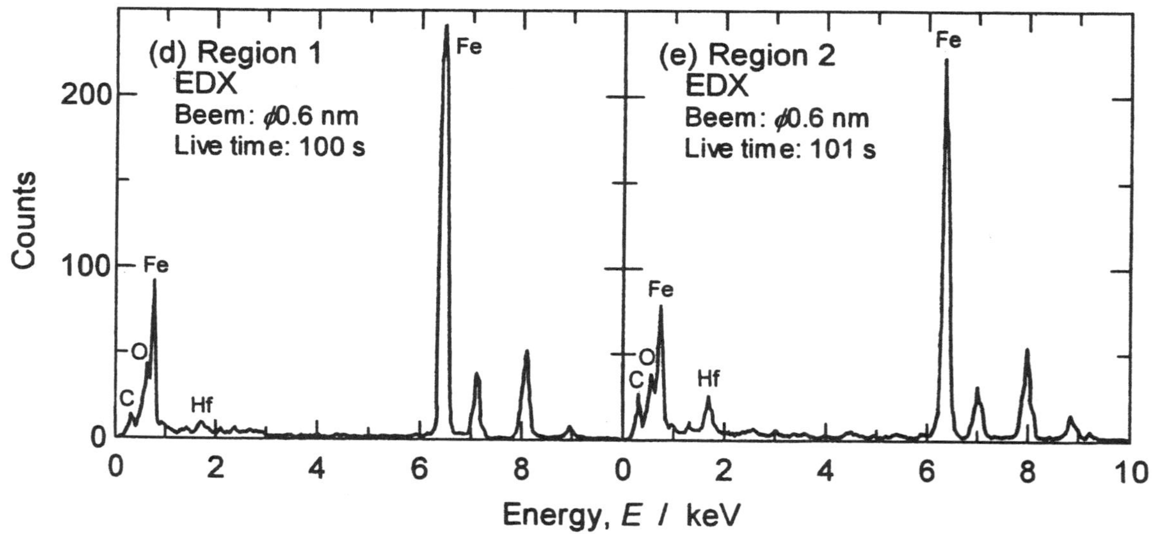

Figure 6. *(Cont'd.)*

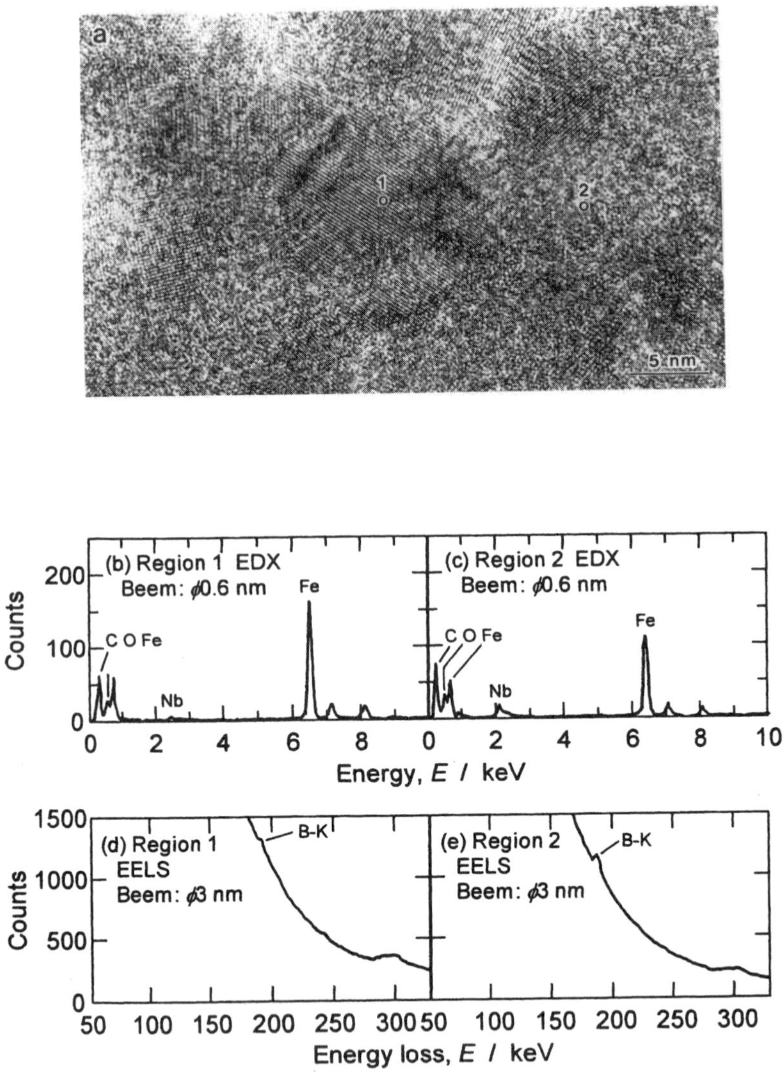

Figure 7. (*a*) High-resolution TEM image, (*b*) and (*c*) EDX spectra, and (*d*) and (*e*) EELS profiles taken from the small regions (1) and (2), respectively, with a diameter of 0.6 nm (EDX) and 3 nm (EELS) for the amorphous $Fe_{84}Nb_7B_9$ alloy annealed for 3.6 ks at 923 K.

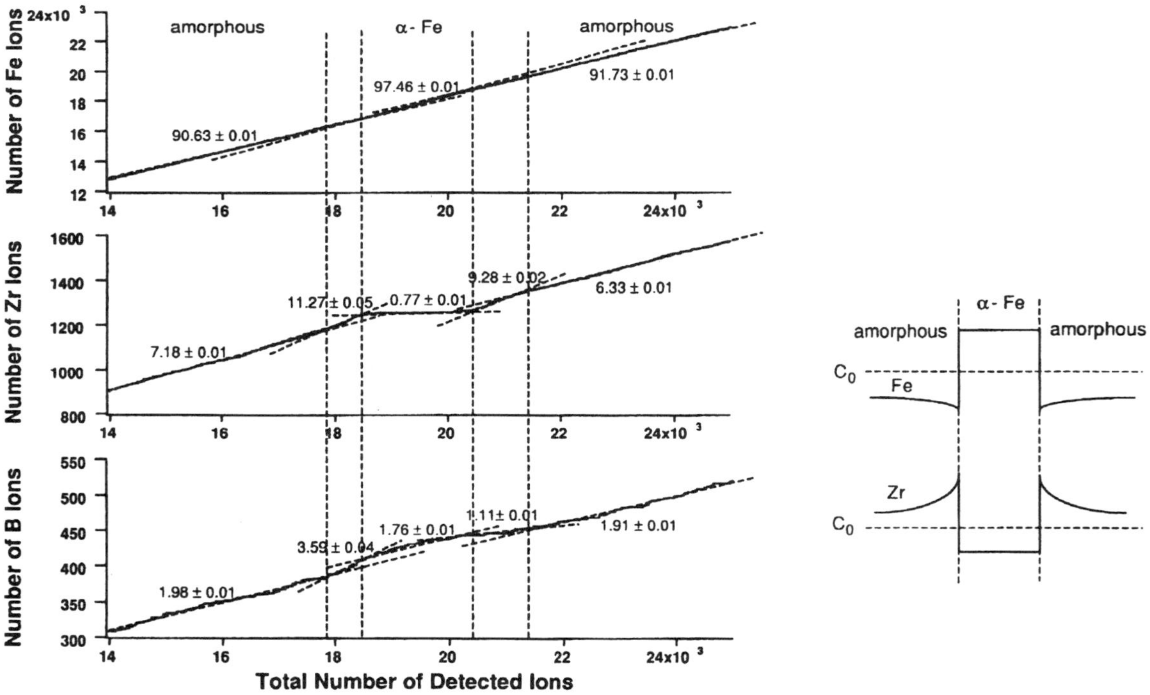

Figure 8. (*a*) Fractions of the numbers of Fe, Zr, and B ions in the total numbers of detected ions and (*b*) concentration profiles of Fe and Zr elements in the region across the interface between amorphous and bcc-Fe phases for the amorphous $Fe_{90}Zr_7B_3$ alloy annealed for 3.6 ks at 723 K.

Makino, et al., proposed[8]–[10] the mechanism for the appearance of the good soft magnetic properties for the nanoscale bcc Fe-M-B alloys. Here, it is important to describe the mechanism because the information is thought to be useful for improving the soft magnetic properties. The good soft magnetic properties have been thought[6]–[11] to result from the simultaneous satisfaction of the following factors:

(1) The achievement of high B_s resulting from the magnetic coupling between the nanoscale bcc particles via ferromagnetic amorphous phase.

(2) The ease of the reversion of magnetization due to the achievement of magnetic homogeneity resulting from the width of magnetic domain walls which is larger than the grain size of the bcc-Fe phase.

(3) The retainment of the nanoscale bcc structure resulting from the residual existence of an amorphous phase where the solute elements are enriched and the thermal stability increases.

(4) The reduction of λ_s resulting from the redistribution of the solute elements between bcc-Fe and remaining amorphous phase.

If this mechanism is appropriate, the soft magnetic properties are expected to be further improved by the modifications of the following four factors:

(1) An increase in the Curie temperature (T_c) for the remaining amorphous phase.

(2) A decrease in the bcc grain size and an increase in the volume fraction of the bcc-Fe phase.

(3) An increase in the thermal stability of the remaining amorphous phase.

(4) An achievement of a nearly zero λ_s value.

4.0 IMPROVEMENT OF SOFT MAGNETIC PROPERTIES BY THE ADDITION OF SMALL AMOUNTS OF SOLUTE ELEMENTS

The improvement of the soft magnetic properties of the Fe-M-B alloys has been tried by modifying the above-described four factors by adding small amounts of various solute elements. The addition of Co has been reported to be the most effective among VI to VIII group transition metals. Figure 9 shows B_s, μ_e, T_c of the remaining amorphous phase, and D values as a function of T_a for the $(Fe_{0.985}Co_{0.015})_{90}Zr_7B_3$ alloy, together with the data for the $Fe_{90}Zr_7B_3$ alloy. Although no appreciable changes in crystallization behavior and D values are seen with the addition of Co, one can notice significant increases in B_s and μ_e, as well as the extension of annealing temperature range leading to high B_s and μ_e. Notice also that the high μ_e values above 20,000 are obtained in the wide T_a range between 823 and 923 K. The extension of the T_a range is also important from an engineering point of view because wide thermal treatment conditions are possible. Considering that T_c of the amorphous phase increases significantly for the Co-containing alloy, the improvement of the soft magnetic properties seems to result from an increase in the degree of magnetic coupling between bcc particles by the increase in the magnetization of the residual amorphous phase containing Co.[9]

The effect of grain size on the soft magnetic properties of the nanoscale bcc alloys has been also examined. Figure 10 shows changes in the D, μ_e at 1 kHz, and H_c with a heating rate (α) up to 923 K for the $Fe_{90}Zr_7B_3$, $Fe_{89}Hf_7B_4$, and $Fe_{84}Nb_7B_9$ amorphous alloys. With increasing α, μ_e increases by 2 to 10 times and H_c decreases by about 1.4 to 4 times, accompanying a significant decrease in D. It is, therefore, concluded that the decrease in the D value is effective for the improvement of soft magnetic properties. The decrease in D with increasing α has been interpreted[8] to originate from the increase in the number of nucleation sites and the suppression of grain growth. The rapid heating treatment is a useful technique for the production of a nanocrystalline alloy with better soft magnetic properties. Besides, the use of the rapid heating technique gives us a unique opportunity to produce a nanocrystalline structure with significantly different D values. In the relation between the logarithms of H_c and D, the slope (n value) for the present bcc alloys has been measured to be

5.2,[8][9] which is nearly the same as the theoretically evaluated value of 6.0.[20] The agreement allows us to consider that the decrease in H_c occurs by the increase in the ease of the reversion of magnetization resulting from the decrease in D.

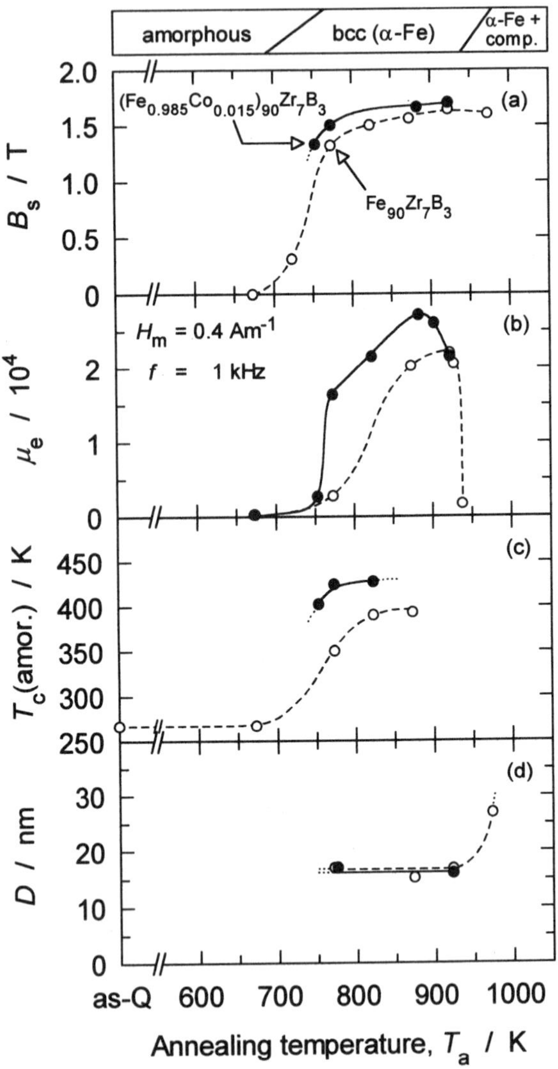

Figure 9. Changes in the structure, B_s, μ_e, T_c (amorphous), and D values with T_a for the amorphous $(Fe_{0.985}Co_{0.015})_{90}Zr_7B_3$ and $Fe_{90}Zr_7B_3$ alloys.

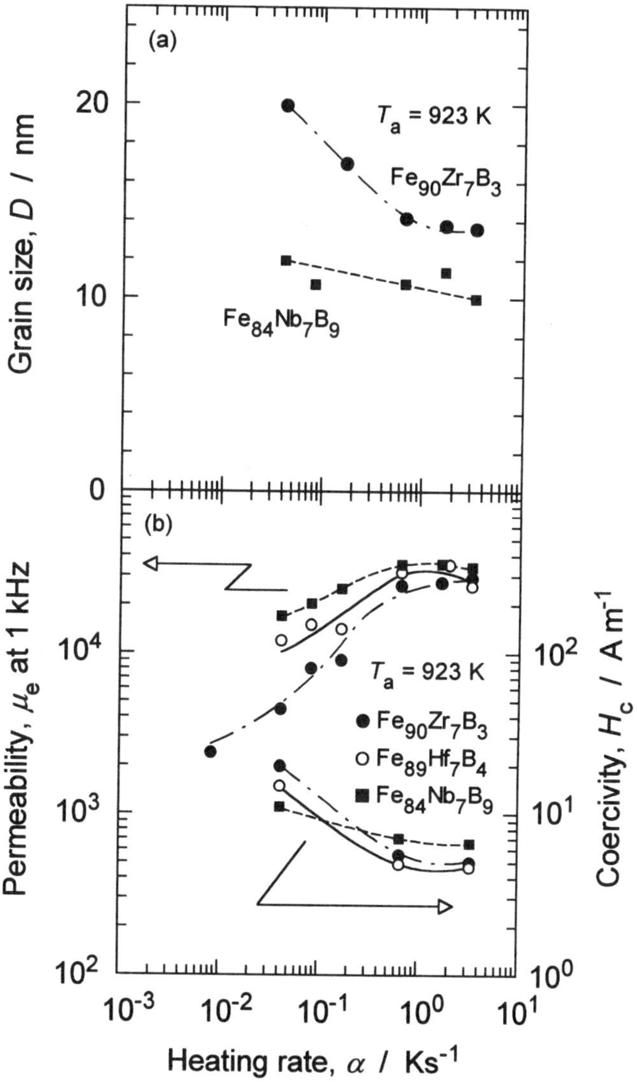

Figure 10. Changes in the D, μ_e, and H_c with α for amorphous $Fe_{90}Zr_7B_3$, $Fe_{89}Hf_7B_4$, and $Fe_{84}Nb_7B_9$ alloys for 3.6 ks at 923 K.

Figure 11 shows B_s, H_c, μ_e, λ_s, and D values as a function of Cu content for the bcc $Fe_{90-x}Zr_7B_3Cu_x$ and $Fe_{84-x}Nb_7B_9Cu_x$ alloys.[21] As Cu content increases from 0 to 2 at%, the B_s, H_c, and D values decrease for both alloys and the λ_s tends to increase. However, no systematic change in μ_e with Cu content is seen. From these changes, it is surmised that the addition of only 1 at% Cu causes a decrease in D, accompanying a decrease in H_c. Similar decreases in D and H_c have also been recognized for Fe-Zr-B alloys containing 1 at% Pd.[8] From the atom-probe field ion microscopic data, it has been confirmed[22] that the decrease in D by the dissolution of Cu is due to an easy heterogeneous nucleation of bcc phase in the Cu-rich region of the amorphous matrix. Furthermore, it is seen that λ_s is negative for the Fe-Zr-B-Cu alloys and positive for the Fe-Nb-B-Cu alloys. The opposite λ_s values indicate the possibility that the coexistence of Zr and Nb causes an achievement of nearly zero λ_s, leading to the improvement of soft magnetic properties. Figure 12 shows the relation among D, λ_s and μ_e or H_c for the bcc Fe-M-B and Fe-M-B-Cu (M = Zr and/or Nb) alloys. One can see a tendency that the highest μ_e and the lowest H_c values are obtained around the slightly positive λ_s of 0.3×10^{-6}. Also notice that the Fe-Zr-Nb-B-Cu alloys have a grain size of 7 nm, slightly positive λ_s of 0.3×10^{-6}, high μ_e above 10^5 at 1 kHz, and low H_c below 2 A/m. The B_s is also as high as 1.53 T. The simultaneous achievement of the high μ_e, B_s, low H_c, and nearly zero λ_s is the first evidence and exceeds those for all soft magnetic materials including Fe- and Co-based amorphous alloys and nanocrystalline Fe-Si-B-Nb-Cu[3] and Fe-P-C-Ga-Cu[23] alloys reported hitherto. The excellent soft magnetic properties are concluded to result from the combination of the grain size refinement by the addition of Cu and nearly zero λ_s by the dissolution of Zr and Nb.

Table 1 summarizes soft magnetic properties of B_s, μ_e, H_c, λ_s, electrical resistivity (ρ) and core losses (W) for the bcc Fe-M-B and Fe-M-B-Cu alloys, together with the data of sample thickness (t) and D, in comparison with those for amorphous Fe-Si-B[24] and Co-Fe-Si-B[24] and nanoscale bcc Fe-Si-B-Nb-Cu[3] alloys. It is confirmed that the soft magnetic properties combined with high B_s and low W for the present nanoscale bcc alloys are superior to those for the other magnetic materials. Based on the data summarized in Table 1, the relation between B_s and μ_e for the soft magnetic materials is shown in Fig. 13, wherein the data of silicon steels and Mn-Zn ferrites are also presented for comparison. It is seen that the Fe-M-B type alloys possess much better soft magnetic properties than obtained for all other soft magnetic materials.

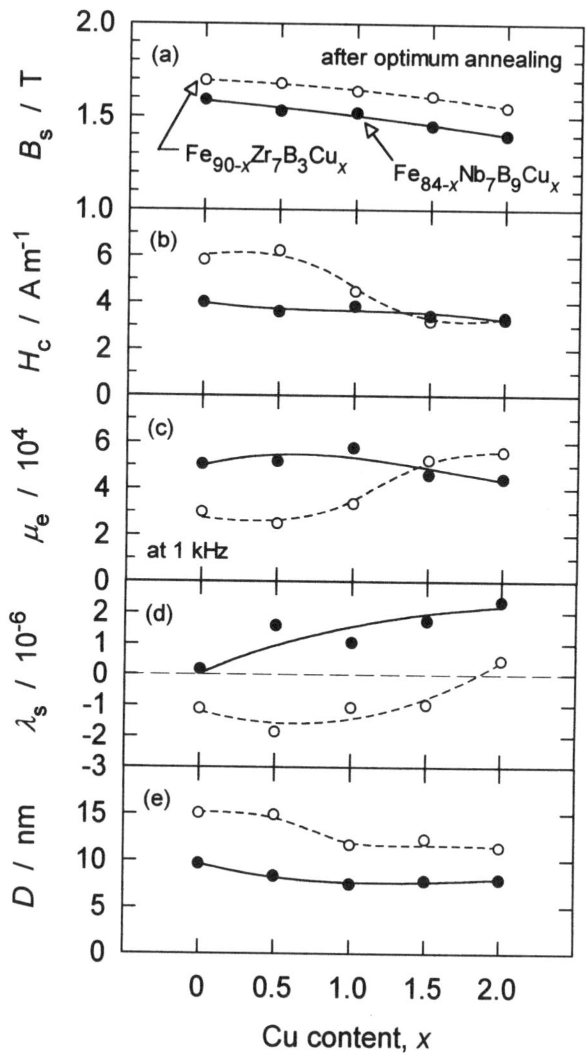

Figure 11. Changes in the B_s, H_c, μ_e, λ_s, and D values with Cu content for amorphous $Fe_{90-x}Zr_7B_3Cu_x$ and $Fe_{84-x}Nb_7B_9Cu_x$ alloys annealed for 1.8 ks at 723–923 K.

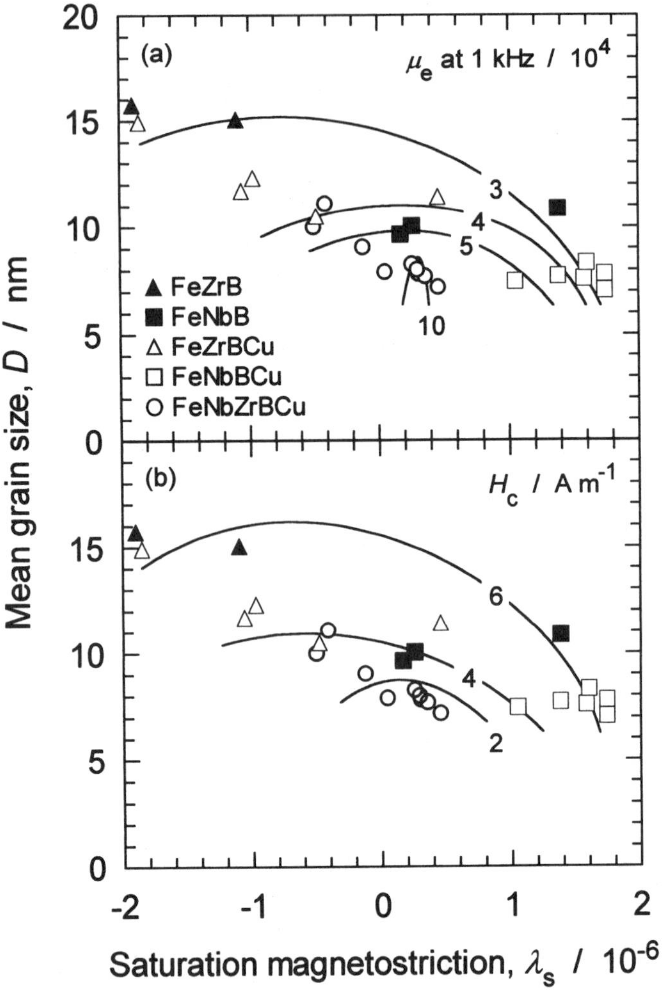

Figure 12. Relation among the D, λ_s, and μ_e, or H_c for the bcc Fe-M-B (M = Zr and/or Nb) alloys.

Table 1. Sample Thickness (T), Grain Size (D), Electrical (r), and Magnetic Properties (B_s, μ_e, H_c, λ_s, and W) for the Nanogranular bcc Fe-M-B Alloys, Fe-Si-B-Nb-Cu Alloys and Amorphous Alloys

	t (μm)	D (nm)	B_s (T)	μ_e*	H_c (A/m)	λ_s (10^{-6})	ρ (μΩm)	$W_{14/50}$** (W/kg)	$W_{2/100k}$** (W/kg)
$Fe_{91}Zr_7B_2$	18	16	1.67	27,000	5.5				
$Fe_{90}Zr_7B_3$	20	16	1.63	29,000	4.2	-1.1	0.44	0.21	79.7
$Fe_{89}Zr_7B_3Cu_1$	20	12	1.64	34,000	4.5	-1.1	0.51		185.4
$(Fe_{0.985}Co_{0.015})_{90}Zr_7B_3$	19	16	1.64	27,000	4.2	0		0.12	63.7
$(Fe_{0.98}Co_{0.02})_{90}Zr_7B_2Cu_1$	22	12	1.70	48,000	4.2	-0.1	0.53	0.08	80.8
$(Fe_{0.995}Ni_{0.005})_{90}Zr_7B_3$	21		1.62	34,000	3.5				
$Fe_{89}Zr_7B_3Pd_1$	20	13	1.63	30,000	3.2				
$Fe_{89}Hf_7B_4$	18	13	1.59	32,000	4.5	-1.2	0.48	0.14	59.0
$Fe_{84}Nb_7B_9$	22	10	1.50	36,000	7.0	+0.1	0.58	0.14	75.7
$Fe_{84}Nb_7B_9$	10		1.55	30,000	7.6				27.5
$Fe_{84}Nb_7B_9$	6								20.8
$Fe_{83}Nb_7B_9Ga_1$	19	10	1.48	38,000	4.8		0.70	0.22	47.0
$Fe_{83}Nb_7B_9Ga_1$	14								35.4
$Fe_{83}Nb_7B_9Ge_1$	24	9	1.47	29,000	5.6	+0.2	0.69		69.2
$Fe_{83}Nb_7B_9Cu_1$	19	8	1.52	49,000	3.8	+1.1	0.64		54.7
$Fe_{86}Zr_{3.25}Nb_{3.25}B_{6.5}Cu_1$	19	9	1.61	110,000	2.0	-0.3	0.56		60.0
$Fe_{85.6}Zr_{3.3}Nb_{3.3}B_{6.8}Cu_1$	18	8	1.57	160,000	1.2	-0.3	0.56	0.05	49.0
$Fe_{84}Zr_{3.5}Nb_{3.5}B_8Cu_1$	19	8	1.53	120,000	1.7	+0.3	0.61	0.06	58.7
$Fe_{78}Si_9B_{13}$	20		1.56	10,000	3.5		1.37	0.28	166.0
$Co_{70.5}Fe_{4.5}Si_{10}B_{15}$	21		0.88	70,000	1.2	0	1.47		62.0
$Fe_{73.5}Si_{13.3}B_9Nb_3Cu_1$	20	10	1.28	85,000	1.1				49.4
$Fe_{73.5}Si_{13.3}B_9Nb_3Cu_1$***	18		1.24	100,000	0.5	-2.1	1.15		39.1

* f = 1 kHz, ** W a/b is the core loss at $a \times 10^{-1}$ T and b Hz, *** Ref. (3)

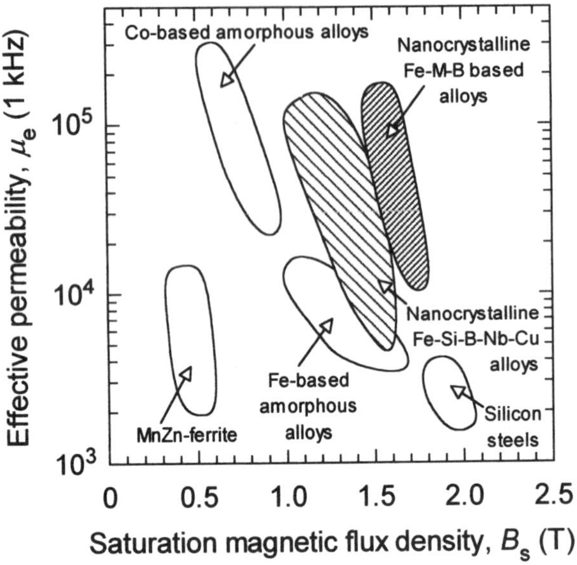

Figure 13. Relation between B_s and μ_e for the bcc Fe-M-B alloys. The data of other soft magnetic materials are also shown for comparison.

5.0 IMPROVEMENT OF HIGH-FREQUENCY PERMEABILITY BY THE DISSOLUTION OF OXYGEN INTO THE SURROUNDING AMORPHOUS PHASE

5.1 As-Sputtered Structure

As described above, the permeability (μ') of the nanocrystalline Fe-M-B alloys ($t = 20$ μm) is as high as 10^5 at 1 kHz, but decreases to about 2,000 at 1 MHz (700 at 3 MHz). For the future development of high-performance and miniaturized electronic devices, it is important to improve the high-frequency permeability characteristics. It is generally known that an increase in electrical resistivity and a decrease in sample thickness are effective for the improvement of high-frequency permeability because these changes cause a decrease in eddy current loss which becomes significant in a high frequency range. As one of the methods to increase the electrical resistivity, the B content in Fe-M-B alloys was increased up to about 20 at%. However, the increase in B content causes the disappearance of the nanocrystalline bcc structure by the simultaneous precipitation of bcc and compound, though an amorphous single phase is obtained in the melt-spun state.[24] Subsequently, Hayakawa, et al.,[25] examined the effect of the addition of oxygen element on the formation and soft magnetic properties of nanogranular bcc alloys for Fe-M base alloys. As an effective method to dissolve a large amount of oxygen and to prepare a thin sample, they used a sputtering method in a mixed atmosphere of argon and oxygen. It has been reported[11][12] that as-deposited $Fe_{46\text{-}88}Hf_{2\text{-}20}O_{7\text{-}41}$ films have four types of structures; they consist of a bcc single phase, mixed bcc and amorphous phases, an amorphous single phase, and an oxide phase, depending on the film composition. Useful soft magnetic properties are obtained in the films which have the mixed bcc and amorphous phases. Figure 14 shows x-ray diffraction patterns of as-deposited Fe-M-O [M = Ti, Zr, Hf, V, Nb, Ta, W, rare earth metals(Re)][26] films. In the Zr-, Hf- and Re-containing alloys, a broad peak due to an amorphous phase and diffraction peaks due to a bcc phase are observed. Contrary to these results, only the diffraction peaks corresponding to bcc phase are observed for the Ti-, V-, Ta-, Nb-, and W-containing alloys and no appreciable broad peak is seen. Furthermore, one can notice that the diffraction angle of bcc (110) peak shifts to a lower angle side than that of pure bcc-Fe marked with a dashed line in Fig. 14, indicating that the bcc phase includes a larger amount of M and oxygen elements.

Figures 15 (*a*) and (*b*) show high-resolution TEM images, nanobeam electron diffraction patterns, and EDX spectra for the as-deposited $Fe_{55}Hf_{11}O_{34}$ and $Fe_{49}Hf_{16}O_{35}$ films, respectively, together with the data of electrical resistivity at room temperature (ρ_{RT}). The electron diffraction patterns and the EDX spectra were taken from the points marked in the figure. Both films are composed of very fine-grained crystals less than 10 nm in diameter, which were surrounded by the amorphous phase. The grains with diameters less than 5 nm in size are smaller and the region of amorphous phase becomes larger for the $Fe_{49}Hf_{16}O_{35}$ film than for the $Fe_{55}Hf_{11}O_{34}$ film. These crystals are identified as a bcc-Fe phase supersaturated with Hf and O from the nanobeam diffraction pattern and the EDX spectrum of the crystal (region 1). The ρ_{RT} value increases from 10.5 to 492 μΩm with an increase in the amorphous region of the film.

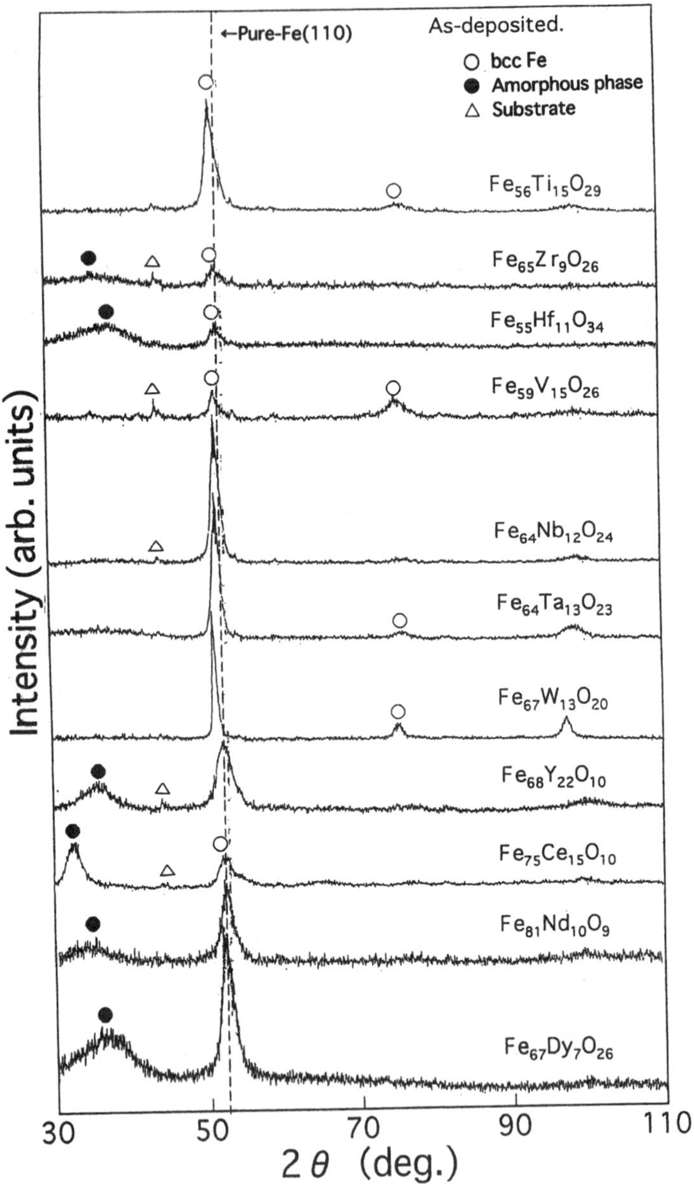

Figure 14. X-ray diffraction patterns of as-deposited Fe-M-O films.

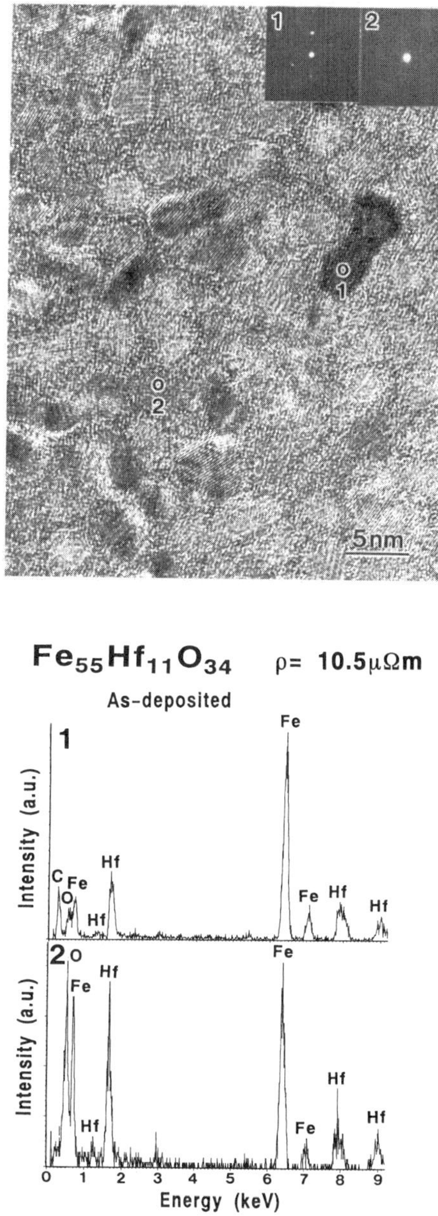

(a)

Figure 15. High-resolution TEM images, nanobeam electron diffraction patterns and EDX spectra taken from each microregion for as-deposited (*a*) $Fe_{55}Hf_{11}O_{34}$ and (*b*) $Fe_{49}Hf_{16}O_{35}$ films.

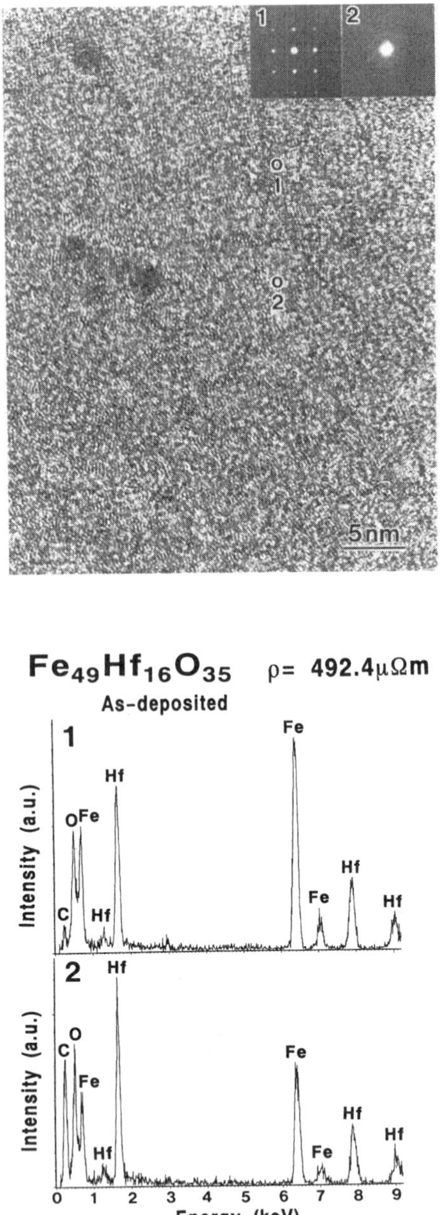

(b)

Figure 15. *(Cont'd.)*

Figure 16 shows an XPS spectra of $Fe_{2p3/2}$, $Hf_{4f7/2}$, $Y_{3d5/2}$, and $Ta_{4f7/2}$ for as-deposited $Fe_{55}Hf_{11}O_{34}$, $Fe_{68}Y_{22}O_{10}$, and $Fe_{55}Ta_{18}O_{27}$ films. Each spectrum is indicated with a solid triangle in the figure. In all systems, the binding energy of $Fe_{2p3/2}$ agrees with that of pure Fe. On the contrary, the binding energies of $Hf_{4f7/2}$ and $Y_{3d5/2}$ are close to those of HfO_2 and Y_2O_3, respectively. Therefore, O preferentially combines with Hf and Y in the $Fe_{55}Hf_{11}O_{34}$ and $Fe_{68}Y_{22}O_{10}$ films consisting of bcc and amorphous phases, as shown in Figs. 14 and 15. Then, these elements are mainly dissolved in the amorphous phase and probably form a M-oxide-like structure. On the contrary, for the $Fe_{55}Ta_{18}O_{27}$ film consisting of a mostly single bcc phase, the binding energy of $Ta_{4f7/2}$ is close to that of metallic Ta and there is no evidence of chemical binding between Ta and O elements. They are presumed to be supersaturated into the bcc phase, which results in the shift of bcc (110) diffraction peak as shown in Fig. 14. As a result, it is considered that the rapid increase in ρ_{RT} results from the high resistive amorphous region including M-O atomic pairs in the Fe-(Hf,Zr,Re)-O films. The formation of the mixed structure is due to the combination of three factors:

1. A low solid solubility limit of Hf in bcc-Fe phase
2. Preferential interaction of oxygen to Hf
3. Large solubility of oxygen in an amorphous phase

5.2 Magnetic Properties

Figure 17 shows the compositional dependence of B_s and H_c of as-deposited Fe-Hf-O films which were sputtered under no magnetic field. Open and solid circles represent the single bcc phase and the single amorphous phase, respectively. Double open and half solid circles indicate the oxide phase and the mixed bcc and amorphous phases, respectively. The B_s tends to decrease with increasing Hf and O contents and has a ridge around the Hf content of 10 to 15 at%. The H_c decreases with increasing Hf and O contents and has a valley around the same Hf content with the ridge of B_s. Therefore, the ridge of B_s approximately agrees with the valley of H_c, and there is a region in which B_s above 1.0 T and H_c below 160 A/m are simultaneously obtained. In this region, the real part of initial permeability (μ') exhibits about 400 even in an as-deposited state and the film structure is composed of fine bcc and amorphous phases as indicated with shaded mark.

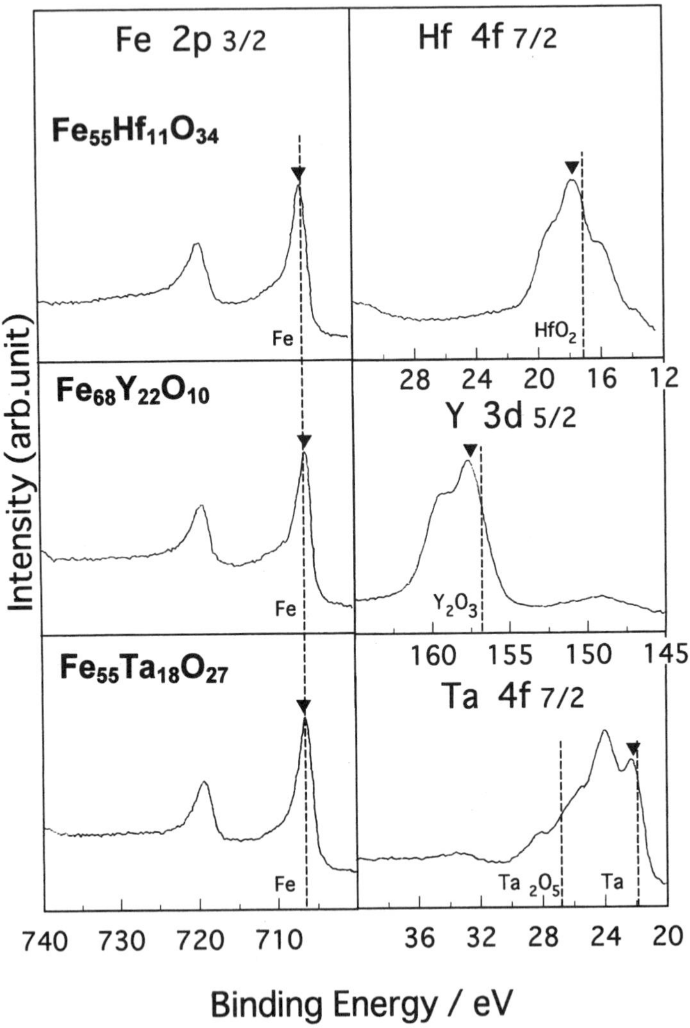

Figure 16. XPS spectra for as-deposited $Fe_{55}Hf_{11}O_{34}$, $Fe_{68}Y_{22}O_{10}$, and $Fe_{55}Ta_{18}O_{27}$ films.

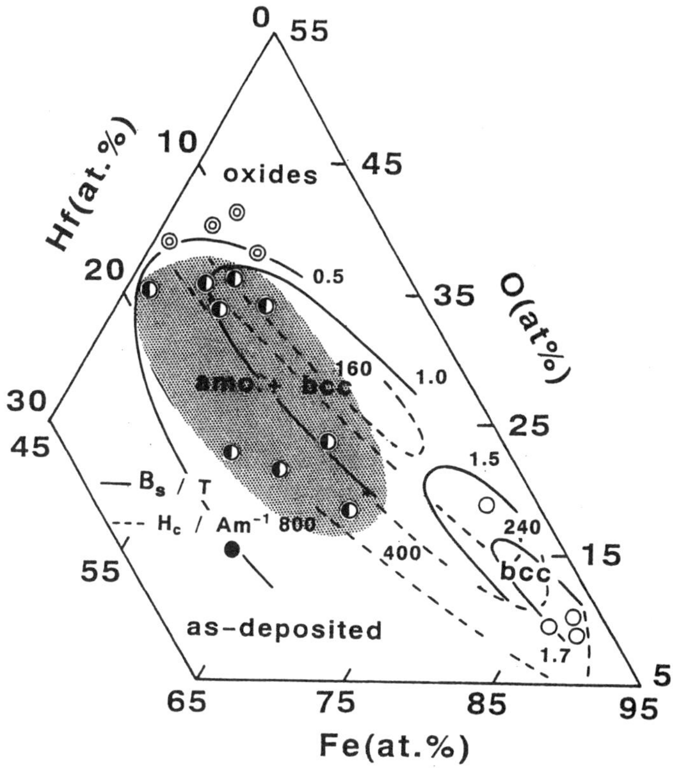

Figure 17. Compositional dependence of B_s and H_c for as-deposited Fe-Hf-O films.

Figure 18 shows the temperature dependence of B_s for as-deposited Fe-Hf-O films in comparison with a Fe-Hf amorphous alloy film. The Fe-Hf-O films with the mixed amorphous and bcc structure exhibit two-stage crystallization behavior. The arrows indicate the first crystallization temperature corresponding to the grain growth of bcc phase, which was measured by DSC. The B_s values of both films decrease with increasing temperature with bending points in the course of changes, at 500 and 750 K for the $Fe_{49}Hf_{16}O_{35}$ and $Fe_{55}Hf_{11}O_{34}$ films, respectively. An amorphous $Fe_{76}Hf_{24}$ film has a low T_c below room temperature owing to the Invar effect.[27] For the $Fe_{49}Hf_{16}O_{35}$ film, the bending point is thought to result from the T_c of the amorphous phase, not from crystallization, because the temperature is lower than that of the first crystallization. It is difficult to conclude, obviously, that the bending point is attributed to the T_c of amorphous phase for the $Fe_{55}Hf_{11}O_{34}$ film, because the temperature of the bending point is close to the first crystallization temperature. However, there is no bending point below that temperature, and the T_c of the amorphous phase for the $Fe_{55}Hf_{11}O_{34}$ film can be regarded to be higher than 700 K. As a consequence, the T_c of the amorphous phase including Fe and Hf increases by the dissolution of O in the Fe-Hf-O films. Furthermore, the T_c of the amorphous phase for the Fe-Hf-O films has been confirmed to increase after annealing. The magnetic properties and ρ_{RT} values for as-deposited Fe-M-O (M = group IV and V group transition metals and Re) films are summarized in Table 2, together with their film structures. The film compositions are equal to those of the films shown in Fig. 14. In all systems, B_s above 0.9 T and high ρ_{RT} above 4 μΩm are simultaneously obtained. The λ_s values were 0.1–2.9 × 10^{-6}. Relatively low H_c values below 400 A/m are obtained in M = Hf, Zr, and Re systems, which have the mixed structure of nanogranular bcc and amorphous phases.

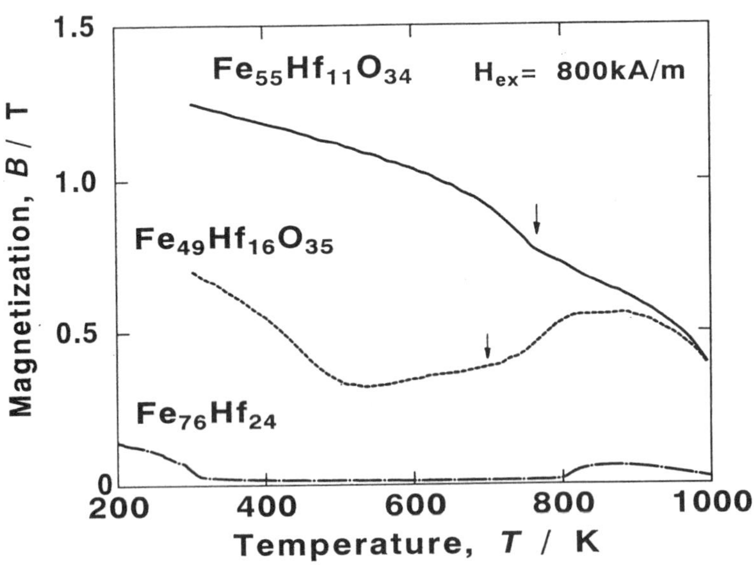

Figure 18. Temperature dependence of B_s for as-deposited $Fe_{76}Hf_{24}$, $Fe_{55}Hf_{11}O_{34}$, and $Fe_{49}Hf_{16}O_{35}$ films. The arrows indicate the first T_x of the Fe-Hf-O films.

Table 2. Magnetic Properties, (B_s, H_c) Electrical Resistivity (ρ), and Film Structure for As-deposited Fe-M-O Films

	Bs(T)	Hc (Am^{-1})	ρ ($\mu\Omega m$)	Structure
$Fe_{56}Ti_{15}O_{29}$	1.2	1040	6.4	bcc
$Fe_{65}Zr_{9}O_{26}$	1.3	211	6.6	amo.+bcc
$Fe_{55}Hf_{11}O_{34}$	1.2	154	9.1	amo.+bcc
$Fe_{59}V_{15}O_{26}$	1.2	···	5.6	bcc
$Fe_{64}Nb_{12}O_{24}$	1.3	168	4.1	amo.+bcc
$Fe_{64}Ta_{13}O_{23}$	1.3	658	4.7	bcc
$Fe_{67}W_{13}O_{20}$	1.3	···	8.3	bcc
$Fe_{68}Y_{22}O_{10}$	0.9	207	22.6	amo.+bcc
$Fe_{75}Ce_{15}O_{10}$	1.2	116	5.9	amo.+bcc
$Fe_{81}Nd_{10}O_{9}$	1.4	286	5.2	amo.+bcc
$Fe_{67}Dy_{7}O_{26}$	1.3	322	9.3	amo.+bcc

As shown in Fig. 14, no amorphous phase is formed in the Ti-, V-, Ta-, and W-containing alloys. In these alloys, H_c is relatively large and sufficient soft magnetic properties are not obtained. Therefore, the formation of the mixed structure consisting of the nanoscale bcc and amorphous phases is required for the achievement of a good magnetic softness in the as-deposited state. The soft magnetic properties in as-deposited Fe-(Hf,Zr,Re)-O films are probably due to the nanoscale grain size and the intergrain ferromagnetic coupling through the high-T_c amorphous phase as shown in Fig. 18, which averages out the magnetocrystalline anisotropy of bcc-Fe phase.[28] Furthermore, it is to be noticed that good soft magnetic properties are obtained even in the films including Re elements, which have a large magnetocrystalline anisotropy and usually inhibit soft magnetic properties. This is attributed to the decrease in magnetic interaction between Fe and Re resulting from the preferential binding with O.

The soft magnetic properties of the Fe-(Hf,Zr,Re)-O films are improved by sputtering under an uniaxial magnetic field, or uniaxial field annealing (UFA) after deposition under no magnetic field. Furthermore, the improvement of the frequency characteristics was tried by the enhancement of uniaxial anisotropy (H_k) with addition of Co into the Fe-M-O films. Figure 19 shows a B-H curve for an as-deposited $Co_{44.3}Fe_{19.1}Hf_{14.5}O_{22.1}$ film, together with the data on the $Fe_{61}Hf_{13}O_{26}$ film. The former was deposited under a static magnetic field and the latter was UFA-treated after deposition under no magnetic field. The H_k value of 1.15 kA/m for the $Fe_{61}Hf_{13}O_{26}$ film is the largest in all the Fe-Hf-O films. However, the H_k of the $Co_{44.3}Fe_{19.1}Hf_{14.5}O_{22.1}$ film exhibits 4.8 kA/m which is approximately four times larger than that of the $Fe_{61}Hf_{13}O_{26}$ film. This film exhibits a high B_s of 1.1 T and a high ρ_{RT} of 15.1 μΩm simultaneously and, moreover, the angle dispersion of magnetic anisotropy becomes small. Therefore, excellent high-frequency characteristics are expected to be obtained for the $Co_{44.3}Fe_{19.1}Hf_{14.5}O_{22.1}$ film owing to its significantly high ρ_{RT} and large H_k values.

Figure 20 shows the frequency dependence of the μ and quality factor ($Q=\mu'/\mu''$) of the Fe-Hf-O and Co-Fe-Hf-O films prepared by various methods, together with the data on other metallic soft magnetic alloy films developed to date. The Q value is a very important factor from a viewpoint of applications. Actually, the conventional soft magnetic films exhibit high permeability at 1 MHz, in particular, approximately 10,000 has been obtained for nanocrystalline soft magnetic films such as Fe-Si-Al-Hf-C film.[29] However, μ' decreases with increasing frequency because of their low ρ_{RT} values. On the other hand, the μ' of the Fe-Hf-O films is lower than that of conventional films in the range below 30 MHz, while in the frequency range higher than that, the Fe-Hf-O films exhibit higher and flatter μ' characteristics over 100 MHz because of their high ρ_{RT} values and moderate magnetic anisotropy field (H_k). The $Fe_{62}Hf_{11}O_{27}$ film exhibits a B_s of 1.3 T and a high μ' of 1400 at 100 MHz in an as-deposited state. Moreover, the Q values are also higher than those of the conventional films. The highest Q value of 61 at 100 MHz is obtained for the $Co_{44.3}Fe_{19.1}Hf_{14.5}O_{22.1}$ film. Needless to say, these films have higher Q values than other metallic films even in a low-frequency range. In the other M systems, similar high-frequency characteristics are observed, so that the Fe-M-O and Co-Fe-Hf-O films can be regarded as low loss soft magnetic films in a wide frequency range from MHz to around GHz.

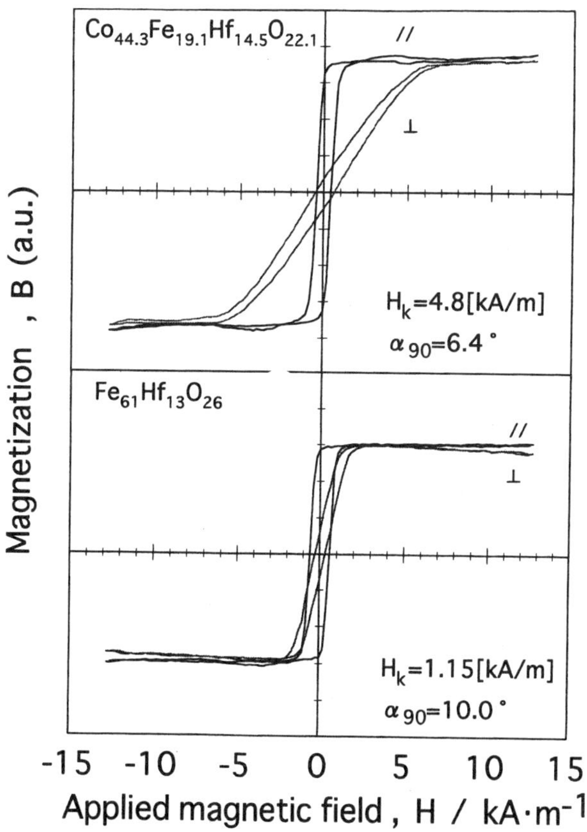

Figure 19. Magnetization curves for an as-deposited $Co_{44.3}Fe_{19.1}Hf_{14.5}O_{22.1}$ film and an $Fe_{61}Hf_{13}O_{26}$ film after UFA at 673 K for 10.8 ks.

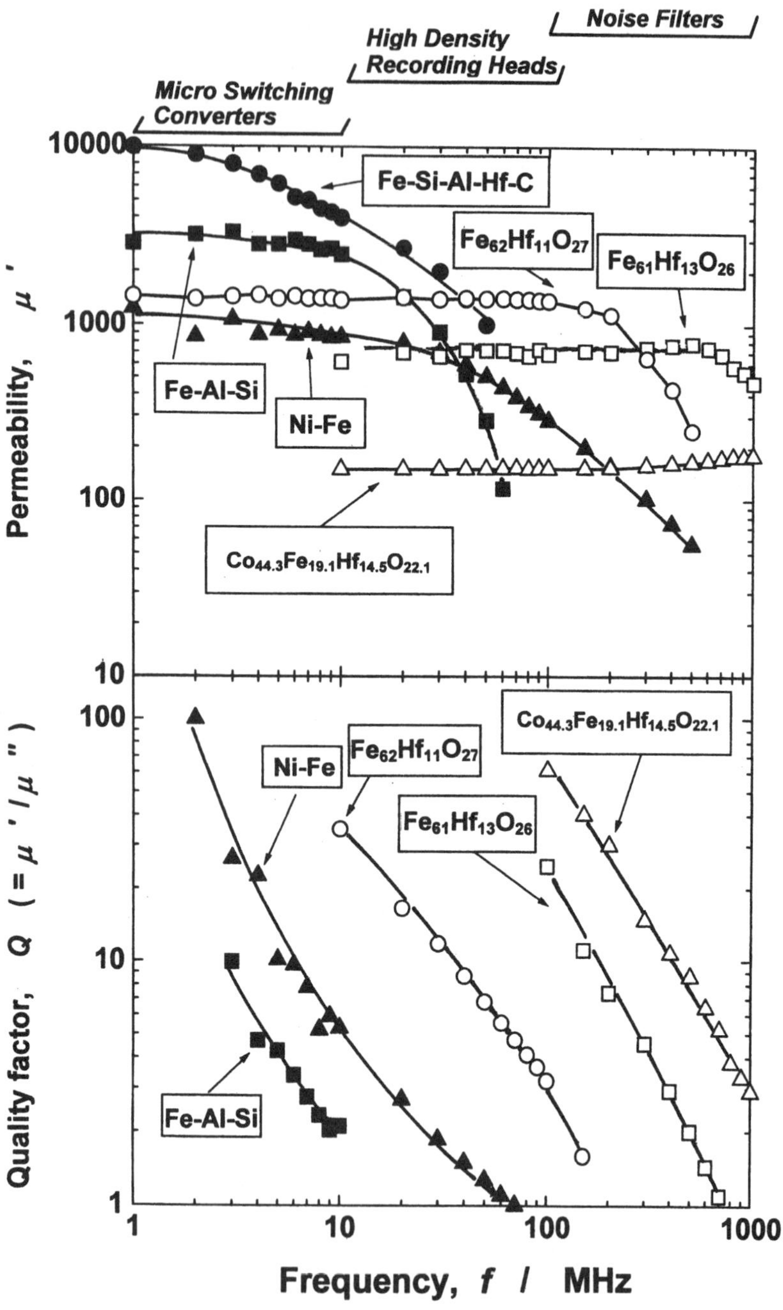

Figure 20. Frequency dependence of the real part of initial permeability μ' and the quality factor $Q(=\mu'/\mu'')$ for an $Fe_{62}Hf_{11}O_{27}$ film (as-deposited), $Fe_{61}Hf_{13}O_{26}$ film (UFA) at 673 K for 10.8 ks, and $Co_{44.3}Fe_{19.1}Hf_{14.5}O_{22.1}$ film (as-deposited) compared with the other soft magnetic films that have ever been reported.

6.0 APPLICATIONS

Figure 21 summarizes expected application fields for the soft magnetic Fe-Zr-Nb-B-Cu alloys, together with the magnetic characteristics which are required for their applications. Application fields include the power transformers, data communication interface components, electromagnetic interference (EMI) prevention components, magnetic heads, sensors, magnetic shielding, and reactors. The expected application to the power transformers comes from the lower core losses over a wide maximum induction range as compared with the oriented Si-steels and amorphous $Fe_{78}Si_9B_{13}$ alloy as shown in Fig. 22.[9][10] In addition, the efficiency of the power transformer was examined as a function of output current for the bcc Fe-Zr-Nb-B-Cu alloys and clarified to be much higher than those for the Fe-Si-B amorphous alloys and oriented Si-steels (Fig. 23),[10] in agreement with the tendency for core losses. We have also evaluated the noise attenuation characteristics which are important for the common mode choke coil. Figure 24 shows that the bcc Fe-Zr-Nb-B-Cu alloys have better noise attenuation values over the whole frequency range compared with Fe-Si-B amorphous alloys.[10] The better performance of the power transformers and common mode choke coils allows us to expect that the newly developed nanocrystalline Fe-M-B alloys are practical soft magnetic materials. Finally, it is important to note that the Fe-Zr-Nb-B-Cu alloys have already been used as pulse transformers in ISDN terminal adapters.

Figure 20 also summarizes typical application items and required characteristics for Fe-M-O and Co-Fe-Hf-O films. In the frequency range around 10 MHz, one can imagine their use in thin-film inductors or transformers for microswitching converters[30] in portable electric equipment. The microswitching dc-dc converters using Co-based amorphous alloy film as the core material of thin-film inductors have already been reported,[31] but they will be operated at higher frequency for further miniaturization and improvement of power supplies. For such high-frequency switching converters operating around 10 MHz, the Fe-M-O films exhibit their ability as a core material.

In the high frequency range up to 100 MHz, the Fe-M-O films are useful for magnetic heads for ultrahigh-density recording exceeding 1 Gbit/in^2. High B_s soft magnetic films such as Fe-M-N[32] have been studied for high-density recording heads, but the high-frequency characteristics around 100 MHz must be developed. In the further high-frequency range around GHz, noise filters, thin-film transformers, or other micromagnetic devices dealing with electromagnetic waves will be proposed.

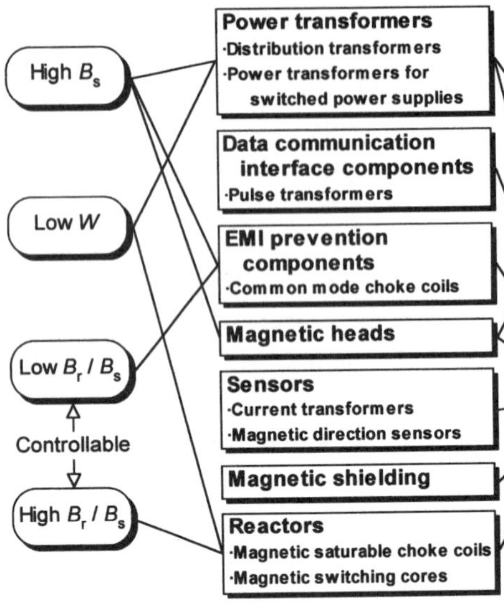

Figure 21. Magnetic characteristics and application fields for the bcc Fe-M-B (M = Zr and/or Nb) alloys.

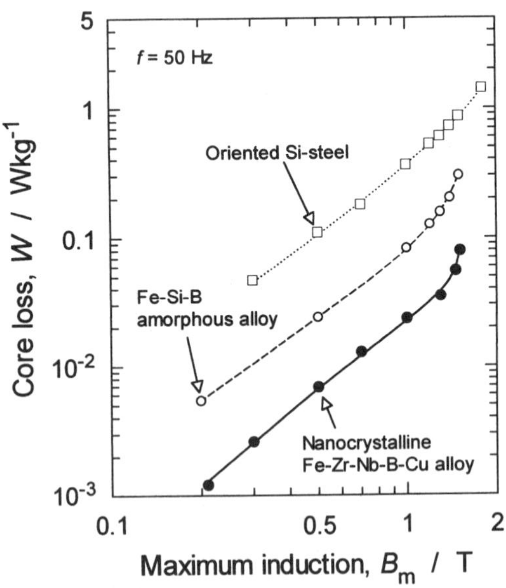

Figure 22. Core losses at 50 Hz as a function of maximum induction for the bcc Fe-Zr-Nb-B-Cu alloy. The data of amorphous $Fe_{78}Si_9B_{13}$ and oriented Si-steel are also shown for comparison.

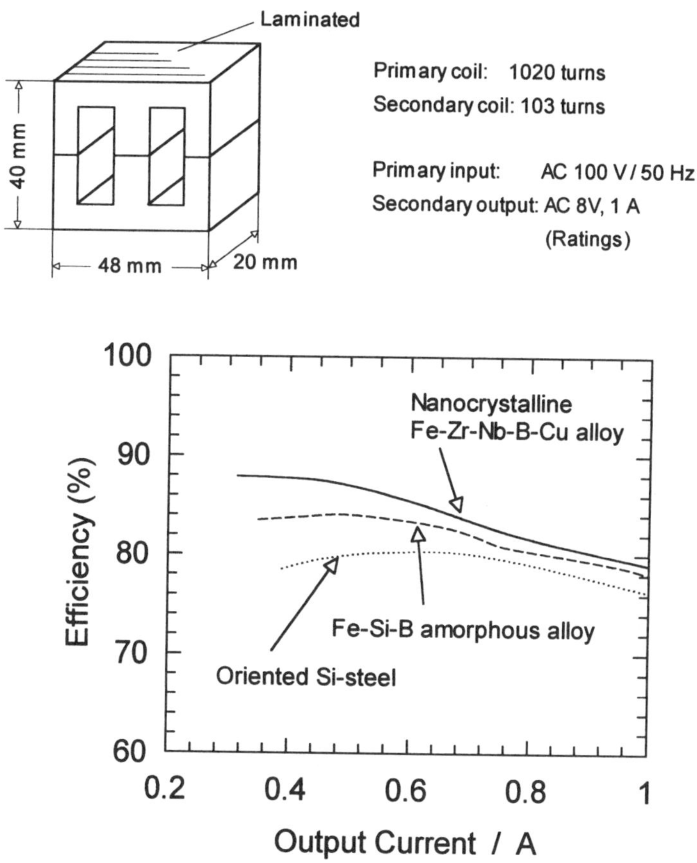

Figure 23. Change in the efficiency of a power transformer with output current for the bcc Fe-Zr-Nb-B-Cu alloy, amorphous Fe-Si-B alloy, and Si steels.

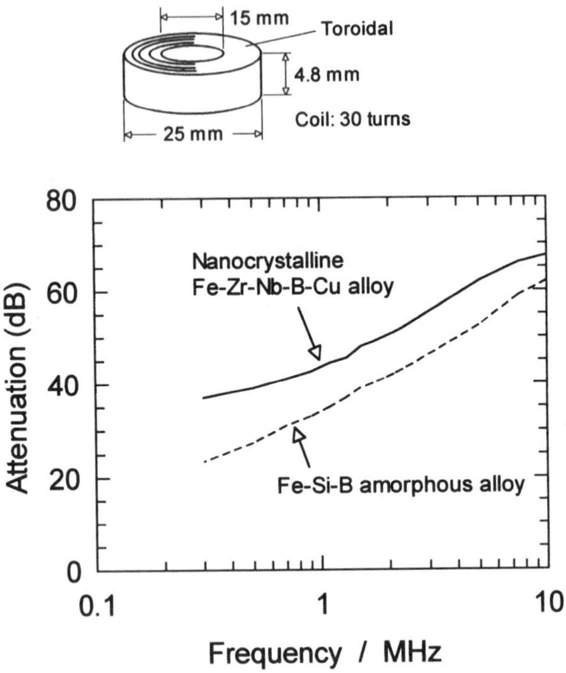

Figure 24. Change in the noise attenuation of a common mode choke coil with frequency for the bcc Fe-Zr-Nb-B-Cu alloy and amorphous Fe-Si-B alloy.

For the other example of their applications, Figs. 25 (a and b) show a schematic illustration of a planar inductor for microswitching converters and the frequency dependence of Q (= $\omega L/R$) values of a planar Cu coil using each magnetic film, respectively. The inductor is made up of a planar coil sandwiched between two magnetic films facing each other as shown in Fig. 25a. In the case of the inductor, one can use one side of the Cu coil as a conductor. As can be seen in Fig. 25b, an inductor using the Fe-Hf-O film shows a maximum. The high Q values of 12.3 at 6 MHz and 21.8 at 20 MHz are obtained for an inductor using the Co-Fe-Hf-O film, therefore, these planar inductors enable higher frequency operation and higher efficiency than the inductors using Co-Ta-Hf films for the microswitching converters, owing to the loss characteristics of the magnetic films.

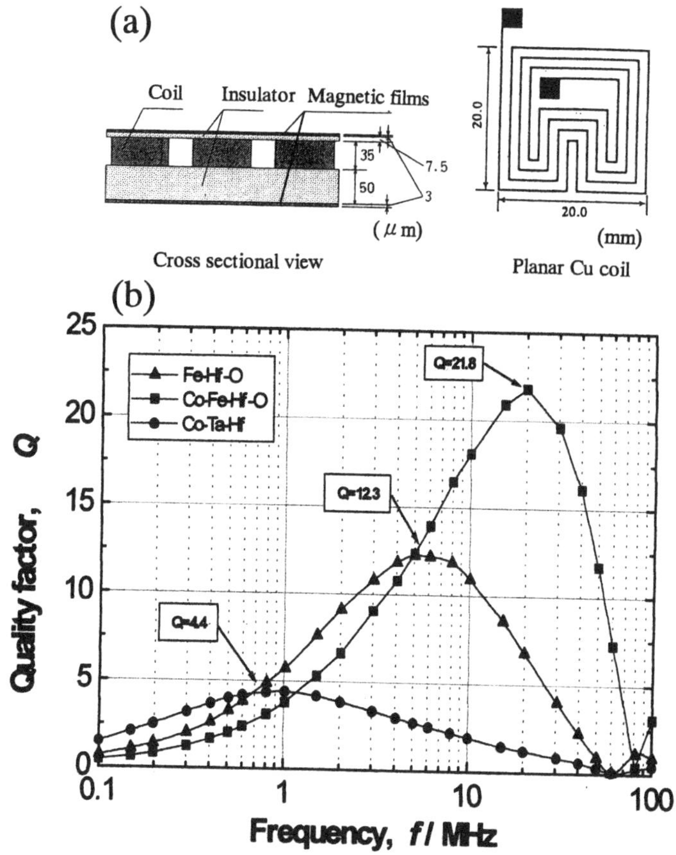

Figure 25. (*a*) Schematic illustration of planar inductor and (*b*) frequency dependence of quality factor Q (= $\omega L/R$) of the inductor using Fe-Hf-O, Co-Fe-Hf-O, and Co-Ta-Hf amorphous films.

7.0 CONCLUSIONS

The development of the new soft magnetic materials with the excellent characteristics of high B_s above 1.5 T combined with high μ_e above 10^5 at 1 kHz was achieved by nanocrystallization of the Fe-Zr-Nb-B-Cu amorphous alloys. In addition, the dissolution of oxygen into the remaining amorphous phase caused the significant improvement of high-frequency permeability of 1000 in the frequency range of 1 to 100 MHz by the drastic increase in ρ_{RT} to 1000 μΩm. Considering that these characteristics had not been previously reported, the nanostructure control is concluded to be a useful method for the development of a new type of high functional material. The future progress of nanostructure-controlled materials is expected to enable the fabrication of a new material exhibiting useful characteristics which have not been achieved for conventional materials.

REFERENCES

1. Masumoto, T., Kimura, H., Inoue, A., and Waseda, Y., *Mater. Sci. Eng.*, 23:141–144 (1976)
2. Croat, J. J., Herbst, J. F., Lee, R. W., and Pinkerton, F. E., *J. Appl. Phys.*, 55:2078–2082 (1984)
3. Yoshizawa, Y., Oguma, S., and Yamauchi, K., *J. Appl. Phys.*, 64:6044–6046 (1988)
4. Hasegawa, N., and Saito, M., *J. Magn. Soc. Jpn.*, 14:313–318 (1990); IEEE Translation, *J. Magn. Jpn,*. 6:91–100 (1991)
5. Inoue, A., Kobayashi, K., Nose, M., and Masumoto, T., *J. Phys.*, (Paris), Collog. C8:831–834 (1980)
6. Suzuki, K., Kataoka, N., Inoue, A., Makino, A., and Masumoto, T., *Mater. Trans.*, JIM, 31:743–746 (1990)
7. Suzuki, K., Makino, A., Inoue, A., and Masumoto, T., *J. Appl. Phys.*, 70:6232–6237 (1991)
8. Makino, A., Inoue A., and Masumoto, T., *Nanostruct. Mater.*, 6:985–988 (1995)
9. Makino, A., Inoue, A., and Masumoto, T., *Mater. Trans.*, JIM, 36:924–938 (1995)
10. Makino, A., Inoue, A., Hatanai, T., and Bitoh, T., *Mater. Sci. Forum*, 235–238:723–728 (1997)
11. Makino, A., Hatanai, T., Inoue, A., and Masumoto, T., *Mater. Sci. Eng.*, A226/228:594–602 (1997)
12. Makino, A., and Hayakawa, Y., *J. Jpn. Inst. Metals*, 57:1301–1309 (1993)
13. Makino, A., and Hayakawa, Y., *J. Magn. Soc. Jpn.*, 18: 411–414 (1994)
14. For example, Naka, M., Masumoto, T., and Chen, H. S., *J. de Phys.*, C8:839–842 (1980)
15. For example, Nose, M., and Masumoto, T., *Sci. Rep.* RITU, A28:232-241 (1980)

16. For example, Forester, D. W., Vittoria, C., Schelleng, J., and Lubitz, P., *J. Appl. Phys.*, 49:1966–1968 (1978)

17. Suzuki, K., Makino, A., Tsai, A. P., Inoue, A., and Masumoto, T., *Mater. Sci. Eng. A*, A179/180:501–505 (1994)

18. Suzuki, K., Makino, A., Inoue, A., and Masumoto, T., *J. Jpn. Inst. Metals*, 57:964–971 (1993)

19. Zhang, Y., Hono, K., Inoue, A., Makino, A., and Sakurai, T., *Acta Metall.*, 44:1497–1510 (1996)

20. Herzer, G., *IEEE Trans. Magn.*, 26:1397–1402 (1990)

21. Makino, A., Bitoh, T., Inoue, A., and Masumoto, T., *J. Appl. Phys.*, 81:2736–2739 (1997)

22. Hono, K., Zhang, Y., Inoue, A., and Sakurai, T., *Mater. Trans.*, JIM, 36:909--917 (1995)

23. Fujii, Y., Fujita, H., Seki, A., and Tomida, T., *J. Appl. Phys.*, 70:6241–6243 (1991)

24. Smith, C. H., *Rapidly Solidified Alloys*, (H. H. Liebermann, ed.), pp. 617–663, Marcel Dekker, New York (1993)

25. Hayakawa, Y., Makino, A., Inoue, A., and Masumoto, T., *Sci. Rep.* RITU A42:115–119 (1996)

26. Makino, A., and Kojima, A., J. *Magn. Soc. Jpn.*, 17:814–819 (1993)

27. Saito, N., Hiroyoshi, H., Fukamichi, K., and Nakagawa, Y., *J. Phys.*, F16:911–919 (1986)

28. Hayakawa, Y., and Makino, A., *Nanostruct. Mater.*, 6:985–988 (1995)

29. Hasegawa, N., Saito M., and Makino, A., *J. Mgn. Soc. Jpn.*, 18:750–758 (1994)

30. Hayakawa, Y., Makino, A., Fujimori H., and Inoue, A., *J. Appl. Phys.*, 81:3747–3752 (1997)

31. Mino, M., Tsukamoto, K., Yanagisawa, K., Tago, A., and Yachi, T., *APEC 96 Proceedings*, pp. 422–426 (1996)

32. Ishiwata, N., Wakabayashi, C., and Urai, H., *J. Appl. Phys.*, 69:5616–5618 (1991)

10

Mechanical Behavior of Nanocrystalline Metals

Julia R. Weertman

1.0 INTRODUCTION

The production of metals and alloys with grain size in the range of a few nanometers to about 50–100 nm led to expectations of materials with extremely high strength. The empirical Hall-Petch equation[1][2] predicts that

Eq. (1) $$\sigma_y = \sigma_o + k/\sqrt{d}$$

where σ_y is the yield strength, σ_o is a friction stress below which dislocations will not move in the material in the absence of grain boundaries, k is a constant, and d is the grain size. However, most measurements seem to indicate that the strength of nanocrystalline metals falls well below that predicted by an extrapolation of Eq. (1) to very fine grain sizes. Not only does k, the slope of a Hall-Petch plot, drop below the coarse-grain value with increasing grain refinement, but in some cases it even becomes negative. Many and varied are the models of deformation in nanocrystalline metals that have been proposed to account for this fall-off in strengthening. Representatives of a number of the different types of models are described in the first part of this chapter. A great spread exists in the strength data reported for the same material by various laboratories, or even by the same laboratory over a period of time. It has become evident that measured values are significantly affected by imperfections in the sample material. Therefore, the second part of the chapter is devoted to characterization of the nanocrystalline material used in the mechanical measurements, particularly characterization of the defects present. Finally, the results of various mechanical measurements are reported.

2.0 MODELS OF MECHANICAL BEHAVIOR OF NANOCRYSTALLINE MATERIALS

The usual textbook explanation of Eq. (1) was given by Cottrell,[3] who calculated the stress ahead of a large array of dislocations piled up at a grain boundary. The length of the pileup is taken to be of the same order of magnitude as the grain size d (Fig. 1a). When the stress from the pileup at a Frank-Reed source in the adjacent grain rises to the value needed to activate this source, yielding will spread across the grain boundary and, thence, throughout the material.

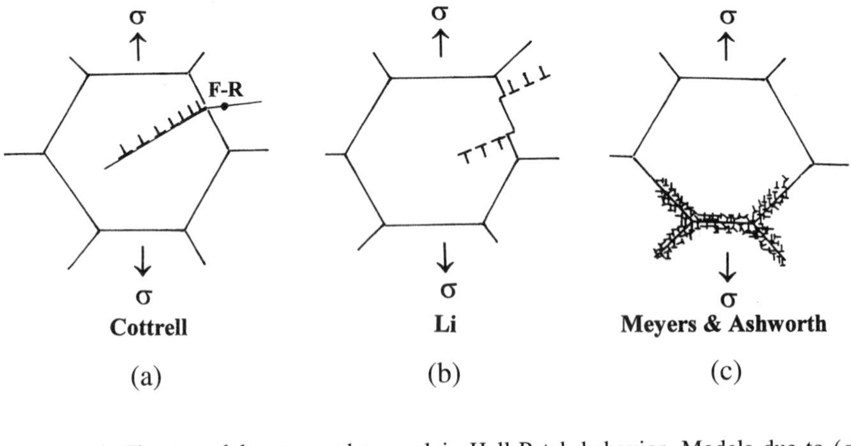

Figure 1. Three models proposed to explain Hall-Petch behavior. Models due to (*a*) Cottrell, (*b*) Li, and (*c*) Meyers and Ashworth.

It is obvious that Eq. (1) cannot be extrapolated to arbitrarily small grain sizes. At very small values of d, the pileup containing the large numbers of dislocations assumed in the Cottrell theory requires applied stresses approaching or exceeding the theoretical strength. A number of models have been proposed to estimate the limit of applicability of Eq. (1). Nieh and Wadsworth[4] used experimental measurements of the yield strength of several nanocrystalline metals to calculate the smallest grain size that could support two dislocations, the minimum number to cc a "pileup." Strength measurements in materials with a grain size be critical value would be likely to remain constant or even drop with increasing grain refinement. Explanations offered for this reduced or negative Hall-Petch slope include onset of fracture at triple points, approach to the amorphous state (which is likely to be weaker than the crystalline material), and the increased importance of grain boundary sliding.[4] More detailed calculations of the dependence of yield stress on grain size in fine-grained material have been made by Armstrong and colleagues.[5]–[8] It is assumed that the concentrated shear stress at the tip of a pileup must attain a critical value (taken to be independent of grain size) before yielding occurs. If the grain size, d, is such that the number, n, of dislocations in the pileup is larger than ~20, the conventional Hall-Petch relationship is closely approximated. At smaller grain sizes, the plot of yield stress σ_y versus $(d)^{-1/2}$ becomes increasingly discontinuous as n drops from 20 to 19, etc., down to one, and the curve falls below the coarse-grain extrapolation (e.g., Fig. 4 in Ref. 6).

Not all explanations of the strengthening effects of grain refinement are based on dislocation pileups at grain boundaries. Guided by observations that dislocations are seen to emanate from grain boundaries in the very early stages of deformation, whereas dislocation pileups are usually absent except in low stacking fault or ordered alloys, Li[9] and Li and Chou[10] proposed that Hall-Petch behavior can be explained by dislocations generated at grain boundary ledges during deformation (Fig. 1b). The flow stress inside a grain can be expressed as proportional to the square root of the dislocation density, ρ. The value of ρ is taken to be proportional to the number of grain boundary ledges per unit volume, assumed to scale with the grain boundary area per unit volume and, thus, with d^{-1}, leading to the familiar Hall-Petch relationship. It is not clear how this model would extrapolate to very small grain sizes. For example, the relationship between flow stress and dislocation density must break down when there are, at most, only a few dislocations per grain. Meyers and Ashworth[11] proposed a model of strengthening by grain refinement that is also based on the generation of dislocations at grain boundaries. In this model, as the material deforms, elastic anisotropy results in stress concentrations in the boundary regions that give rise to the generation of dislocations that effectively form a hardened reinforcing second-phase network (Fig. 1c). General yielding occurs when the applied stress becomes sufficient to push dislocations through the hardened layers into the grain interior. In order to insure a $d^{-1/2}$ dependence of σ_y at large grain sizes, Meyers and Ashworth set the thickness, t, of the hardened layer proportional to $d^{1/2}$. The resulting expression for σ_y contains not only the usual $d^{-1/2}$ term but also a second term, proportional to d^{-1}, that becomes important at small values of d, and acts to lower the slope of a Hall-Petch plot in this grain size region. Using experimental values for yield stress versus grain size for Cu and Fe, Meyers, Benson, and Fu[12] calculated that the maximum in σ_y occurs at a grain size of about 10 nm in these materials.

Many of the models of the mechanical behavior of fine grain materials are based on a two-component or multi-component concept. In a few,[11]–[13] the components are taken to be the hardened regions adjacent to the grain boundaries and the relatively soft grain interiors. However, in most models, the two components are the grain boundaries themselves and the crystalline grains. Following this idea, Carsley, Ning, Milligan, Hackney, and Aifantis[14] carried out a straightforward calculation of the strength (as measured by hardness) of nanocrystalline metals as a function of grain size. Hall-Petch behavior is assumed for the grains down to the finest grain sizes while the hardness of the boundaries, which are regarded as being in the amorphous state, is estimated from a simple relationship between hardness and shear modulus in the glassy state (namely, the shear modulus is approximately 6 times the hardness). The shear modulus of the glassy form of the metal is taken to be about one half of the crystalline value. A simple rule of mixtures is used to obtain the hardness of the combined phases. A typical Hall-Petch plot from Ref. 14 is shown for nickel in Fig. 2. A more complex model[15] includes the influence of triple lines and quadruple nodes as well as that of the grain boundaries.

A problem in two-component models is the choice of the effective width of the phase associated with the grain boundary, whether it is that of the grain boundary itself as in Ref. 14 or the hardened region of the grains adjacent to the boundaries.[11]–[13] In the former case, Gleiter and colleagues[16][17] argue that the grain boundaries in nanocrystalline materials are essentially different from those in conventional materials; they are of low density and in a gas-like state of disorder. A number of direct observations in the electron microscope[18][19] have indicated that the grain boundaries are similar in nanocrystalline and coarse-grained metals. Values ranging from 0.5–1 nm, or about 2–5 atomic widths, are frequently chosen in two-phase deformation models, such as in Ref. 14.

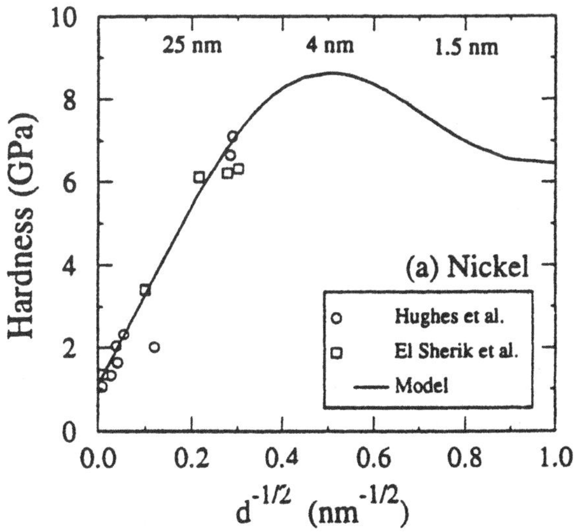

Figure 2. Hall-Petch plot of the model of Carsley, et al.,[14] for Ni, calculated with a grain boundary width of 1 nm. *(Data points from Refs. 56 and 61.)*

Masumura, Hazzledine, and Pande[20] proposed a two-component model that takes into account the dispersion in grain sizes that characterizes real nanocrystalline samples (as opposed to model material that conveniently is made up of grains of a single size). These authors assume that the stress, $\sigma_{hp} = \sigma - \sigma_o$, required to deform those grains in the sample that are larger than a certain critical size, d^*, follow Hall-Petch behavior, i.e., $\sigma_{hp} = \sigma - \sigma_o = kd^{-1/2}$, whereas the strength, σ_c, of the grains smaller than d^* derives from a Coble creep[21] mechanism. The expression for σ_c is taken to be: $\sigma_c = A/d + Bd^3$. The threshold term A/d, which is suggested by experimental results, is presumed to be related to the stress, $\approx Gb/d$, required by a dislocation loop pinned at the grain boundary nodes to climb. The vacancies created and destroyed in the course of the climbing are needed in the Coble diffusion process. (Here, G is the shear modulus and b is the Burgers vector of the dislocation.) The dispersion in grain size in the nanocrystalline metal is assumed to follow a log-normal distribution, which has been found to give a good description of actual distributions.[22][23] The yield stress corresponding to a given average grain size, $\bar{d}$, is found by a simple volume averaging of the grains, using Hall-Petch behavior for grains greater than d^* and the Coble relationship for $d < d^*$. The authors introduce a parameter, $p = (A/d^*)/B(d^*)^3$, which indicates the relative value at the transition grain size, d^*, of the threshold stress compared to the usual grain boundary diffusional creep term. Figure 3 shows a plot of normalized stress versus normalized average grain size, $\bar{d}$, for various values of p (and a fixed value for the standard deviation of the log-normal distribution). It can be seen that, with a suitable adjustment of parameter p, this model predicts that the slope of a Hall-Petch plot at small grain sizes levels off, or even becomes negative.

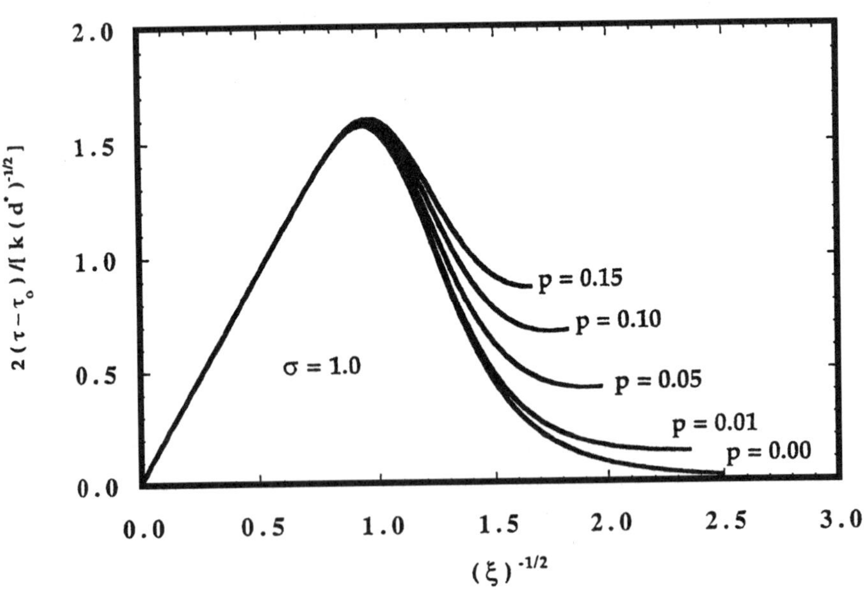

Figure 3. Plot of normalized stress vs $\xi^{-1/2}$ for various values of the parameter p. ξ is the normalized average grain size $\bar{d}/d^*$. See text for explanation of symbols. *(From Ref. 20.)*

The influence of a dispersion in grain size on the strength of nanocrystalline metals has been considered in several papers.[20][22][24] A deformation model, with assumptions appropriate to many actual nanocrystalline samples, has been developed by Morita and colleagues.[24] Again, a log-normal dispersion in grain size is assumed. High density samples of nanocrystalline metals, at least many of those made in clean systems by inert gas condensation (IGC)[16] and compaction, tend to have few grains small enough that grain boundary sliding is likely to be important.[23][25] Therefore, it was assumed that, for a given applied stress, all grains larger than a critical size (determined by the applied stress) undergo plastic deformation while the smaller grains remain elastic. Morita, et al., take into account the internal stresses that arise as grains in an elastic "matrix" deform plastically. The micromechanics of inclusions[26][27] were used to solve this problem. An interesting result of the calculations is the influence on overall strength of the spread in grain sizes. Figure 4 shows calculated stress-strain curves for a fixed average grain size and varying widths of the grain size dispersion. It can be seen that the apparent 0.2% offset yield strength is strongly affected by the magnitude of the dispersion.

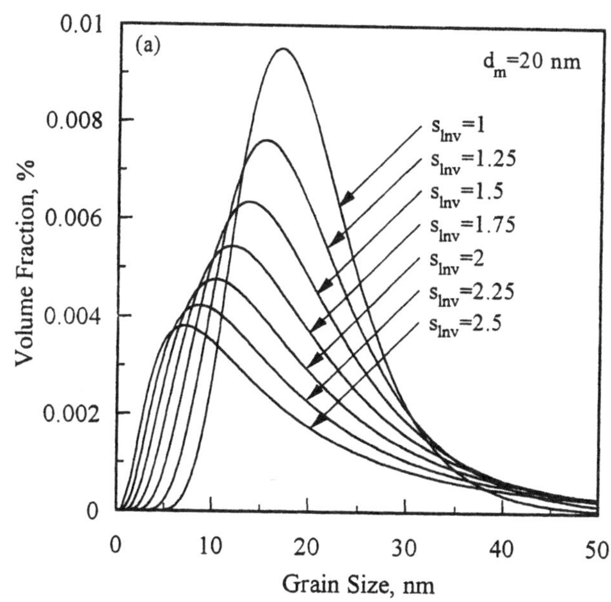

Figure 4. The effect of the size of the standard deviation l_{inv} on (*a*) the grain size distribution expressed in volume fractions and (*b*) the stress-strain curves. The mean grain size d_m is constant at 20 nm. *(From Ref. 24.)*

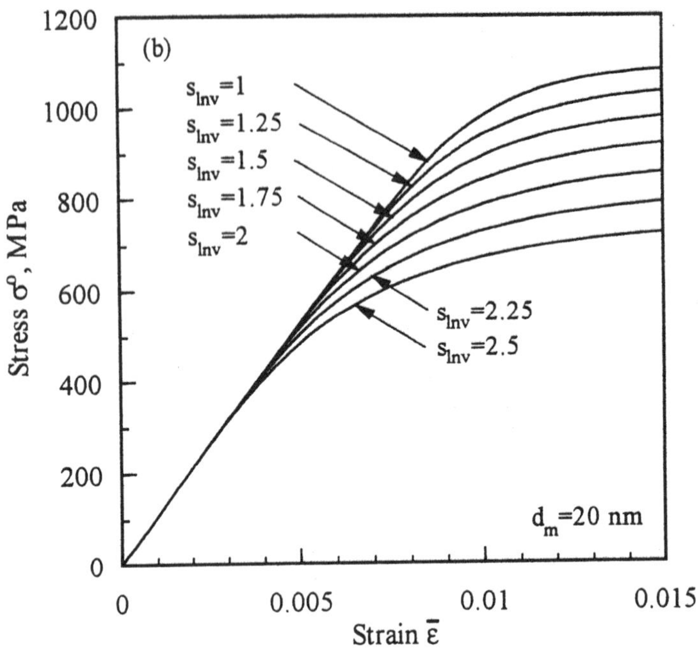

Figure 4. *(Cont'd.)*

In the preceding models it was tacitly assumed that dislocations, when they enter into the picture, are exactly the same as the dislocations in large-grain material. However, it is evident that as the grain size begins to approach the core radius, the standard dislocation theory must be modified. Scattergood and Koch[28] emphasized the drop in dislocation line tension, at very small grain sizes, in their model to explain negative Hall-Petch slopes. The authors argue that, at large grain sizes, cutting of dislocation forests is the easiest way for glide dislocations to get past these obstacles. But at sufficiently small sizes, the expression for dislocation line tension, T,

Eq. (2) $$T = (Gb^2/4\pi)\ln R/r_o$$

becomes sensitive to the value of R, the long-range stress field screening distance. (Here, r_o is the dislocation core cutoff radius.) The authors scale R with the grain size d. As a result, the stress required for a glide dislocation to get past the obstacles by Orowan looping, which is proportional to T, becomes comparable to the cutting stress at a critical grain size and is the favored mechanism at still smaller grain sizes. The continued drop of the line tension with decreasing grain size leads to a negative Hall-Petch slope. When the theory is fitted to limited experimental data, values of R, the screening distance, generally are found to lie between $d/2$ and $d/4$.

Attempts have been made by the authors of the various models to validate their calculations by comparing predicted grain size dependence of some measure of strength (usually hardness) with experimental data. As shown below, the experimental results are so diverse that any such comparison is highly suspect.

3.0 CHARACTERIZATION OF NANOCRYSTALLINE METALS

Before valid comparisons can be made between measurements of mechanical properties of nanocrystalline metals and the predictions of various models, it is necessary to be sure that the data represent the inherent behavior of the material and are not compromised by sample imperfections. Most methods for producing nanocrystalline samples tend to introduce various flaws in the materials such as pores and microcracks, high internal stresses, and impurities and trapped gases. While not exactly a flaw, the broad dispersion in grain size of many samples has a strong effect on the overall mechanical properties, as seen in the previous section, and must be taken into account when comparing experimental results with theoretical expectations. It is evident that careful characterization of the samples to be used in measurements of mechanical properties is essential.

A second factor compromising mechanical test results is the small size of most nanocrystalline samples. Inadequate measurement techniques, especially in strain measurements, have led to erroneous results that were only corrected as the techniques improved.

Density, Pores, and Microcracks. In the early days of nanocrystalline research, density measurements of samples often gave values ranging from only about 70% to >90% of the single crystal density.[16][29] The density shortfall has been interpreted variously as the consequence of nanocrystalline grain boundaries having extremely low densities[16] or as caused primarily by the presence of pores.[30] Molecular Dynamics computer simulations[25] indicate that the density of Ni and Cu with grain sizes down to 5 nm only falls to about 96% of the single crystal value. Since porosity is known to have a strong effect on the elastic moduli and on other mechanical properties as well, it is important to have knowledge of the size and number density of any pores in the mechanical test specimens.

Schaefer and colleagues,[31][32] using positron spectroscopy, identified three size classes of voids in nanocrystalline samples. The smallest are about a lattice vacancy in size and are presumed to be located at grain interfaces. The largest are identified with "missing grain" pores and are of the order of the grain size. The middle-sized voids correspond to clusters of about 10 vacancies and have been identified by Schaefer[31] as residing at grain triple junctions. It is likely that the two larger classes make the major contribution to the density shortfall in many samples. These larger voids lie in a size range accessible to measurement by small-angle neutron scattering (SANS).

Figure 5 compares the total void volume fractions in a series of nanocrystalline Pd samples obtained by precision density measurements with the results from the SANS data.[33] The good agreement between the two methods lends support to the interpretation of the SANS data on the basis of scattering by small pores. Figure 6 shows that the void population can be decreased by improving vacuum conditions in the synthesis operation and by using an elevated temperature to compact the powders into specimens.[33] Since pore size tends to scale with grain size, it can be seen from this figure that the decrease in porosity with warm compaction comes at the expense of grain growth.

Small-angle neutron scattering becomes insensitive to pores or flaws much above 0.1 μm. Larger flaws, which are especially damaging to strength properties, can result from imperfect bonding during compaction between agglomerations of nanocrystalline powders produced by IGC and compaction. An example of potential crack nucleation sites is shown in Fig. 7.[34] The "roll-ups" (Fig. 7a) shown in this polished and etched transverse section through a nanocrystalline Cu specimen produced by IGC and compaction probably come from the rolls of nanocrystalline powder scraped from the cold finger during evaporation. The TEM micrograph of part of a roll-up (Fig. 7b) shows that the compaction operation was not able to create a perfect bonding between its layers. Trapped gas in nanocrystalline samples can stabilize voids and flaws against sintering under pressure, even at elevated temperatures.[34]

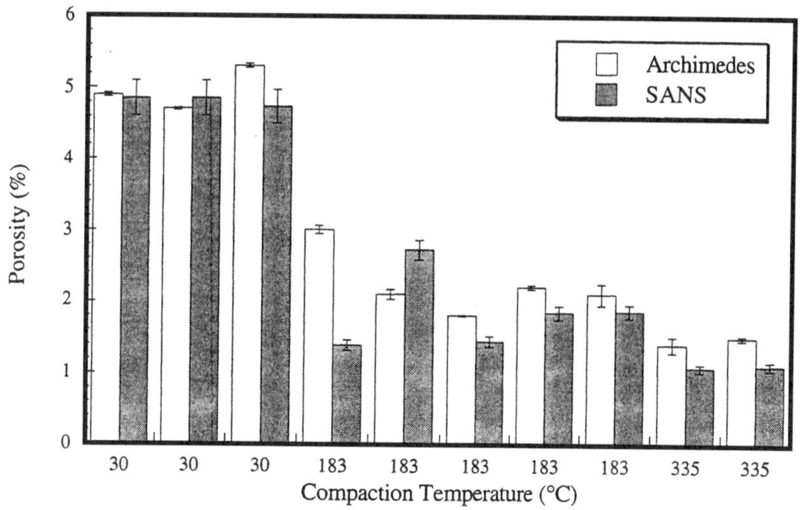

Figure 5. Correlation between the porosity in nanocrystalline Pd obtained from Archimedes density and SANS as a function of compaction temperature. *(From Ref. 33.)*

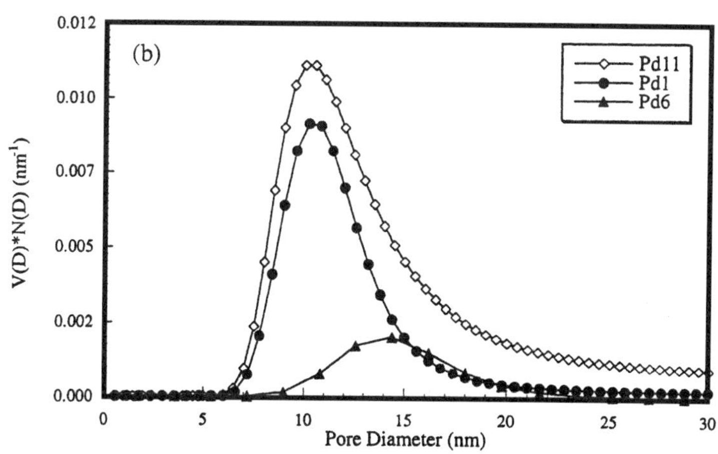

Figure 6. Effect of synthesis conditions and compaction temperature on the pore volume fraction in nanocrystalline Pd. *Open symbols* are from a sample made before improvements to the vacuum in the synthesis chamber and filled symbols are from samples after the improvements, compacted either at room temperature (*circles*) or $T_m/4$ (*triangles*). *(From Ref. 33.)*

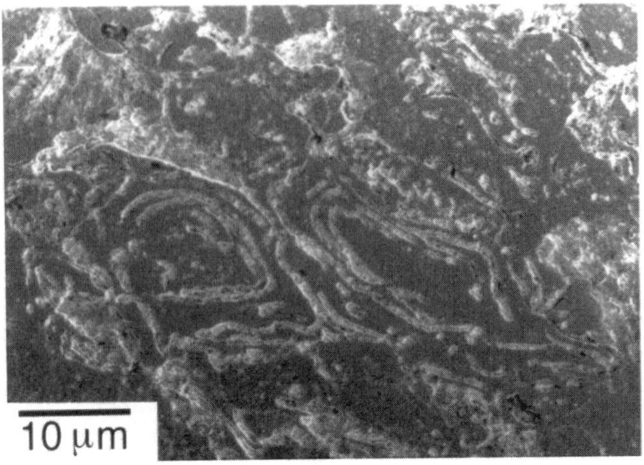

(a)

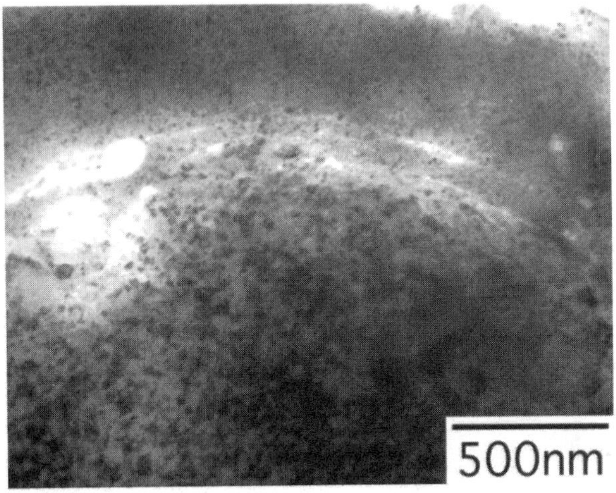

(b)

Figure 7. (*a*) SEM micrograph of a polished and etched transverse section of a nanocrystalline Cu sample made by IGC and compaction showing "roll-ups" (see text). (*b*) TEM micrograph of an ion beam-etched sample similar to that in (*a*) showing the spiral morphology of a roll-up. Different naoncrystalline grain sizes are apparent on either side of the roll-up boundary. *(From Ref. 34.)*

Grain Size Dispersion and Internal Stresses. In Hall-Petch plots of nanocrystalline metals it is customary to plot some measure of strength against an "average" grain size. Various x-ray diffraction (XRD) techniques have been developed for determination of an average grain size. These include analysis of Bragg peaks using the Scherrer equation,[35] integral breadth analysis,[36] and methods derived by Warren and Averbach[37] and Williamson and Hall.[38] The various methods lead to different average grain sizes. For example, Nieman,[39] while measuring the grain sizes of nanocrystalline Cu, Pd, and Ag samples prepared by IGC, found that the values estimated from the Scherrer equation were 1 to 4 times larger than those obtained by the Warren-Averbach method. Measurements by TEM[23][40] show that the grain size is anything but monodisperse in actual nanocrystalline samples. The internal structure is poorly described by a single grain size. Several of the models described in the first section of this chapter show the importance of the width of the grain size distribution in influencing mechanical behavior.[20][22][24] This finding must be borne in mind when interpreting the usual Hall-Petch plot of nanocrystalline materials. Ungár and colleagues[41][42] further modified peak profile analysis methods of Warren-Averbach and Williamson-Hall to estimate grain size distributions as well as dislocation density and dislocation arrangements. Comparisons of grain size distributions calculated by XRD and measured by TEM are shown in Ref. 23.

According to the analysis of x-ray measurements of nanocrystalline compacts of powders produced by ball milling or inert gas condensation, the material is highly strained. The average RMS value of these internal microstrains typically is several tenths of a percent.[40][43][44] In contrast, the internal strains in nanocrystalline Cu produced by electrodeposition[45] are only 0.03%, an order of magnitude smaller. This small strain increases in value upon deformation of the sample to become comparable to that in the compacted powders.[45]

4.0 MECHANICAL BEHAVIOR

Experimental studies of the mechanical properties of nanocrystalline metals have concentrated on the grain size dependence of strength (usually determined by hardness measurements) and, to a lesser extent, elastic behavior.

Elastic Properties of Nanocrystalline Metals. Early studies[16][29] of the elastic behavior of nanocrystalline metals indicated that their elastic moduli are likely to be only a fraction of the coarse-grain values. Gleiter[16] reported that several different techniques for measuring Young's modulus, *E*, led to a value of 88 GPa for a Pd sample with a grain size of 8 nm. (The coarse-grain value is 123 GPa.) However, the modulus of a Mg sample (grain size of 12 nm) was found to be close to the conventional value. The initial slopes of stress-strain curves of nanocrystalline Cu and Pd samples measured by Nieman, et al.,[29] also suggested a substantial decrease in Young's modulus in these materials. Kristic, Erb, and Palumbo[46] obtained a value for *E* in 7 nm grain size Ni-1.5%P that is similar to the coarse-grain modulus. These authors attributed the measured shortfalls in *E* described above to the presence in the samples of pores with cracks growing out of them. In both Refs. 16 and 29, the samples were made by compacting nanocrystalline powders obtained by inert gas condensation to somewhat less than 100% density whereas the Ni-1.5%P material was produced by electrodeposition and reported to be fully dense.

Shen, Koch, Tsui, and Pharr[47] used a nanoindentation technique to investigate the elastic moduli in nanocrystalline Fe, Cu, and Cu-Ni alloys made by mechanical milling and mechanical alloying. The nanoindentation was carried out on individual particulates in compacts, away from interpowder porosity. The average grain size in the Cu, Ni, and Cu-Ni samples ranged from 17 to 26 nm and down to 7 nm in the ball-milled iron. Only at the smallest grain sizes was an appreciable deviation from the coarse-grain value found and, even in the case of the 7 nm Fe, this deviation was only 5%. Shen, et al.,[47] modeled the grain size dependence of *E* on the basis of a two-component system consisting of crystalline grains and interfacial regions. A comparison of the model's predictions and experimental results is shown in Fig. 8.

The results of the various modulus studies implicate porosity as an important factor in producing low *E* values. In the case of the modulus results deduced from stress-strain curves,[29] a poor technique for measuring the strain contributed to the apparent low values. The later use of a miniature strain gage glued to the small tensile specimens led to reasonable results.[48] More accurate measurements were obtained using a pulse echo technique operating at 50 MHz, then correcting for porosity effects in high density samples. Figure 9 (from Ref. 48) shows that the extrapolation to zero porosity yields modulus values in good agreement with the usual values.

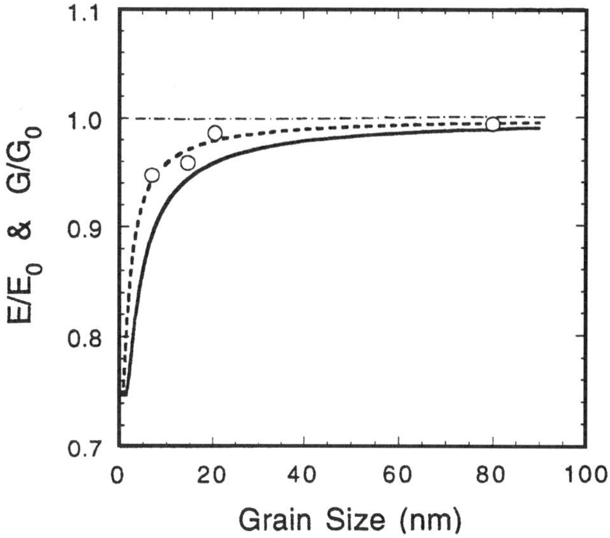

Figure 8. Calculated ratios of Young's (E) and shear (G) moduli of nanocrystalline material to those of polycrystals (E_o, G_o) as a function of grain size d. The dashed and solid curves correspond to a grain boundary thickness of 0.5 and 1.0 nm, respectively. The open circles show the E/E_o values of nanocrystalline Fe vs grain size. The horizontal dotted-dashed line represents the E/E_o and G/G_o ratios for materials having an infinite grain size. *(From Ref. 47.)*

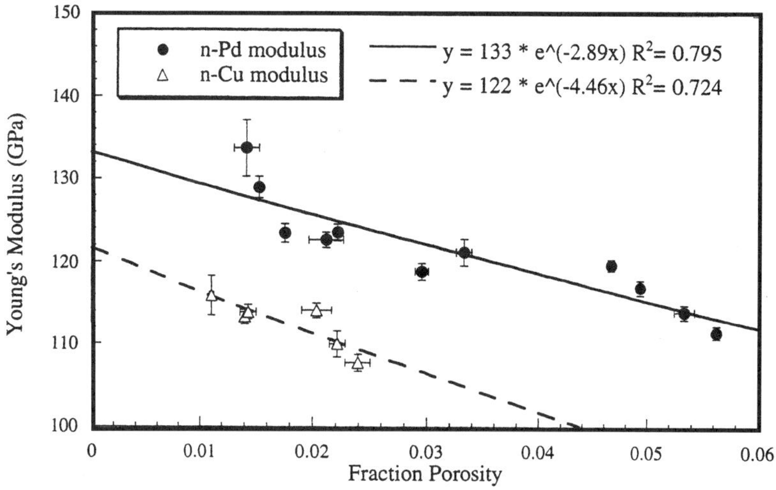

Figure 9. Young's Modulus as a function of porosity for nanocrystalline Pd and Cu. *(From Ref. 48.)*

Relaxation effects in nanocrystalline metals at room temperature have been noticed by a number of investigators.[49]–[51] The modulus measurements described above that were carried out at high frequencies obviously give unrelaxed values, but other techniques could yield results involving some degree of relaxation. Indeed, in a study of the elastic properties of 100% dense nanocrystalline gold, Sakai, et al.,[51] found a strain rate effect in the measured values of E. Below about 220 K, values obtained from a vibrating reed technique are similar to results from stress-strain curves, and the temperature dependence of both is similar to that of ordinary gold. Above 220 K, the vibrating reed results continue to track the coarse-grain values, but modulus values from the stress-strain curves, obtained at a very low strain rate, decrease with increasing temperature at a much faster rate. A stress-strain plot taken at 77 K at a strain rate of 10^{-4}/s is linear over a wide strain range, but at 293 K the plot continuously curves away from the initial slope, indicating the presence of creep. The authors conclude that experimental values of E in nanocrystalline material depend on the strain rate of the measurements.

Hardness, Yield Strength, and Tensile/Compressive Strength. Just as sample porosity affects modulus values in nanocrystalline metals, the presence of voids, cracks, and other defects seriously impacts strength measurements. This problem is well illustrated in Fig. 10. Figure 10a[52] shows the large increase in hardness in nanocrystalline Cu after sample synthesis procedures were improved. Figure 10b demonstrates that imperfections are still present. Cutting the sample gage volume by a factor of 250, and thus decreasing the chances of a large flaw being present in this part of the sample, substantially increases the yield and tensile stresses and strain-to-failure.[53] As previously stressed, measured values of mechanical properties of nanocrystalline metals may be more representative of the flaws present than of the material itself.

Because of the small size of most nanocrystalline samples, studies of strength usually are confined to hardness measurements. Typically, the results are displayed as Hall-Petch plots. While error bars are frequently given for the strength values, they seldom appear for the grain sizes although it is not a single grain size that determines the strength but rather a whole range. Even when grain size distributions are obtained, they are usually plotted as number frequencies. When converted to the more realistic volume frequencies, the influence of a relatively few large grains becomes apparent.

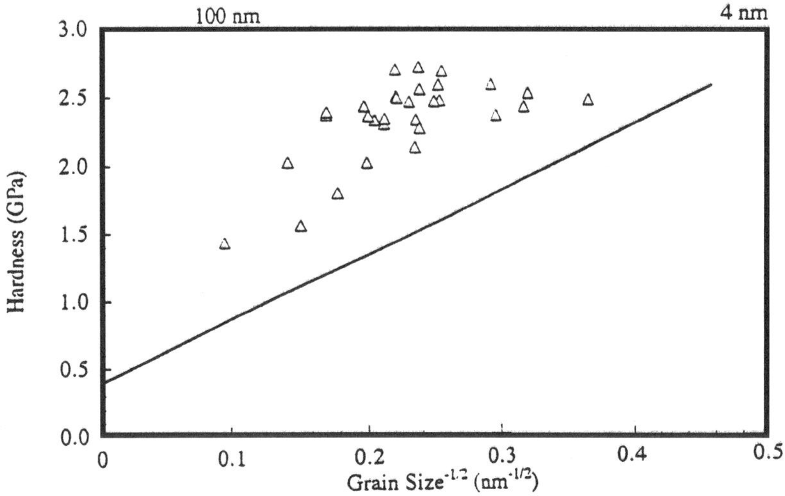

(a)

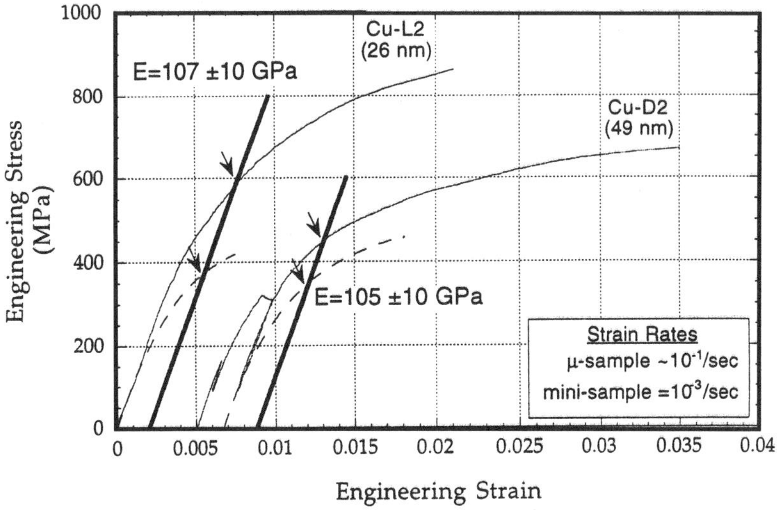

(b)

Figure 10. (*a*) Hall-Petch plot for nanocrystalline Cu. Straight line: best fit to data before synthesis improvements. Data points: after improvements. *(From Ref. 52.)* (*b*) Increase in strength and strain-to-failure in two nanocrystalline samples when the volume of the gage section is decreased by a factor of 250 ("micro" sample vs "mini" sample). Grain sizes are nominal x-ray-measured values. Note relaxation in sample D2 when the sample was partially unloaded and left for 10 minutes. *(From Ref. 53.)*

Some of the earlier work on the influence of grain size on strength was carried out on Ni by Wilcox and Clauer[54] and by Thompson and Saxton.[55] The grain size range later was extended down to 12 nm by Hughes, Smith, Pande, Johnson, and Armstrong.[56] Following the pioneering work of Gleiter[16] in opening up the field of nanostructured materials, a large number of strength versus grain size measurements were reported. (An idea of the spread in reported hardness values just in nanocrystalline Cu alone is given in Fig. 2 of Ref. 20.) One of the earliest measurements, by Chokshi, Rosen, Karch, and Gleiter,[57] aroused much interest with their finding of a negative Hall-Petch slope measured over a narrow range of grain sizes (~7–16 nm) in nanocrystalline Cu and Pd (Fig. 11). Grain refinement softening had been predicted by Gleiter[16] as the result of the dominance of Coble creep at very fine grain sizes and this mechanism was used by the authors to explain their results. Later, Molecular Dynamic simulations[25] showed that grain refinement softening occurs in very fine-grained material as the result of grain boundary sliding. Hardness versus grain size measurements by Nieman, Weertman, and Siegel,[58] also on nanocrystalline Cu and Pd and made by IGC and compaction (as in Ref. 57), showed a small but generally positive Hall-Petch slope down to the smallest grain sizes measured. A fundamental difference between the experiments of Chokshi, et al.,[57] and Nieman, et al.,[58] involves the method of producing a variety of grain sizes. In the latter case, all data points correspond to individual samples in the as-prepared state, whereas the grain size in Ref. 57 was changed by repeated annealing of a sample with originally a very small grain size. Later, it was shown that nanocrystalline metals, at least those made by IGC and compaction or by intense deformation, are substantially strengthened by a short heat treatment.[44][59] At longer annealing times, the decrease in grain refinement strengthening with grain growth appears to overshadow the strengthening effect of the heat treatment and the Hall-Petch slope becomes positive. The cause of the strengthening is likely to be associated with a reduction in internal strains[44] and in dislocation content[60] produced by the annealing. The drop in internal friction seen in nanocrystalline Ni and Fe after the first temperature scan also is probably a reflection of the changes produced by heating.[49]

While the strengthening effect of annealing and the common practice of changing grain size by a heat treatment may account for some of the apparent negative Hall-Petch slopes at very fine grain sizes, not all cases can be explained in this manner. In contradiction to the results of Hughes, et al.,[56] on electrodeposited Ni, which showed a linear Hall-Petch plot down to 12 nm grain size, Erb and colleagues,[61][62] in an extensive study of electrodeposited Ni, found a decided deviation from linearity below ~25 nm grain size and a negative Hall-Petch slope between 11 and 6 nm (Fig. 12). The results of hardness and compressive yield stress measurements on nanocrystalline Cu made by IGC and compaction (Fig. 13) also hint at a deviation from linearity at the smallest grain sizes measured. Explanations for negative Hall-Petch slopes usually depend on some version of Coble creep[20][57] or grain boundary sliding.[25] A different mode of deformation was observed by Carsley, Milligan, Hackney, and Aifantis[63] in a nanocrystalline Fe-Cu alloy prepared by ball milling. Compression testing led to intense shear banding with marked offsets. This deformation behavior is similar to that seen in amorphous metals.

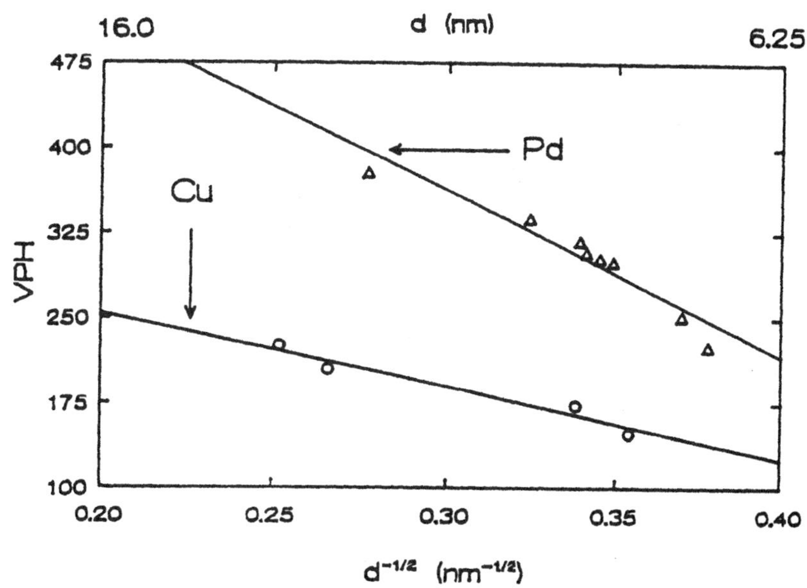

Figure 11. Variation in Vicker's hardness with $d^{-1/2}$ for nanocrystalline Cu and Pd. *(From Ref. 57.)*

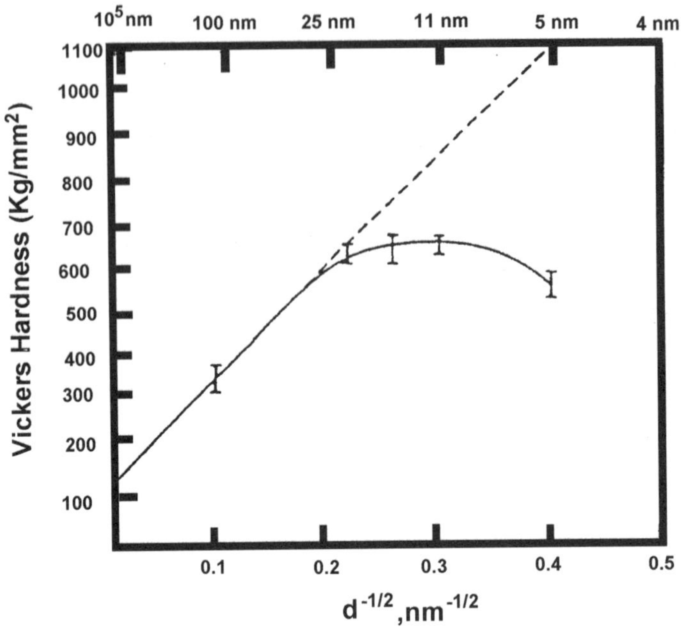

Figure 12. Hall-Petch plot for bulk electrodeposited Ni. *(From Ref. 62.)*

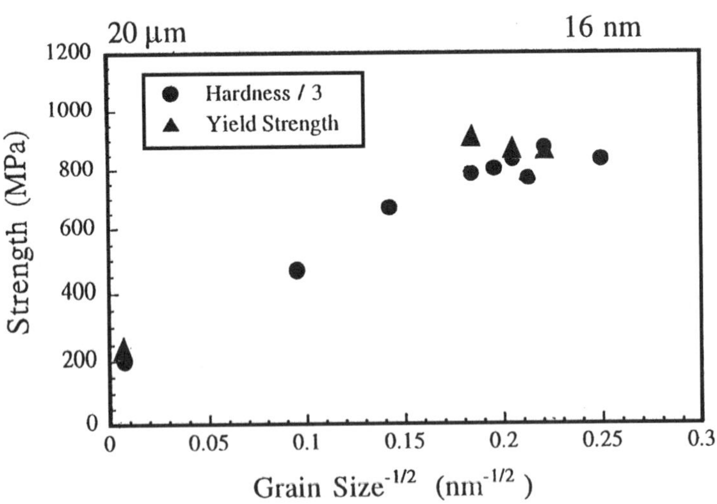

Figure 13. Hall-Petch plot for nanocrystalline Cu samples after synthesis improvements. *Circles*: hardness/3. *Triangles*: compressive yield stress. *(After Ref. 52 plus additional points.)*

Despite the various imperfections found in most nanocrystalline metals, very high strengths have been observed, especially in compression where internal cracks are less damaging. Figure 13 shows yield strengths in pure nanocrystalline Cu that approach 1 GPa, while ordinary Cu yields around 50 MPa or less. Nanocrystalline Ni with an x-ray-measured grain size of 21 nm failed in compression at > 2.1 GPa,[53] a value close to the theoretical shear strength of Ni. The nanocrystalline Cu of Fig. 13, deformed in compression, exhibited considerable ductility, whereas deformation in the Ni was almost entirely elastic in both tension and compression.[53]

Most nanocrystalline metals have limited ductility, especially in tension. Lu and colleagues[45][64] used an electrodeposition technique to produce nanocrystalline Cu that, as mentioned in Sec. 2.0, has internal strains an order of magnitude smaller than those usually measured in nanocrystalline metals made by IGC and compaction or by ball milling. This electrodeposited Cu was found to undergo extensions greater than 5000% under cold rolling.[45] During the deformation, the internal strains gradually increase to a value comparable to other nanocrystalline metals and then remain constant. Hardness values increase with deformation to about 1.2 GPa, well below the values seen in nanocrystalline Cu made by inert gas condensation and compaction (e.g., Fig. 13) and even lower than found in ordinary cold-rolled Cu. The authors observed an appreciable creep strain at room temperature which depends linearly on stress. This linear stress dependence together with the order of magnitude of the creep rate led to an identification of the creep process with a Coble mechanism. Stress-strain tests in tension[64] showed a strain-to-failure of about 30%, remarkably high for nanocrystalline metals but far short of superplastic behavior which, by definition, must be seen in tension. The unusual behavior of this electrodeposited Cu probably is related to the microstructure, which consists of large grains or "domains" of a few micrometers in extent that are subdivided into "grains" ranging from a few nanometers in size to about 80 nm.[64] These nanocrystalline grains are separated from one another by boundaries with misorientation angles of ~ 1–10°. It was observed, during rolling,[45] that these low misorientation angles increase with deformation and that dislocation density builds up at the grain boundaries.

Not all electrodeposited nanocrystalline metals behave in a fashion similar to that of the Cu in Refs. 45 and 64. As shown Fig. 12, electrodeposited nanocrystalline Ni[61][62] attains high values of hardness. (These hardness values are equivalent to the compressive failure stress and similar to the hardness results in Ni made by IGC and compaction observed by Elliott.[53])

5.0 CONCLUSIONS

The high hardness values and high compressive strengths obtained in nanocrystalline metals indicate the promising potential of these materials. Examples of extensive deformation in compression and occasionally even in tension give hope that elimination of flaws and other defects will lead to nanocrystalline metals with acceptable ductility, especially if material with adequate strain hardening to prevent early plastic instabilities can be developed. Clearly synthesis of quality material in useful quantities and at competitive costs is a top priority. From the scientific point of view, pinning down the mechanism (or mechanisms) of deformation at grain sizes below ~10–20 nm is of great interest. Knowledge of the mode of deformation in this grain size regime may be useful in the design of nanocrystalline alloys and composites.

REFERENCES

1. Hall, E. O., *Proc. Roy. Soc. London B,* 64:474 (1951)
2. Petch, N. J., *J. Iron Steel Inst.*, 174:25 (1953)
3. Cottrell, A. H., *Trans. TMS-AIME*, 212:192 (1958)
4. Nieh, T. G., and Wadsworth, J., *Scripta Met. Mat.*, 25:955 (1991)
5. Armstrong, R. W., Chou, Y. T., Fisher, R. M., and Louat, N., *Phil. Mag.*, 14:943 (1966)
6. Pande, C. S., Masumura, R. A., and Armstrong, R. W., *Nanostruct. Mater.*, 2:323 (1993)
7. Smith, T. R., Armstrong, R. W., Hazzledine, P. M., Masumura, R. A., and Pande, C. S., in: *Grain Size and Mechanical Properties: Fundamentals and Applications,* (M. A. Otooni, R. W. Armstrong, N. J. Grant, and K. Ishizaki, eds.), 362:31, Mater. Res. Soc., Pittsburgh, PA (1995)

8. Armstrong, R. W., and Smith, T. R., in: *Processing and Properties of Nanocrystalline Materials*, (C. Suryanarayana, J. Singh, and F. H. Froes, eds.) p. 345, The Minerals, Metals, and Materials Soc., Warrendale, PA (1996)
9. Li, J. C. M., *Trans. TMS-AIME*, 227:239 (1963)
10. Li, J. C. M., and Chou, Y. T., *Metall. Trans.*, 1:1145 (1970)
11. Meyers, M. A., and Ashworth, E., *Phil. Mag. A*, 46:737 (1982)
12. Meyers, M. A., Benson, D. J., and Fu, H. H., in: *Advanced Materials for the 21st Century*, (Y. W. Chung, D. C. Dunand, P. K. Liaw, and G. B. Olson, eds.), p. 499, The Minerals, Metals, and Materials Society, Warrendale, PA (1999)
13. Sangal, S., and Tangri, K., *Scripta Met.*, 23:2079 (1989)
14. Carsley, J. E., Ning, J., Milligan, W. W., Hackney, S. A., and Aifantis, E. C., *Nanostruct. Mater.*, 5:441 (1995)
15. Wang, N., Wang, Z., Aust, K. T., and Erb, U., *Acta Met. Mat.*, 43:519 (1995)
16. Gleiter, H., *Progress in Mats. Sci.*, 33:223 (1989)
17. Gleiter, H., *Acta Mat.*, 48:1 (2000)
18. Thomas, G. J., Siegel, R. W., and Eastman, J. A., *Scripta Met. Mat.*, 24:201 (1990)
19. Kung, H., Sanders, P. G., and Weertman, J. R., in: *Advanced Materials for the 21st Century*, (Y. W. Chung, D. C. Dunand, P. K. Liaw, and G. B. Olson, eds.), p. 455, The Minerals, Metals, and Materials Society, Warrendale, PA (1999)
20. Masumura, R. A., Hazzledine, P. M., and Pande, C. S., *Acta Mat.*, 46:4527 (1998)
21. Coble, R. L., *J. Appl. Phys.*, 34:1679 (1963)
22. Lian, J., Baudelet, B., and Nazarov, A. A., *Mater. Sci. Eng. A*, 172:23 (1993)
23. Mitra, R., Ungar, T., Morita, T., Sanders, P. G., and Weertman, J. R., in: *Advanced Materials for the 21st Century*, (Y. W. Chung, D. C. Dunand, P. K. Liaw, and G. B. Olson, eds.), p. 553, The Minerals, Metals, and Materials Society, Warrendale, PA (1999)
24. Morita, T., Weertman, J. R., Weertman, J., Youngdahl, C. J., and Mitra, R., unpublished data. See also Ref. 23.
25. Van Swygenhoven, H., Spaczer, M., and Caro, A., *Acta Mat.*, 47:3117 (1999)
26. Mura, T., *Micromechanics of Defects in Solids*, pp. 79, 204–211, 216, Martinus Nijhoff Publishers, Dordrecht (1987)

27. Mori, T., and Tanaka, K., *Acta Metall.*, 21:571 (1973)
28. Scattergood, R. D., and Koch, C. C., *Scripta Met. Mat.*, 27:1195 (1992)
29. Nieman, G. W., Weertman, J. R., and Siegel, R. W., *J. Mater. Res.*, 6:1012 (1991)
30. Sanders, P. G., Weertman, J. R., Barker, J. G., and Siegel, R. W., *Scripta Met. Mat.*, 29:91 (1993)
31. Schaefer, H. E., in: *Mechanical Properties and Deformation Behavior of Materials Having Ultra-Fine Microstructures*, (M. Nastasi, D. M. Parkin, and H. Gleiter, eds.), p. 81, Dordrecht, Kluwer (1993)
32. Würschum, R., Scheytt, M., and Schaefer, H. E., *Phys. Status Solidi. A*, 102:119 (1987)
33. Sanders, P. G., Eastman, J. A., and Weertman, J. R., *Acta Mat.*, 46:4195 (1998)
34. Agnew, S. R., Elliott, B. R., Youngdahl, C. J., Hemker, K. J., and Weertman, J. R., *Mater. Sci. Eng. A*, 285:391 (2000)
35. Scherrer, P., Nachr. Ges. Wiss. Goettingen, *Math. Phys.*, K1(2):98 (1918)
36. Klug, H. P., and Alexander, L. E., *X-Ray Diffraction Procedures*, p. 643, John Wiley and Sons, New York (1974)
37. Warren, B. E., and Averbach, B. L., *J. Appl. Phys.*, 21:595 (1950)
38. Williamson, G. K., and Hall, W. H., *Acta Metall.*, 1:22 (1953)
39. Nieman, G. W., Ph.D. Thesis, Northwestern University, Evanston, IL, (1991)
40. Jiang, H. G., Rühle, M., and Lavernia, E. J., *J. Mater. Res.*, 14:549 (1999)
41. Ungár, T., and Borbély, A., *Appl. Phys. Lett.*, 69(21):3173 (1996)
42. Ungár, T., Ott, S., Sanders, P. G., Borbély, A., and Weertman, J. R., *Acta Mat.*, 46:3693 (1998)
43. Sanders, P. G., Witney, A. B., Weertman, J. R., Valiev, R. Z., and Siegel, R. W., *Mater. Sci. Eng. A*, 204:7 (1995)
44. Volpp, T., Göring, E., Kuschke, W.-M., and Arzt, E., *Nanostruct. Mater.*, 8:855 (1997)
45. Lu, L., Sui, M. L., and Lu, K., *Science*, 287:1463 (2000)
46. Kristic, V., Erb, U., and Palumbo, G., *Scripta Met. Mat.*, 29:1501 (1993)
47. Shen, T. D., Koch, C. C., Tsui, T. Y., and Pharr, G. M., *J. Mater. Res.*, 10:2892 (1995)
48. Sanders, P. G., Eastman, J. A., and Weertman, J. R., *Acta Mat.*, 45:4019 (1997)
49. Bonetti, E., Campari, E. G., Del Bianco, L., Pasquini, L., and Sampaolesi, E., *Nanostruct. Mater.*, 11:709 (1999)

50. Huang, Z., Gu, L. Y., and Weertman, J. R., *Scripta Mat.*, 37:1071 (1997)

51. Sakai, S., Tanimoto, H., and Mizubayashi, H., *Acta Mat.,* 47:211 (1999)

52. Sanders, P. G., Youngdahl, C. J., and Weertman, J. R., *Mater. Sci. Eng. A*, 234–236:77 (1997)

53. Elliott, B. R., PhD. Thesis, Northwestern University, Evanston, IL (1998)

54. Wilcox, B. A., and Clauer, A. H., *Acta Metall.*, 20:743 (1972)

55. Thompson A. W., and Saxton, H. J., *Metall. Trans.*, 4:1599 (1973)

56. Hughes, G. D., Smith, S. D., Pande, C. S., Johnson, H. R., and Armstrong, R. W., *Scripta Met.*, 20:93 (1986)

57. Chokshi, A. H., Rosen, A., Karch, J., and Gleiter, H., *Scripta Met.*, 23:1679 (1989)

58. Nieman, G. W., Weertman, J. R., and Siegel, R. W., *Scripta Met.*, 23:2013 (1989)

59. Weertman, J. R., and Sanders, P. G., *Solid State Phenomena*, 35–36:249 (1994)

60. Huang, J. Y., Wu, Y. K., and Ye, H. Q., *Acta Mat.,* 44:1211 (1996)

61. El-Sherik, A. M., Erb, U., Palumbo, G., and Aust, K. T., *Scripta Met. Mat.*, 27:1185 (1992)

62. Erb, U., *Nanostruct. Mater.*, 6:533 (1995)

63. Carsley, J. E., Milligan, W. W., Hackney, S. A., and Aifantis, E. C., *Met. Mat. Trans. A*, 26A:2479 (1995)

64. Lu, L., Wang, L. B., Ding, B. Z., and Lu, K., *J. Mater. Res.*, 15:270 (2000)

11

Structure Formation and Mechanical Behavior of Two-Phase Nanostructured Materials

Jürgen Eckert

1.0 INTRODUCTION

The development of nanostructured materials has led to a new class of materials that are single- or multiphase polycrystals with microstructural features, i.e., particle or grain sizes, layer thicknesses, or domain sizes, in the nanometer range (typically less than 100 nm at least in one dimension). Owing to the extremely small dimensions, nanostructured materials have an appreciable fraction of their atoms in defect environments such as grain or interface boundaries. For example, nanocrystalline material with an average grain size of 5 nm has about 50% of its atoms within the first two nearest-neighbor planes of a grain boundary, in which distinct atomic displacements from the normal lattice sites are exhibited. These unique features have an important impact on their physical and chemical properties, which may be significantly different compared to conventional coarse-grained polycrystals of the same chemical composition.[1]–[3] For example, nanostructured materials may exhibit enhanced diffusivity, superior soft or hard magnetic properties, enhanced catalytic activity, ultrahigh strength or hardness, or improved ductility and toughness in comparison with conventional polycrystals. This not only offers a new scope for applications,[4] but also opens up fundamental questions regarding the basic understanding of the structure-property relations of materials as the microstructural scale is reduced to nanometer dimensions.

While considerable progress has been achieved in the basic understanding of the structures of nanophase materials and their interplay with different physical and chemical properties, the shift from basic science to the design of nanostructured materials with optimum mechanical properties has been rather limited. One of the reasons for this is the fact that it is relatively easy to produce powders, thin films, or ribbons, with nanoscale structure, but it is rather difficult to obtain nanostructured materials in bulk form, which is required for structural applications utilizing mechanical properties.

It has been known for many years that the mechanical strength of crystalline metals or alloys is largely controlled by the grain size, *d*. The well-known empirical Hall-Petch equation[5][6] relates the yield strength, σ_y, to the average grain size, *d*, according to

Eq. (1) $$\sigma_y = \sigma_0 + k\, d^{-1/2}$$

where σ_0 is the friction stress and *k* is a constant. A similar relation exists between the hardness and the grain size. Consequently, reducing *d* down to the nanometer regime increases the strength considerably. However, the limits of the conventional description of yielding and of new mechanisms that may come into play at these small dimensions need to be explored and studied in much more detail. Besides, the intrinsically high interface-to-volume ratio of nanostructured materials may enhance interface-controlled processes to extend the strain to failure or the plasticity. However, other structural features such as pores or larger flaws, grain boundary junctions, and other crystal lattice defects that can depend on the preparation technique and the processing conditions also play an important role. It has become increasingly clear during recent years that all these structural aspects have to be carefully considered in order to fully understand the mechanical properties of nanostructured materials.[7]–[10] This chapter will refer to some of the material presented in previous reviews but will emphasize the more recent work on the mechanical properties of bulk nanostructured materials consisting of at least two different phases, in particular, nanostructured materials derived from bulk glass forming alloys which contain amorphous and nanocrystalline phases.

In view of the importance of processing methods on materials' structures and properties, this overview starts with a description of the methods and the process variables typically employed for producing bulk nanostructured materials. Experimental data for the resulting nanostructures, determining their mechanical behavior at room and elevated temperatures, will be described for selected materials, and the toughness and the ductility increase due to the nanoscale structure will be critically assessed. Consideration of the further development, as well as a discussion of the outstanding questions and challenges, will also be addressed at the end of this chapter.

2.0 METHODS OF PREPARATION

Detailed overviews on the different processing techniques employed for nanostructure formation can be found in several review articles or books, such as in Refs. 1, 3, and 11. Here, only processes are considered that are relevant for nanostructured two-phase materials, i.e., nanostructured metallic materials, whose common characteristic is to consist of at least two different phases and to have a microstructural scale which is in the nanometer regime. A brief presentation of the most common synthesis routes and the typical microstructures obtained will first be given, as such basic information is necessary for an understanding of the microstructures and for the mechanical behavior discussed later on.

2.1 Rapid Solidification Techniques

The advent of the "gun" technique, by Duwez and coworkers in 1960,[12] to rapidly solidify metallic melts at a cooling rate of about 10^6 Ks^{-1}, has most strongly influenced materials science by vastly expanding the processing window for a variety of existing materials and enabling the synthesis of entirely new compositions and alloys. Since then, rapid solidification processing has come to be recognized to result in characteristic constitutional and microstructural changes. The constitutional changes include extension of solubility limits, formation of new non-equilibrium crystalline or quasicrystalline phases, and preparation of metallic glasses. The microstructural effects include changes in the morphology and the refinement of the dimensions of microstructural features such as size, shape, and location of grains and phases. In general, the change is towards a more uniform and finer microstructure with a large reduction of solute segregation effects. Based on these effects, rapidly solidified materials have been extensively investigated over the last three decades. Several books, reviews and conference proceedings give an overview on the state-of-the-art situation on structure, properties and applications of rapidly solidified materials.[13]–[15]

High enough cooling rates during solidification can be accomplished when some important requirements are satisfied. The solidification rate, $\dot{T}$, during cooling is related to the section thickness, z, through the relation

Eq. (2) $$\dot{T} = 10^4 z^{-2}$$

suggesting that the solidification rate increases by two orders of magnitude when the section thickness decreases by one order of magnitude. Hence, the molten metal must be delivered in a thin enough stream, at least in one dimension, and must have a large surface area-to-volume ratio to allow rapid heat removal. This can be achieved by maximizing the contact area between the melt and the cooling medium by rapidly increasing the surface area of the liquid alloy, which can be realized by spreading the melt as a thin layer on a substrate. Alternatively, an increase of the surface area can be effected by disintegrating the melt into small droplets, as is the case for atomization techniques.

Chill Methods. The most commonly used method for rapid solidification is the melt-spinning technique. This process involves directing a molten metal jet onto a cold, moving heat sink where the jet is reshaped and solidified.[13] Thin ribbons or foils of up to 40–60 μm thickness are produced by this technique. The cooling rates can reach up to 10^6 Ks^{-1}, thus making it possible to produce a variety of non-equilibrium phases and a large number of alloy systems, not only on a laboratory scale, but also for industrial production.

The die method involves forcing a liquid metal into copper or steel chill-mold cavities of small cross-sections through the use of vacuum,[16] gravity,[17] or pressure plus vacuum.[18] The advantage of this method is that specimens with predetermined cross-sections (mostly circular) can be obtained. The cooling rates achieved strongly depend on the thickness of the specimen, ranging from about 10^6 to 10^4 Ks^{-1} for thicknesses of 0.2 or 0.7 mm, respectively.[19] Hence, the cooling rates achieved, and the critical thickness of the sample, are strongly coupled or, in other words, the usually high critical cooling rate necessary for metastable phase formation usually limits the thickness of the specimen to thin strips or foils significantly thinner than 1 mm.

In principle, die casting into chill-mold cavities also enables direct preparation of three-dimensional bulk specimens and can be easily scaled up to industrial scale for appropriate materials and processing conditions. For example, copper mold casting, high pressure die casting, or suction casting methods can be used for production of multicomponent bulk metallic glasses with dimensions in the centimeter range.[20][21] In this case, the excellent glass forming ability of these alloys permits glass formation at cooling rates as low as 1–100 Ks^{-1}, thus enabling the preparation of bulk specimens by die casting methods. Combined with devitrification, such materials can be easily transformed into fully dense bulk nanostructured materials.

Spray Methods. In these methods, a continuous stream of liquid metal is atomized, i.e., broken down into fine droplets by means of a gas (e.g., argon) or a liquid (e.g., water).[13][15] The resultant product after solidification is a powder with typical particle size in the micrometer range. The mechanism of achieving atomization and the means of cooling can be different in different techniques. In particular, atomization techniques are applied for large-scale production of rapidly solidified metal and alloy powders for commercial application. The typical solidification rates achieved are on the order of 10^6–10^2 Ks^{-1}.

2.2 Mechanical Attrition

Besides rapid quenching techniques, the formation of metastable phases and also, in particular, nanostructured materials, can be achieved within the solid state. Since the 1970s, mechanical attrition (MA) of powders as a method for materials synthesis has been developed as an industrial process to successfully produce new alloys, phases, and phase mixtures. Powder metallurgy allows the preparation of alloys and composites which cannot be synthesized via conventional casting or rapid solidification routes. Some examples are uniform dispersions of ceramic particles in a metallic matrix or alloys of metals with rather different melting points with the aim of improved strength and corrosion resistance.[22][23] Over the years, mechanical attrition has gained a lot of interest as a non-equilibrium processing technique, resulting in solid-state alloying beyond the equilibrium solubility limit and the formation of amorphous, quasicrystalline, or nanostructured materials for a broad range of alloys, intermetallic compounds, ceramics, and composites. The details of the process of MA, the microstructure development, alloy formation, and the properties of the different classes of materials have been described in a variety of overviews, book chapters, special issues in journals, etc.[24]–[31]

Powder Synthesis. Mechanical attrition is usually carried out in high energy mills. For this purpose, a variety of different types of mills with different characteristics has been developed, including attrition mills, shaker mills, planetary ball mills, vibratory mills, etc.[25] Powders with typical particle diameters of 50–150 μm are placed together with a number of balls in a sealed container, which is violently agitated at high frequencies or rotational velocities. The energy of the milling media (typically steel or tungsten carbide) depends on the internal mechanics of the specific mill, the power supplied for agitation, and the mass, the size, and the number of the balls. Mostly, the milling is carried out under dry conditions in an inert gas atmosphere (e.g., argon), but also milling is possible under air, reactive gases with or without liquid or solid process control agents, depending on the nature of the milled powders, in order to prevent sticking of the materials to the milling tools.

The Process of Mechanical Attrition. The central event of the milling process is the ball-powder-collision. Powder particles are trapped between the colliding balls during milling. They undergo repeated severe plastic deformation and fracture processes, leading to incorporation of lattice defects and to a continuous refinement of the initial structure of the powder particles to the nanometer regime. The nature of these processes strongly depends on the mechanical behavior of the powder components, their phase equilibria, and the stress state during milling, thus being different for pure metals or combinations of ductile/ductile, ductile/brittle, or brittle/brittle components.[25] Hence, the behavior of the different classes of materials during MA depends on the intrinsic properties of the materials and, for alloys, is governed either by the stable or metastable phase equilibria. The temperature that the powders experience during milling also depends on the type of mill used and the actual milling parameters. However, experimental observations and modeling of the mechanics, kinetics, and the energy transfer during collision suggest that the temperature rise during MA is about ≤ 100–200 K.[24][25] Hence, (metastable) alloy formation by MA is driven by intermixing, solid-state interdiffusion, and supersaturation beyond the equilibrium solubility limit and chemical reaction, whereby the individual processes are governed by the high defect density and the severe plastic deformation during MA, resulting in a highly non-equilibrium state of the powder particles.[24][25][32][33]

Influence of Milling Parameters - Milling Induced Phase Transitions. The actual milling parameters, i.e., the kinetic energy transferred and the local temperature during a collision event, may control phase formation and the final structure of the material for a given composition.[24][34] Adjusting or changing the milling conditions can affect the phase formation or can lead to phase transitions. Depending on the chosen experimental conditions, an amorphous phase, as well as a quasicrystalline or a crystalline phase, can be synthesized directly from the composite of the starting elemental powder mixture. Which phase forms depends on the actual milling intensity used for the experiment (i.e., energy input, ambient milling temperature). In addition, the different phases can be transformed into each other by additional milling at higher or lower milling intensity. In particular, this has been demonstrated for the crystal-to-quasicrystal transition, the crystal-to-amorphous transition, and the quasicrystal-to-amorphous transition.[24][35]–[37]

Nanostructure Formation upon Mechanical Attrition. Besides the variability with respect to metastable phase formation, one other important feature inherent to MA is the development of nanoscale microstructures. As described above, during milling the powder particles are subjected to severe mechanical deformation from the collisions with the milling tools. Consequently, plastic deformation at high strain rates ($\sim 10^3$–10^4 s^{-1}) occurs in the particles and a high level of internal strain is created due to the large number of dislocations and other deformation faults introduced. The dislocations rearrange into cell structures and deformation shear bands and finally evolve into high-angle grain boundaries by recovery and recrystallization processes. At this stage, no further grain refinement is possible but only grain boundary sliding can occur, which does not refine the microstructure any further.[34][38] Such results were first observed for pure metals and intermetallics[38]–[40] and have since then been reported for a variety of materials including supersaturated solid solutions[41][42] and multicomponent alloys.[43][44] Detailed overviews on nanostructure formation by MA for a variety of systems were recently given by Koch[25] or Fecht.[32]

Nanoscale Phase Mixtures. Besides preparation of homogeneous nanocrystalline materials, mechanical attrition can also be applied to produce nanoscale phase mixtures of materials with distinctly different ductility. For example, 10 nm Ge particles can be embedded in a ductile Pb or Sn matrix.[45] Similarly, ultra-fine dispersions have been produced for different systems such as Fe-W, Cu-Ta, TiNi-C, and Ag-Fe.[46][47] These results link the field of nanocrystalline materials to the well-known oxide-dispersion strengthened materials, where nanometric dispersions of oxides are embedded in a metallic matrix.[22][23] Hence, MA allows the production of different metastable phases and the utilization of mechanically induced phase transitions in order to either fully transform one phase into another, or to achieve phase mixtures of different metastable phases by properly adjusting the milling conditions. Moreover, it simultaneously offers the opportunity to achieve nanoscale microstructures due to the mechanical deformation of the material.

Consolidation of Powders. Consolidation of nanocrystalline powders into fully dense bulk specimens is of primary interest for the development of near-net shape parts for technological applications. The essence of all compaction techniques is to apply high pressure for densification, and rather high temperature to soften the material so that plastic deformation allows better filling, and material flow by diffusion helps to remove the remaining porosity.[48] In addition, sliding or shearing of powder particles over each other will help to remove oxide and other surface layers to assist bonding. Acting to render consolidation into essentially fully-dense specimens with complete bonding between the initial particles difficult is the fact that the nanocrystalline powders of interest are typically hard. Moreover, diffusional processes will not only assist densification, but will also allow grain coarsening to occur. As such, the temperature and the time span of the consolidation process have to be adjusted carefully in order to achieve a balance between densification and grain growth. With respect to the pressure requirements, as a rule of thumb, a pressure of about three times the material flow stress, that is on the order of the hardness of the material ($p \sim H$), will be required for continued densification by deformation.[48] However, using high pressure for consolidation bears the problem of cracking as the pressure is released.[49]–[53] At the other extreme of the temperature-pressure requirements are processes of pressureless sintering, hot isostatic pressing, or uniaxial hot pressing, where rather high temperatures are required to compensate for low pressures.[50] However, high temperatures bear the risk of significant structural coarsening or even phase transitions, such as crystallization of metastable nano-quasicrystalline phases. Therefore, there is a strong interest in using as low temperatures as possible in order to avoid structural changes.

The difficulties associated with the use of high pressures and the excessive structural coarsening for using high temperatures[50] render that commercial processes such as Hot Isostatic Pressing (HIP) at reasonably high pressures, e.g., about 200–300 MPa, and moderate temperatures on the order of 0.5–0.65 the melting temperature, are most frequently used for consolidation of nanostructured powders. This allows one to obtain highly dense (above 97%) specimens with final grain sizes after compaction on the order of 50–150 nm. [54]–[57] While methods using such pressure and temperature conditions are most common and easy to use, there are considerable advantages to using consolidation techniques that incorporate large amounts of shear deformation, which assists cracking of surface layers and particle deformation to enhance interparticle bonding and densification. Particularly interesting are forging and extrusion methods, where high pressures and shear strains are imposed at moderate temperatures to achieve good densification and interparticle bonding.[58][59]

In many cases, improvements in density and particle bonding not only require optimized consolidation techniques and parameters, but also great care has to be taken with respect to clean powder handling and compaction conditions. This often requires the use of closed-loop processing or ultrahigh vacuum systems to ensure a high purity and good dynamic degassing.[58] In particular, restricting oxygen contamination and reducing the water and hydrogen content of powders seems to play a major role in achieving high densities, preventing residual porosity, and promoting good bonding of particle surfaces.

The most significant problems of powder consolidation stem from the risks of possible phase transformations and structural coarsening, i.e., grain growth or second phase or particle growth, during high temperature exposure. Grain growth can usually be described by an equation of the type

Eq. (3) $$d^n - d_0^n = K \cdot t$$

which relates the growth from an initial grain size, d_0, to a larger grain size, d, during the annealing time, t. Simple theory suggests a value of $n = 2$, but larger values are frequently found. The kinetic term, K, is sensitive to the annealing temperature, but also to such factors as the chemistry (cleanliness) and structure of the grain and particle boundaries. For nanocrystalline materials, the driving force for coarsening is high, because of the high surface area, and growth can occur even at temperatures as low as room temperature.[58][60][61] Grain growth can be hindered by a narrow size distribution,[61] and by impurities.[58]

Almost any deviation from pure single-phase material will reduce the tendency for grain growth. This includes residual pores,[62] impurities and solutes,[63] and second phase particles,[64] all slowing down grain growth. Pinning by fine particles is, in fact, a very efficient way of slowing or stopping grain coarsening, with the Zener drag process giving a relationship between the grain size, d, the volume fraction, f, and the size of particles, D, as approximately $d \propto D/f$.[64] Hence, incorporating a significant volume fraction of fine second phase particles enables one to retain a fine grain size both during and after processing.[59][65]–[67]

2.3 Devitrification of Metallic Glasses

Alternatively to directly synthesizing metastable phases with nanocrystalline microstructure as it can be achieved by rapid quenching or mechanical attrition techniques, controlled devitrification (crystallization) of amorphous solids can be employed to produce large quantities of nanocrystalline material. The basic principle for the devitrification method is to control the crystallization kinetics by optimizing the heat treatment conditions (annealing temperature and time, heating rate, etc.) so that the amorphous phase partially or completely transforms into nanocrystalline material.

Nucleation and Crystal Growth. Crystallization of metallic glasses is known to generally proceed by nucleation and growth processes. The phases and morphologies of crystallization products are dominated by the transformation mechanism, which is closely related to the chemical composition of the amorphous state and the thermodynamic properties of the corresponding phases. Both crystal nucleation, as well as the growth rate, depend strongly on temperature. Since the driving force for crystallization increases with decreasing temperature, but the diffusivity decreases with decreasing temperature, the highest growth rate is observed close to the melting temperature. The maximum nucleation rate is observed at significantly lower temperatures. Prerequisites for nanocrystallization, i.e., for the formation of material with a fine-grained microstructure upon crystallization, is a very high nucleation rate, I, combined with a low growth rate, u. For designing a very fine microstructure, one has to choose an annealing temperature for which I/u is maximum,[68]–[71] i.e., crystallization has to proceed at temperatures just below the temperature for the highest nucleation rate. The nucleation rate is given by[68][69]

Eq. (4) $$I = I_0 \cdot \exp\left(-\frac{Q}{R \cdot T}\right) \cdot \exp\left(-\frac{\Delta G_c}{R \cdot T}\right)$$

where Q is the activation energy for diffusion and ΔG_c the nucleation barrier. For homogeneous nucleation the nucleation barrier is given by

Eq. (5) $$\Delta G_c = \frac{\gamma^3}{(\Delta G)^2}$$

In general, the nucleation rate can be influenced by the interfacial energy, γ, the driving force, ΔG, and the atomic mobility (diffusivity or viscosity) of the system. Most efficient for increasing the nucleation rate is a reduction of the interfacial energy, γ.[70] It is known that γ is smaller for metastable phases or can be reduced by small additions of other elements. For example, in $Co_{50}Zr_{50}$ metallic glasses, the shape of the stable crystalline CoZr phase changes significantly when the oxygen content of the glass increases to about 4 at%. This observation has been explained by a reduction of the interfacial energy by oxygen;[70] a significant increase of the driving force, ΔG, due to the small oxygen content, is less likely.

Whereas, in $Co_{50}Zr_{50}$ metallic glass, no metastable phases have been observed during crystallization, amorphous $Co_{67}Zr_{33}$ crystallizes first into a metastable cubic $CoZr_2$ phase before transforming into another metastable hexagonal phase and finally into the stable tetragonal $CoZr_2$ phase.[72] The lower driving force, ΔG, for the formation of metastable phases might be responsible for a reduced growth rate. The lower interfacial energy, γ, which is typical for metastable phases, however, will reduce the nucleation barrier, ΔG_c, significantly, thus increasing the nucleation rate. Due to the reduced nucleation barrier, crystallization of a metastable phase should always lead to a much finer microstructure than direct crystallization into the equilibrium phase.

Types of Transformation. During conventional crystallization of amorphous solids, three different types of transformation are identified depending on the composition of the respective alloy:[68]–[71] polymorphous, eutectic, and primary crystallization. All these types of transformations have been observed in nanocrystallization of amorphous alloys in different alloy systems. Figure 1 shows a hypothetical free energy diagram to illustrate the different reactions occurring upon crystallization of amorphous phases, displaying the change of free energy with composition of the glass and different nanocrystalline phases (e.g., two nanophases, a solid solution, α, and an intermetallic compound, β) at a given temperature. Possible metastable equilibria between the different phases are marked by dashed tangents, the equilibrium coexistence tangent is represented by the solid line. Depending on composition, the transition of the metastable amorphous phase into the crystalline phases can proceed by one of the following reactions:

Polymorphous Crystallization. An amorphous phase crystallizes into a single crystalline phase without any change in composition. The crystallization product can be a supersaturated solid solution or a metastable or stable crystalline compound. This type of reaction can occur only in composition ranges near compounds or pure elements (C1 or C2 in Fig. 1). As far as supersaturated phases (solid solutions) form during this reaction they decompose by subsequent reactions. A metastable intermetallic compound will undergo a phase transition into the stable thermodynamic equilibrium phases.

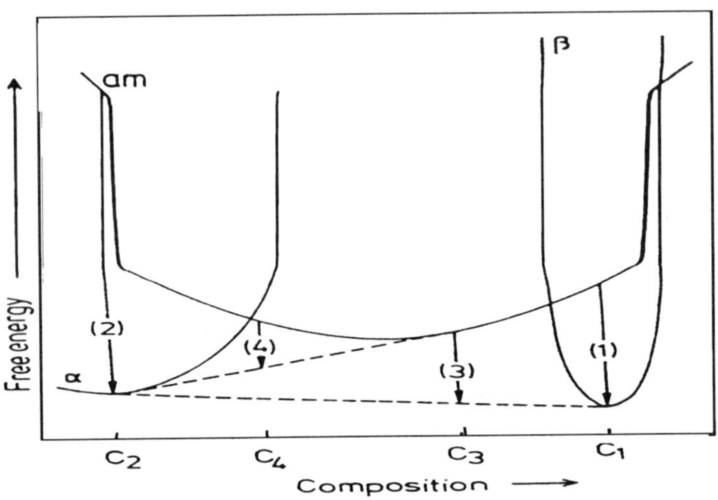

Figure 1. Hypothetical diagram of the free energy of the amorphous phase (*am*), a terminal solid solution (α) and an intermetallic phase (β), illustrating the nanocrystallization of amorphous solids. The numbers (1) to (4) correspond to the crystallization reactions mentioned in the text.[68]

Eutectic Crystallization. An amorphous phase crystallizes simultaneously into two nanocrystalline phases (e.g., reaction 3: $am \rightarrow \alpha + \beta$) by a discontinuous reaction. By this, two nanophases grow in a coupled fashion analogously to the eutectic crystallization of liquids. This type of reaction has the largest driving force and can occur in the concentration range between the two stable phases. The overall composition of the two phases remains the same as that of the initial amorphous phase. There is no concentration difference across the reaction front. However, in the reaction front the components have to separate into the different phases. Therefore, this type of reaction is initially slower compared to a polymorphous crystallization without any separation of the components.

Primary Crystallization. An amorphous phase with a composition deviating from that for either polymorphous or eutectic reaction crystallizes in a first step into a primary nanocrystalline phase (either a supersaturated solid solution or an intermetallic compound) embedded in the amorphous matrix [reaction 4: $am\,(C4) \rightarrow \alpha + am\,(C3)$]. During this reaction the residual amorphous phase will change its composition until further crystallization is stopped by reaching the metastable equilibrium given by the dashed line. The residual amorphous phase (concentration C3) can transform in a second step (i.e., later or at higher temperatures) into nanophases in the mechanism of either eutectic or polymorphous nanocrystallization. The dispersed primary crystallized phase may act asa preferred nucleation site for the subsequent crystallization of the residual amorphous matrix.

In any case, the finest nanocrystalline structures should be obtained during primary crystallization. Primary crystallization is governed by long-range diffusion of the constituents and proceeds by a time-dependent growth rate. As the crystals grow, their growth rate is further decreasing until the diffusion fields of the individual crystals overlap. This long-range diffusion controlled growth rate is significantly smaller than growth during polymorphous or eutectic reaction processes, which are controlled by interfacial diffusion processes. However, in a lot of cases a clear distinction between the different types of crystallization reactions is not unambiguously possible. This is due to the possibility of formation of intermediate metastable crystallization products. For example, oxygen-induced metastable phase formation makes it very difficult to clearly distinguish between primary or polymorphous crystallization for $(CoZr_2)_{1-x}O_x$ metallic glasses.[70]

Amorphous Phase Separation. A large glass forming ability is generally related to deep eutectics, indicating a strongly negative enthalpy of mixing of the constituents.[73] However, there is also evidence for the existence of systems which show, at least in some concentration range, a different behavior, i.e., a positive enthalpy of mixing, leading to a miscibility gap which may be due to singularities in the structure of the amorphous state. Therefore, a concentration range exists where the free energy of a mixture of two coexisting amorphous phases is lower than that of a single amorphous phase. Amorphous materials in this concentration range not only have the possibility of transforming by one of the crystallization reactions mentioned above, but may also decompose into two amorphous phases, i.e., may undergo amorphous phase separation.

Amorphous phase separation occurring by nucleation and growth processes, or even by spinodal decomposition without any nucleation process, are well-known from oxide glasses.[74][75] There, extremely fine-grained glass ceramics (partially crystallized glasses) can be produced, which is generally assumed to be a result of amorphous phase separation followed by subsequent crystallization. As illustrated in Fig. 2, in the case of a miscibility gap there exists a concentration range between *a* and *b*, where phase separation into two amorphous phases can occur by nucleation and growth or in a narrower concentration range between *c* and *d* by spinodal decomposition. The spinodal is the boundary between the metastable and the unstable state of the phase. In the metastable state, a finite fluctuation, i.e., a nucleus of a new phase, is required for phase transformation. In contrast, beyond the spinodal, a single phase is unstable with respect to infinitesimal fluctuations in composition and begins to separate into two related phases with different compositions without nucleation. In a binary system, the condition for metastability is given by $\partial^2 G/\partial c^2 < 0$, where G is the molar free enthalpy and c is the mole fraction.[70] The driving force for increasing the composition fluctuations is the gain in free energy upon decomposition. Outside this concentration range decomposition into two amorphous phases will proceed by nucleation and growth processes. The kinetics of spinodal decomposition and the characteristic wavelength for a given annealing temperature have been determined by Huston, et al.,[76] Typical wavelengths at temperatures not too far below the critical temperature, T_c, are in the range of 5 to 10 nm. The decomposition rate is zero at the temperature of the spinodal and passes through a maximum with decreasing temperature.

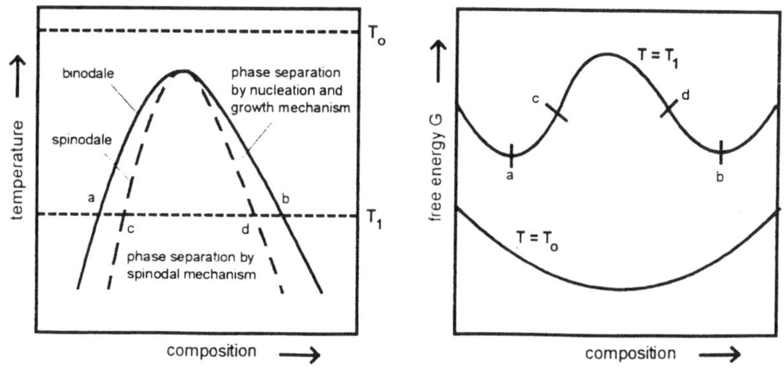

Figure 2. Spinodal decomposition:[70] *(a)* schematic phase diagram showing the phase boundary and the spinodal of a two-liquid immiscibility region, *(b)* free energy vs composition diagrams for the temperatures given in *(a)*.

Amorphous phase separation is expected to have a strong potential for increasing the nucleation rate, and to decrease the growth rate of crystals. Hence, it promotes nanocrystal formation. For example, an increase in the nucleation rate may be due to an increase in the diffusivity of the respective element(s), a decrease in the crystal/glass interfacial energy, a local increase in the thermodynamic driving force, or heterogeneous nucleation of crystals at the amorphous/amorphous interface. Nucleation can start only after amorphous phase separation, thus leading to significant incubation times for crystallization. On the other hand, phase separation is expected to reduce the crystal growth rates, due to composition shifts induced by the phase separation and/or by interference between the growing crystals and the amorphous regions.

Phase Separation in Bulk Glass Forming Alloys. Evidence for amorphous phase separation in metallic glasses was reported more than twenty years ago, in $Pd_{74}Au_8Si_{18}$,[77] $(Pd_{0.5}Ni_{0.5})_{81}P_{19}$,[78] or $Be_{40}Ti_{24}Zr_{36}$[79][80] metallic glasses—either prior to crystallization or even in the as-quenched state. Only very recently, strong evidence for amorphous phase separation was also found for new bulk glass forming alloys. In $Zr_{41.2}Ti_{13.8}Cu_{12.5}Ni_{10}Be_{22.5}$ metallic glass, Busch, et al.,[81] observed significant composition fluctuations in Be and Zr concentrations in as-quenched bulk specimens by field ion microscopy. Schneider, et al.,[82][83] concluded from small-angle neutron scattering (SANS) data that a second miscibility gap opens in the undercooled liquid near the glass transition temperature, primarily involving Ti and Cu. They revealed the formation of

Ti-rich regions, which in later stages promote nanocrystallization. Depending on the annealing temperature, the wavelength of phase separation has been determined to be about 13–35 nm, following a relation, $\lambda^{-2} \propto T$, as suggested by Cahn's theory of spinodal decomposition. Furthermore, decomposition triggers nanocrystal nucleation within the decomposed zones. The growth of these nanocrystals is limited by the size of the domains in which they nucleate.

Similar observations were made for Mg-Y-Cu-Li bulk metallic glasses,[84] also showing evidence for phase-separated domains in high-resolution TEM images. Additional small-angle x-ray scattering (SAXS) and anomalous small-angle x-ray scattering (ASAXS) studies of such alloys also gave evidence for phase separation into Cu-rich and Cu-poor regions.[85] Again, these phase-separated domains act as preferred nucleation sites for nanocrystals.[85][86] Hence, the precipitation of primary nanocrystalline phases in these bulk glass forming systems is governed by the initial chemical composition of the undercooled liquid, and chemical decomposition controls their crystallization behavior.

By choosing alloys with different compositions, e.g., preparing an alloy having the same composition as the Ti- and Cu-poor decomposed regions in the Zr-Ti-Cu-Ni-Be system, allows to obtain bulk metallic glasses without tendency for decomposition. For example, a $Zr_{46.75}Ti_{8.25}Cu_{7.5}Ni_{10}Be_{27.5}$ glassy alloy shows no evidence of phase separation in SANS.[87] Hence, the occurrence of phase separation and nanocrystallization strongly depends on the actual composition of the alloy, giving the unique opportunity to specially design bulk nanostructured materials by appropriate choice of alloy composition and optimization of annealing conditions.

3.0 PHENOMENOLOGY OF NANOSTRUCTURE FORMATION AND TYPICAL MICROSTRUCTURES

The nanostructured materials being discussed here may contain crystalline, quasicrystalline, or amorphous phases of the metallic constituents. In addition, nanostructured composites based on metallic glasses will be considered. The materials can be produced by fabrication routes based on rapid quenching or slow cooling from the melt, or powder metallurgy, respectively. These synthesis routes may directly lead to nanostructured materials or, in other cases, have to be combined with an appropriate heat treatment to obtain the desired nanoscale microstructure, such as in the case of (partial) devitrification of metallic glasses. Whereas crystallization of bulk metallic glasses can directly yield bulk nanostructured materials, subsequent consolidation of rapidly quenched powders and ribbons, or compaction of mechanically attrited powders has to be performed in order to obtain bulk samples.

3.1 Rapidly Solidified Materials

Substantial increases in strength, along with good ductility, have been observed in a number of alloys with multiphase nanoscale microstructures. Such "nanophase composites" can be made directly by liquid quenching or by annealing of initially fully-amorphous material.[88][89] In addition, devitrification can also be induced by mechanical working.[90] Most of the work on multiphase nanoscale microstructures has concentrated on Al-, Mg-, and Zr-based multicomponent systems with the goal of developing new materials with high specific strength (and thus, low density) and high ductility. By the control of alloy composition and quenching conditions, several different types of alloys containing non-equilibrium phases and microstructures have been fabricated. In particular, great attention has been paid to non-periodic structure alloys consisting of amorphous and quasiperiodic phases.

Al-Based Nanostructured Alloys. The first success in synthesizing Al-based high specific strength material with good ductility was achieved for amorphous Al-Ni-Si alloys by Inoue, et al., in 1987.[91] Since then, a large number of Al-based amorphous alloys have been found.[92][93] These alloys can be classified into Al-metal and Al-metalloid systems.[94] Among these alloy systems, the Al-Ln-LTM and Al-ETM-LTM [Ln = lanthanide metal, ETM = early transition metal (groups IV–VI), LTM = late transition metal (groups VII and VIII)] amorphous alloys are particularly important because of the simultaneous achievement of high strength and good bending ductility. Subsequently, it was found[90] that a homogeneous dispersion of nanoscale Al particles into an amorphous matrix causes a strong increase in mechanical strength. This fine-mixed structure consisting of nanoscale fcc Al particles with 3–5 nm in size embedded in an amorphous phase is typically found in Al-Ln-LTM alloys, and can be produced either by annealing the amorphous single phase[95] or by decreasing the cooling rate upon quenching from the liquid.[88] More recently, similar microstructures have also been observed in alloys not containing any lanthanides.[96] In the historical progress of the development of high-strength Al-based alloys a drastic increase in tensile fracture strength was found for the more recent non-equilibrium base alloys including amorphous or quasicrystalline phases. Figure 3 summarizes the features of the microstructure and the mechanical strengths of the different types of non-equilibrium alloys developed so far.[94][97]–[99] The non-equilibrium structures for Al-based alloys can be classified into the following six types.

1) An amorphous single phase.
2) A nanostructure of Al and intermetallic compounds obtained by devitrification of an initial amorphous phase.
3) A partially crystallized structure of nanoscale Al particles embedded in an amorphous matrix.
4) A nano-quasicrystalline structure consisting of nanoscale icosahedral particles, which are surrounded by a nanoscale fcc Al phase without grain boundaries.
5) A phase mixture of coexisting nanogranular amorphous and Al phases.
6) A nanogranular Al phase surrounded by an amorphous network phase.

Amorphous Single-Phase Alloys. The amorphous alloys obtained by melt spinning can be divided into metal-metal and metal-metalloid systems. Among these systems, both Al-Ln-TM[93][100] and Al-ETM-LTM[101] systems are more important because of the achievement of higher fracture strength. In addition to melt-spun thin ribbons, also Al-based amorphous wires with diameters in the range of 40 to 120 µm, which exhibit high strength and good ductility, have been produced by a melt-extraction method.[102] No bulk amorphous single-phase alloys in Al-based systems have been obtained by any kind of solidification method up to date. This is due to the relatively low glass forming ability of these systems. However, amorphous alloys with Al concentrations above about 85 at% exhibit a glass transition phenomenon, followed by a rather narrow supercooled liquid region of about 20 K prior to crystallization.[103] This promises that by utilizing the viscous flow of the supercooled liquid, bulk amorphous samples can be produced by extrusion of pieces of melt-spun ribbons or from gas atomized powders with particle sizes below about 25 µm.[104] For example, first results for the extrusion of $Al_{85}Ni_{10}Ce_5$ amorphous powders at temperatures between 443 and 493 K yielded mostly single-phase bulk amorphous samples with diameters of up to 20 mm.[105]

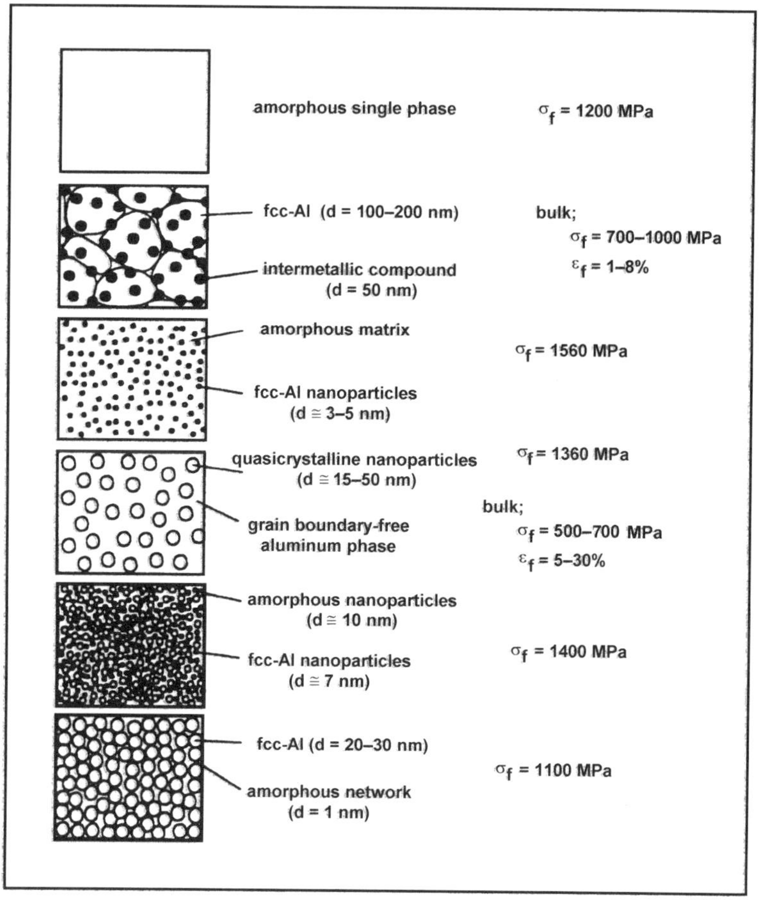

Figure 3. Microstructures and mechanical properties of non-equilibrium Al-based alloys.[99]

Partially Crystallized Alloys. The phase formation from the highly undercooled liquid (or the amorphous phase) has the following features:[106]

1) Homogeneous nucleation with a high nucleation rate and a low growth rate.

2) A large concentration gradient of solute elements resulting from the low atomic diffusivity of the solute atoms and a highly dense-packed structure at the liquid/solid interface.

3) Formation of metastable phases with new compositions

4) Formation of a residual amorphous phase with high solute concentration.

5) Precipitation of almost ideally spherical defect-free nano-sized particles with nanoscale interparticle spacings.

These features enable the synthesis of new nanoscale mixed phase alloys upon rapid solidification or during subsequent annealing.

In general, Al-rich amorphous alloys with high Al contents above about 85 at% crystallize through two stages upon heating to elevated temperatures. The first exothermic reaction is due to the precipitation of fcc Al and the second stage results from the decomposition of the residual amorphous phase into intermetallic compounds.[107] Appropriate control of the cooling rate during quenching from the melt allows one to obtain a nanostructure consisting of 3–5 nm Al particles embedded in an amorphous matrix in melt-spun ribbons.[107] The volume fraction of precipitates can be adjusted by changing the rotation speed of the wheel. The hardness, the yield strength, and the Young's modulus of the material increase for volume fractions of up to about 25%. For larger volume fractions the yield strength decreases again, which is attributed to the embrittlement of the residual amorphous phase due to structural relaxation and enrichment of solute elements above a critical level. Similar microstructures form upon devitrification of initially amorphous ribbons, i.e., for annealing in the first-stage exothermic reaction range.

Bulk Nanocrystalline Alloys. Raising the annealing or extrusion temperature to temperatures corresponding to the second-stage exothermic reaction allows to produce fully-dense nanoscale multiphase alloys consisting of fine intermetallic compounds of about 50 nm in size, which are embedded in an fcc Al matrix with typical grain sizes of about 100–200 nm.[106] This yields a unique microstructure in which intermetallic phases are dispersed homogeneously within the Al matrix and on the grain boundaries. In particular, the density of the dispersed compound particles is not high enough to suppress all dislocation activity and a significant fraction of dislocations can interact with the grain boundaries of the Al phase. Such nanocrystalline Al-based alloys have been commercialized under the name of "GIGAS,"[99] and have already reached some technological importance.

Nanoscale Quasicrystalline Alloys. A finely mixed structure consisting of nanoscale icosahedral (I) quasicrystalline particles with sizes of 20–50 nm surrounded by an fcc Al phase, with a thickness of about 5–10 nm, was found in melt-spun Al-(Mn, Cr)-Ce,[108]–[110] Al-(Mn, Cr)-Ce-TM,[111][112] Al-V-Ce-TM,[113] and Al-(V, Cr, Mn)-TM[114] systems (TM = Fe, Co, Ni, or Cu). These alloys were chosen because Al-Mn and

Al-Cr alloys are well-known I-phase forming systems and Al-Ln-based alloys possess a good glass-forming ability (see above). Figure 4 illustrates the compositional dependence of structure formation for melt-spun Al-Mn-Ce alloys as a typical example for this type of material. Depending on the actual alloy composition, different phase mixtures of amorphous, quasi-crystalline, and fcc Al phases form directly upon quenching.[115] X-ray diffraction (XRD) analysis and transmission electron microscopy (TEM) investigations proved the formation of a mixed structure consisting of nanoscale icosahedral and fcc Al phases in the Al-rich composition range above 92 at% Al, as displayed in Fig. 5 for an $Al_{92}Mn_6Ce_2$ alloy.[116][117] The inset in Fig. 5 shows a typical DSC trace for the metastable phase mixture, revealing that the I-phase starts to decompose into Al_6Mn at about 720 K. The nanoscale I-phase particles, with diameters of about 50–100 nm, have an almost spherical morphology and random orientation, and are homogeneously dispersed in the Al phase without high-angle grain boundaries (Fig. 6). The volume fraction of the I-phase is as high as 60–70%. A similar microstructure forms for other rapidly solidified Al-(Mn, Cr, V)-Ce-TM alloys and also in Al-Fe-Cr-Ti alloys without any lanthanide metal.[118]

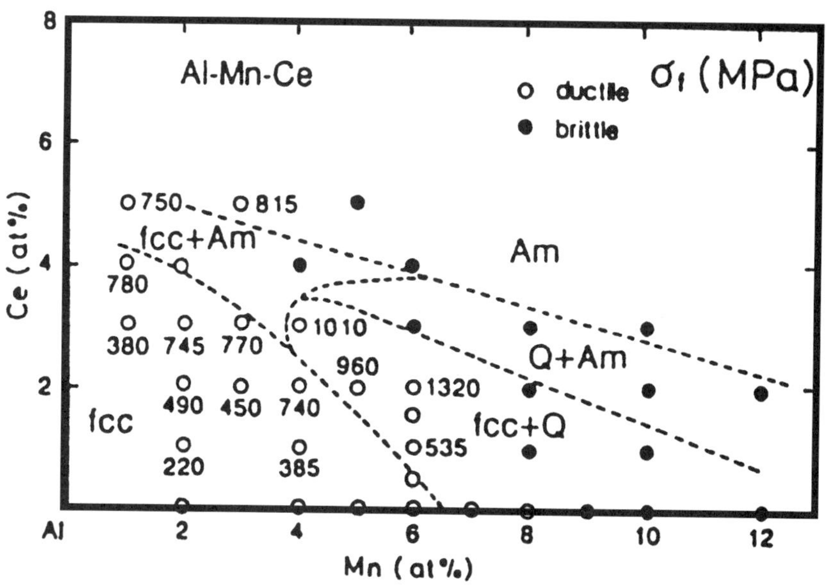

Figure 4. Compositional dependence of structure, tensile strength, σ_f, and ductility of melt-spun Al-Mn-Ce alloys.[99]

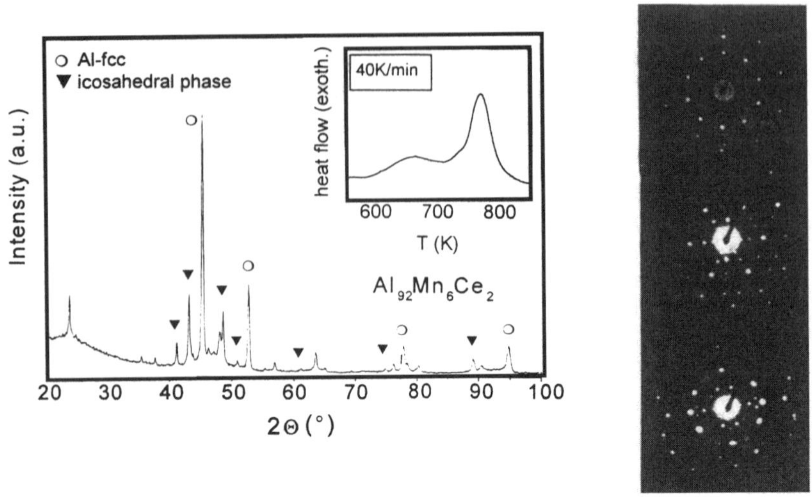

Figure 5. X-ray diffraction pattern of a melt-spun $Al_{92}Mn_6Ce_2$ alloy. The selected-area electron diffraction patterns confirm the formation of an icosahedral phase. The DSC trace (inset) reveals a transition of the metastable icosahedral phase starting at about 720 K.[116]

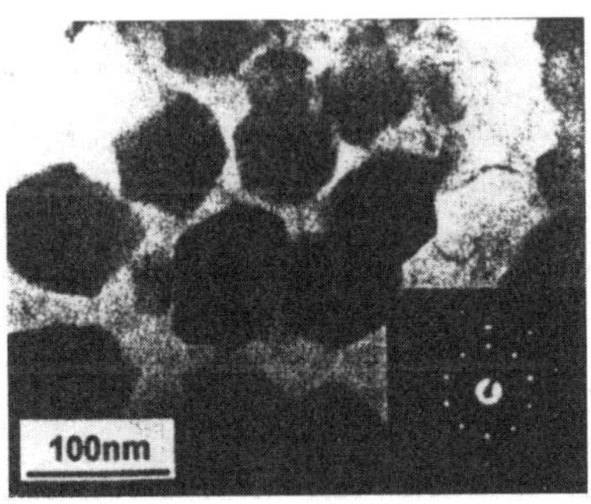

Figure 6. Brightfield TEM micrograph of a melt-spun $Al_{92}Mn_6Ce_2$ alloy and selected-area electron diffraction pattern taken from one of the particles.[117]

In addition to melt-spun ribbons, a mixed microstructure of nano-scale I-phase surrounded by an Al phase is also obtained for atomized Al-Mn-Ln and Al-Cr-Ln-TM powders, as well as in Al-Mn-TM and Al-Cr-TM systems.[98] The microstructural scale of the different phases is slightly larger, i.e., 50–200 nm, than in the case of melt-spun ribbons. In addition, some intermetallic phases, such as Θ'-$Al_{13}Cr_2$ and similar compounds, may also be present. Both the slightly coarser microstructure, as well as the appearance of additional compounds, is due to the lower quenching rate achievable by atomization compared to melt spinning. Extruding the atomized powders at temperatures below 673 K, i.e., below the decomposition temperature of the icosahedral phase, renders fully dense bulk specimens which maintain the mixed microstructure of the initial powder and show no extensive growth of the quasicrystalline domains or the fcc Al phase.

Nanogranular Two-Phase Alloys. In addition to the Al-Mn/Cr- and Al-Fe-based alloys, amorphous and quasicrystalline phases also form in melt-spun Al-V- and Al-Ti-based systems containing solute elements from the groups IV and V of the periodic table. For example, melt-spun $Al_{94}V_4Fe_2$ consists of nanogranular amorphous grains with a size of about 10 nm in diameter coexisting with fcc Al grains with a size of about 7 nm.[119] The nanoscale amorphous phase crystallizes through two stages upon heating to elevated temperatures, i.e., first transforms into an icosahedral phase, which in a second step crystallizes into an $Al_{11}V$ compound. A similar nanogranular amorphous microstructure is observed for $Al_{95}Ti_3Fe_2$.[120] In contrast, melt-spun $Al_{93}Ti_5Fe_2$ with a slightly different composition consists of fcc Al grains with a size of about 30–40 nm, which are surrounded by an amorphous phase network with a thickness of about 7–10 nm.[120] This indicates that the Al phase precipitates as primary phase and the amorphous network phase subsequently forms from the remaining liquid. This change in solidification mode causes the disappearance of the nanogranular amorphous phase observed for larger Al contents, revealing that already small changes in composition may have a rather drastic effect on the solidification behavior and the resulting microstructure of the alloys.

Mg-Based Alloys. From the interest of developing high strength materials with light weight, Mg-based alloys have attracted a lot of attention because of the low density of Mg. Since amorphization of metallic alloys by rapid quenching causes a significant increase in mechanical strength compared with the corresponding crystalline alloys, the formation of amorphous Mg-based alloys as new structural materials with higher specific strength has been investigated extensively through recent years. This has led to the development of new Mg-based alloys with high tensile strength in a number of alloy systems, such as Mg-Ln-TM (TM = Ni, Cu, or Zn),[121][122] Mg-Y-Ln,[123] Mg-Y-Al,[123] Mg-Ca-Al,[124] Mg-Zn-Al,[125] and Mg-Al-Ga[126] ternary systems. These alloy systems can be divided into two types: metal-metal systems and metal-metalloid systems. Table 1 summarizes the various alloy systems in which an amorphous phase containing more than 50 at% Mg can be obtained by single-roller melt spinning. In particular, the alloys of the Mg-Ln-TM group are of importance because of their high tensile fracture strength (see Sec. 4.2). In addition, the Mg-Ca-Al, Mg-Y-Al and Mg-Y-Ln systems are also attractive for achieving a high corrosion resistance.[125] In general, glass formation in these ternary Mg-based alloy systems is observed over a rather broad range of compositions. This is exemplified in Fig. 7 for the Mg-Ni-Y and Mg-Cu-Y systems.[127] An amorphous phase forms in almost the whole composition range when the solute concentration is above 12 at% in both alloy systems. Similar data are observed for other ternary Mg-based alloys.[127]

Glass Transition and Crystallization in Mg-Based Alloy Systems. The crystallization temperatures of the Mg-based amorphous alloys are in the range of 440–600 K. There is a clear tendency for the crystallization temperature, T_x, to increase with increasing solute content,[125] and T_x is rather insensitive to the actual solute element introduced into the amorphous phase. However, only the Mg-Ln-TM amorphous alloys exhibit a distinct glass transition and a rather wide supercooled liquid region before crystallization as well as a large glass-forming ability.[128] Figure 8 shows differential scanning calorimetry (DSC) traces for Mg-Ni-Y and Mg-Cu-Y amorphous alloys as typical examples for the thermal stability behavior of such alloys.[128] The temperature interval, ΔT_x, between the glass transition temperature, T_g, and the onset of crystallization starting at T_x ($\Delta T_x = T_x - T_g$), i.e., the extension of the supercooled liquid region, is as large as 69 K for $Mg_{65}Cu_{25}Y_{10}$ (at a heating rate of 40 K/min). The glass transition phenomenon is observed for a variety of compositions, e.g., from 3–5 at% Y and 5–50 at% Cu.[127] Similar data was found for the Mg-Ni-Y system.[127] The supercooled liquid crystallizes through one exothermic crystallization event, leading to the simultaneous formation of nanoscale Mg_2Cu and $Mg_{24}Y_5$ precipitates.[84][123][129][130]

Table 1. Mg-Based Amorphous Alloys Produced by Melt Spinning[125]

Metal-Metal Systems	Metal-Metalloid Systems
Mg-Ca-Al, Mg-Ca-Li, Mg-Ca-M, Mg-Sr-M, Mg-Al-Ln, Mg-Al-Zn, Mg-Ni-Ln, Mg-Cu-Ln, Mg-Zn-Ln	Mg-Ca-Si, Mg-Ca-Ge, Mg-Ni-Si, Mg-Ni-Ge, Mg-Cu-Si, Mg-Cu-Ge, Mg-Zn-Si, Mg-Zn-Ge

Ln = Lanthanide Metal; M = Transition Metal (Ni, Cu, Zn)

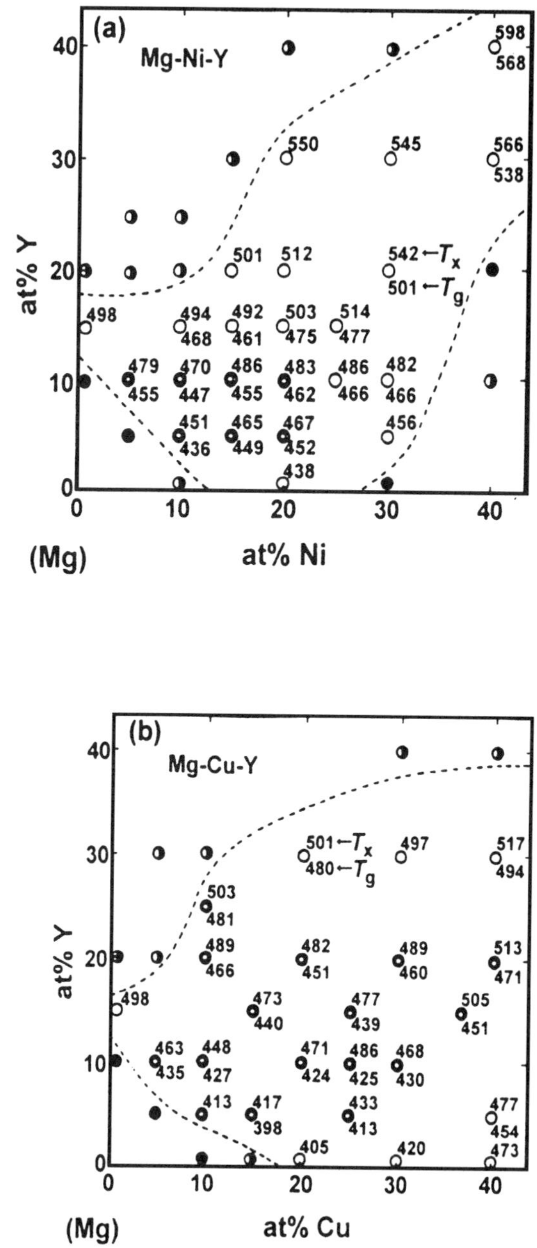

Figure 7. Composition ranges in which an amorphous phase is formed by melt spinning for (*a*) Mg-Ni-Y and (*b*) Mg-Cu-Y systems.[127]

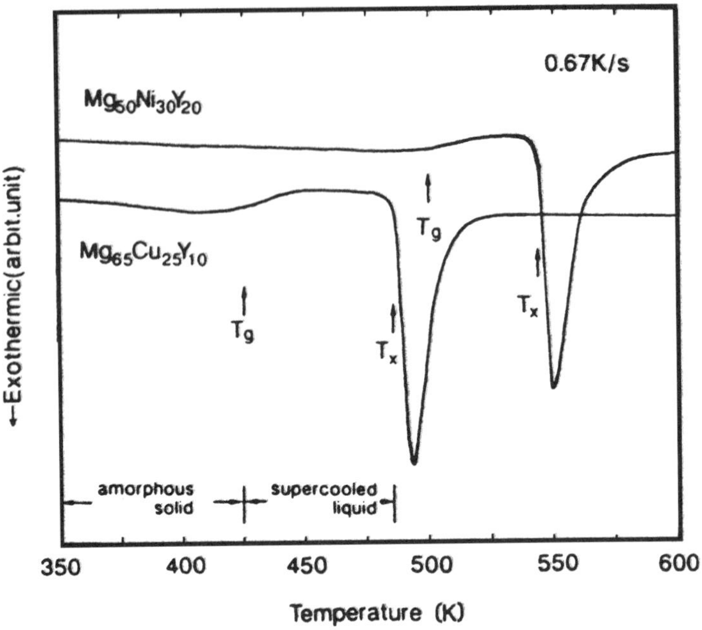

Figure 8. Typical DSC traces of Mg-Ni-Y and Mg-Cu-Y amorphous alloys.[125]

Mg-rich alloys with more than 80 at% Mg crystallize through several stages upon heating to elevated temperatures.[123][131][132] The first exothermic reaction is due to precipitation of hcp Mg nanocrystals, coexisting with a residual amorphous matrix/supercooled liquid with changed composition. Subsequent crystallization proceeds by decomposition of the residual amorphous phase into intermetallic compounds. The volume fraction of the precipitates can be controlled by adjusting the annealing temperature and time properly. Hence, similar to Al-based alloys, fine-scale mixed structures consisting of homogeneously dispersed nanoscale hcp Mg precipitates in an amorphous matrix can be obtained by partial devitrification of the amorphous phase.

Bulk Nanostructured Samples by Extrusion. Melt-spinning or gas-atomization combined with extrusion at elevated temperatures in the supercooled liquid state or near the onset of crystallization can also be applied to Mg-based powders, similar to Al-based amorphous alloys (see above). Since Mg-based powders are extremely reactive, this requires very stringent control of the oxygen and moisture contents in the atmosphere during atomization, and also during consolidation of the resulting fine powders. Hence, the production of bulk Mg-based alloys from atomized powders is typically carried out in a closed-loop processing system, where melting of the master alloy, atomization, powder collection, sieving, as well as precompaction and sealing into Al or Cu cans for subsequent extrusion are conducted in high-purity controlled argon with less than 1 ppm oxygen in a single chamber.[133] Using this type of equipment, Mg-rich atomized powders (e.g., $Mg_{85}Y_{10}Cu_5$) were successfully produced for different alloy systems, such as Mg-Cu-Y, Mg-Ni-Y, Mg-Ca-Al, Mg-Al-Zn, etc.[125] The atomized powders exhibit a spherical morphology and the average particle size is about 20 µm. After atomization, the powders consist mainly of an amorphous phase. Besides, they contain a small amount of a fine hcp Mg-phase,[133] which can not be suppressed due to the lower cooling rate upon atomization compared to melt-spinning. The typical grain size of the hcp Mg particles is on the order of 10 nm and their interparticle spacing is also of about the same scale.[132] The thermal stability data and the features of the DSC scan for the atomized powders are almost the same as for single-phase amorphous melt-spun ribbons. For Mg-Cu-Y alloys, extruding at temperatures between about 460 and 630 K[134] yields dense bulk compacts without significant structural transformation,[133] i.e., the bulk alloy also consists of an amorphous matrix phase with embedded hcp Mg particles.

Using higher extrusion temperatures leads to the formation of nanostructured phase mixtures of hcp Mg and intermetallic compounds, e.g., $Mg_{24}Y_5$ and Mg_2Cu phases, as found for extruded bulk $Mg_{85}Y_{10}Cu_5$.[133] TEM investigations revealed a microstructure consisting of Mg grains with diameters of about 150–200 nm, in which $Mg_{24}Y_5$ particles are homogeneously distributed. The Mg_2Cu precipitates mainly reside at the grain boundaries. A similar fine mixed structure was obtained in a bulk $Mg_{70}Ca_{10}Al_{20}$ alloy produced by extrusion at 673 K.[124] The extruded alloy exhibits a hcp matrix with a grain size of about 100 nm, and the particle size and the interparticle spacing of the Al_2Ca compound are about 80 nm and 50 nm, respectively. Even larger grain and particle sizes were found for a Mg-8.3 wt% Al- 8.1 wt% Ga alloy, where the average grain size of the hcp Mg matrix is about 600 nm and that of the intermetallic $Mg_{17}Al_{12}$ and Mg_5Ga_2 compounds ranges from 100 to 300 nm.[126][135] The Mg grains are almost equiaxed and the intermetallic compounds are evenly distributed throughout the grains and at grain boundaries, yielding a rather homogeneous microstructure after extrusion without any significant texture.

Zr-Based Alloys. Multicomponent Zr-based alloys (e.g., Zr-Al-Cu-Ni or Zr-Ti-Al-Cu-Ni)[136]–[139] exhibit exceptional glass forming ability and belong to the best glass-forming systems known so far. They form a glass at cooling rates of under 100 K/s with critical casting thicknesses ranging upward toward several centimeters thickness.[20] This is illustrated in Fig. 9, showing the critical cooling rate for glass formation, R_c, and the maximum sample thickness for glass formation, t_{max}, as a function of the reduced glass transition temperature, $T_{rg} = T_g/T_m$ (T_m is the melting temperature of the alloy). However, the glass forming ability and the maximum attainable sample thickness strongly depend on a suitable alloy composition and the requirement of high purity starting materials and processing conditions. In particular, the overall oxygen content of the alloy is a key parameter for bulk glass formation. For example, Lin, et al.,[138] studied the effect of oxygen on the crystal nucleation and growth for a Zr-Ti-Cu-Ni-Al alloy upon cooling from the melt. Over the range of oxygen contents studied (300 to 5000 at. ppm), the time-temperature-transformation curves vary roughly by two orders of magnitude along the time axis. In other words, oxygen contamination ranging up to 0.5 at% can increase the necessary critical cooling rate for glass formation by two orders of magnitude. Similar behavior has been found for other Zr-based alloys, such as quaternary Zr-Al-Cu-Ni alloys.[140] This effect is related to oxygen-induced formation of metastable phases which, in turn, can act as heterogeneous nucleation sites for further crystallization upon solidification from the melt. This, of course, strongly affects bulk glass formation and limits the production of bulk specimens to high-purity material and optimum compositions with a large resistance against oxygen-induced crystal nucleation.

To circumvent the limitations of casting processes at low cooling rates, there have also been efforts to produce bulk samples through consolidation of rapidly quenched ribbons or atomized powders,[141]–[143] since the nucleation of the competing metastable crystalline phases can be suppressed by fast enough cooling from the melt. Choosing appropriate consolidation parameters yields dense bulk samples without crystallization and almost identical thermal stability as that of melt-spun ribbons. In particular, the material retains an extended supercooled liquid region before crystallization, which can be used for further viscous flow deformation for shaping of three-dimensional parts.

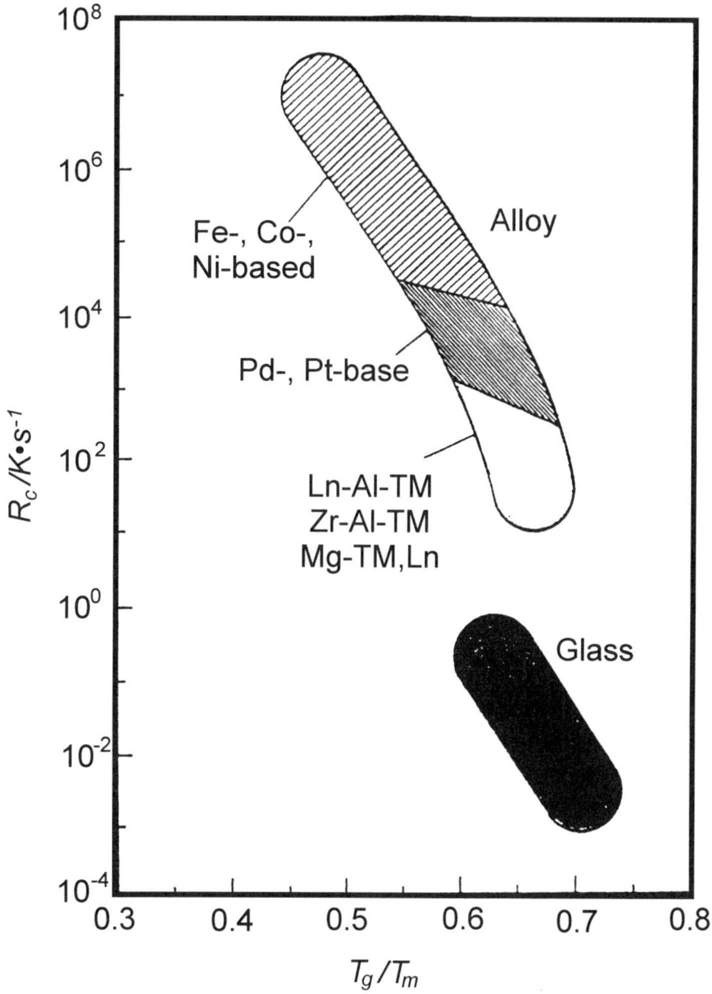

Figure 9. Relation between the critical cooling rate for glass formation, R_c, the maximum sample thickness, t_{max}, and the reduced glass transition temperature, T_g/T_m, for amorphous alloys.[127]

Oxygen-Induced Precipitation of Nanoscale Compound Particles upon Crystallization. Oxygen impurities not only affect the solidification of multicomponent Zr-based alloys upon quenching from the melt, but also strongly influence the crystallization behavior during heating to elevated temperatures or upon isothermal annealing. This has been studied in detail for $(Zr_{0.65}Al_{0.075}Cu_{0.175}Ni_{0.10})_{100-x}O_x$ rapidly quenched amorphous ribbons (x = 0.2, 0.4 and 0.8).[144] X-ray diffraction and TEM proved that all the samples are fully amorphous after melt-spinning. The DSC scans of the ribbons show an endothermic heat event characteristic of the glass transition, followed by exothermic heat release events due to crystallization (Fig. 10). For example, amorphous $(Zr_{0.65}Al_{0.075}Cu_{0.175}Ni_{0.10})_{99.8}O_{0.2}$ exhibits a wide supercooled liquid region of more than 100 K and crystallizes via one sharp exothermic peak. Most prominent is the change of the DSC signal from one sharp exothermic crystallization peak for x = 0.2 to two or three more or less well resolved crystallization peaks for samples with higher oxygen content. This indicates a successive stepwise transformation from the supercooled liquid to the equilibrium crystalline intermetallic compounds at different temperatures. The glass transition temperature, T_g, slightly increases and the onset of crystallization, T_x, decreases with increasing oxygen content, thus, leading to a reduction of the extension of the supercooled liquid region, ΔT_x, for oxygen-rich samples.

The phases formed upon devitrification of the $(Zr_{0.65}Al_{0.075}Cu_{0.175}Ni_{0.10})_{100-x}O_x$ amorphous alloys were studied by time and temperature resolved XRD and TEM measurements.[144] Figure 11 shows schematic phase formation diagrams for the $(Zr_{0.65}Al_{0.075}Cu_{0.175}Ni_{0.10})_{100-x}O_x$ ribbons with different oxygen content. The details of phase formation strongly depend on the annealing conditions. For x = 0.2, crystallization is governed by simultaneous precipitation of quasicrystalline, tetragonal $CuZr_2$, and hexagonal Zr_6NiAl_2 phases, which are embedded in the residual amorphous matrix (Fig. 11a). A metastable fcc $NiZr_2$-type phase forms as an intermediate crystallization product. At elevated temperatures, the quasicrystals transform into $CuZr_2$, and the fcc phase transforms into the stable Zr_6NiAl_2 compound. The grain sizes of the different phases vary between about 10 and 70 nm, depending on the actual annealing conditions. Further details are given in Ref. 144.

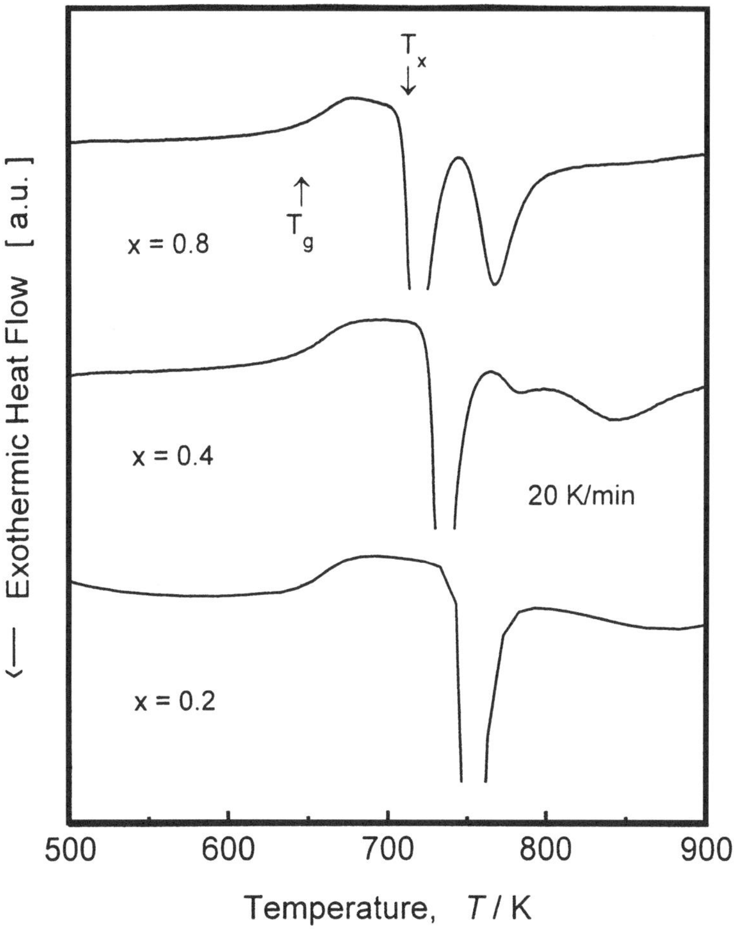

Figure 10. DSC traces of amorphous $(Zr_{0.65}Al_{0.075}Cu_{0.175}Ni_{0.10})_{100-x}O_x$ melt-spun ribbons (x = 0.2, 0.4 and 0.8).[144]

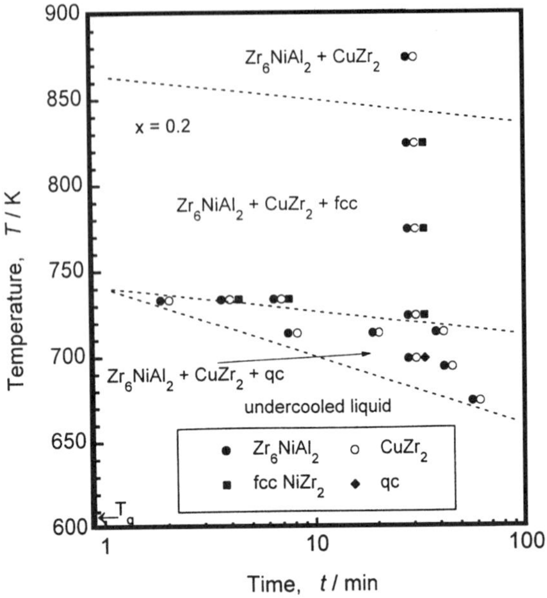

(a)

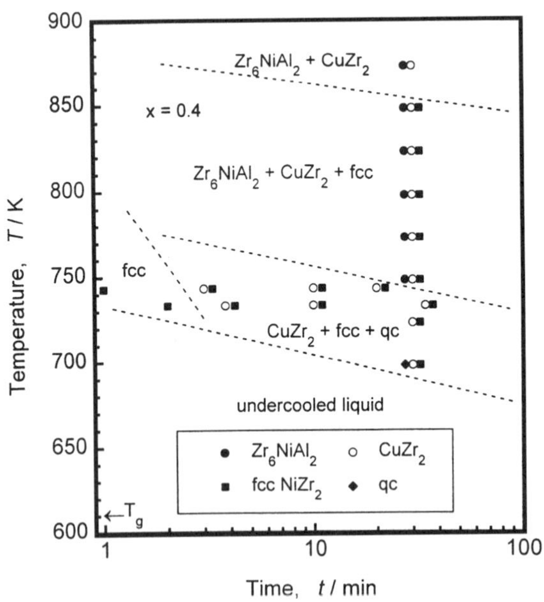

(b)

Figure 11. Schematic phase formation diagram for $(Zr_{0.65}Al_{0.075}Cu_{0.175}Ni_{0.10})_{100-x}O_x$ ribbons with different oxygen content: (*a*) x = 0.2, (*b*) x = 0.4 and (*c*) x = 0.8. The dashed lines are only guidelines for the eye illustrating the temperature-time regimes where the different phases were detected.[144]

(c)

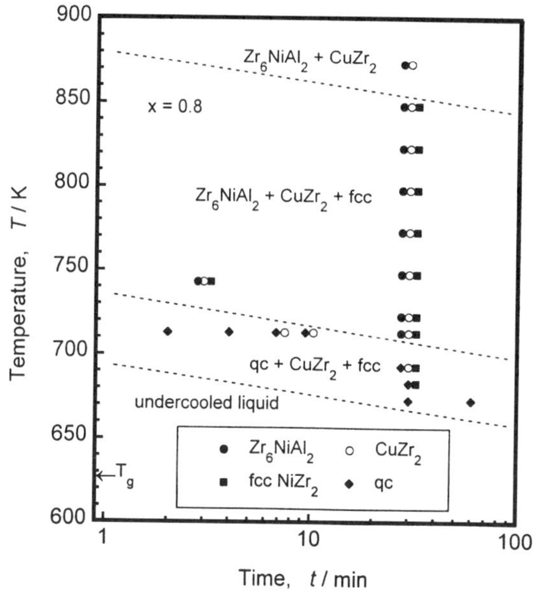

Figure 11. *(Cont'd.)*

Sequential transformations from the supercooled liquid to the intermetallic compounds are also observed for $(Zr_{0.65}Al_{0.075}Cu_{0.175}Ni_{0.10})_{99.6}O_{0.4}$ and $(Zr_{0.65}Al_{0.075}Cu_{0.175}Ni_{0.10})_{99.2}O_{0.8}$ (Figs. 11b and 11c). With increasing oxygen content, the existence regions of the metastable fcc and quasicrystalline phases become more extended. This reveals that the presence of oxygen promotes the formation of these intermediate phases and stabilizes the metastable structures against transformation into the equilibrium compounds. Hence, the phase formation and stability regions of the oxygen-stabilized metastable phases depend systematically on the oxygen content of the material and the actual annealing conditions. Therefore, the alloy, in fact, has to be viewed as a 5-component system in which the dilute component oxygen plays a decisive role for the characteristics of crystallization.

Similar results were obtained by different groups for multicomponent Zr-based alloys of different composition.[138][145]–[147] On one hand, the effect of oxygen lowers the thermal stability of the material against crystallization and is, therefore, detrimental. On the other hand, this leads to phases and microstructures which cannot be achieved in the pure metallic glass, and allows to synthesize amorphous alloys with nanoscale compound particles in an amorphous matrix by proper annealing. Therefore, oxygen or other trace elements are useful for obtaining new nanostructured composites from amorphous Zr-based alloys.

Nanoscale Compound Formation in Zr-Based Alloys with Different Composition. The effect of successive stepwise transformation from the supercooled liquid to the crystalline equilibrium phases is not limited to the influence of trace elements such as oxygen. One example is the multicomponent alloy system $Zr_{60}Al_{10}Cu_{30-x}Pd_x$.[148][149] The crystallization of the alloy with x = 10 occurs also through two stages: amorphous → Zr_2(Cu, Pd) + amorphous′ → Zr_2(Cu, Pd) + Zr_2Al. For this alloy, annealing for 1 hour at 726 K between T_g and T_x induces precipitation of a large volume fraction (about 70%) of Zr_2(Cu, Pd) nanocrystals with a diameter of about 5–10 nm, which are embedded in a remaining amorphous matrix with changed composition, occupying a volume fraction of about 30%. Similar results were found for a variety of Zr-based alloys with different alloying elements, e.g., Zr-Al-Cu-Ni-Fe,[150][151] Zr-Al-Cu-Pd-Fe,[149] Zr-Al-Ti-Cu-Ni,[139][152][153] Zr-Al-Cu-Ni-Pd,[154] Zr-Al-Ni-Cu-M (M = Ag, Au, Pt).[155] In all cases, a sequential crystallization process was observed, involving precipitation of a nanoscale intermetallic or quasicrystalline phase in the first step of transformation, which coexists with an amorphous phase with changed composition. At higher temperatures, the residual amorphous phase also crystallizes and the material becomes fully nanocrystalline.

The details of precipitation and transformation strongly depend on the actual composition and the annealing conditions.[151] As an example, Fig. 12 shows the average crystallite size, *D*, of the primary nanocrystals with metastable fcc $NiTi_2$ structure precipitated during the first crystallization step of $(Zr_{0.65}Al_{0.075}Cu_{0.175}Ni_{0.10})_{100-x}Fe_x$ alloys. The average crystallite size fits well to $\sqrt[3]{x}$ functional behavior, indicating that the number of fcc crystallites, *N*, is proportional to the iron content in the alloy ($N \sim x$). Figure 13 shows the behavior of the average crystallite size versus annealing time for a $(Zr_{0.65}Al_{0.075}Cu_{0.175}Ni_{0.10})_{80}Fe_{20}$ alloy for the annealing temperature $T_A = 748$ K. The crystallites grow to a saturation value of only about 4 nm upon annealing for up to 16 hours. For longer annealing times, the fcc phase starts to transform into the stable equilibrium phases.[151] These results prove that the composition of the alloy, which is directly linked to the nucleation and growth characteristics of nanocrystal formation, determines the nature and the size of the primary crystals precipitating from the amorphous phase. Hence, optimizing the stoichiometry and the annealing conditions enables synthesis of nanostructured two-phase Zr-based materials with extremely fine nanometer-scale precipitates in a residual amorphous matrix.

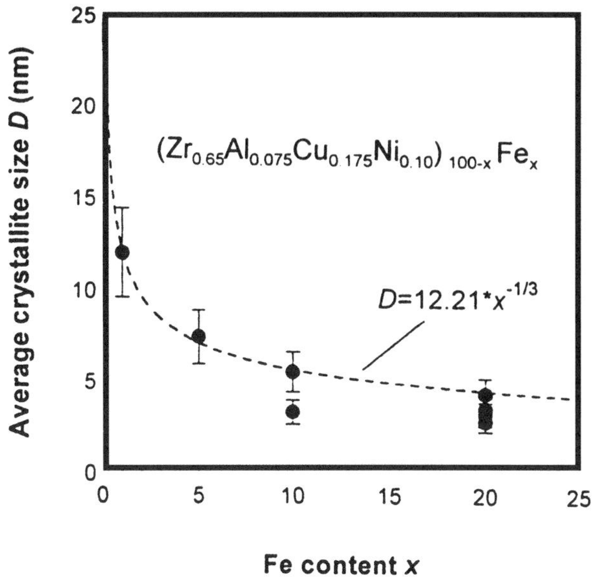

Figure 12. Average crystallite size, *D*, of the primary fcc nanocrystals formed upon annealing of amorphous $(Zr_{0.65}Al_{0.075}Cu_{0.175}Ni_{0.10})_{100-x}Fe_x$ alloys at 773 K for 30 min vs Fe content *x*.

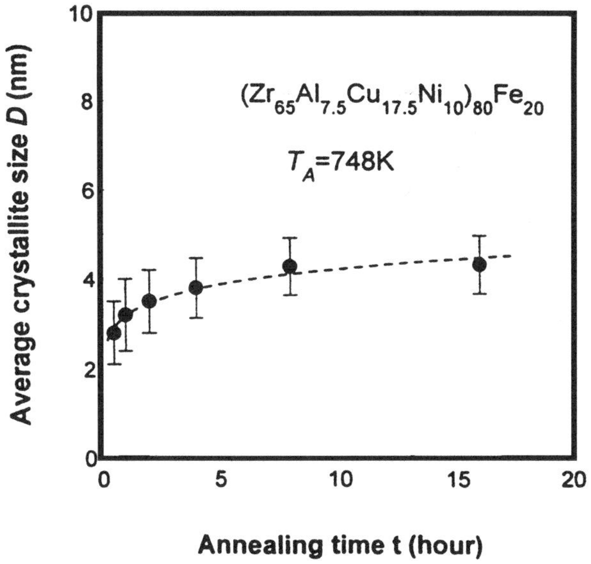

Figure 13. Average crystallite size, *D*, of the primary fcc nanocrystals formed upon annealing of amorphous $(Zr_{0.65}Al_{0.075}Cu_{0.175}Ni_{0.10})_{80}Fe_{20}$ at 748 K vs annealing time.

3.2 Conventional Solidification and Devitrification of Bulk Samples

As described above, rapid solidification techniques, in some cases combined with proper annealing conditions, enable the preparation of nanoscale multiphase Al-, Mg- and Zr-based alloys. Combined with consolidation at elevated temperatures, this can yield bulk specimens with dimensions in the millimeter or centimeter range. However, in most cases, phase transitions and microstructural coarsening during consolidation cannot be completely suppressed, which may deteriorate the desired mechanical properties of the material compared to the as-quenched state. Moreover, the consolidation of rapidly quenched ribbons or powders involves several processing steps, which have to be carefully optimized and carried out under high purity conditions. Hence, alternative processing techniques based on conventional casting techniques, such as die casting, copper mold casting, suction or squeeze casting, are highly desirable for large scale production of such materials. Due to the lower cooling rates of these casting techniques, they may not be applicable for multiphase nanostructure formation in all cases, but have to be combined with additional annealing treatments to produce a nanoscale microstructure, e.g., through crystallization of a bulk amorphous precursor obtained during slow cooling from the melt. In the following, selected examples for bulk multiphase microstructures in different alloy systems will be reviewed, which were prepared through slow cooling methods.

Al-Based Alloys Containing Quasicrystalline Particles. Bulk Al-based alloys containing a dual-phase microstructure of quasicrystalline particles embedded in an fcc Al matrix can be produced by squeeze casting the melt into a water-cooled copper mold.[156] This allows one to obtain bulk rod-shape samples of several millimeters diameter and about 50 mm in length (this geometry is determined by the copper mold used). Although the attainable quenching rate depends on the size of the sample and is limited to some 100 Ks^{-1} and, therefore, is more than three orders of magnitude smaller than the typical cooling rates achievable upon melt spinning or atomization, this quenching rate is sufficient for quasicrystal phase formation in Al-Mn-Ce and Al-Mn-Fe alloys.[156] Figure 14 shows an x-ray diffraction pattern and the corresponding DSC scan for an $Al_{92}Mn_6Ce_2$ sample as a typical example, revealing the coexistence of an icosahedral phase, fcc Al and a small volume fraction of the competing Al_6Mn intermetallic equilibrium phase. The DSC scan, recorded at a heating rate of 40 $Kmin^{-1}$, reveals the decomposition of the metastable I-phase at around 750 K. Both the XRD pattern as well as the thermal stability data are very similar to the results for melt-spun ribbons of the same composition (Fig. 5).

Figure 15 shows a scanning electron microscopy (SEM) image taken from the cross-section of the squeeze cast rod, revealing a homogeneous distribution of globular-shaped quasicrvstals with sizes between 0.5 and 5 µm embedded in the Al matrix. This two-phase structure is coarser than the nanoscale microstructure observed in melt-spun ribbon samples. This is due to the lower cooling rate achieved upon squeeze casting, which is even lower than that achievable upon atomization which, in turn, changes the solidification behavior with respect to nucleation and growth of the different phases. This also leads to a gradient of the microstructural scale originating from slightly different cooling conditions in the center of the rod and at the outer surface layer. Although this type of slowly cast bulk multiphase material does not exhibit a truly nanoscale microstructure, the dimensions of the different phases and their morphology are quite similar to the microstructure obtained for extruded bulk specimens prepared from atomized powders,[98][157]–[159] i.e., it is also characterized mainly by a rather inhomogeneous microstructure consisting of submicrometer- to micrometer-sized quasicrystalline and intermetallic particles embedded in an Al matrix.

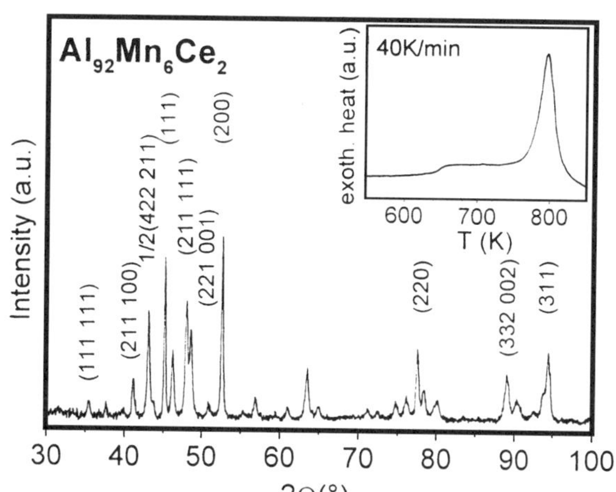

Figure 14. X-ray diffraction pattern of a squeeze-cast $Al_{92}Mn_6Ce_2$ bulk sample, revealing the coexistence of an icosahedral phase, fcc Al and a small volume fraction of the competing Al_6Mn equilibrium compound. The corresponding DSC trace shows the decomposition of the metastable icosahedral phase around 750 K.

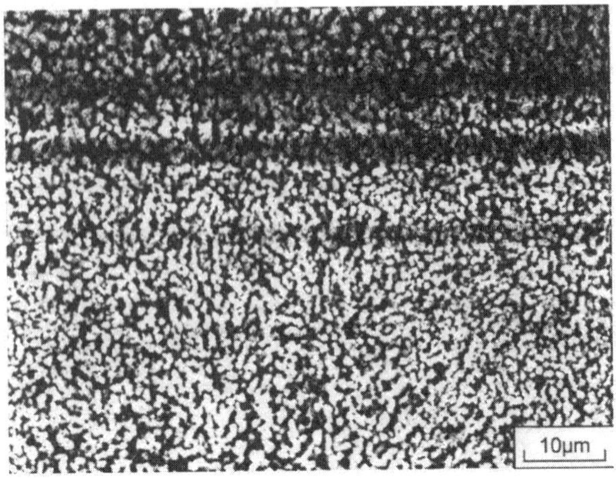

Figure 15. SEM micrograph of the cross-section of a squeeze-cast $Al_{92}Mn_6Ce_2$ bulk sample, showing homogeneously distributed quasicrystalline particles in the fcc Al matrix.

Mg-Based Nanostructured Alloys From Bulk Metallic Glasses. It was shown in Sec. 3.1 that the family of Mg-based ternary alloys exhibits a high glass-forming ability upon quenching from the liquid state. Moreover, the Mg-Ln-TM amorphous alloys show a distinct supercooled liquid region before crystallization, indicating that the phase transition into a crystalline phase or phase mixture is rather difficult. This result also implies that the supercooled liquid obtained by quenching the melt from temperatures above the melting point, T_m, also has a high resistance against nucleation and growth of crystalline phases upon cooling to room temperature. While systematically studying rapid solidification in Mg-based alloys, Inoue, et al.,[20][131][160] developed Mg-based alloys which can be cast into fully glassy rods or bars with cast thicknesses of several millimeters. Figure 16 shows the compositional dependence of the maximum diameter for formation of an amorphous phase in $Mg_{90-x}Cu_xY_{10}$ cylinders prepared by a high-pressure die casting method.[131] The maximum diameter, d, increases with increasing Cu content from about 3 mm for $Mg_{80}Cu_{10}Y_{10}$ to 7 mm for $Mg_{65}Cu_{25}Y_{10}$. A further increase in Cu content causes a decrease of the maximum achievable diameter. The significant compositional dependence of d reflects the stability of the supercooled liquid against crystallization and reveals that the ability of bulk glass formation upon casting is closely correlated with the optimum composition of the glass close to the deep eutectics of the phase diagram.[21]

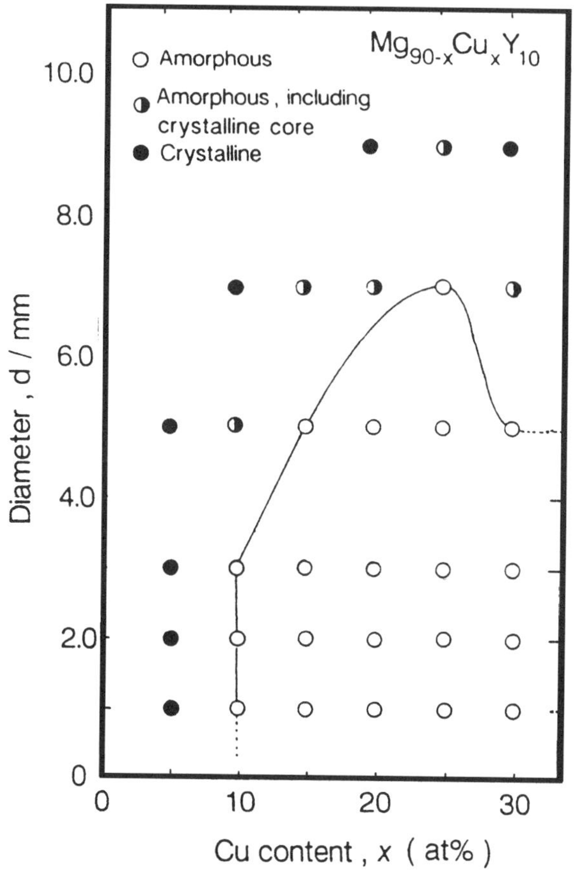

Figure 16. Maximum diameter for the formation of an amorphous phase for cast $Mg_{90-x}Cu_xY_{10}$ cylinders.[131]

The thermal stability data of the cast bulk glasses reveal no appreciable difference between the bulk samples and rapidly-quenched melt-spun ribbons, as long as the bulk samples are fully amorphous.[20][160] Hence, partial or complete devitrification of the cast specimens by annealing at temperatures in the supercooled liquid region can be used to produce nanostructured bulk specimens. There is evidence that the as-cast bulk specimens undergo phase separation upon quenching from the melt, at least for certain alloy compositions.[84][130] Figure 17 shows a high-resolution TEM micrograph of a Mg-Y-Cu-Li alloy obtained by Liu, et al.,[130] The micrograph gives evidence of phase-separation into Cu-rich and Cu-poor domains, which was furthercorroborated by small-angle x-ray scattering

(SAXS) and anomalous small-angle x-ray scattering (ASAXS) studies.[85] These phase-separated domains appear to be preferred sites for nucleation of bcc Mg_6Li nanocrystals.[85][86] Hence, both preceding phase separation in the supercooled liquid, perhaps already during cooling from the melt or after annealing, and the nucleation and growth characteristics of the competing crystalline phases seem to contribute to the microstructure evolution of bulk nanostructured materials prepared from bulk glassy specimens.

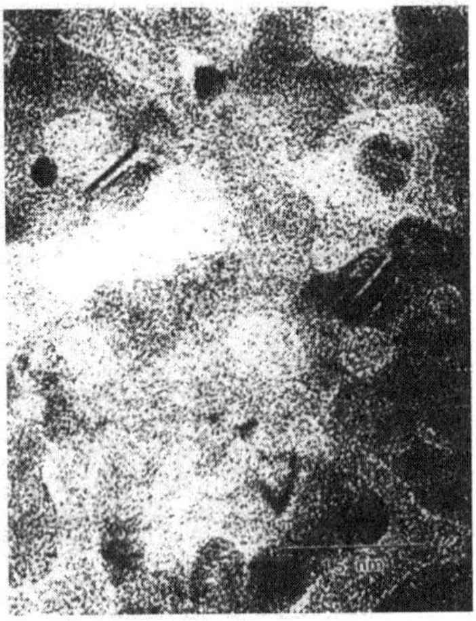

Figure 17. High-resolution brightfield TEM micrograph of a cast $Mg_{62}Li_3Cu_{25}Y_{10}$ alloy.[131]

Zr-Based Bulk Alloys with Nanoscale Precipitates. The high glass forming ability of multicomponent Zr-based alloys can be used for the production of completely glassy bulk specimens with dimensions in the millimeter to centimeter range, as well as be exploited for the formation of bulk nanostructured materials. For this, the as-cast glassy specimens are annealed at temperatures within the supercooled liquid region or close to the onset of crystallization. This procedure is based on the results first obtained for rapidly quenched thin ribbons, where sequential crystallization was observed for a variety of Zr-based bulk glass forming alloys such as Zr-Cu-Al-Ni-Ti,[139][152][153] Zr-Al-Cu-Ni-Pd,[154] Zr-Al-Ni-Cu-M (M = Ag, Au, Pt),[155] etc. This stepwise devitrification behavior leads to primary precipitation of nanoscale intermetallic or quasicrystalline phases from the supercooled liquid, which are embedded in an amorphous phase with changed composition.

Figure 18 shows the DSC traces for $Zr_{55}Cu_{30}Al_{10}Ni_5$ and $Zr_{57}Cu_{20}Al_{10}Ni_8Ti_5$ fully amorphous rods with 5 mm diameter[161] as typical examples for bulk glass-forming Zr-based alloys. Both alloys exhibit an extended supercooled liquid region before crystallization. Whereas $Zr_{55}Cu_{30}Al_{10}Ni_5$ appears to crystallize through a single exothermic event, $Zr_{57}Cu_{20}Al_{10}Ni_8Ti_5$ exhibits a stepwise crystallization behavior from the supercooled liquid into crystalline phases, indicating a primary-type of precipitation upon devitrification. This change in crystallization mode is attributed to the Ti addition. Figure 19 displays the DSC traces for $Zr_{62-x}Cu_{20}Al_{10}Ni_8Ti_x$ bulk amorphous cylinders with different Ti content. The alloy without Ti shows just one exothermic peak. When Ti is added, the DSC scans exhibit two exothermic peaks. With increasing Ti content, the first peak shifts to lower temperatures and the enthalpy of the second exothermic peak decreases.[162] Isothermal annealing experiments together with systematic x-ray and TEM investigations revealed,[162] that for alloys with low Ti content ($x < 5$), the first exothermic event is related to primary precipitation of a nanoscale icosahedral quasicrystalline particles (Fig. 20). Hence, similar to Al-based alloys, also in Zr-based multicomponent alloys, bulk nanostructured quasicrystal-based two-phase materials can be produced. Similar results were found recently for other Zr-based 5-component alloy systems.[154][163][164] With increasing Ti content, a metastable fcc-type phase forms besides the quasicrystals (Fig. 20). Moreover, the size of the nanoscale grains precipitated during the first crystallization step decreases with increasing Ti content, yielding extremely fine-grained material with grain sizes of about 2 nm for $x = 10$. At higher temperatures the metastable icosahedral and fcc-type phases successively transform into equilibrium compounds.

Figure 21 shows isothermal DSC traces for the $Zr_{57}Cu_{20}Al_{10}Ni_8Ti_5$ bulk alloy annealed at 673 K (curve *a*) and 703 K (curve *b*), respectively. Three significantly different regions can be distinguished. Initially, there is a short period of incubation without exothermic heat flow, followed by a sharp drop of the curve indicating pronounced exothermic heat flow and rapid crystallization. Finally, a third part with small exothermic heat flow is observed, which reveals the continuing crystallization of the residual amorphous phase with lower crystallization rate. As shown in curve *a*, annealing at 673 K yields an incubation time of about 4 min and a duration of rapid crystallization of about 40 min. Increasing the annealing temperature to 703 K (curve *b* in Fig. 21), enhances diffusion and chemical redistribution, leading to reduced times for incubation (1 min) and rapid crystallization (7 min). This demonstrates that a desired volume fraction of crystallites can be obtained by adjusting the annealing temperature and time for an amorphous alloy of given composition.[152][161][162]

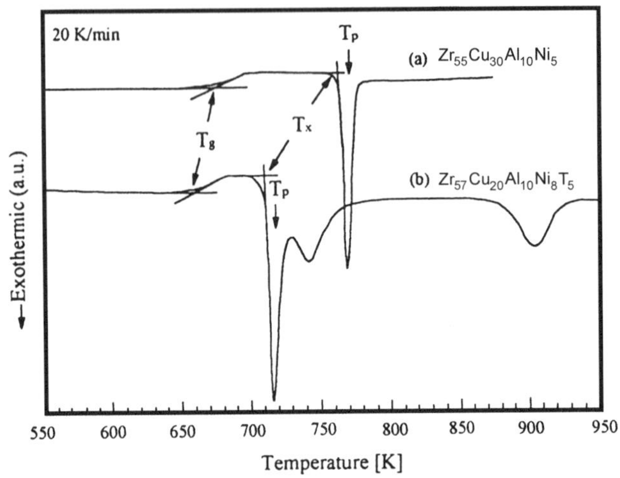

Figure 18. DSC traces of (*a*) amorphous $Zr_{55}Cu_{30}Al_{10}Ni_5$ and (*b*) amorphous $Zr_{57}Cu_{20}Al_{10}Ni_8Ti_5$.[161]

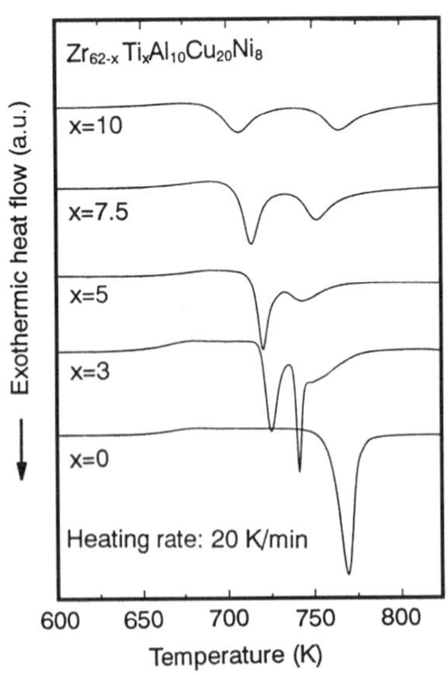

Figure 19. DSC traces of $Zr_{62\text{-}x}Ti_xCu_{20}Al_{10}Ni_8$ amorphous cylinders with different Ti content *x*.

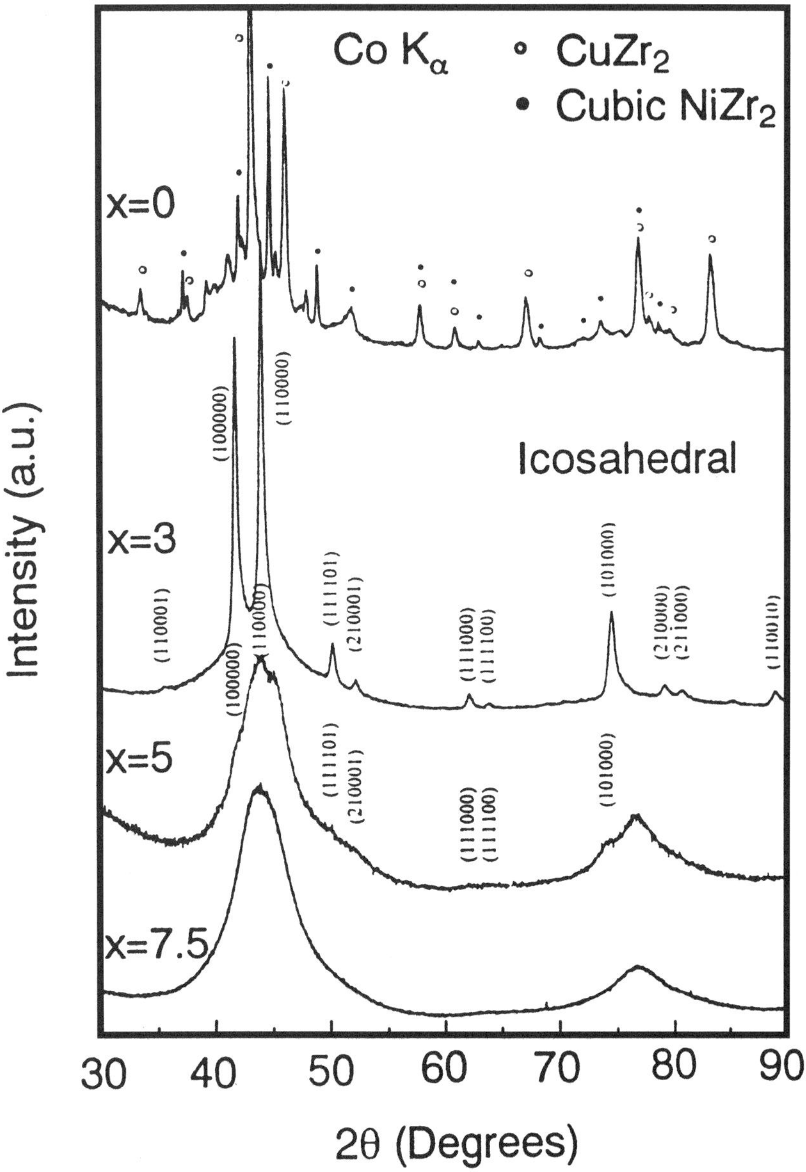

Figure 20. X-ray diffraction patterns of $Zr_{62-x}Ti_xCu_{20}Al_{10}Ni_8$ amorphous cylinders with different Ti content x after primary crystallization.

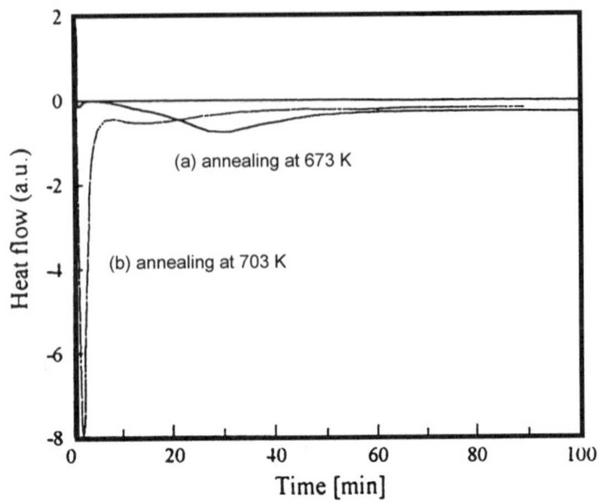

Figure 21. Isothermal DSC curves of amorphous $Zr_{57}Cu_{20}Al_{10}Ni_8Ti_5$ upon annealing at (*a*) 673 K and (*b*) 703 K.[161]

Figure 22 displays a high-resolution brightfield TEM image of a $Zr_{57}Cu_{20}Al_{10}Ni_8Ti_5$ bulk specimen after annealing at 673 K for 40 min as a typical example for the microstructure obtained upon annealing. The annealed sample consists of nanosized grains with sizes between 2 and 10 nm which are embedded in the residual amorphous matrix. Due to the small grain size, the structure of the phases cannot be determined unambiguously. Presumably, both nanoscale quasicrystalline and metastable fcc-type phases coexist for the alloy with x = 5.[162] The volume fraction of nano(quasi)crystals is about 40%, as estimated from the remaining crystallization enthalpy of the partially crystallized sample compared with the total crystallization enthalpy of the fully amorphous as-cast material.

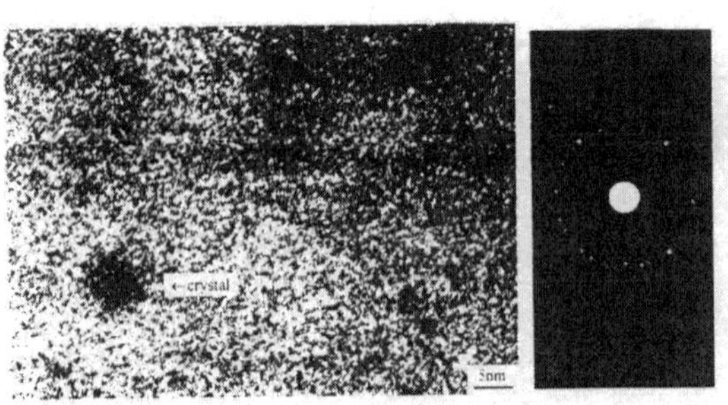

Figure 22. High-resolution brightfield TEM micrograph of amorphous $Zr_{57}Cu_{20}Al_{10}Ni_8Ti_5$ annealed at 673 K for 30 min.[152]

The $Zr_{55}Cu_{30}Al_{10}Ni_5$ alloy, which exhibits just one single exothermic crystallization event, can also be partially crystallized by proper annealing, resulting in nanocrystalline precipitates with particle sizes between 50 and 100 nm which are embedded in the residual amorphous matrix.[161] Curve (I) in Fig. 23a reveals the common broad amorphous x-ray diffraction maxima for the as-cast specimen. In contrast, curve (II) exhibits additional superimposed diffraction peaks due to the crystalline particles embedded in the amorphous matrix. The brightfield TEM image in Fig. 23b shows the corresponding microstructure of the annealed partially crystallized sample (Fig. 23a, curve II). Elliptical crystallites were found with a length of up to 200 nm. Where possible, their aspect ratio (*AR*) was evaluated to yield $AR = 2$. From images such as Fig. 23b the average size of the nanocrystalline precipitates was determined to be ≈ 50 nm. Hence, formation of nanocrystals embedded in a residual amorphous matrix is also observed for this alloy, which appears to crystallize through one single exothermic event and not through a successive stepwise transformation into the equilibrium compounds as in the case of the $Zr_{57}Cu_{20}Al_{10}Ni_8Ti_5$ glassy alloy. However, the size of the precipitating nanocrystals (2–10 nm) is much smaller for the $Zr_{57}Cu_{20}Al_{10}Ni_8Ti_5$ alloy with primary-type crystallization than for the $Zr_{55}Cu_{30}Al_{10}Ni_5$ alloy (50–100 nm), indicating that the Ti-containing 5-component alloy exhibits a higher nucleation rate and a smaller growth rate than the quaternary Zr-Cu-Al-Ni metallic glass. Again, this reveals that the alloy composition and the details of nucleation and growth play a decisive role for the formation of ultrafine-scaled microstructure formation of two-phase nanostructured materials.

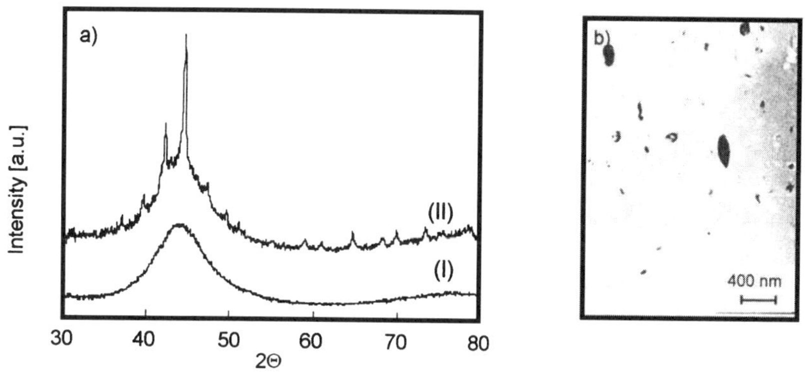

Figure 23. (*a*) X-ray diffraction patterns of (I) amorphous and (II) partially nanocrystallized $Zr_{55}Cu_{30}Al_{10}Ni_5$ (volume fraction of 25% crystallites); (*b*) brightfield TEM micrograph of the annealed $Zr_{55}Cu_{30}Al_{10}Ni_5$ sample showing elliptical nanosize precipitates (cf. curve II).

3.3 Mechanically Attrited Powders

Metastable phase formation and the development of nanostructures in multicomponent alloy systems can also be achieved by solid state processing, e.g., through mechanical alloying of elemental powder mixtures.[35][43][44][116][117][165] Similar to liquid quenched specimens, the mechanically alloyed powders can be subsequently annealed and partially nanocrystallized in order to achieve nanoscale two- or multiphase microstructures,[44][166] which are almost identical to that of Al-, Mg-, or Zr-based melt-spun ribbons or slowly cooled bulk samples.[117][144][166][167] Moreover, mechanical alloying also allows direct synthesizing of nanostructured composites by blending a metallic glass with insoluble metallic or ceramic particles which usually induce heterogeneous crystallization upon cooling from the melt.[168] This is due to the different reaction temperatures and timescales for the phase formation by solid state processing compared to quenching from the melt.[24][34][166]

Al-Based Amorphous or Quasicrystalline Alloys. The work on mechanical alloying of Al-based alloys is triggered by the attempt to explore the possibilities of metastable phase formation in the solid state compared to what is already known from rapid solidification of such alloys, with special focus on the similarities and differences of phase formation under completely different thermodynamic and kinetic conditions. Moreover, mechanical alloying as an alternative method of metastable phase formation, combined with appropriate consolidation techniques, should circumvent the limitations of melt-spinning with respect to the restrictions in geometry and size of the resulting nanostructured multiphase ribbons and, simultaneously, may also enable the production of bulk material without microstructural coarsening as observed for consolidated atomized powders or squeeze-cast bulk samples (see Secs. 3.1 and 3.2).

Al-Rich Amorphous Alloys. The first attempts to synthesize Al-rich high-strength ternary and quaternary amorphous alloys with more than 80 at% Al for compositions close to that known for good glass forming ability upon rapid solidification, were reported by Dougherty, et al.,[43] and Benameur, et al.,[169][170] in 1994. For Al-Ni-Fe-Gd alloys, a fully amorphous phase could only be achieved for a maximum Al content of 80 at%, i.e., for powder corresponding to an overall composition of $Al_{80}Ni_8Fe_4Gd_8$.[43] No complete amorphization could be achieved for larger Al contents, but the powders consist of an amorphous phase, fcc Al phase mixture, most likely also coexisting with some intermetallic compound(s). Nevertheless, the thermal stability of the amorphous phase is comparable to that of melt-spun ribbons. Similar results were found for mechanically alloyed Al-Ni-

Co-(Zr) powders[169][170] as well as for Al-Y-Ni-Co/Fe alloys.[171] These data indicated that the composition range for optimum amorphization is different for rapid quenching and mechanical alloying techniques.

However, more recent experiments[116] employing a systematic variation and optimization of the milling parameters, i.e., using proper milling conditions such as interval-milling at low intensity corresponding to a rather low kinetic energy and resulting ambient milling temperature during milling, revealed that for $Al_{85}Y_8Ni_5Co_2$ complete amorphization can be achieved and no indications for remaining crystalline phases were found by x-ray diffraction and TEM (Fig. 24). From energy-dispersive x-ray analysis the composition of the amorphous phase was proved to agree with the nominal composition within the accuracy of the measurement. This, together with the thermal stability data obtained from DSC measurements (Fig. 25), which are almost identical to that of melt-spun ribbons,[169][172] reveal a good homogeneity of the amorphous phase. The first sharp exothermic DSC peak at about 620 K corresponds to precipitation of fcc Al nanocrystals, whereas the second transition step is related to the formation of intermetallic phases from the remaining amorphous phase. Again, this is almost identical to what is known for the crystallization of melt-spun ribbons.[169][172] This seems to be the first example for complete amorphization of Al-rich multicomponent alloys by mechanical alloying and, combined with proper annealing, promises a further extension of the mechanical alloying technique for the preparation of Al-rich multiphase nanostructured materials.

Nanostructured Quasicrystalline Alloys. Based on previous investigations on quasicrystal formation by mechanical attrition,[35]–[37] Al-rich Al-Mn-Ce and Al-Mn-Fe ternary alloys with more than 85 at% Al and a nanoscale microstructure consisting of nanoscale quasicrystalline particles embedded in an fcc Al matrix were successfully prepared by mechanical alloying of elemental powder mixtures, as well as by ball milling of pre-alloyed ingots.[116][117][156] For example, $Al_{92}Mn_6Ce_2$ or $Al_{91}Mn_7Fe_2$ powders with grain sizes of the I-phase of about 10–30 nm and a large volume fraction of I-phase particles of up to about 70% form upon milling. This yields powders with similar microstructural features and thermal stability as reported for melt-spun samples.[108][114] Compared to amorphization of multicomponent Al-based alloys, higher milling intensities, i.e., a higher kinetic energy during milling, has to be used for quasicrystal formation. This is consistent with previous results for metastable quasicrystal formation by mechanical alloying,[37] and can be understood in terms of the thermodynamic relations between the metastable amorphous and quasicrystalline phases.

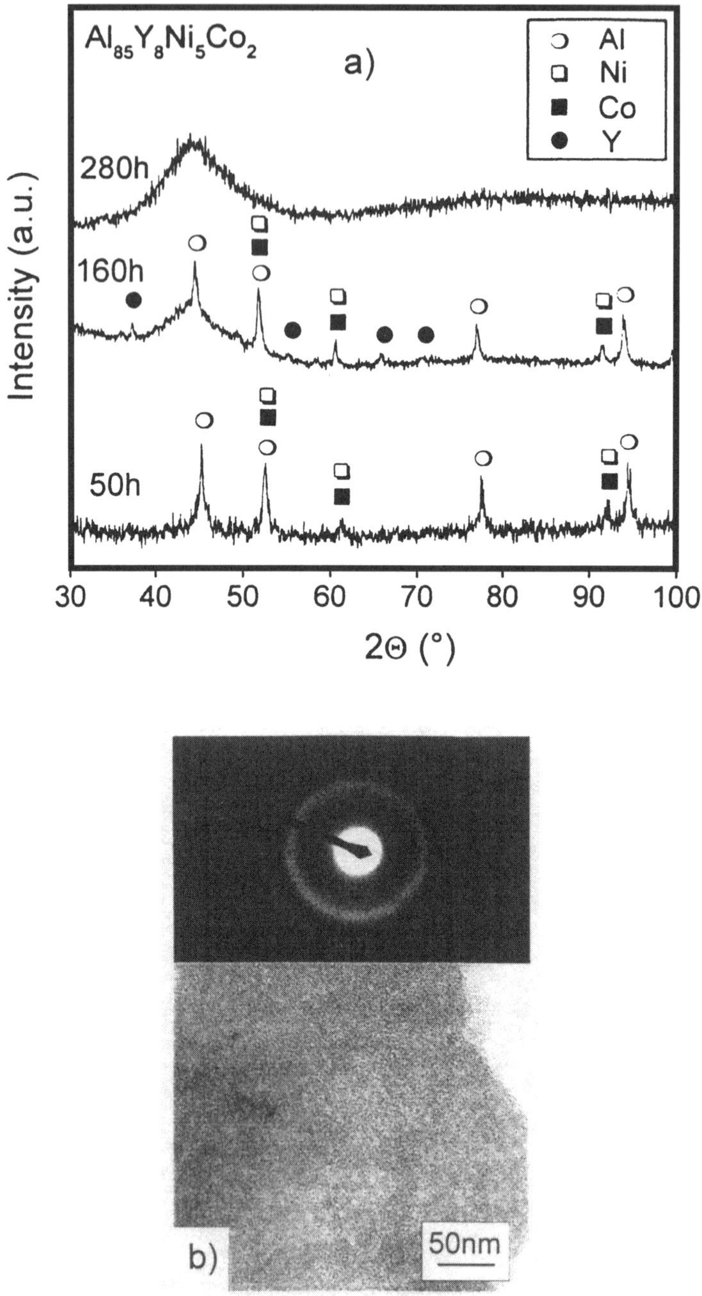

Figure 24. (*a*) X-ray diffraction patterns of mechanically alloyed $Al_{85}Y_8Ni_5Co_2$ after different milling times; (*b*) brightfield TEM micrograph and corresponding electron diffraction pattern of the powder after 280 h of milling.[116]

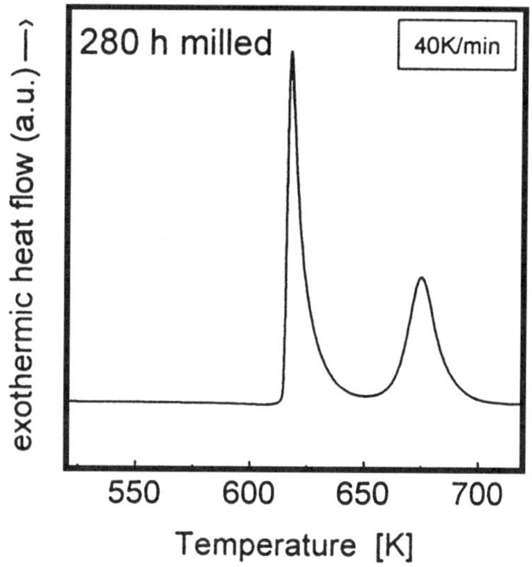

Figure 25. DSC traces of mechanically alloyed amorphous $Al_{85}Y_8Ni_5Co_2$ powder after 280 h of milling.[116]

Besides formation of metastable icosahedral phases in Al-Mn-based systems, mechanical alloying can also be applied to alloy systems known to form thermodynamically stable quasicrystals.[173]–[175] Figure 26 illustrates the progress of phase formation for a mechanically alloyed elemental powder mixture with overall composition of $Al_{70.5}Mn_{8.5}Pd_{21}$.[156] After about 20 h of milling a quasicrystalline phase with face-centered (fci) structure shows up in the x-ray diffraction pattern. The quasicrystal grain size at this stage is about 20 nm. The quasicrystal formation and the resulting microstructure have been corroborated by TEM. Milling for up to 100 h does not lead to any structural changes.

In order to artificially create a dual-phase microstructure of coexisting nanoscale fci particles and fcc Al phase, the initially formed quasicrystalline $Al_{70.5}Mn_{8.5}Pd_{21}$ powder was blended with an excess of pure aluminum powder and subsequently milled for an extended time (Fig. 26). The desired phase mixture of fci and Al phases can be obtained after blending with 20 vol% of Al and milling for an additional 5 h. However, additional SEM and TEM investigations revealed a rather poor distribution of quasicrystalline particles in the Al matrix after short time milling. The homogeneity of the fci phase distribution improves with increasing milling, but care has to be taken to properly adjust the milling parameters in order not to lead to significant alloying of Al with the quasicrystalline phase upon extended milling. This would shift the composition of the quasicrystalline phase towards a more Al-rich stoichiometry, followed by the formation of the competing Al_3Pd_2 phase which was reported as a milling product in an earlier study.[176] Despite the fact that further, more detailed, investigations are necessary to fully elaborate the potential of mechanical alloying for artificially creating nanoscale microstructures based on stable quasicrystalline phases, this synthesis route is believed to be an interesting alternative method for the preparation of such novel microstructures.

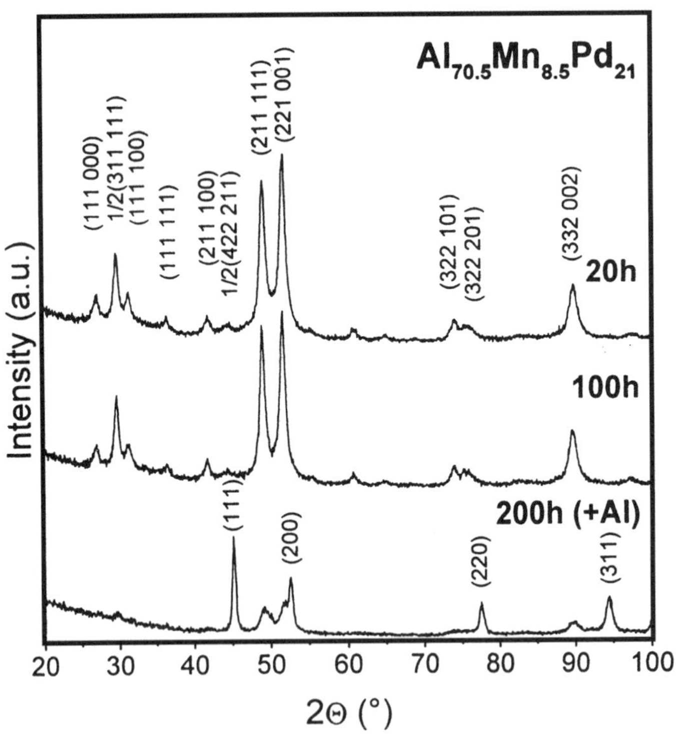

Figure 26. X-ray diffraction patterns of mechanically alloyed $Al_{81}Mn_{14}Fe_5$ powder after different milling times. After comparatively short milling time, the stable quasicrystalline fci phase forms. The quasicrystalline powder is then blended with pure Al powder and further milled for up to 200 h.

Mg-Based Metallic Glasses and Metallic Glass Composites. Upon quenching Mg-based alloys from the melt, special care has to be taken to account for the high reactivity of these alloys in order to prevent contamination-induced heterogeneous crystal formation, which would deteriorate the castability and the thermal stability of the material. In contrast, phase formation by mechanical alloying occurs on a completely different timescale than for quenching from the melt.[24][166] Thus, mechanical alloying is expected to allow glass formation at low temperatures, which may help to circumvent the limitations of quenching techniques with respect to unwanted crystallization effects.

Based on this idea, a systematic study of glass formation by mechanical alloying of elemental powder mixtures of Mg-based alloys has been carried out.[129][166][177][178] Figure 27 shows typical x-ray diffraction patterns for as-milled Mg-Y-Cu powders, revealing the formation of an amorphous phase after 60 h of milling. Besides, some traces of unreacted Mg, Y, and Cu together with Y_2O_3 are visible. No complete amorphization was observed even after long milling times, but XRD and TEM investigations proved the formation of an amorphous phase coexisting with nanoscale crystals of about 10–20 nm in size.[177] This inhomogeneity of the material is related to the high reactivity of the constituents. Especially Mg and Y tend to readily react with oxygen traces stemming from the starting materials or from the argon inert gas atmosphere used for milling. Moreover, thin oxide layers covering the grain boundaries and particle interfaces of the starting powders are present, preventing complete alloying upon milling.[166][177] Hence, mechanically alloyed Mg-Y-Cu alloys have to be considered as nanostructured metallic glass matrix composites which form in-situ during processing. These powders exhibit a supercooled liquid region of about 40 to 50 K, being almost identical to that of high-purity melt-spun ribbons of comparable composition,[177]–[179] which can be used to easily extrude the powders into bulk specimens. Combined with partial or complete devitrification, this yields bulk nanostructured composites consisting of two or more nanoscale phases.[180]

Since the high reactivity of the powders leads to nanostructured metallic glass matrix composites already during milling, it is tempting to further intentionally introduce oxide particles into the Mg-Y-Cu metallic glass matrix in order to synthesize oxide-dispersion strengthened Mg-based metallic glasses.[181] Figure 28 shows x-ray diffraction patterns for an alloy with nominal composition of $Mg_{55}Y_{15}Cu_{30}$ in comparison with the results obtained for composites blended with 5 vol% MgO, CeO_2, Cr_2O_3, or Y_2O_3

oxide particles after 60 h of mechanical alloying. The XRD patterns of the composite powders reveal peaks of the oxides superimposed on the broad diffraction maxima of the amorphous phase. Besides, traces of residual unreacted material and Y_2O_3 are present (see above). No other crystalline phases were detected by XRD or TEM. The average grain size of the oxide particles as estimated from x-ray line broadening and TEM is about 10 to 20 nm after milling.

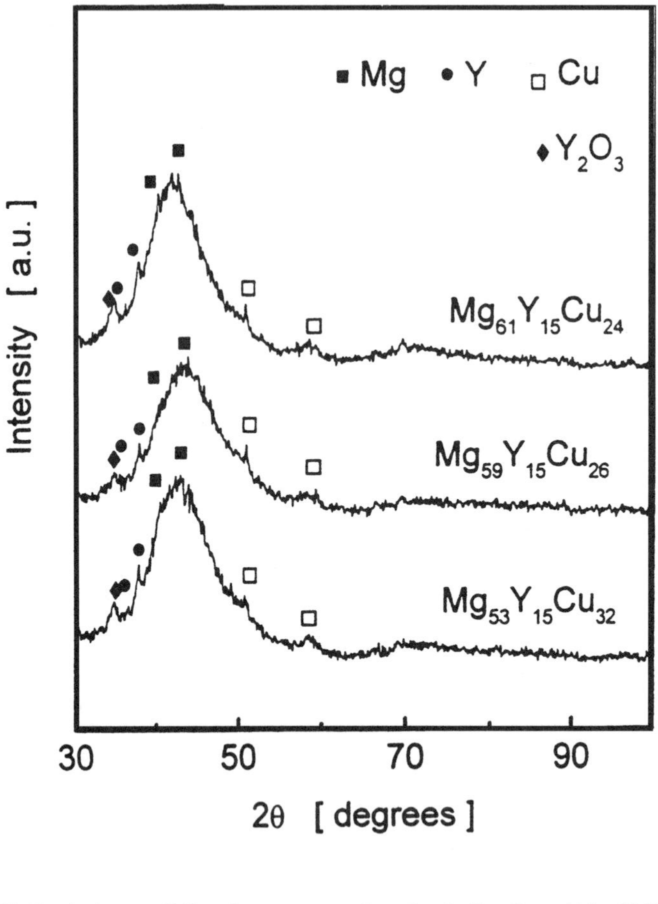

Figure 27. Typical x-ray diffraction patterns of mechanically alloyed Mg-Y-Cu powders, showing the formation of an amorphous phase after 60 h of milling. Besides, some traces of unreacted Mg, Y or Cu, and of Y_2O_3 impurities are visible.[177]

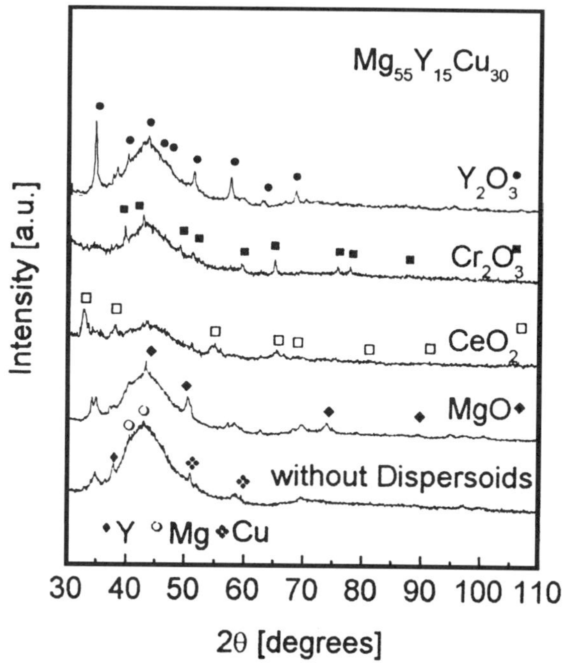

Figure 28. X-ray diffraction patterns of mechanically alloyed $Mg_{55}Y_{15}Cu_{30}$ powders without dispersoids and with 5 vol% MgO, CeO_2, Cr_2O_3, or Y_2O_3 after 60 h of milling.[181]

The addition of oxide particles is not limited to a small volume fraction.[181] Figure 29 shows the DSC traces for $Mg_{55}Y_{15}Cu_{30}$ powders blended with up to 30 vol% Y_2O_3 particles. All samples exhibit a glass transition before crystallization, revealing that mechanical alloying can yield metallic glass matrix composites with an extended supercooled liquid region for up to rather large volume fractions of particles. Detailed investigations[182] proved that the thermal stability of the glassy matrix alloy is not markedly affected by the presence of the nanoscale Y_2O_3 dispersoids.

Zr-Based Multicomponent Alloys. Glass formation by mechanical alloying has been observed for several multicomponent Zr-based systems, such as Zr-Al-Ni,[182][183] Zr-Al-Cu-Ni,[44][166] Zr-Ti-Cu-Ni,[166][184] Zr-Al-Ti-Cu-Ni,[182] Zr-Al-Cu-Ni-Co,[165] etc. Complete amorphization over a wide range of compositions is achieved for such alloy systems. The mechanically alloyed powders show a glass transition and a rather wide supercooled liquid region before crystallization, as exemplified in Fig. 30 showing the DSC traces for $Zr_{65}Al_{10}Cu_{25-x}Ni_x$ alloys. In particular, Zr-Ti-Cu-Ni metallic glasses can be formed by mechanical alloying also for compositions in the central part of the pseudoternary phase diagram (Fig. 31) where glass formation by slow cooling from the melt is prevented due to the existence of a very stable low-melting Laves phase.[185] These results reveal that the glass forming ranges accessible by mechanical alloying or by liquid quenching can differ significantly.

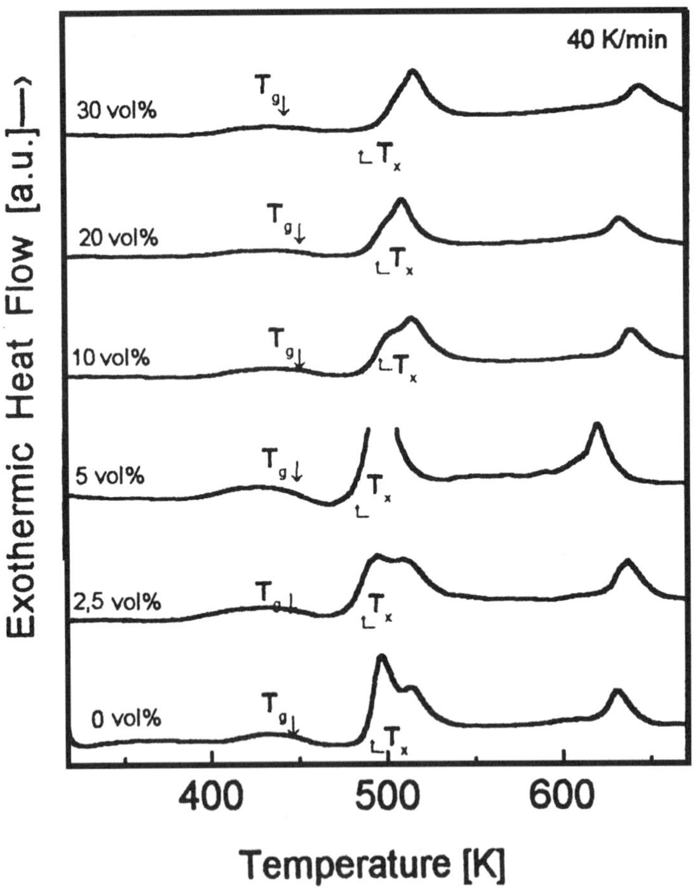

Figure 29. DSC traces of mechanically alloyed $Mg_{55}Y_{15}Cu_{30}$ powders without dispersoids and with up to 30 vol% Y_2O_3 dispersoids after 60 h of milling.

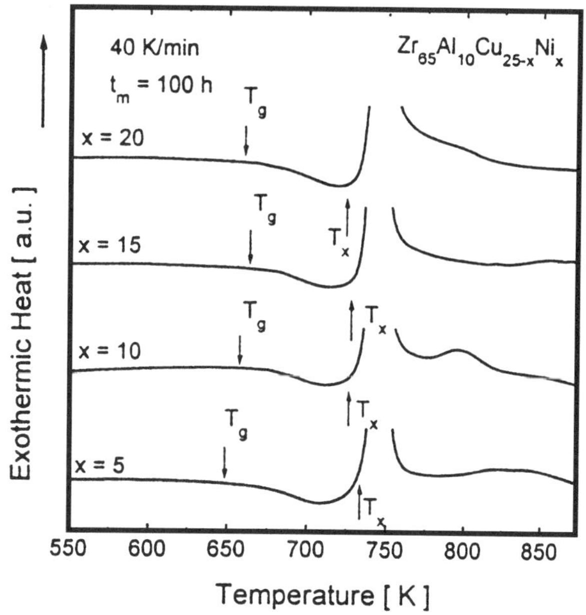

Figure 30. DSC scans for mechanically alloyed amorphous $Zr_{65}Al_{10}Cu_{25-x}Ni_x$ powders with different Ni content x.[166]

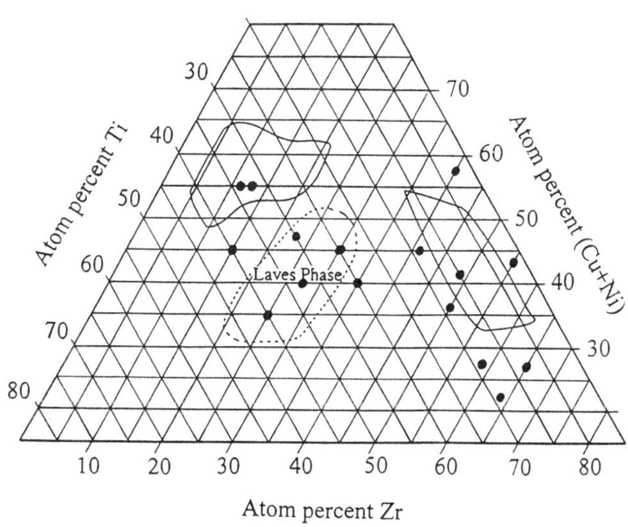

Figure 31. Formation ranges of amorphous Zr-Ti-Cu-Ni: The lines surround the regions where bulk samples can be prepared by slow cooling from the melt (*after Ref. 185*). The dots represent mechanically alloyed glassy powders with extended supercooled liquid region.[166]

Influence of Oxygen on Thermal Stability and Nanostructure Formation. Contamination effects are always a concern in powder metallurgy due to the high reactivity of the fine powders produced by milling. The contamination level due to the wear debris of the milling tools (WC or steel vials and balls) is typically smaller than 0.5 at%.[166] Detailed investigations proved that the influence of such low values on the properties of the material is negligible.[44][186] The oxygen content after milling mostly stems from the contamination of the starting powders, and can reach values above 1 at%, even for careful processing under inert gas atmosphere, because of the large surface area and the high reactivity of the fine powder particles. Although some oxides are present in as-milled material with large oxygen content, e.g., for material with up to about 3 at% oxygen,[187] most of the oxygen is alloyed in the glassy phase, thus changing the actual composition and the short-range order of the alloy compared to the pure material.[144] The rather high oxygen content in general changes T_g, T_x, and ΔT_x compared to high purity melt-quenched alloys, as exemplified in Fig. 32 for $Zr_{65}Al_{7.5}Cu_{17.5}Ni_{10}$ alloys with different oxygen content. This was investigated in detail for mechanically alloyed $Zr_{65}Al_{7.5}Cu_{17.5}Ni_{10}$ and $Zr_{55}Al_{10}Cu_{30}Ni_5$ alloys.[144][166][187][188] Similar to melt-spun ribbons with rather large oxygen content (see Sec. 3.1), the crystallization of mechanically alloyed powders is determined by oxygen-triggered crystallization of a metastable fcc $NiZr_2$-type phase.[182] Nevertheless, despite the rather high oxygen content the milled powders are in a truly glassy state without indications for the formation of intermetallic phases. This is in striking contrast to slowly cooled bulk specimens where, already, a much smaller oxygen content prevents the formation of fully glassy material.[140] Hence, mechanical alloying allows a much larger oxygen level for glass formation than slow cooling from the melt. This is due to the completely different mechanisms of glass formation for the different techniques. Whereas, cooling from the melt is governed by the thermodynamics and kinetics of nucleation and growth at rather high temperatures,[21][189] mechanical alloying is a low-temperature process involving different constraints for phase formation, thus enabling metallic glass formation also for rather large impurity contents.

Considering that oxygen-triggered formation of metastable phases upon crystallization not only changes the thermal stability and crystallization sequence, but also allows sequential precipitation and transformation of nanoscale crystalline phases upon devitrification, this offers interesting perspectives for nanostructure design in such materials, yielding unique nanostructured two- or multiphase materials, which cannot be obtained by any other processing technique. When the annealing treatment is combined with powder consolidation, i.e., by carefully adjusting the temperature-time regime applied for extrusion or hot pressing, this easily enables the preparation of bulk nanostructured composites from mechanically alloyed powders.

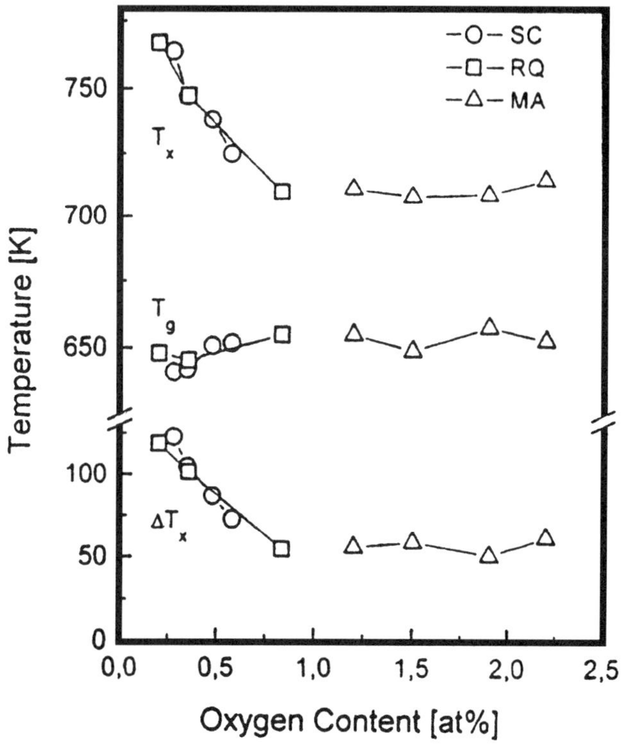

Figure 32. T_g, T_x and ΔT_x (heating rate 40 Kmin^{-1}) for $Zr_{65}Al_{7.5}Cu_{17.5}Ni_{10}$ slowly cooled bulk samples (SC), rapidly quenched ribbons (RQ) and mechanically alloyed powders (MA) vs oxygen content.[187]

Zr-Based Metallic Glass Matrix Composites. Nanostructured Zr-based metallic glass composites can also be prepared directly by mechanical alloying, by blending the metallic constituents of the glass with insoluble oxide[168][190] carbide,[191] or nitride[192] particles. For Zr-based alloys with various compositions, the phase formation and the development of the microstructure upon milling was characterized by XRD, TEM, and

DSC measurements. The XRD patterns of composite powders show the diffraction peaks of the particles superimposed on the broad amorphous maxima.[168][186] Figure 33 shows a brightfield TEM image and the corresponding diffraction pattern for $Zr_{65}Al_{7.5}Cu_{17.5}Ni_{10}$ with 5 vol% CaO particles as a typical example. The diffraction pattern shows a diffuse halo besides some diffraction spots of the crystalline oxide particles, proving the formation of an amorphous phase coexisting with the CaO dispersoids. The brightfield image reveals a uniform distribution of 5–20 nm dispersoids embedded in the glassy matrix. Similar results were obtained for other composite samples with different volume fraction of particles, V_f, and also for Zr-Al-Cu-Ni glassy alloys containing insoluble W particles.[168][186]

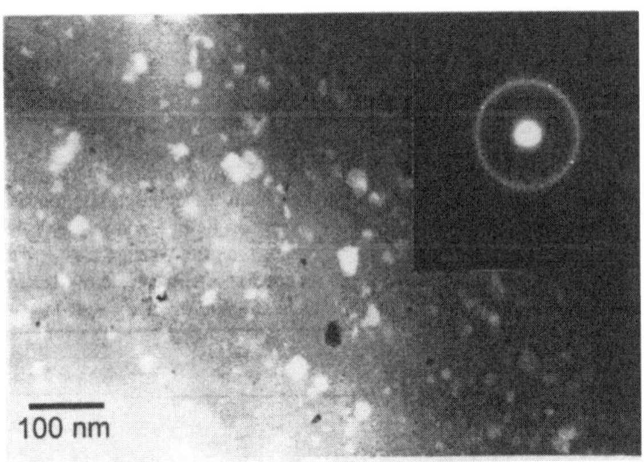

Figure 33. Brightfield TEM micrograph and corresponding electron diffraction pattern of mechanically alloyed $Zr_{65}Al_{7.5}Cu_{17.5}Ni_{10}$ powder containing 5 vol% CaO particles after 100 h of milling, revealing a uniform distribution of the dispersoids in the amorphous matrix.[168]

Figure 34 shows DSC traces for glassy $Zr_{55}Al_{10}Cu_{30}Ni_5$ powders blended with different volume fractions of W particles as a typical example of the thermal stability data of mechanically alloyed composite powders. The scans exhibit a glass transition at low temperatures, followed by an exothermic crystallization peak. The addition of W particles into the glassy $Zr_{55}Al_{10}Cu_{30}Ni_5$ matrix leads to a shift of the crystallization peak to slightly higher temperatures. This indicates that a small amount of W might be alloyed into the matrix and, therefore, slightly changes the overall composition of the glass. A careful examination of the thermal stability data reveals that T_g and T_x, as well as the incubation time for crystallization under isothermal conditions, remain almost unchanged upon particle addition.[186] However, crystallization of the composites proceeds faster than for the particle-free material. Hence, nanosized W particles do not significantly deteriorate the stability of the supercooled liquid against crystallization, but promote formation of a nanoscale microstructure upon crystallization due to interface controlled nucleation at the particle/supercooled liquid interfaces.[186] Similar data were obtained for composites containing nanoscale oxide particles with volume fractions of up to 30%.[168][190] Again, the overall thermal stability of the metallic glass remains almost unchanged compared to the particle-free material (Fig. 35), but the presence of nanoscale oxides promotes nanocrystal formation upon crystallization.

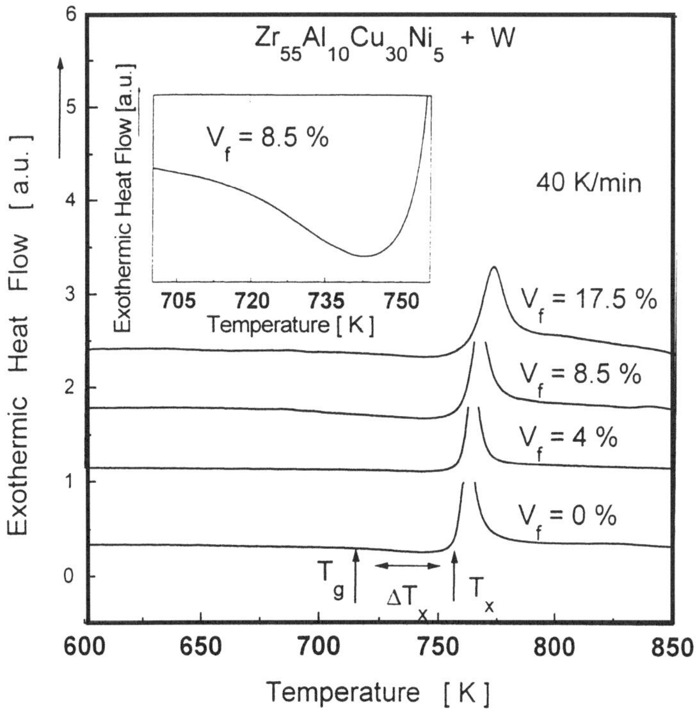

Figure 34. DSC traces for mechanically alloyed $Zr_{55}Al_{10}Cu_{30}Ni_5$ with different volume fractions of W particles, V_f. The inset displays an enlarged part of the DSC scan for the sample with V_f = 8.5% as a typical example for the glass transition behavior of the composite powders.[186]

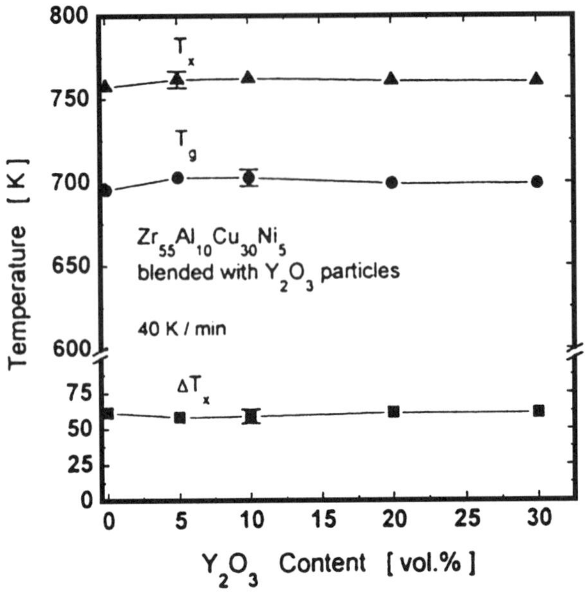

Figure 35. Thermal stability data (heating rate 40 $Kmin^{-1}$) for mechanically alloyed amorphous $Zr_{55}Al_{10}Cu_{30}Ni_5$ and for composites with different volume fraction of Y_2O_3 particles.[190]

4.0 MECHANICAL PROPERTIES AT ROOM AND ELEVATED TEMPERATURES

On one hand, the Hall-Petch relationship predicts large strength increases as the grain size approaches the nanometer regime.[5][6] On the other hand, theory indicates that a high volume fraction of interfacial regions leads to enhanced deformation due to grain boundary sliding in the low end of the nanometer range.[193] Despite a significant strength increase at room temperature and the observation of superplastic flow at elevated temperatures,[9] the room temperature ductility of single-phase nanocrystalline metals, alloys, and compounds is typically rather poor.[10] In this sense, their deformation and fracture behavior is quite similar to that of amorphous alloys, i.e., there seems to be an analogy of the deformation behavior of nanocrystalline and non-crystalline structures. Detailed descriptions of the mechanical behavior of "conventional" nanostructured materials can be found in Refs. 7 and 8. Here, the deformation behavior of recently developed two- or multiphase alloys with nanoscale microstructure will be described for selected Al-, Mg-, and Zr-based multicomponent alloys.

4.1 Al-Based Two-Phase Nanostructured Alloys

Conventional high-strength aluminum alloys make use of different strengthening mechanisms, such as solid solution strengthening, precipitation hardening, grain size refinement, dispersion strengthening, work hardening, and fiber reinforcement. Utilizing these different mechanisms leads to an upper limit for the tensile strength of about 600 MPa at room temperature for conventional crystalline Al alloys. For developing new types of high-strength Al-based alloys with further improved strength levels, the use of different strengthening mechanisms is required. This can be realized by fabricating alloys with non-periodic structures, i.e., amorphous alloys without long-range periodicity or quasicrystalline alloys. In such alloys, there are no specific slip planes and, therefore, there is no plastic deformation via dislocation motion at low stress levels. However, this, in turn, may limit the room temperature ductility of the material. To overcome this problem, different types of two-phase nanostructured materials have been developed.

Two-Phase Nanostructured Alloys with Amorphous Matrix. The starting point for the development of new high-strength Al alloys was the discovery of high tensile strengths, σ_f, exceeding 1200 MPa for melt-spun Al-Ln-TM[93][100] and Al-ETM-LTM[101] amorphous ribbons or amorphous wires prepared by melt-extraction.[102] Such strength levels are about twice as high as for conventional crystalline Al alloys. The achievable σ_f and hardness values depend, of course, on the actual composition of the respective alloy. Similar high hardness data was found for mechanically alloyed powders and consolidated bulk specimens. For example, Vickers hardness (H_v) values of about 430 were found for mechanically alloyed and consolidated $Al_{85}Y_8Ni_5Co_2$.[116] Neglecting residual porosity in the bulk samples, the fracture strength is estimated to be about 1400 MPa. Dougherty, et al.,[43] report even higher H_v values (625–650) for mechanically alloyed $Al_{80}Ni_8Fe_4Gd_8$, corresponding to an estimated fracture strength of about 2000 MPa. This very high value indicates the presence of additional intermetallic precipitates and/or oxides in the material.

Even higher strength levels are observed for nanostructured two-phase alloys with nanoscale fcc Al particles in an amorphous matrix. Such nanoscale mixed-phase materials can be made directly upon rapid quenching or by partial devitrification of the melt-spun amorphous ribbon. They exhibit tensile fracture strengths reaching up to 1560 MPa for an $Al_{88}Ni_9Ce_2Fe_1$ alloy,[107] which is considerably larger than that of the corresponding single-phase amorphous alloy (1100 MPa). The tensile strength increases almost linearly with the volume fraction of particles, V_f, reaching its maximum value for $V_f = 25\%$. Simultaneously, the Vickers hardness increases from 280 to 400, and the Young's modulus increases from 63 to 71 GPa with increasing V_f.[107] Similar data was found for a variety of alloys with different compositions,[89][194][195] including two-phase nanogranular amorphous alloys, such as $Al_{94}V_4Fe_2$,[119] or Al-Ti-Fe alloys with an amorphous network phase.[120]

The increase in σ_f by the dispersion of nanoscale fcc Al particles in the amorphous matrix is due to a combination of different effects.[99] The very small fcc Al nanograins of about 3–5 nm in size are too small to contain dislocations[7][8] and have a perfect crystalline structure. In addition, the particles precipitate from a supercooled liquid/amorphous phase. Thus, the interfacial energy of the amorphous/Al particle interface is lower, by about an order of magnitude,[196][197] than the solid/solid interfacial energy in the case of the precipitation of a crystalline phase from a crystalline fcc Al alloy. Consequently, the amorphous/Al particle interface has a highly dense-packed structure without excess vacancies or voids.[196] This interfacial structure suppresses crack generation at the interface, favoring the transmittance of the applied load between the different phases. In addition, the nanoscale of the fcc Al particles and the interparticle spacing are smaller than the width of the shear bands which carry deformation in the amorphous phase. Hence, the Al particles can act as an effective barrier against the shear deformation of the material. Finally, chemical redistribution and solute enrichment of the remaining amorphous matrix after precipitation might also contribute to the observed strength and hardness increase.[90][197]–[199]

However, the tensile strength of these alloys is too high to produce bulk alloys by extrusion at low temperatures, which would be required to maintain the extremely fine nanoscale phase mixture obtained after solidification. Moreover, some of the alloys tend to significantly embrittle upon annealing at elevated temperatures, yielding fracture toughness values on the order of 1–5 MPam$^{1/2}$.[43][92][103] This is believed to be due to relaxation of the amorphous phase coupled with solute redistribution.[90][197] This limits the practical use of such two-phase nanostructured alloys containing an amorphous matrix phase.

Fully Nanocrystalline Bulk Alloys. Extruding atomized amorphous powders or pieces of melt-spun ribbons at temperatures near the onset of crystallization results in bulk nanostructured alloys consisting of intermetallic compounds (about 50 nm in size) and fcc Al with a grain size of up to 200 nm.[106] Alloys with this type of microstructure exhibit a tensile strength above 900 MPa, a Young's modulus of about 100 GPa, and a rotating-beam fatigue strength of about 300–350 MPa at room temperature.[200] The high strength values result from the combination of dispersion strengthening and grain refinement. Such alloys show elongations of up to 10%, indicative of a good room temperature ductility. This is attributed to the presence of dislocations, in particular, in the submicrometer-sized Al phase. On the other hand, the dispersion of intermetallics, together with the grain boundaries of the Al phase, effectively hinder dislocation motion, thus, yielding the observed strength increase. Besides, these alloys exhibit a high strain rate sensitivity exponent $m = 0.7$ as well as large elongations of 600–700% at a strain rate of 1 s^{-1} at 885 K, revealing that they can be deformed by high strain rate superplasticity. This makes them also attractive for shaping operations at moderate to elevated temperatures.

Alloys with Quasicrystalline Phases. Micrometer-sized quasicrystalline poly- or single crystals exhibit high strength and Young's modulus, but are brittle at room temperature.[201]–[203] They also show increased ductility at elevated temperatures.[204][205] Due to their room temperature brittleness, their practical relevance has been limited to low-friction or thermal barrier coatings so far.[206][207] Hence, it is of major importance to increase the ductility of quasicrystal-based Al alloys for exploiting their potential as high-strength materials. This goal can be realized in various Al-(Mn, Cr)-Ce,[108]–[110] Al-(Mn, Cr)-Ce-TM,[111][112] Al-V-Ce-TM,[113] and Al-(V, Cr, Mn)-TM[114] alloys (TM = Fe, Co, Ni, or Cu), forming a nanoscale phase mixture of quasicrystalline icosahedral particles coexisting with a ductile fcc Al phase upon quenching from the melt or through solid state reaction. This type of mixed structure typically forms for more than about 90 at% Al and exhibits good ductility and high room temperature tensile strength of about 1000–1340 MPa.[99][112][115] (See Fig. 4.) These properties originate from the fine distribution of globular quasicrystalline particles in an Al matrix on a nanometer scale, where the quasicrystals act as strength-bearing components, while the Al matrix supplies ductility. The existence of a crystalline approximant phase at the interface between the I-phase particles and the fcc Al matrix improves interfacial bonding between the different phases and is, thus, important for the combination of high strength and good ductility without failure at the interface.[91]

Although melt-spun ribbons exhibit ultimate strengths exceeding those of conventional crystalline Al alloys by a factor of 2–3, their small size prevents engineering applications. For that reason, powder metallurgical methods such as gas atomization or mechanical attrition have been employed to create powder particles with the desired microstructure, which in a second step can be consolidated into nearly full density bulk specimens by extrusion at temperatures below the decomposition temperature of the nanoscale I-phase.[93][116][208] Figure 36 illustrates the relation between the tensile strength and the plastic elongation at room temperature for consolidated bulk I-phase-containing alloys[99] in comparison with data for conventional crystalline Al alloys obtained from the Japanese Institute of Steels (JIS) handbook. The tensile strength and the elongation of the bulk I-phase-based alloys with different composition are in the range of 500–850 MPa and 5–25%, respectively. Hence, both the strength and the ductility values of these alloys are superior to those of conventional Al alloys. The bulk alloys also have a high Young's modulus of about 100 GPa and exhibit a rather high elevated temperature strength, i.e., about 350 MPa at 473 K and 200 MPa at 573 K.[105] This elevated temperature strength can even be maintained after quite long temperature exposure, e.g., 1000 h at 573 K, indicating a high stability of the fine-scale microstructure against phase transition or coarsening. Similar data were also reported for Al-Mn-LTM (LTM = Fe, Co, Ni, or Cu) alloys.[200] Their impact fracture energy reaches values of about 180 kJ/m^2, which is significantly larger than the values for typical age-hardened crystalline Al alloys after T6 annealing treatment.

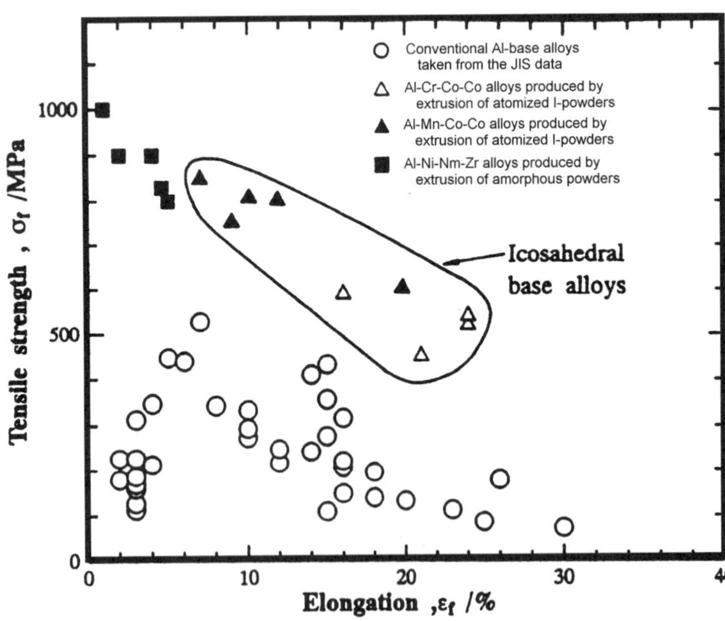

Figure 36. Relation between tensile strength and plastic elongation for extruded bulk I-phase-containing alloys. Data for conventional Al-based alloys are also shown for comparison.[99]

Bulk I-phase containing Al-based alloys can also be obtained directly by squeeze casting without any additional consolidation step as is necessary in the case of atomized or mechanically alloyed powders (see Sec. 3.2). Although the microstructural scale of such specimens is larger than for extruded material and, therefore, this material has to be regarded as a "submicron-to-micron-sized" composite rather than a "nanosized" two-phase mixed alloy, this type of microstructure exhibits similar mechanical properties as extruded bulks. This is exemplified in Fig. 37, comparing the room temperature deformation behavior under constant true strain rate compression conditions for squeeze-cast $Al_{92}Mn_6Ce_2$ and $Al_{91}Mn_7Fe_2$ alloys with that of gas-atomized and extruded Al-Cr-Cu-Mn.[156] The combination of a rather high strength of 565 MPa and good ductility of about 13% for the $Al_{92}Mn_6Ce_2$ cast rod is comparable to the data for the extruded bulk specimens, despite the coarser, but also rather homogeneous, globular microstructure of the cast material. The $Al_{91}Mn_7Fe_2$ sample shows a lower strength, and failure already occurs after less than 6% plastic deformation. This is attributed to a dendritic microstructure formed upon casting, leading to inhomogeneous deformation and extensive slip band formation.[156] Hence, care has to be taken to optimize alloy composition and microstructure of the cast alloys. Nevertheless, the further optimization of the squeeze casting technique as a rather simple single-step process is believed to be promising for the production of Al-based bulk alloys containing quasicrystalline particles for strengthening, at least for specimens with not too large dimensions.

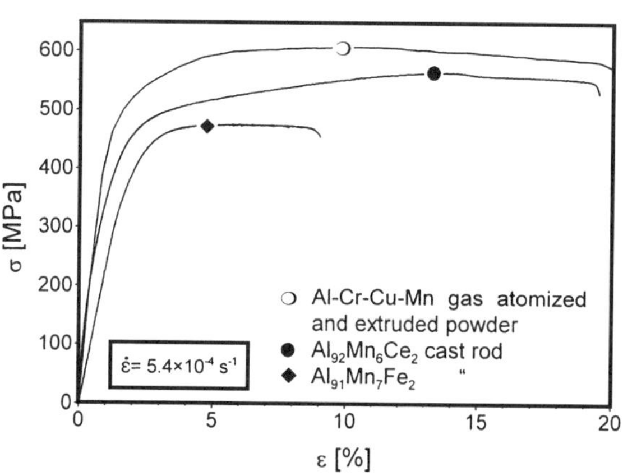

Figure 37. Compressive true stress – true strain curves of squeeze-cast $Al_{92}Mn_6Ce_2$ and $Al_{91}Mn_7Fe_2$ bulk samples in comparison with a gas-atomized and extruded Al-Cr-Cu-Mn sample.

4.2 Mg-Based Amorphous and Nanostructured Alloys

Rapidly Quenched and Cast Materials. The development of high-strength two-phase nanostructured Mg-based alloys makes use of a similar concept as already described for Al-based alloys: nanoscale microstructures, consisting of amorphous and nanocrystalline phases or of nanocrystalline phases with different strength and ductility, are combined to achieve a high room temperature strength together with good ductility and good deformability at elevated temperatures. This will be described in the following paragraphs for different types of materials and samples, i.e., rapidly quenched ribbons, slowly cooled bulk samples, as well as bulk samples prepared from gas-atomized powders.

Deformation Behavior of Amorphous Alloys. Most Mg-based amorphous alloys containing more than about 80 at% Mg have good bending ductility when prepared as rapidly quenched thin ribbons and can be bent through 180° without fracture. The tensile fracture strength, σ_f, the Young's modulus, Y, the Vickers hardness, H_v, the specific strength, σ_f/ρ, and the fracture elongation, ε_f, for different Mg-based alloys with good bending ductility are summarized in Table 2. Alloys belonging to the Mg-Ln-TM group exhibit higher σ_f values than the other amorphous systems. The highest σ_f value for the Mg-Ln-TM alloys reaches 830 MPa, which is about twice as high as for conventional crystalline Mg-based alloys.[125] The highest Y and H_v values found for this group of alloys are 46 and 224 GPa, respectively. Although the σ_f values of Mg-Ca-Al amorphous alloys are considerably lower than those of Mg-Ln-TM alloys, there is no distinct reduction in the specific strength of the Mg-Ca-Al alloys, because of their very low densities. The fracture surface under tensile loading conditions typically shows a smooth region due to shear band sliding and a vein pattern due to the final fracture after sliding. This feature is the same as for other conventional amorphous alloys with good bending ductility.[125] Most of the ternary Mg-based amorphous alloys known so far exhibit a good ductility in the as-quenched state even for high concentrations of alloying elements, such as Y or Cu. However, the amorphous alloys with less than 80 at% Mg embrittle during aging and the degree of embrittlement increases with decreasing Mg content.[125] This effect is attributed to the development of short-range ordering and rearrangement of Mg and the other alloying elements upon aging.[121]

Table 2. Mechanical Properties of Mg-Based Amorphous Alloys[125]

Alloy (at%)	σ_f (MPa)	Y (GPa)	H_v	σ_f/ρ (10^5 Nmkg^{-1})	ε_f (%)
$Mg_{70}Ca_{10}Al_{20}$	670	35	199	3.7	1.9
$Mg_{90}Ca_{2.5}Ni_{7.5}$	670	40	182	3.4	1.7
$Mg_{87.5}Ca_5Ni_{7.5}$	720	47	176	3.7	1.5
$Mg_{80}Y_5Ni_{7.5}$	830	46	224	2.5	1.8
$Mg_{85}Y_{10}Cu_5$	800	44	205	2.3	1.8
$Mg_{80}Y_{10}Cu_{10}$	820	46	218	2.5	1.8
$Mg_{91}Y_5Cu_4$	550	30	140	2.3	1.8

Besides melt spinning, the high glass forming ability of Mg-based alloys, in particular, of Mg-Cu-Y or Mg-Ni-Y alloys, enables the production of bulk amorphous specimens by high pressure die casting or suction casting.[130][131][160] Figure 38 shows nominal tensile stress—elongation curves for bulk amorphous $Mg_{80}Cu_{10}Y_{10}$ obtained at different temperatures between 273 and 473 K.[131] The deformation behavior can be divided into three different types. At room temperature (293 K) the material has a high fracture stress and shows no appreciable elongation under tensile conditions. In the temperature range between 353 and 393 K a high yield stress combined with a rather small elongation is observed. In contrast, the material exhibits a distinct yield phenomenon, followed by a significant decrease in nominal flow stress accompanying large elongations at temperatures above 423 K. The fracture stress reaches 630 MPa at 293 K and remains at this level for temperatures up to 353 K. For higher temperatures the fracture stress decreases rapidly with increasing temperature. Hence, the material keeps a mechanical strength exceeding 400 MPa for temperatures up to about 373 K. On the other hand, the elongation is below 7% for temperatures lower than 393 K, and the material reveals inhomogeneous deformation and fracture behavior without distinct work hardening in this temperature range. Above 400 K, the elongation increases rapidly to about 35%. This is due to the viscous flow of the material in this temperature range. Similar results were found for other cast bulk Mg-based amorphous specimens,[125][131] as well as for extruded amorphous samples prepared from atomized powders.[133][134]

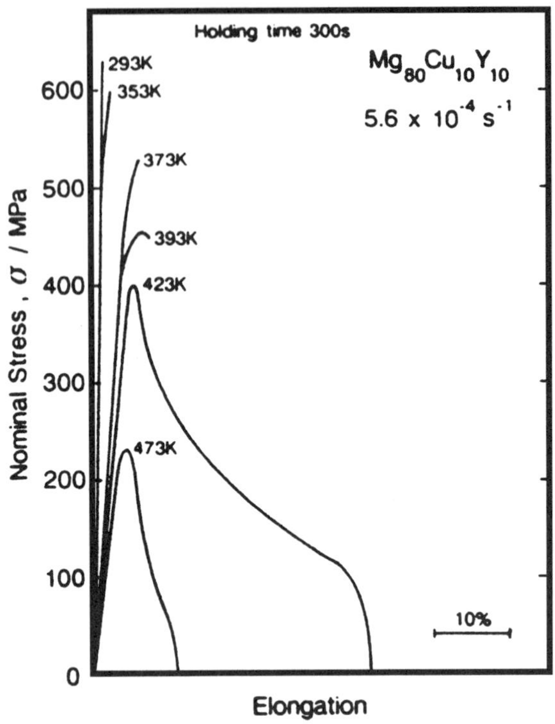

Figure 38. Temperature dependence of the nominal tensile stress – elongation curve of amorphous $Mg_{80}Cu_{10}Y_{10}$ bulk samples.[131]

Strength of Amorphous Alloys Containing Nanoscale Particles. Partially crystalline Mg-rich alloys with more than 80 at% Mg consisting of amorphous and nanoscale hcp Mg phases can be prepared either directly by quenching from the melt or through partial crystallization.[123][131][132] About 5–10 nm sized hcp particles are dispersed homogeneously in the amorphous matrix with a typical interparticle spacing of about 3–10 nm.[131][132] Compared to the fully amorphous alloy, the ultimate tensile strength, the Vickers hardness, and the Young's modulus increase for the mixed phase alloys. They also exhibit good room temperature ductility. For example, for Mg-Zn-La alloys the ultimate room temperature tensile strength increases from about 600 MPa for the amorphous phase to about 700 MPa for the mixed structure containing nanoscale hcp Mg particles.[132] In particular, the nanostructured two-phase $Mg_{85}Zn_{12}La_3$ alloy exhibits a pronounced yielding phenomenon with a maximum at about 3% plastic strain. This indicates that the homogeneous dispersion of ultrafine hcp particles embedded in the amorphous matrix suppresses the generation of local shear deformation and assists homogeneous plastic deformation as long as the amorphous matrix phase exhibits some ductility. The ductilization effect has been attributed to the small size of the Mg precipitates (5–10 nm). They are smaller than the width of the inhomogeneously deformed regions (10–20 nm), which allows the second phase to act as an effective barrier for shear deformation by increasing the flow stress on the shear plane by the presence of the nanoscale particles. As a result, subsequent deformation on the same shear plane is suppressed, leading to the increase in plastic elongation at room temperature. With increasing temperature, the mixed phase nanostructured alloys exhibit a decrease in yield strength and an increase in the elongation, similar to that observed for fully amorphous alloys.

Deformation Behavior of Nanostructured Alloys. Crystallization of the amorphous phase or extrusion at higher temperatures produces nanostructured phase mixtures consisting of hcp Mg grains and intermetallic compounds with a size of about 100–200 nm.[124][133] Figure 39 shows the temperature dependence of the tensile yield strength, $\sigma_{0.2}$, for bulk nanostructured $Mg_{87.5}Cu_5Y_{7.5}$[133] and $Mg_{70}Ca_{10}Al_{20}$[124] alloys. For comparison, the data for the heat-resistant commercial crystalline WE54-T6 alloy[209] are also shown in Fig. 39. The tensile yield strength at room temperature is 740 MPa for the $Mg_{87.5}Cu_5Y_{7.5}$ alloy and 600 MPa for $Mg_{70}Ca_{10}Al_{20}$. Although the $\sigma_{0.2}$ values decrease with increasing temperature, both alloys keep a high tensile yield strength exceeding that of the commercial crystalline alloy up to rather high temperatures. Simultaneously to the decrease in yield strength the elongation increases with increasing temperature, similar to that shown above for fully amorphous or partially crystalline alloys of comparable composition. The specific tensile yield strength, $\sigma_{0.2}/\rho$, of the Mg-Cu-Y and Mg-Ca-Al alloys is as high as 3.3×10^5 Nmkg^{-1}, which is much higher than the values of typical commercial Mg-based alloys.[210]

The significant increase in elongation with increasing temperature points to the possibility of superplastic deformation characteristics of the nanostructured Mg-based alloys. In fact, this has been observed for some of the alloys. For example, evidence for superplastic behavior was found for a Mg-8.3 wt% Al-8.1 wt% Ga alloy, which was produced by rapid quenching and extrusion.[135] A maximum elongation to failure of 1080% was obtained at 573 K at a relatively high strain rate of 10^{-2} s^{-1}. This has been attributed to the fine-scale equiaxed microstructure, consisting of a hcp Mg matrix with a grain size of about 2 μm and intermetallic compounds with sizes of 100–300 nm at 573 K, the optimum temperature for superplastic deformation. The superplastic strain rate is higher than for conventional superplastic aluminum and magnesium alloys. This opens perspectives for investigations of possible high strain-rate superplasticity in other Mg-based two-phase materials.

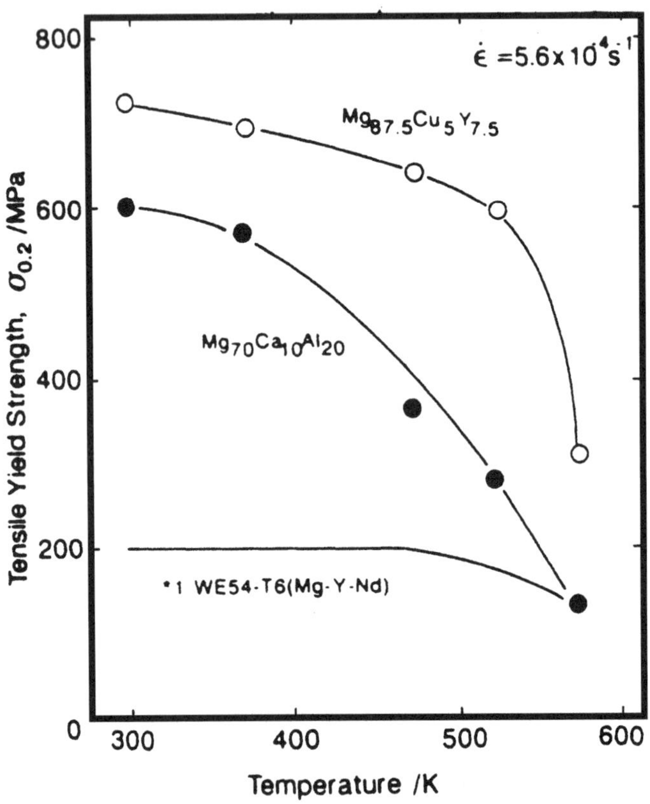

Figure 39. Temperature dependence of the tensile yield stress, $\sigma_{0.2}$, for extruded $Mg_{87.5}Cu_5Y_{7.5}$ and $Mg_{70}Ca_{10}Al_{20}$ alloys annealed for 2 h at each testing temperature.[125] Data for the commercial WE54-T6 (Mg-Y-Nd) alloy taken from Ref. 209 are also shown for comparison.

Mechanically Alloyed Powders. Mechanically alloyed Mg-based alloy powders typically contain nanoscale unreacted elemental material and traces of Y_2O_3 homogeneously embedded in the amorphous matrix even after extended milling.[166][177] Nevertheless, the thermal stability of the powders is comparable to rapidly quenched or cast fully amorphous specimens, and reveals a distinct supercooled liquid region. Heating through the glass transition is accompanied by a softening of the material, leading to viscous flow behavior for temperatures above T_g. The typical behavior of mainly amorphous powders under static compression is shown in Fig. 40 for $Mg_{55}Y_{15}Cu_{30}$ as a typical example.[178] Figure 40 compares the DSC scan for the mechanically alloyed powder with the deformation data obtained from thermomechanical analysis (TMA). Above the glass transition temperature a distinct contraction of the sample, i.e., a penetration of the probe, occurs due to viscous flow. This allows an easy consolidation by hot pressing or extrusion in the viscous state above T_g. Figure 41 shows an amorphous $Mg_{55}Y_{15}Cu_{30}$ plate prepared by hot compaction. The spherical indent in the middle of the plate is 1.5 mm deep and was produced by additional die-upset deformation of the initially flat plate above T_g, showing that the as-consolidated dense bulk sample can be easily deformed subsequently without crack formation or crystallization.[129]

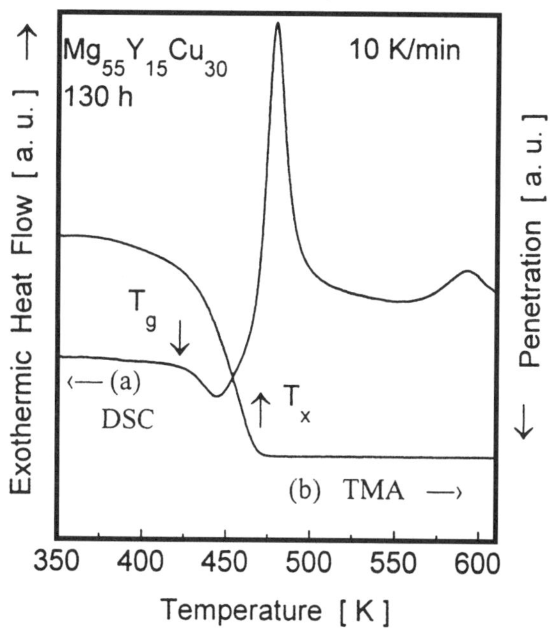

Figure 40. (*a*) DSC trace and (*b*) corresponding TMA scan of mechanically alloyed $Mg_{55}Y_{15}Cu_{30}$ powder.[178]

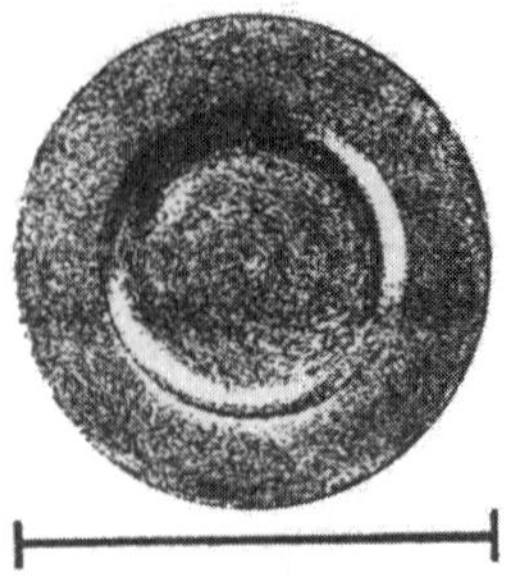

Figure 41. Example of a bulk sample of an amorphous $Mg_{55}Y_{15}Cu_{30}$ alloy prepared by consolidation in the supercooled liquid region.[178]

The Vickers hardness of such consolidated samples is about 345–385 H_v for samples with different composition.[129][178] Using the relation $H_v = 3\sigma_y$,[186] this points to a fracture strength of about 1150–1280 MPa for the mechanically attrited material. These data are significantly larger than the values for as-quenched amorphous or partially crystallized ribbons or cast bulk specimens.[125][130] Thus, the nanocrystalline metallic particles or oxides present after mechanical attrition and also after consolidation effectively increase the mechanical strength of the powder metallurgically prepared specimens.

4.3 Zr-Based Alloys

Cast Material. High strengths and good ductility have been observed in a number of multicomponent Zr-based alloys belonging to the group of bulk metallic glass forming systems, e.g., Zr-Cu-Al-Ni-Ti[152][153][162] or Zr-Al-Cu-Ni-Pd,[154] etc. Already the bulk metallic glasses are known to reveal outstanding mechanical properties at room temperature, namely a beneficial combination of very high strength, relatively low Young's modulus, some microplasticity, and high wear resistance.[211]–[215] In contrast, crystalline specimens of the same nominal composition are brittle and show significantly lower values of fracture strength.[211] Primary crystallization of such metallic glasses further increases the strength of the material due to the precipitation of nanoscale particles while maintaining a good ductility as long as the volume fraction of precipitates does not exceed a critical limit.[161][162][216]

In contrast, fully nanocrystallized material fails in a completely brittle manner.[161] Since good ductility is observed when the alloy contains an amorphous phase, the good ductility of the material is attributed to the ductile nature of the remaining amorphous phase, which seems to retain a significant amount of free volume even after partial crystallization. The following sections focus on some details regarding the effect of nanoscale precipitates on the room temperature and elevated temperature mechanical behavior of bulk glass forming Zr-based alloys, comparing data for fully amorphous and partially crystallized $Zr_{57}Al_{10}Cu_{20}Ni_8Ti_5$ and $Zr_{55}Al_{10}Cu_{30}Ni_5$ alloys as examples for this class of materials.

Room Temperature Mechanical Properties. For the compression tests at room temperature a strain rate of $\dot{\varepsilon} = 1 \times 10^{-3}$ s^{-1} was employed. Figure 42 compares typical compressive stress-strain curves for as-cast fully amorphous $Zr_{57}Al_{10}Cu_{20}Ni_8Ti_5$ and for partially crystallized samples with different volume fraction of nanosized particles prepared by isothermal annealing at 673 K for different times.[162][217] Fully amorphous samples exhibit a combination of high strength of up to 1800 MPa, a relatively low Young's modulus of about 70 GPa, which is comparable to Al-based alloys, and some microplasticity of up to 1%. Their fracture toughness amounts to about 55 $MPam^{1/2}$, which is sufficient for engineering applications.[211] While the elastic behavior of the partially crystallized two-phase samples is almost identical to that of the fully amorphous alloy regardless of the amount of nanoparticles, only samples with a crystalline volume fraction of less than 45% show microplasticity of up to 1%. The deformation proceeds by inhomogeneous deformation localized in thin shear bands.[161][216][218] The precipitation of the fine and homogeneously dispersed nanocrystals increases the hardness and the flow stress significantly. For example, the $Zr_{57}Al_{10}Cu_{20}Ni_8Ti_5$ sample containing a volume fraction of 40% nanoparticles [Fig. 42, curve (b)] appears to provide the best compromise between high strength and good ductility. This shows that a homogeneous dispersion of nanosized particles in the amorphous matrix can lead to significant strengthening of the material. Detailed TEM studies proved that the nanoscale precipitates are free of dislocations, which may give rise to the observed high mechanical strength.[217] A further increase in the volume fraction of nanocrystals to 45% [Fig. 42, curve (c)] and to 68% [Fig. 42, curve (d)] mainly favors brittleness. Hence, despite a high hardness, the fracture stress decreases considerably for samples containing larger volume fractions of nanocrystals than about 45%. These results agree well with data reported by several authors.[187][219]–[221]

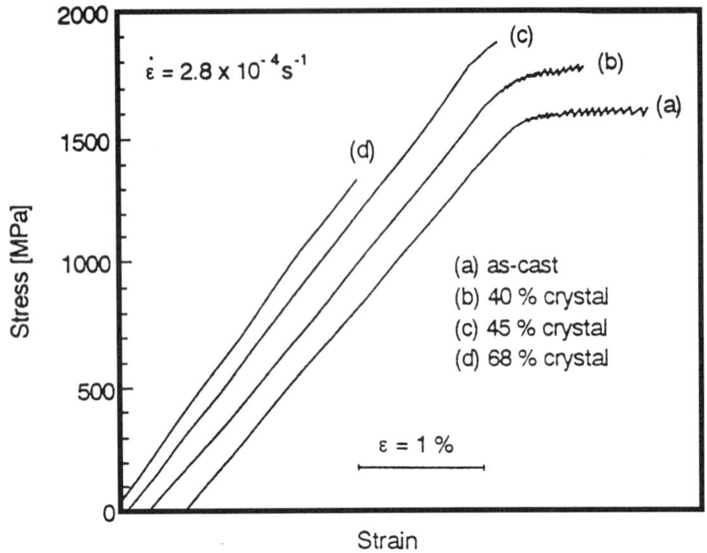

Figure 42. Compressive stress – strain curve of the amorphous and partially crystallized $Zr_{57}Al_{10}Cu_{20}Ni_8Ti_5$ alloy: (*a*) as-cast, (*b*) 40 vol% nanocrystals, (*c*) 45 vol% nanocrystals and (*d*) 68 vol% nanocrystals.[161]

The fracture morphology of partially crystallized samples in comparison with fully amorphous samples remains unchanged. Figure 43 shows a typical fracture surface for a partially crystallized $Zr_{55}Al_{10}Cu_{30}Ni_5$ specimen containing about 25 vol% of nanocrystals as a typical example. The fracture surface reveals a well-developed vein pattern, indicating that the deformation mechanism is governed by the amorphous phase and not by the nanocrystals. Similar findings were obtained for samples containing up to about 50 vol% of nanocrystals. Assuming the viscous shear band deformation mechanism discussed in Ref. 218 is the dominant deformation mechanism at room temperature, precipitated crystallites interacting with the shear bands do not initiate fracture provided their grain size is smaller than the thickness of the shear band. When the volume fraction of nanosized precipitates becomes larger than about 50%, the brittle nature of the precipitated intermetallic compounds is likely to dominate the overall mechanical behavior, leading to the observed decrease in ductility since the deformation is no longer governed by the deformation mechanism of the amorphous phase. This is supported by the brittle deformation behavior of the sample containing 68 vol% nanocrystals [Fig. 42, curve (d)].

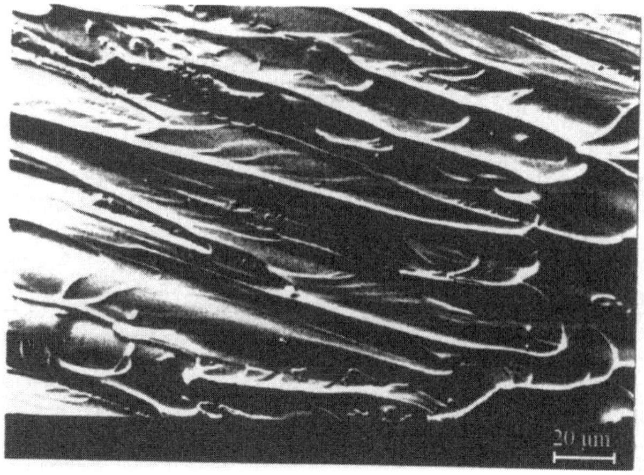

Figure 43. SEM micrograph showing the fracture surface of a $Zr_{57}Al_{10}Cu_{20}Ni_8Ti_5$ sample containing 25 vol% of nanocrystals embedded in an amorphous matrix.

Deformation Behavior at Elevated Temperatures. In order to investigate the deformation characteristics of the supercooled liquid, compression tests at the onset of the glass transition, T_g, as well as at the point of inflection of the glass transition were carried out at constant true strain rates ranging from $\dot{\varepsilon} = 3 \times 10^{-3}$ to 1×10^{-5} s^{-1}. Because of the well known dependence of the glass transition temperature on the heating rate,[222] a heating rate of 20 Kmin^{-1} for all creep specimens was applied prior to testing and the test temperature was set to the corresponding T_g value to ensure that the deformation occurs in the supercooled liquid state. Figure 44 (a and b) shows typical examples for true stress-true strain curves of fully amorphous and partially crystallized $Zr_{55}Al_{10}Cu_{30}Ni_5$ samples at the onset of T_g at strain rates of $\dot{\varepsilon} = 1 \times 10^{-3}$ and 3×10^{-5} s^{-1}, respectively. In contrast to the room temperature measurements, the overall stress level is much lower. Moreover, predominantly homogeneous flow is observed, resulting in extended plasticity without failure. Obviously, the plastic deformation starts with a stress maximum (stress overshoot), which is also observed in other metallic glasses at temperatures below and around T_g[223][224] and then decreases to an almost constant (steady-state) flow level at plastic strains of up to 30%. The underlying reason for the presence of the stress overshoot in these bulk metallic glassy systems is still a matter of debate. The most likely explanation for this phenomenon was given by De Hey, et al.,[223] They measured in tensile creep of ribbons of Pd-based amorphous alloys at temperatures below T_g an increase in the free volume produced by the continuously ongoing shear deformation. Consequently, the viscosity in the deformed specimen, i.e., within the volume subjected to shear, decreases, and, thus, the stress level decays in a test under constant strain rate (Fig. 44 a and b). Recent experiments on $Zr_{55}Al_{10}Cu_{30}Ni_5$ bulk samples verified the same effect to result in an increase of the creep rate in constant stress creep tests.[225]

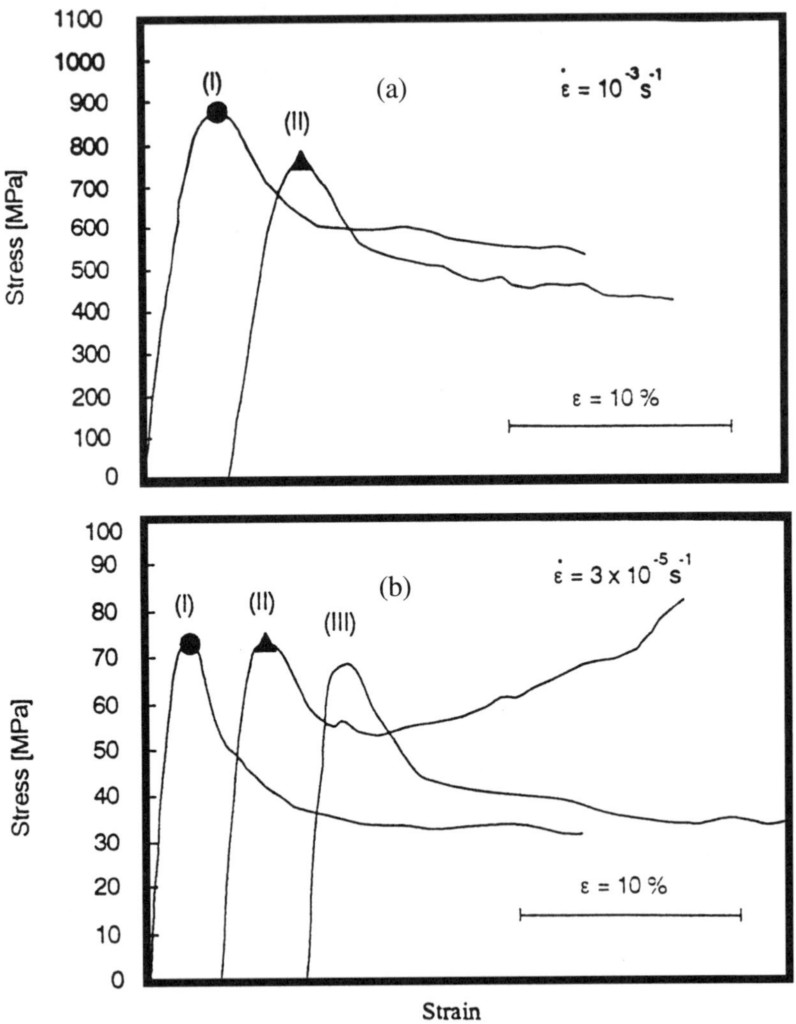

Figure 44. Compressive true-stress–true-strain curves of (I) amorphous, (II) and (III) partially nanocrystallized samples with 25 and 15 vol% of nanocrystals at different strain rate, $\dot{\varepsilon}$.

Comparing deformation curves at different strain rates (Fig. 44, a and b) reveals a mere trend that the peak stress decreases from the fully amorphous state (I) to the partially crystallized material (II), if at all. A peculiar behavior is observed for the partially nanocrystallized two-phase material at the lowest strain rate of $\dot{\varepsilon} = 3 \times 10^{-5}\ s^{-1}$ (Fig. 44b). At this low strain rate the material reveals a significant stress increase after the stress plateau, even exceeding the initial peak stress value. This increase is attributed to the formation of additional nanocrystallites during the long-term thermal exposure upon slow compression. Obviously, the volume fraction of precipitated nanocrystals has increased substantially (Fig. 45). Whether there is a connected change in the viscosity of the remaining amorphous matrix which may superimpose this strengthening effect needs further clarification.

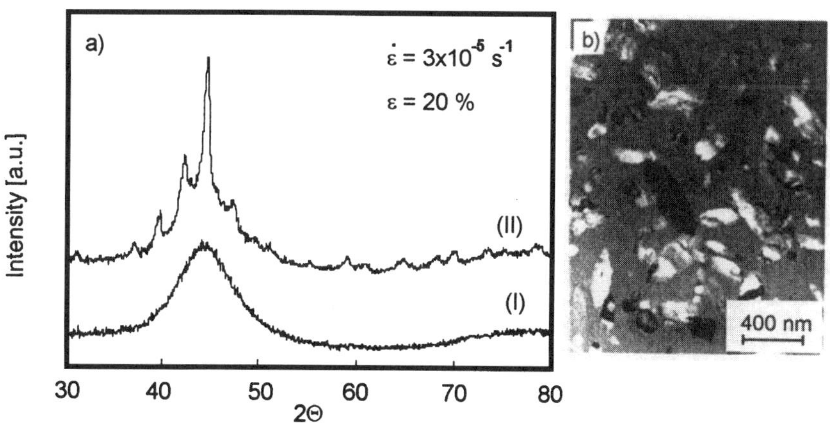

Figure 45. *(a)* X-ray diffraction patterns of (I) amorphous, (II) partially nanocrystallized $Zr_{55}Al_{10}Cu_{30}Ni_5$ samples; *(b)* brightfield TEM micrograph of the specimen corresponding to the diffraction pattern (II) after creep deformation at the onset of T_g [cf. curve (II) in Fig. 44b].

In order to check the occurrence of viscous Newtonian flow, the data points at the stress peak (full symbols) are adopted from tests such as those displayed in Fig. 44 (a and b) and plotted in Fig. 46 on a log of $\dot{\varepsilon}$ - log σ scale for two temperatures, i.e., the onset of T_g (circles) and the point of inflection (squares). Figure 46 comprises results from different testing modes, namely constant strain rate tests (full symbols), constant stress (creep) tests (half filled symbols), and from viscosity measurements using parallel plate rheometry (open symbols). At both temperatures the data points can be successfully explained applying transition state theory to the flow stress which predicts a strain rate–stress relation of the form[226][227]

Eq. (6) $$\dot{\varepsilon} = \dot{\varepsilon}_0 \sinh \frac{\sigma \gamma_0 \Omega_f}{M k_B T}$$

where γ_0 is the local strain produced by the shear site of volume Ω_f. Both, the product of $\gamma_0 \times \Omega_f$ as well as the reference strain rate $\dot{\varepsilon}_0$ are used as fitting parameters for adapting model curves (solid lines in Fig. 46) to the experiments. As expected from theory, Fig. 46 reveals a linear relationship between the strain rate and the stress in the low stress regime, revealing purely Newtonian viscous flow with a stress exponent n $(= \Delta \log \dot{\varepsilon} / \Delta \log \sigma) = 1$. At high stresses, transition to nonlinear stress-dependent flow occurs with n-values of 4 and higher.

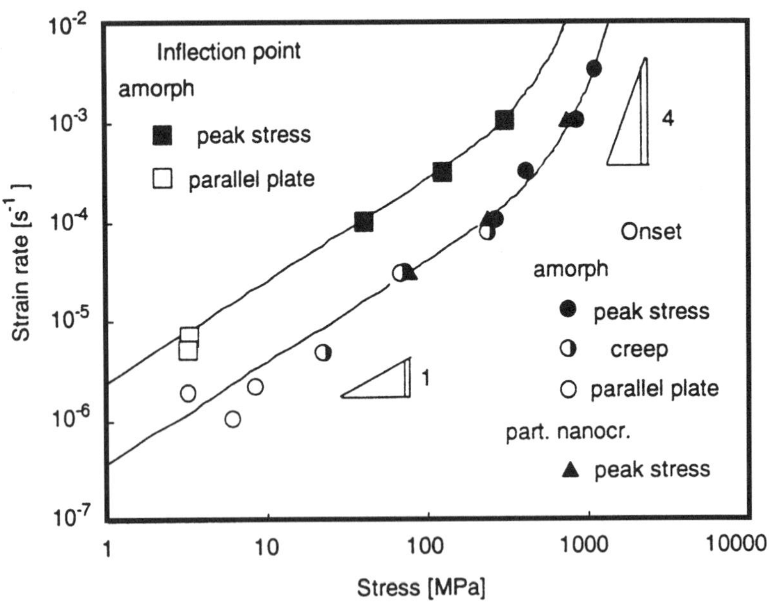

Figure 46. Strain rate–stress relationship (Norton plot) of fully amorphous and partially crystallized $Zr_{55}Al_{10}Cu_{30}Ni_5$ alloys.

Consolidated Bulk Specimens from Amorphous Powders. The easy deformation of metallic glasses at temperatures above T_g allows consolidation of fully dense bulk specimens starting from atomized powders, utilizing the Newtonian flow behavior of the supercooled liquid (further details regarding the viscous flow behavior of amorphous powders and nanostructured composites based on metallic glasses will be given in Sec. 4.4). For this, the extrusion conditions, e.g., the extrusion temperature, the extrusion ratio, and the pressure have to be optimized in order to achieve a high density of the compact without excessive temperature rise due to deformation-induced heating. For example, gas atomized amorphous $Zr_{55}Al_{10}Cu_{30}Ni_5$ powder extruded at an extrusion ratio $R = 5$ at 673 K yields bulk compacts with a density of 98.7%.[141]–[143] The tensile strength of 1520 MPa, the fracture strain of 2%, and the Young's modulus of 80 GPa of the compacted material are nearly the same as the values measured for cast bulk and melt-spun samples. Moreover, the fracture surface of the extruded bulk sample consists of a well-developed vein pattern without indications for interparticle fracture,[142][143] revealing a good bonding between the initial powder particles. Similar data are possible for consolidated bulk specimens when using mechanically alloyed amorphous powders as starting materials.[184]

Increasing the extrusion temperature to 693 K leads to partial crystallization, due to the heat generated by friction. The Young's modulus of the partially crystallized nanostructured composite material increases to 117 GPa and the achieved density also further increases to 99.7%. However, the precipitation of the nanocrystalline compounds embrittles the material, leading to a reduced fracture strain of 0.9% and a fracture strength of 930 MPa. This behavior is consistent with the findings for partially crystallized cast bulk samples, also showing embrittlement when the volume fraction of precipitates exceeds a critical value (see above), and stresses the importance of carefully adjusting the extrusion parameters in order to optimize the mechanical properties of the compacts. In addition, the samples have to be quenched from the extrusion temperature to reintroduce a significant amount of free volume for ductilization.[154] Nevertheless, optimized consolidation promises to allow for the production of high-strength Zr-based two-phase nanostructured materials for alloys exhibiting a successive stepwise crystallization behavior through the powder metallurgy route.

4.4 Mechanically Attrited Composites

The thermal stability investigations of as-milled amorphous powders and composites reveal a distinct glass transition followed by a wide supercooled liquid region. Together with the results obtained for the deformation behavior of cast bulk specimens at elevated temperatures, this points to an extended temperature region where the material exhibits a deformation regime governed by Newtonian flow. This viscous flow behavior is not only of fundamental scientific interest, but also opens the possibility to consolidate and shape such alloys and composites into bulk parts. A better insight into the flow behavior of the supercooled liquid can be derived from viscosity measurements using parallel plate rheometry at a constant heating rate and under isothermal conditions. Details are given in Ref. 228.

Flow Behavior of the Supercooled Liquid and Consolidation. Figure 47 shows viscosity measurements for dispersoid-free $Mg_{55}Y_{15}Cu_{30}$ and for composites with 5 vol% MgO, Cr_2O_3, or Y_2O_3 particles as typical examples for the viscosity data of mechanically alloyed powders. The measured viscosity, η, is an effective viscosity due to the presence of residual unreacted material and the dispersion of the nanosized oxide particles in the Mg-Y-Cu supercooled liquid. For all samples, the viscosity first decreases with increasing temperature upon passing through the glass transition into the supercooled liquid regime. Crystallization at higher temperatures strongly increases the viscosity. The observed course of the viscosity data is consistent with the results obtained for cast Mg-Y-Cu bulk glasses.[84] However, looking more closely reveals that the mechanically attrited metallic glass composites show some differences compared to the dispersoid-free material. Slight differences were also observed in thermal stability investigations of these powders, and are related to (partial) dissolution of some of the dispersoids.[182] Nevertheless, the $Mg_{55}Y_{15}Cu_{30}$ matrix alloy and the composites exhibit a similar temperature dependence of viscosity in the supercooled liquid region.

As stated above, the measured viscosities are effective viscosities due to the combination of hard particles of unreacted material, together with the oxide particles, which are embedded in the supercooled liquid. An estimate for the contribution of the oxide particles to the effective viscosity is possible using the Einstein equation for the flow of mixtures.[229]

Eq. (7) $$\eta_{\text{eff}} = \eta \times (1 + 2.5\, V_f)$$

where η_{eff} is the effective viscosity of the mixture, η the viscosity of the $Mg_{55}Y_{15}Cu_{30}$ supercooled liquid, and V_f the volume fraction of oxide dispersoids. In a first approximation, η is assumed to be identical with or without dispersoids. According to Eq. (7), 5 vol% oxide particles should increase η_{eff} by 12.5%. For all Mg-Y-Cu composites the viscosity changes are larger than this value.

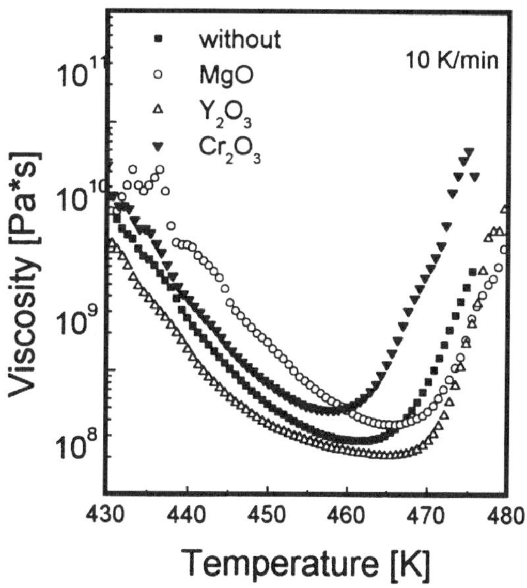

Figure 47. Viscosity values, η, of mechanically alloyed $Mg_{55}Y_{15}Cu_{30}$ powders without dispersoids and with 5 vol% MgO, Cr_2O_3, or Y_2O_3 vs temperature.[181]

A similar effect was observed for Zr-based composites[186][220] and can be attributed to compositional changes of the supercooled liquid. Normalizing η_{eff} of composite samples with the viscosity, η, of the glassy matrix alloy gives the relative viscosity

Eq. (8) $$\eta_{rel} = \eta_{eff}/\eta$$

which directly reveals the changes in viscosity due to the nanocrystalline particles introduced into the glass. Figure 48 shows a plot of the minimum relative viscosity, $\eta_{rel,min}$, in the supercooled liquid just before the beginning of crystallization for $Zr_{55}Al_{10}Cu_{30}Ni_5$ blended with different V_f of nanocrystalline W particles. According to Eq. (8), volume fractions of up to 17.5 vol% are expected to increase the effective viscosity by less than 45% if the viscosity of the metallic liquid remains the same in the presence of the particles. However, the data show a considerably larger increase in $\eta_{rel,min}$ for the W-containing composites than predicted by Eq. (7) (dashed line in Fig. 48). Similar trends were also reported for cast bulk $Zr_{55}Al_{10}Cu_{30}Ni_5$ metallic glass containing micrometer-sized ZrC particles[220] (this data is also included in Fig. 48b for comparison) and for partially crystallized $Zr_{41.2}Ti_{13.8}Cu_{12.5}Ni_{10}Be_{22.5}$.[230] In the latter alloy this was attributed to a significant change in stoichiometry of the glass towards a composition with a higher equilibrium viscosity at a given temperature.

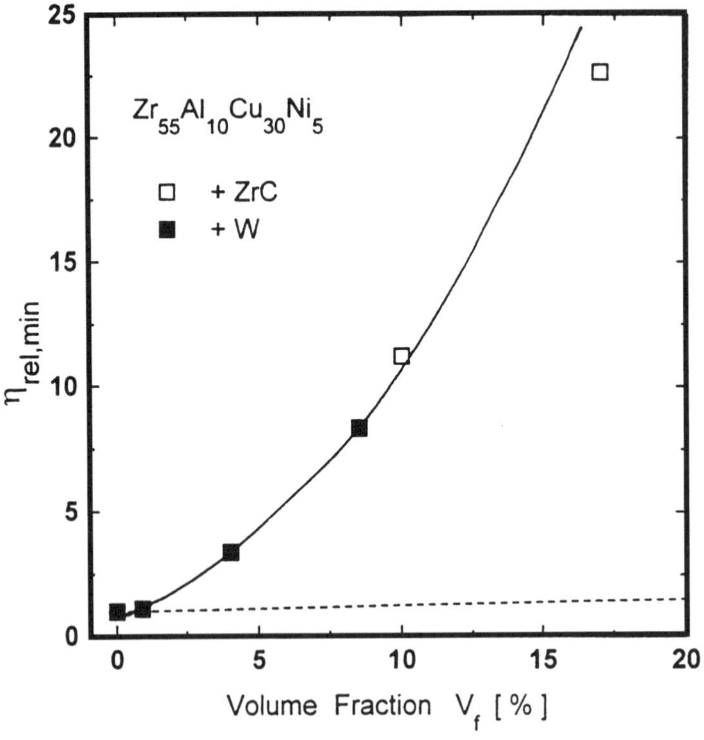

Figure 48. Changes in the minimum relative viscosity of the supercooled liquid, $\eta_{rel,min}$, for mechanically alloyed $Zr_{55}Al_{10}Cu_{30}Ni_5$ composites with different V_f in comparison with data for composites containing micrometer-sized ZrC particles[220] in the same glassy matrix. The expected values according to the Einstein Eq. (7) are also shown for comparison (dashed line).[186]

Comparing the present viscosity data with those for cast bulk composites indicates that the viscosity increase with increasing V_f is mainly related to slight changes in the composition of the glassy matrix phase due to (partial) dissolution of particles. The viscosity increase due to alloying can be described in the framework of the free volume theory of viscous flow[231]–[233] Density fluctuations in the supercooled liquid relate to the viscosity according to:[231]

Eq. (9) $$\eta = \eta_0(T)\cdot \exp\left(\frac{bv_m}{v_f}\right)$$

with the initial viscosity, η_0, the average free volume per atom, v_f, and the critical volume for flow, bv_m.[231] Taking bv_m to be independent of temperature and considering a temperature dependent free volume v_f[232] describes the viscosity as a function of temperature very well.[234] Assuming slight changes in packing density and short-range order by dissolution of particles suggests that the dissolved atoms reduce the amount of free volume of the glass, leading to a viscosity increase. Therefore, the viscosity changes of metallic glass composites can be rationalized by considering changes in the free volume of the glassy matrix together with a minor contribution from the particles hindering the flow of the supercooled liquid.

The free volume model also describes the temperature dependence of the equilibrium viscosity. Rewriting Eq. (9) yields the Vogel-Fulcher-Tammann (VFT) equation[233][235]

Eq. (10) $$\eta = \eta_0 \cdot \exp\,[D^*\cdot T_0/(T - T_0)]$$

where D^* is the fragility parameter and T_0 is the VFT temperature where the barriers, with respect to flow, would go to infinity. D^* is a measure for the fragility of the glass in terms of the fragility concept, classifying material according to the temperature dependence of their kinetics in the liquid and supercooled liquid state: the larger D^*, the stronger is the glass.[235] Taking $A = \ln\eta_0$, $B = bv_m$, and $D^* = B/T_0$ as constants gives a modified form of the VFT relation:[233]

Eq. (11) $$\ln\eta = A + B/(T - T_0)$$

which was used for fitting the data (*A, B, and* T_0 are fitting parameters). The VFT fit yields $D^* = 27$ and $T_0 = 363$ K for the particle-free matrix alloy. Similar fragility parameters were found for other bulk metallic glass forming alloys.[168][228][236]–[238] The determined VFT temperatures are much lower than the calorimetric T_g observed in the DSC which is characteristic for strong liquid behavior.[234]

Although the fragility concept is strictly valid only for homogeneous liquids,[235] one can describe the flow behavior of the nanostructured composites in an analogous way. The presence of nanocrystals in the matrix causes only a small viscosity change according to the Einstein relation [see Eq. (7)]. Such a small change can be neglected when compared to the overall change in viscosity upon the changes in free volume of the matrix. VFT fits for the composites, therefore, primarily describes the viscous flow behavior of the matrix. The VFT parameters derived for the Zr-based W-containing composites are $D^* = 27$–28 and $T_0 = 336$–359 K for the samples with different volume fractions of particles, thus being almost the same as the values found for the matrix alloy. There is no clear correlation between the VFT parameters and the amount of particles present in the supercooled liquid. Hence, the observed changes in the effective viscosity do not significantly affect the T_0 values or the fragility parameters D^* for samples with different V_f. This indicates that the changes in effective viscosity do not drastically alter the overall flow behavior of the material, and that the composites behave, overall, as equally strong liquids as the fully glassy $Zr_{55}Al_{10}Cu_{30}Ni_5$ alloy.

In Fig. 49 the viscosity data are compared with other glass forming alloys[230][231][239] in a fragility plot. This plot enables a true comparison of different systems because the viscosity is normalized to the temperature where all the materials exhibit the same viscosity of 10^{12} Pas,[235] which is taken as the laboratory glass transition temperature, T_g. Strong glasses such as SiO_2 are characterized by an Arrhenius-type temperature dependence of the viscosity. They exhibit a small VFT temperature far below the glass transition temperature and high melt viscosities.[235][239] In contrast, fragile glasses have low melt viscosities and a VFT temperature close to T_g. The fragility plot reveals that the $Zr_{55}Al_{10}Cu_{30}Ni_5$ alloy, as well as the composite samples, behaves closer to the strong than to the fragile glasses. Similar data are known for other bulk metallic glass forming alloys,[84][239] which are also much more viscous than "conventional" metallic glasses like $Au_{77.8}Ge_{13.8}Si_{8.4}$.

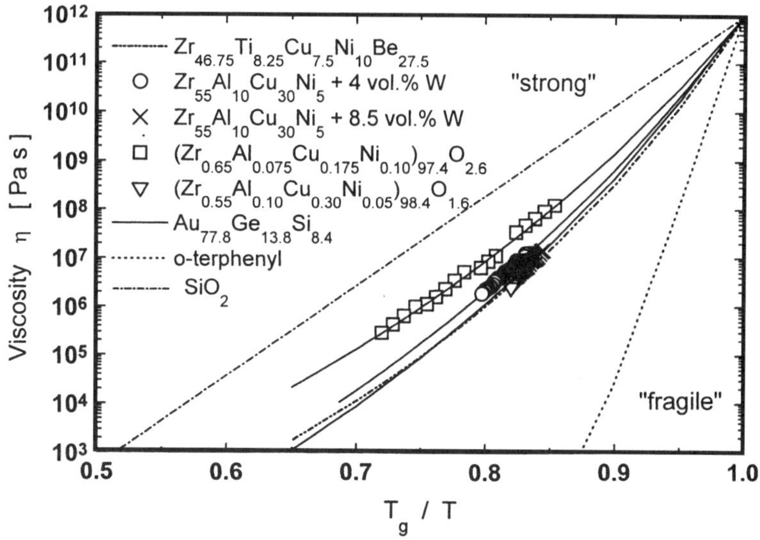

Figure 49. Fragility plot comparing mechanically alloyed $Zr_{55}Al_{10}Cu_{30}Ni_5$ composites with different volume fraction of W particles, V_f, with other "strong" and "fragile" glasses.[186]

However, the changes in effective viscosity for the composites do not drastically affect the overall flow behavior of the material, thus, enabling the consolidation of dense bulk composite specimens from mechanically alloyed powders by hot pressing at temperatures above T_g. This is exemplified in Fig. 50 for a consolidated $Zr_{55}Al_{10}Cu_{30}Ni_5$ composite with 4 vol% W particles. X-ray diffraction and TEM[186] gave no hint for crystallization during compaction. The features of the DSC scans and the heat release upon crystallization of the bulk samples are essentially the same as for the as-milled powders. The bulk samples are about 99.5% dense and can be further deformed or shaped above T_g without crack formation or crystallization. Similar results were obtained for Mg- and Fe-based alloys.[177][182]

Room Temperature Mechanical Properties. The mechanical behavior of consolidated bulk composite specimens prepared from mechanically attrited powders was investigated by room temperature Vickers microhardness, H_v, measurements. No cracks were observed radiating from the corners or sides of the Vickers indentation, indicating a rather high fracture toughness of the material. High hardness values were determined for a variety of Zr- and Mg-based alloys and composites. For example, a hardness of 5.6 ± 0.1 GPa was determined for $Zr_{55}Al_{10}Cu_{30}Ni_5$ glassy specimens. This value increases to about 6.1 ± 0.2 GPa when introducing 17.5 vol% nanocrystalline W particles. This hardness increase is similar to cast $Zr_{55}Al_{10}Cu_{30}Ni_5$ samples containing micrometer-sized ZrC particles,[220] or partially crystallized Zr-Al-Ni-Cu-Ag[240] or Zr-Cu-Pd-Al glasses[148] with nanoscale precipitates. Using the relation $H_v = 3\sigma_y$[186] and neglecting residual porosity in the bulk samples, the yield strength, σ_y, is estimated to be about 1870 MPa for $Zr_{55}Al_{10}Cu_{30}Ni_5$ which increases to about 2000 MPa for a volume fraction of 17.5% nanocrystalline W particles. This is in good agreement with the results from room temperature compressive tests on $Zr_{57}Al_{10}Cu_{20}Ni_8Ti_5$ composite alloys as discussed in the context of Fig. 42.

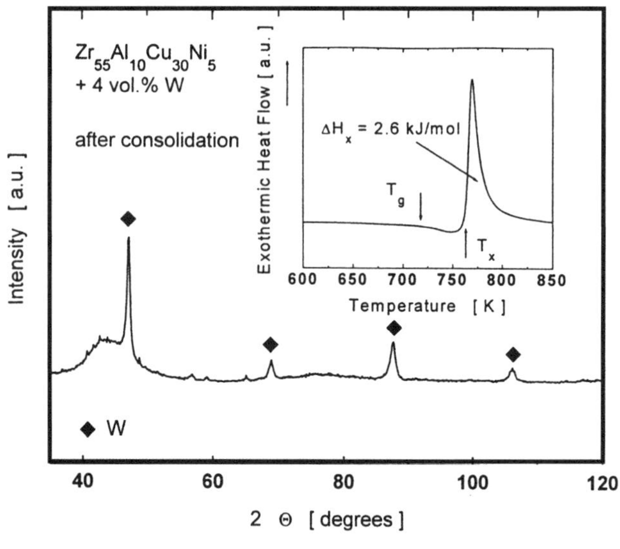

Figure 50. X-ray diffraction pattern for a consolidated $Zr_{55}Al_{10}Cu_{30}Ni_5$ composite with 4 vol% W particles taken from the cross-sectional surface of a bulk sample. The inset shows the corresponding DSC scan (heating rate 40 $Kmin^{-1}$).[186]

Also for $Mg_{55}Y_{15}Cu_{30}$-based composites the hardness increases considerably upon blending with nanosized oxide particles. The H_v values vary between 3.45 GPa for the dispersoid-free material and 3.68–3.88 GPa for the samples with 5 vol% of different oxide particles, corresponding to an estimated yield strength of about 1150 MPa for $Mg_{55}Y_{15}Cu_{30}$ and of

1.23–1280 MPa for the composites. These data are larger than the hardness values of 2.0–3.2 GPa for as-quenched amorphous ribbons and are comparable to the hardness of 3.9–4.3 GPa reported for partially or fully crystallized ribbons of same or comparable composition.[178][179] This suggests that both the nanometer-sized crystals of unreacted constituents and oxide additions increase the mechanical strength of the mechanically alloyed Mg-composites.

To get a better insight into the general trend of the hardness data for different metallic glass composites containing nanoscale particles, the measured hardness data of the composites were normalized to the hardness of the particle-free glassy matrix alloy according to

Eq. (12) $$H_{v,norm} = H_v / H_{v,0}$$

where H_v is the measured hardness of the composite and $H_{v,0}$ is the hardness of the glassy matrix alloy. This normalizing allows the direct comparison of hardness changes between different composites with different matrices and different types of particles. In contrast to the results obtained for composites containing micrometer-sized particles,[220] the normalized hardness of the mechanically alloyed composites containing nanosized particles indicates a steady (linear) increase with V_f (Fig. 51). While the hardness increase for the $Zr_{55}Al_{10}Cu_{30}Ni_5$-based composites is rather moderate, a strong effect is found for the $Mg_{55}Y_{15}Cu_{30}$-based and the $Zr_{65}Al_{7.5}Cu_{17.5}Ni_{10}$-based composites containing nanoscale oxide particles.

The TEM microstructure investigations revealed that the mechanically alloyed composite samples contain uniformly distributed nanocrystals embedded in the glassy matrix. It is hardly conceivable that any dislocation motion takes place in the composites and, therefore, no hardening in the different phases of the composite is expected.[7][8][241] Assuming no special interaction between the nanoparticles and the glassy phase except the force balance, finite element analysis of the unit cell model[197] suggests that the overall hardness of such a composite can be described by a rule of mixtures based on the volume fraction and the hardness of each phase

Eq. (13) $$H_v = V_{f,m} \cdot H_{v,m} + V_{f,W} \cdot H_{v,W}$$

where $V_{f,i}$ is the volume fraction of the respective phase and $H_{v,i}$ is the hardness of the glassy matrix alloy and the nanocrystalline particles, respectively. As shown in Fig. 51, the normalized $H_{v,norm}$ data fit well to a linear relation as expected from a rule of mixtures (dashed and solid lines in Fig. 51). This indicates that the room temperature hardness of the nanoscale composites is determined by both phases, suggesting that nanocrystalline particles are more effective for improving the room temperature strength of metallic glasses than micrometer-sized reinforcements.

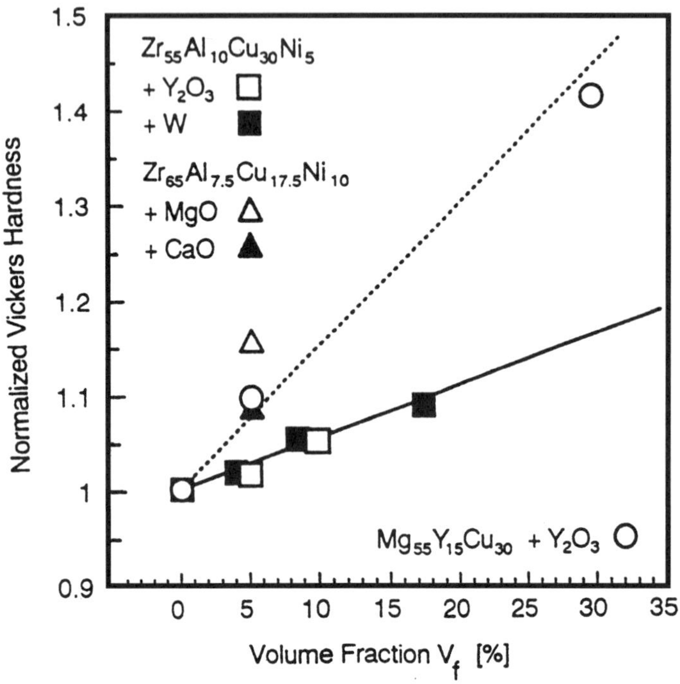

Figure 51. Changes in normalized Vickers hardness, $H_{v,norm}$, for mechanically alloyed $Zr_{55}Al_{10}Cu_{30}Ni_5$-, $Zr_{65}Al_{7.5}Cu_{17.5}Ni_{10}$- and $Mg_{55}Y_{15}Cu_{30}$-based composites as a function of volume fraction of different nanosized particles, V_f.

However, this simple approach neglects effects like partial dissolution of particles into the glassy matrix or composition changes due to the formation of reaction layers/concentration gradients around the particles. Moreover, possible interactions between the shear bands propagating in the glassy matrix and the particles are not considered. Such phenomena are supposed to contribute to the strengthening effect of nanocrystalline metallic precipitates in the amorphous matrix produced by partial crystallization.[148][240] Furthermore, it does not elucidate why the effectiveness of strengthening strongly differs for the various alloy composite systems investigated here, cf. the different slopes of the dashed and the solid line in Fig. 51. Hence, a simple rule of mixtures approach can only give a first estimate for the resulting strength. Further investigations regarding the strengthening mechanisms of metallic glass composites are necessary for a better understanding of their mechanical properties. This will be the topic of future work, also, with respect to their tensile and compression deformation behavior at different strain rates and temperatures.

5.0 SUMMARY AND OUTLOOK

In this overview, phenomenological results concerning the formation of nanostructured two- or multiphase materials in Al-, Mg-, and Zr-based alloy systems and the resulting mechanical properties were presented. Such nanostructured materials may contain crystalline, quasicrystalline, or amorphous phases and are characterized by an ultrafine microstructure, where the different phases are intimately mixed on a nanoscale. As processing techniques, both rapid quenching from the melt or solid state reaction can be utilized. These synthesis routes may directly lead to a two-phase nanostructure. In other cases, additional heat treatment has to be employed in order to create or optimize the desired nanostructure, such as in the case of devitrification of metallic glasses. Whereas crystallization of bulk glass forming alloys can directly yield bulk nanostructured samples with the desired microstructure, rapidly quenched ribbons, gas atomized powders, or mechanically attrited powders and composites have to be subsequently consolidated to achieve dense bulk specimens. When the precursor material is a metallic glass, this can be easily done at relative low temperatures and pressure by using the viscous flow of the supercooled liquid. At least for two-phase materials containing nanoscale crystalline or quasicrystalline particles in a residual amorphous matrix, quenching to room temperature is favorable to achieve a decent ductility. However, special care has to be taken with respect to clean processing and accurate control of consolidation parameters in order to achieve dense bulk specimens without inducing microstructural coarsening or possible phase transitions of the metastable phases building up the unique microstructure of these materials. Hence, despite the possibility of producing bulk material from powders or ribbons, it is highly desirable to obtain such nanostructures directly in bulk, e.g., through casting. In particular, the multicomponent glass forming systems are attractive candidates for creating two-phase nanostructures because of their castability as large-scale bulk specimens and parts, and the ease of inducing nanostructure formation by precipitation of second phase particles by simple partial devitrification of the amorphous phase/supercooled liquid.

The mechanical properties of multicomponent two- or multiphase alloys with nanoscale microstructure are very encouraging regarding the combination of high tensile strength and good ductility at room temperature. For example, Al-, Mg-, or Zr-based alloys containing amorphous/nanocrystalline, amorphous/quasicrystalline, or nanocrystalline/crystalline phases can exhibit very high room temperature strength together with good ductility. At elevated temperatures, two-phase mixtures containing a residual glassy phase can be deformed by viscous flow at rather low temperatures. In addition, some of the nanocrystalline/crystalline or nanocrystalline/quasicrystalline nanostructure variants indicate superplastic behavior at rather high strain rates. Hence, these materials seem to be superior to single-phase nanocrystalline materials which typically exhibit little ductility in tension for grain sizes less than about 25 nm.

This class of materials provides many interesting topics for study of microstructure-property relations, offering both the possibility of discovering and developing new materials and properties, and a way to test models and understand mechanical deformation in ultrafine composites consisting of different phases with different mechanical properties, such as high strength in one phase and good ductility in the other. For example, materials with extremely small grain size and high density of second phase particles, i.e., a small interparticle spacing, provide many challenges to the understanding of deformation theory. Also, procedures for arresting or blunting deformation bands to increase strength, ductility, or toughness have to be further developed and coupled with a detailed knowledge of the actual microstructure of the material, including investigations regarding the nature of the interfaces between the different phases. This is expected to further improve the materials properties and to lead to a basic understanding of how to design new alloys for specific applications.

The development of multicomponent alloys with nanoscale microstructure as technological materials not only requires a basic understanding of the role of chemistry and structure on determining properties, but also an understanding of how to create such structures during large-scale fabrication. Such basic understanding is now developing, and interest in these nanostructured materials is growing. Fabrication processes for large volume production are already being developed, and with the careful optimization of procedures for preparing specific materials for specific applications, the importance of this new class of materials is bound to grow.

ACKNOWLEDGMENTS

The author gratefully acknowledges the support of the German Science Foundation (DFG Schwerpunktprogramm "Unterkühlte Metallschmelzen" and "Quasikristalle;" Grants Ec 111/7, Ec 111/8, Ec 111/9, and He 1872) and of the Alexander von Humboldt Foundation. Special thanks are given to many of my coworkers, in particular, K. Buchholz, S. Deledda, A. Gebert, A. Gümbel, W. Gude, M. Heilmaier, N. Ismail, A. Kübler, U. Kühn, W. Löser, N. Mattern, A. Reger-Leonhard, N. Schlorke-de Boer, F. Schurack, H. Schulze, M. Seidel, B. Weiß, and L. Q. Xing who have all contributed in one way or another to the results presented in this overview. Finally, I would like to thank R. Busch, A. Inoue, J. Z. Jiang, W. L. Johnson, C. C. Koch, J. H. Perepezko, K. Samwer, and L. Schultz for many valuable discussions.

REFERENCES

1. Gleiter, H., *Prog. Mater. Sci.*, 33:223 (1989)
2. Siegel, R. W., in: *Mechanical Properties and Deformation Behavior of Materials Having Ultrafine Microstructures,* (M. Nastasi, D. M. Parkin, and H. Gleiter, eds.), 233:509, NATO ASI Series, Kluwer, Dordrecht (1993)
3. Edelstein, A. S., and Cammarata, R. C., (eds), *Nanomaterials: Synthesis, Properties and Applications*, IOP Publishing, Bristol (1996)
4. Andres, R. P., Averback, R. S., Brown, W. L., Brus, L. E., Goddard, W. A. III, Kaldor, A., Louie, S. G., Moscovits, M., Peercy, P. S., Riley, S. J., Siegel, R. W., Spaepen, F., and Wang, Y., *J. Mater. Res.*, 4:704 (1989)
5. Hall, E. O., *Proc. Phys. Soc. London B*, 64:747 (1951)
6. Petch, N. J., *J. Iron Steel Inst.*, 174:25 (1953)

7. Weertman, J. R., and Averback, R. S., in: *Nanomaterials: Synthesis, Properties and Applications*, (A. S. Edelstein and R. C. Cammarata, eds.), Ch. 13, p. 323, IOP Publ., Bristol (1996)

8. Morris, D. G., *Mechanical Behaviour of Nanostructured Materials*, Materials Science Foundations, Vol. 2, Trans. Tech. Publ., Zürich (1998)

9. Weertman, J. R., Farkas, D., Hemker, K., Kung, H., Mayo, M., Mitra, R., and Van Swygenhoven, H., *MRS Bull.,* 24(2):44 (1999)

10. Koch, C. C., Morris, D. G., Lu, K., and Inoue, A., *MRS Bull.*, 24(2):55 (1999)

11. Hadjipanayis, G. C., and Siegel, R. W., (eds), *Nanophase Materials: Synthesis - Properties - Applications*, Kluwer, Dordrecht (1994)

12. Klement, W., Jr., Willens, W. H., and Duwez, P., *Nature*, 187:869 (1960)

13. Jones, H., *Rapid Solidification of Metals and Alloys*, The Institute of Metallurgists, London (1982)

14. Anantharaman, T. R., and Suryanarayana, C., *Rapidly Solidified Metals: A Technological Overview*, Trans. Tech. Publ., Zürich (1987)

15. Suryanarayana, C., and Jones, H., in: *Materials Science and Technology*, (R. W. Cahn, P. Haasen, and E. J. Kramer, eds.), 15:55, VCH Verlagsgesellschaft, Weinheim (1991)

16. Groeber, H., and Hanemann, H., *Arch. Eisenhüttenwesen*, 11:199 (1937)

17. Serita, Y., Ishikawa, T., and Kimura, F., *J. Jpn. Inst. Light Metals*, 20:1 (1970)

18. Hinesly, C. P., and Morris, J. G., *Metall. Trans.*, 1:1476 (1970)

19. Armstrong, G. R., and Jones, H., in: *Solidification and Casting of Metals*, p. 454, The Metals Society, London (1979)

20. Inoue, A., *Mater. Trans., JIM*, 36:866 (1995)

21. Johnson, W. L., in: *Metastable, Mechanically Alloyed and Nanocrystalline Materials,* (R. Schulz, ed.), 225–227:35, *Mater. Sci. Forum,* Trans. Tech. Publ., Zürich (1996)

22. Benjamin, J. S., *Sci. Am.*, 234:40 (1976)

23. Benjamin, J. S., *Metall. Trans.*, 1:2943 (1970)

24. Schultz, L., and Eckert, J., in: *Glassy Metals III, Topics in Applied Physics*, (H. J. Güntherodt and H. Beck, eds.), 72:69, Springer, Berlin (1994)

25. Koch, C. C., in: *Materials Science and Technology*, (R. W. Cahn, P. Haasen, and E. J. Kramer, eds.), 15:193, VCH, Verlagsgesellschaft, Weinheim (1991)

26. Shingu, P. H., (ed.), Mechanical Alloying, Vol. 88–90, *Mater. Sci. Forum*, Trans. Tech. Publ., Zürich (1992)

27. Yavari, A. R., (ed.), Mechanically Alloyed and Nanocrystalline Materials, Vol. 179–181, *Mater. Sci. Forum*, Trans. Tech. Publ., Zürich (1995)

28. Schulz, R., (ed.), Metastable, Mechanically Alloyed and Nanocrystalline Materials, Vol. 225–227, *Mater. Sci. Forum*, Trans. Tech. Publ., Zürich (1996)

29. Fiorani, D., and Magini, M., (eds.), Synthesis and Properties of Mechanically Alloyed and Nanocrystalline Materials, Vol. 235–238, *Mater. Sci. Forum*, Trans. Tech. Publ., Zürich (1997)

30. Baró, M. D., and Suriñach, S., (eds.), Mechanically Alloyed, Metastable and Nanocrystalline Materials, Vol. 269–272, *Mater. Sci. Forum*, Trans. Tech. Publ., Zürich (1998)

31. Calka, A., and Wexler, D., (eds.), Metastable, Mechanically Alloyed and Nanocrystalline Materials, Vol. 312–314, *Mater. Sci. Forum*, Trans. Tech. Publ., Zürich (1999); *J. Metastable and Nanocrystalline Materials*, 2–6 (1999)

32. Fecht, H. J., in: *Nanomaterials: Synthesis, Properties and Applications*, (A. S. Edelstein and R. C. Cammarata, eds.), Ch. 5, p. 89, IOP Publ., Bristol (1996)

33. Schwarz, R. B., in: Mechanically Alloyed, Metastable and Nanocrystalline Materials, (M. D. Baró and S. Suriñach, eds.), 269–272:665, *Mater. Sci. Forum*, Trans. Tech. Publ., Zürich (1998)

34. Koch, C. C., *Nanostructured Mater.* 2:109 (1993)

35. Eckert, J., Schultz, L., and Urban, K., *Appl. Phys. Lett.*, 55:117 (1989)

36. Eckert, J., Schultz, L., and Urban, K., *Acta Metall. Mater.*, 39:1497 (1991)

37. Eckert, J., Schultz, L., and Urban, K., *Europhys. Lett.*, 13:349 (1990)

38. Eckert, J., Holzer, J. C., Krill, C. E., III, and Johnson, W. L., *J. Mater. Res.*, 7:1751 (1992)

39. Hellstern, E., Fecht, H. J., Fu, Z., and Johnson, W. L., *J. Appl. Phys*, 65:305 (1989)

40. Fecht, H. J., Hellstern, E., Fu, Z., and Johnson, W. L., *Metall. Trans. A*, 21:2333 (1990)

41. Benghalem, A., and Morris, D. G., *Acta Metall. Mater.*, 42:4071 (1994)

42. Eckert, J., Holzer, J. C., Krill, C. E., III, and Johnson, W. L., *J. Appl. Phys.*, 73:2794 (1993)

43. Dougherty, G. M., Shiflet, G. J., and Poon, S. J., *Acta Metall. Mater.*, 42:2275 (1994)

44. Seidel, M., Eckert, J., and Schultz, L., *J. Appl. Phys.*, 77:5446 (1995)
45. Yang, J. S. C., and Koch, C. C., *J. Mater. Res.*, 5:325 (1990)
46. Schlump, W., and Grewe, H., in: *New Materials by Mechanical Alloying Techniques*, (E. Arzt and L. Schultz, eds.), p. 307, DGM Informationsgesellschaft, Oberursel (1989)
47. Shingu, P. H., Ishihara, K. N., and Kuyama, J., in: *Proc. 34th Japan Congress on Materials Research*, p. 19, Society of Materials Science, Kyoto (1991)
48. Swinkels, F. B., Wilkenson, D. S., Arzt, E., and Ashby, M. F., *Acta Metall. Mater.*, 31:1829 (1983)
49. Gutmanas, E. Y., Trusov, L. I., and Gotman, I., *Nanostruct. Mater.*, 4:893 (1994)
50. Sadangi, R. K., Kear, B. H., and McCandlish, L. E., *Nanostruct. Mater.*, 2:563 (1993)
51. Cline, C. F., and Hopper, R. W., *Scr. Metall. Mater.*, 11:1137 (1977)
52. Schwarz, R. B., Kasiraj, P., Vreeland, T., and Ahrens, T. J., *Acta Metall. Mater.*, 32:1243 (1984)
53. Jain, M., and Christman, T., *Acta Metall. Mater.*, 42:1901 (1994)
54. Suryanarayana, C., *Int. Mater. Rev.*, 40:21 (1995)
55. Vance, R. R., and Courtney, T. H., *Scr. Metall. Mater.*, 26:1435 (1992)
56. He, L., and Ma, E., *Mater. Sci. Eng. A,* 204:240 (1995)
57. Shaik, G. R., and Milligan, W. W., *Metall. Mater. Trans. A*, 28:895 (1997)
58. Sanders, P. G., Fougere, G., Thompson, L. J., Eastman, J. E., and Weertman, J., *Nanostruct. Mater.*, 8:243 (1997)
59. Morris, D. G., and Morris, M. A., *Acta Metall. Mater.*, 39:1763 (1991)
60. Birringer, R., *Mater. Sci. Eng. A*, 117:33 (1989)
61. Gertsman, V. Y., and Birringer, R., *Scr. Metall. Mater.*, 30:577 (1994)
62. Hahn, H., Logas, J., and Averback, R. S., *J. Mater. Res.*, 5:609 (1990)
63. Averback, R. S., Höfler, H. J., and Tao, R., *Mater. Sci. Eng. A*, 166:169 (1993)
64. Zener, C., quoted by Smith, C. S., *Trans. AIME,* 175:47 (1948)
65. Lu, K., *Nanostruct. Mater.*, 2:643 (1993)
66. Grundmann, U., Heilmaier, M., and Schultz, L., *Scr. Metall. Mater.*, 37:103 (1997)
67. Heilmaier, M., Saage, H., and Eckert, J., *Mater. Sci. Eng. A*, 239–240:652 (1997)
68. Lu, K., *Mater. Sci. Eng. R*, 16:161 (1996)
69. Köster, U., and Herold, U., in: *Glassy Metals I, Topics in Applied Physics*, (H. J. Güntherodt and H. Beck, eds.), 46:225, Springer, Berlin (1981)

70. Köster, U., in: Synthesis and Properties of Mechanically Alloyed and Nanocrystalline Materials, (D. Fiorani and M. Magini, eds.), 235–238:377, *Mater. Sci. Forum*, Trans. Tech. Publ., Zürich (1997)

71. Köster, U., Rüdiger, A., and Meinhardt, J., in: Advances in Nanocrystallization, (A. R. Yavari, ed.), 307:9, *Mater. Sci. Forum*, Trans. Tech. Publ., Zürich (1997); *J. Metastable and Nanocrystalline Materials*, 1:9 (1999)

72. Köster, U., Spassow, T., and Sutton, M., in: *Proc. 3rd Int. Workshop on Non-Crystalline Solids, Matalascañas 1991*, (A. Conde, C. F. Conde, and M. Millán, eds.), p.149, World Scientific, Singapore (1992)

73. Marcus, M., and Turnbull, D., *Mater. Sci. Eng.*, 23:211 (1976)

74. Ohlberg, S. M., Golob, H. R., and Stickler, D. W., in: *Nucleation and Crystallization in Glasses and Melts*, p. 55, The American Ceramic Society, Columbus, OH (1962)

75. Ramsden, A. H., and James, P. F., *J. Mater. Sci.*, 19:1406 (1984)

76. Huston, E. L., Cahn, J. W., and Hilliard, J. E., *Acta Metall. Mater.*, 14:1053 (1966)

77. Chou, C. P. and Turnbull, D., *J. Non-Cryst. Solids*, 17:169 (1975)

78. Chen, H. S., *Mater. Sci. Eng.*, 23:151 (1976)

79. Tanner, L. E., and Ray, R., *Scr. Metall. Mater.*, 14:657 (1980)

80. Pelton, A. R., and Tanner, L. E., in: *Rapidly Quenched Metals*, (S. Steeb and H. Warlimont, eds.), p. 747, North-Holland Physics Publ., Amsterdam (1985)

81. Busch, R., Schneider, S., and Johnson, W. L., *Appl. Phys. Lett.*, 67:1544 (1995)

82. Schneider, S., Geyer, U., Thiyagarajan, P., Busch, R., Schulz, R., Samwer, K., and Johnson, W. L., in: Metastable, Mechanically Alloyed and Nanocrystalline Materials, (R. Schulz, ed.), 225–227:59, *Mater. Sci. Forum*, Trans. Tech. Publ., Zürich (1996)

83. Schneider, S., Geyer, U., Thiyagarajan, P., and Johnson, W. L., in: Synthesis and Properties of Mechanically Alloyed and Nanocrystalline Materials, (D. Fiorani and M. Magini, eds.), 235–238:337, *Mater. Sci. Forum*, Trans. Tech. Publ., Zürich (1997)

84. Busch, R., Liu, W., and Johnson, W. L., *J. Appl. Phys.* 83:4134 (1998)

85. Liu, W., Johnson, W. L., Schneider, S., Geyer, U., and Thiyagarajan, P., *Phys. Rev. B*, 59:11755 (1999)

86. Liu, W., Ph.D. Thesis, California Institute of Technology, Pasadena (1998)

87. Busch, R., and Johnson, W. L., *Appl. Phys. Lett.*, 72:2695 (1998)

88. Kim, Y. H., Inoue, A., and Masumoto, T., *Mater. Trans. JIM*, 31:747 (1990)
89. Chen, H., He, Y., Shiflet, G. J., and Poon, S. J., *Scr. Metall. Mater.*, 25:1421 (1991)
90. He, Y., Shiflet, G. J., and Poon, S. J., *Acta Metall. Mater.*, 43:83 (1995)
91. Inoue, A., Yamamoto, M., Kimura, H. M., and Masumoto, T., *J. Mater. Sci. Lett.*, 6:194 (1987)
92. Inoue, A., Ohtera, K., Tsai, A. P., and Masumoto, T., *Jpn. J. Appl. Phys.*, 27:L479 (1988)
93. He, Y., Poon, S. J., and Shiflet, G. J., *Science*, 241:1640 (1988)
94. Inoue, A., and Masumoto, T., in: *Encyclopedia of Materials Science and Engineering,* (R. W. Cahn, ed.), 2:660, Pergamon Press, Oxford (1990)
95. Inoue, A., Nakazato, K., Kawamura, Y., and Masumoto, T., *Mater. Sci. Eng. A*, 179/180:654 (1994)
96. Kimura, H., Sasamori, K., and Inoue, A., *Mater. Trans. JIM,* 37:1722 (1996)
97. Inoue, A., *Mater. Sci. Eng. A*, 179/180:57 (1994)
98. Inoue, A., and Kimura, H. M., in: Synthesis and Properties of Mechanically Alloyed and Nanocrystalline Materials, (D. Fiorani and M. Magini, eds.), 235–238:873, *Mater. Sci. Forum*, Trans. Tech. Publ., Zürich (1997)
99. Inoue, A., and Kimura, H. M., in: *Quasicrystals*, (J. M. Dubois, P. A. Thiel, A. P. Tsai, and K. Urban, eds.), 553:495, *Mater. Res. Soc. Symp. Proc.,* Materials Research Society, Warrendale, PA (1999)
100. Inoue, A., Ohtera, K., Tsai, A. P., and Masumoto, T., *Jpn. J. Appl. Phys.,* 27:L280 (1988)
101. Tsai, A. P., Inoue, A., and Masumoto, T., *Metall. Trans. A*, 19:1369 (1988)
102. Inoue, A., Amiya, K., Yoshii, I., Kimura, H. M., and Masumoto, T., *Mater. Trans. JIM*, 35:485 (1994)
103. Inoue, A., Ohtera, K., Tsai, A. P., Kimura, H. M., and Masumoto, T., *Jpn. J. Appl. Phys.*, 27:L1579 (1988)
104. Inoue, A., Kita, K., Ohtera, K., and Masumoto, T., *J. Mater. Sci. Lett.*, 7:1287 (1988)
105. Inoue, A., Kimura, H. M., and Kita, K., in: *New Horizons in Quasicrystals*, (A. I. Goldman, D. J. Sordelet, P. A. Thiel, and J. M. Dubois, eds.), p. 256, World Scientific, Singapore (1997)
106. Ohtera, K., Inoue, A., Terabayashi, T., Nagahama, H., and Masumoto, T., *Mater. Trans. JIM,* 33:775 (1992)
107. Kim, Y. H., Inoue, A., and Masumoto, T., *Mater. Trans., JIM,* 32:599 (1991)

108. Inoue, A., Watanabe, M., Kimura, H. M., Takahashi, F., Nagata, A., and Masumoto, T., *Mater. Trans. JIM,* 33:723 (1992)

109. Watanabe, M., Inoue, A., Kimura, H. M., Aiba, T., and Masumoto, T., *Mater. Trans. JIM*, 34:162 (1993)

110. Inoue, A., Watanabe, M., Kimura, H. M., and Masumoto, T., *Sci. Rep. Res. Inst. Tohoku Univ. A*, 38:138 (1993)

111. Kimura, H. M., Sasamori, K., Watanabe, M., Inoue, A., and Masumoto, T., *Mater. Sci. Eng. A*, 181/182:845 (1994)

112. Inoue, A., Kimura, H. M., Sasamori, K., and Masumoto, T., *Mater. Trans. JIM*, 36:6 (1995)

113. Kimura, H. M., Inoue, A., Sasamori, K., Kim, Y. H., and Masumoto, T., *Mater. Trans. JIM*, 36:1004 (1995)

114. Inoue, A., Kimura, H. M., Sasamori, K., and Masumoto, T., *Mater. Trans. JIM*, 37:1287 (1996)

115. Inoue, A., Kimura, H. M., Sasamori, K., and Masumoto, T., *Mater. Trans. JIM*, 35:85 (1994)

116. Schurack, F., Börner, I., Eckert, J., and Schultz, L., in: Metastable, Mechanically Alloyed and Nanocrystalline Materials, (A. Calka and D. Wexler, eds.), 312–314:49, *Mater. Sci. Forum*, Trans. Tech. Publ., Zürich (1999); *J. Metastable and Nanocrystalline Materials*, 2–6:49 (1999)

117. Schurack, F., Eckert, J., and Schultz, L., *Nanostruct. Mater.*, 12:107 (1999)

118. Inoue, A., Kimura, H. M., and Sasamori, K., *Advanced Mater.*, 4:91 (1998)

119. Inoue, A., Kimura, H. M., and Masumoto, T., *Nanostruct. Mater.*, 7:363 (1996)

120. Kimura, H. M., Sasamori, K., and Inoue, A., *Mater. Trans. JIM*, 37:1722 (1996)

121. Inoue, A., Ohtera, K., Kita, K., and Masumoto, T., *Jpn. J. Appl. Phys.*, 27:L2248 (1988)

122. Inoue, A., Kohinata, M., Ohtera, K., Tsai, A. P., and Masumoto, T., *Mater. Trans. JIM*, 30:378 (1989)

123. Kim, S. G., Inoue, A., and Masumoto, T., *Mater. Trans. JIM*, 31:929 (1990)

124. Kato, A., Horikiri, H., Inoue, A., and Masumoto, T., *Mater. Sci. Eng. A,* 179/180:707 (1994)

125. Inoue, A., and Masumoto, T., *Mater. Sci. Eng. A*, 173:1 (1993)

126. Shibata, T., Kawanishi, M., Nagahara, J., Inoue, A., and Masumoto, T., *Mater. Sci. Eng. A*, 179/180:632 (1994)

127. Inoue, A., *Bulk Amorphous Alloys*, Vol. 4, Materials Science Foundations, Trans. Tech. Publ., Zürich (1998)

128. Inoue, A., and Masumoto, T., *Mater. Sci. Eng. A*, 133:6 (1991)

129. Eckert, J., Schlorke, N., and Schultz, L., in: *Synthesis/Processing of Lightweight Materials II*, (C. M. Ward-Close, F. H. Froes, D. J. Chellman, and S. S. Cho, eds.), p. 383, The Minerals, Metals, and Materials Society, Warrendale, PA (1997)

130. Liu, W., and Johnson, W. L., *J. Mater. Res.*, 11:2388 (1996)

131. Inoue, A., Nakamura, T., Nishiyama, N., and Masumoto, T., *Mater. Trans. JIM*, 33:937 (1992)

132. Inoue, A., Nishiyama, N., Kim, S. G., and Masumoto, T., *Mater. Trans. JIM*, 33:360 (1992)

133. Kato, A., Suganuma, T., Horikiri, H., Kawamura, Y., Inoue, A., and Masumoto, T., *Mater. Sci. Eng. A*, 179/180:112 (1994)

134. Inoue, A., Horikiri, H., Kato, A., and Masumoto, T., *Mater. Trans. JIM*, 35:79 (1994)

135. Uoya, A., Shibata, T., Higashi, K., Inoue, A., and Masumoto, T., *J. Mater. Res.*, 11:2731 (1996)

136. Inoue, A., Zhang, T., and Masumoto, T., *Mater. Trans. JIM*, 31:425 (1990)

137. Inoue, A., Zhang, T., Nishiyama, N., Ohba, K., and Masumoto, T., *Mater. Trans. JIM*, 34:1234 (1993)

138. Lin, X. H., Johnson, W. L., and Rhim, W. K., *Mater. Trans. JIM*, 38:473 (1997)

139. Xing, L. Q., Eckert, J., Löser, W., and Schultz, L., *Appl. Phys. Lett.*, 73:2110 (1999)

140. Gebert, A., Eckert, J., and Schultz, L., *Acta Metall. Mater.*, 46:5475 (1998)

141. Kawamura, Y., Kato, H., Inoue, A., and Masumoto, T., *Appl. Phys. Lett.*, 67:2008 (1995)

142. Kato, H., Kawamura, Y., and Inoue, A., *Mater. Trans. JIM*, 37:70 (1996)

143. Kawamura, Y., Kato, H., and Inoue, A., *Advanced Mater.*, 4:4861 (1998)

144. Eckert, J., Mattern, N., Zinkevitch, M., and Seidel, M., *Mater. Trans. JIM*, 39:623 (1998)

145. Köster, U., Meinhardt, J., Roos, S., and Liebertz, H., *Appl. Phys. Lett.*, 69:179 (1996)

146. Frankwicz, P. S., Ram, S., and Fecht, H. J., *Mater. Lett.*, 28:77 (1996)

147. Chen, M. W., Inoue, A., Sakurai, T., Ping, D. H., and Hono, K., *Appl. Phys. Lett.*, 74:812 (1999)

148. Fan, C., and Inoue, A., *Mater. Trans. JIM*, 38:1040 (1997)

149. Fan, C., Takeuchi, A., and Inoue, A., *Mater. Trans. JIM*, 40:42 (1999)

150. Eckert, J., Mattern, N., Seidel, M., and Schultz, L., in: *Structure and Dynamics of Glasses and Glass Formers,* (A. Angell, K. L. Ngai, J. Kieffer, T. Egami, and G. U. Nienhaus, eds.), 455:465, *Mater. Res. Soc. Symp. Proc.,* Materials Research Society, Warrendale, PA (1997)

151. Mattern, N., Roth, S., Bauer, H. D., Henninger, G., and Eckert, J., *Mater. Sci. Eng. A*, 304–306:311 (2001)

152. Xing, L. Q., Eckert, J., Löser, W., Schultz, L., and Herlach, D. M., *Phil. Mag. A*, 79:1095 (1999)

153. Fan, C., Louzguine, D. V., Li, C., and Inoue, A., *Appl. Phys. Lett.*, 75:340 (1999)

154. Inoue, A., Zhang, T., Saida, J., Matsushita, M., Chen, M. W., and Sakurai, T., *Mater. Trans. JIM*, 40:1137 (1999)

155. Inoue, A., Zhang, T., Saida, J., Matsushita, M., Chen, M. W., and Sakurai, T., *Mater. Trans. JIM*, 40:1181 (1999)

156. Schurack, F., Eckert, J., and Schultz, L., *Mater. Sci. Eng. A*, 294–296:164 (2000)

157. Haas, V., Cho, M. I., Ishii, H., and Inoue, A., in: *Advances in Nanocrystallization*, (A. R. Yavari, ed.), 307:203, *Mater. Sci. Forum*, Trans. Tech. Publ., Zürich (1997); *J. Metastable and Nanocrystalline Materials*, 1:203 (1999)

158. Haas, V., Cho, M. I., Ishii, H., and Inoue, A., *Nanostruct. Mater.*, 12:829 (1999)

159. Büchler, E. H., Watanabe, E., and Kazama, N. S., *Int. J. Non-Equilibrium Processing*, 10:35 (1997)

160. Inoue, A., Kato, A., Zhang, T., Kim, S. G., and Masumoto, T., *Mater. Trans. JIM*, 32:609 (1991)

161. Leonhard, A., Xing, L. Q., Heilmaier, M., Gebert, A., Eckert, J., and Schultz, L., *Nanostruct. Mater.*, 10:805 (1998)

162. Xing, L. Q., Eckert, J., Löser, W., and Schultz, L. *Appl. Phys. Lett.*, 74:664 (1999)

163. Chen, M. W., Zhang, T., Inoue, A., Sakai, A., and Sakurai, T., *Appl. Phys. Lett.*, 75:1697 (1999)

164. Saida, J., Matsushita, M., Zhang, T., Inoue, A., and Sakurai, T., *Appl. Phys. Lett.*, 75:3497 (1999)

165. Sagel, A., Wunderlich, R. K., Perepezko, J. H., and Fecht, H. J., *Appl. Phys. Lett.*, 70:580 (1997)

166. Eckert, J., *Mater. Sci. Eng. A*, 226–228:364 (1997)

167. Schlorke, N., Eckert, J., and Schultz, L., *Mater. Sci. Eng. A*, 226–228:425 (1997)

168. Eckert, J., Seidel, M., Kübler, A., Klement, U., and Schultz, L., *Scr. Metall. Mater.*, 38:595 (1998)

169. Benameur, T., Inoue, A., and Masumoto, T., *Mater. Trans. JIM*, 35: 451 (1994)

170. Benameur, T., Inoue, A., and Masumoto, T., *Nanostruct. Mater.*, 4:303 (1994)

171. Seidel, M., Eckert, J., Bauer, H. D., and Schultz, L., in: *Grain Size and Mechanical Properties - Fundamentals and Applications*, (M. A. Otooni, R. W. Armstrong, N. J. Grant, and K. Ishizaki, (eds.), 362:239, *Mater. Res. Soc. Symp. Proc.,* Materials Research Society, Warrendale, PA (1995)

172. Kawamura, Y., Inoue, A., Sasamori, K., Katoh, A., and Masumoto, T., *Mater. Trans. JIM*, 34:969 (1993)

173. Tsai, A. P., Aoki, K., Inoue, A., and Masumoto, T., *J. Mater. Res.*, 8:5 (1993)

174. Mizutani, U., Takeuchi, T., and Fukunaga, T., *Mater. Trans. JIM*, 34:102 (1993)

175. Schurack, F., and Eckert, J., unpublished results.

176. Asahi, N., *Mater. Sci. Eng. A*, 226–228:67 (1997)

177. Schlorke, N., Eckert, J., and Schultz, L., in: Mechanically Alloyed, Metastable and Nanocrystalline Materials, (M. D. Baró and S. Suriñach, eds.), 269–272:761, *Mater. Sci. Forum*, Trans. Tech. Publ., Zürich (1998)

178. Schlorke, N., Eckert, J., and Schultz, L., in: *Magnesium Alloys and Their Applications*, (B. L. Mordike and K. U. Kainer, eds.), p. 533, DGM Informationsgesellschaft, Frankfurt (1998)

179. Inoue, A., Kohinata, M., Tsai, A. P., and Masumoto, T., *Mater. Trans. JIM*, 30:378 (1990)

180. Eckert, J., Schlorke, N., and Schultz, L., in: *Proc. 1998 Powder Metallurgy World Congress, Granada*, 1:311, European Powder Metallurgy Association, Shrewsbury (1998)

181. Eckert, J., Schlorke-de Boer, N., Weiß, B., and Schultz, L., *Z. Metallkde.*, 90:908 (1999)

182. Seidel, M., Ph.D. Thesis, University of Technology, Dresden (1998)

183. Kuhnast, F. A., Held, O., Regnier, F., and Illeková, E., *Mater. Sci. Eng. A*, 226–228:463 (1997)

184. Eckert, J., Seidel, M., Schlorke, N., Kübler, A., and Schultz, L., in: *Synthesis and Properties of Mechanically Alloyed and Nanocrystalline Materials*, (D. Fiorani and M. Magini, eds.), 235–238:113, Mater. Sci. Forum, Trans. Tech. Publ., Zürich (1997)

185. Lin, X. H., and Johnson, W. L., *J. Appl. Phys.*, 78:6514 (1995)

186. Eckert, J., Kübler, A., and Schultz, L., *J. Appl. Phys.*, 85:7112 (1999)

187. Eckert, J. L., in: Metastable, Mechanically Alloyed and Nanocrystalline Materials, (A. Calka and D. Wexler, eds.), 312–314:3, *Mater. Sci. Forum*, Trans. Tech. Publ., Zürich (1999); *J. Metastable and Nanocrystalline Materials*, 2–6:3 (1999)

188. Kübler, A., Gebert, A., Eckert, J., and Schultz, L., *J. Appl. Phys.*, 83:3438 (1998)

189. Inoue, A., *Mater. Sci. Eng. A*, 179/180:57 (1994)

190. Kübler, A., Eckert, J., and Schultz, L., *Nanostruct. Mater.*, 12:443 (1999)

191. Moelle, C., Lu, I. R., Sagel, A., Wunderlich, R. K., Perepezko, J. H., and Fecht, H. J., in: Mechanically Alloyed, Metastable and Nanocrystalline Materials, (M. D. Baró and S. Suriñach, eds.), 269–272:47, *Mater. Sci. Forum*, Trans. Tech. Publ., Zürich (1998)

192. Zhou, C. R., and Xu, J., *J. Mater. Res.*, in press

193. Coble, R. L., *J. Appl. Phys.*, 34:1679 (1963)

194. Greer, A. L., Zhong, Z. C., Jiang, X. Y., Rutherford, K. L., and Hutchings, I. M., in: *Chemistry and Physics of Nanostructures and Related Non-Equilibrium Materials*, (E. Ma, B. Fultz, R. Shull, J. Morral, and P. Nash, eds.), p. 3, The Mineral, Metals and Materials Society, Warrendale, PA (1997)

195. Inoue, A., Nakazato, K., Kawamura, Y., Tsai, A. P., and Masumoto, T., *Mater. Trans. JIM*, 35:95 (1994)

196. Chalmers, B., *Principles of Solidification*, p. 64, John Wiley and Sons, New York (1964)

197. Kim, H. S., Suryanarayana, C., Kim, S. J., and Chun, B. S., *Powder Metall.*, 41:217 (1998)

198. Greer, A. L., *Curr. Opin. Solid State & Mater. Sci.*, 2:412 (1997)

199. He, Y., Dougherty, G. M., Shiflet, G. J., and Poon, S. J., *Acta Metall. Mater.*, 41:337 (1993)

200. Inoue, A., *Curr. Opin. Solid State & Mater. Sci.*, 2:305 (1997)

201. Yokoyama, Y., Inoue, A., and Masumoto, T., *Mater. Trans. JIM*, 34:135 (1993)

202. Takeuchi, S., in: *Quasicrystals*, (J. M. Dubois, P. A., Thiel, A. P. Tsai, and K Urban, eds.), 553:283, *Mater. Res. Soc. Symp. Proc.,* Materials Research Society, Warrendale, PA (1999)

203. Wittmann, R., Urban, K., Schandl, M., and Hornbogen, E., *J. Mater. Res.*, 6:1165 (1991)

204. Feuerbacher, M., Klein, H., Schall, P., Bartsch, M., Messerschmidt, U., and Urban, K., in *Quasicrystals*, (J. M. Dubois, P. A. Thiel, A. P. Tsai, and K. Urban, eds.), 553:307, *Mater. Res. Soc. Symp. Proc.*, Materials Research Society, Warrendale, PA (1999)

205. Feuerbacher, M., Metzmacher, C., Wollgarten, M., Urban, K., Baufeld, B., Bartsch, M., and Messerschmidt, U., *Mater. Sci. Eng. A*, 233:103 (1997)

206. Sordelet, D. J., Kim, J. S., and Bessier, M. F., in: *Quasicrystals*, (J. M. Dubois, P. A. Thiel, A. P. Tsai, and K. Urban, eds.), 553:459, *Mater. Res. Soc. Symp. Proc.*, Materials Research Society, Warrendale, PA (1999)

207. Sánchez, A., Garcia de Blas, F. J., Algaba, J. M., Alvarez, J., Valles, Garcia-Poggio, M. C., and Agüero, A., in: *Quasicrystals*, (J. M. Dubois, P. A. Thiel, A. P. Tsai, and K. Urban, eds.), 553:447, *Mater. Res. Soc. Symp. Proc.*, Materials Research Society, Warrendale, PA (1999)

208. Inoue, A., Horio, Y., Kim, Y. H., and Masumoto, T., *Mater. Trans. JIM*, 32:669 (1992)

209. Unsworth, W., *Mater. Product. Tech.*, 4:359 (1989)

210. Gjsetland, H., *Light Metal Age*, 47(7):19 (1989)

211. Gilbert, C., Ritchie, R. O., and Johnson, W. L., *Appl. Phys. Lett.*, 71:476 (1997)

212. Kawamura, Y., Kato, H., Inoue, A., and Masumoto, T., *Appl. Phys. Lett.*, 67:2008 (1995)

213. Kawamura, Y., Shibata, T., Inoue, A, and Masumoto, T., *Appl. Phys. Lett.*, 69:1208 (1996)

214. Inoue, A., Zhang, T., and Masumoto, T., *Mater. Trans. JIM*, 36:391 (1995)

215. Bruck, H. A., Christman, T., and Johnson, W. L., *Scr. Metall. Mater.*, 30:429 (1994)

216. Xing, L. Q., Eckert, J., and Schultz, L. *Nanostruct. Mater.*, 12:503 (1999)

217. Xing, L. Q., Bertrand, C., Dallas, J. P., and Cornet, M., *Mater. Sci. Eng. A*, 241:216 (1998)

218. Spaepen, F., *Acta Metall. Mater.*, 25:407 (1977)

219. Kim, Y. H., Inoue, A., and Masumoto, T., *Mater. Trans. JIM*, 31:747 (1990)

220. Kato, H., and Inoue, A., *Mater. Trans. JIM*, 38:793 (1997)

221. Inoue, A., Zhang, T., and Kim, Y. H., *Mater. Trans. JIM*, 38:749 (1997)

222. Leonhard, A., Heilmaier, M., Eckert, J., and Schultz, L., in: *Creep Behavior of Advanced Materials for the 21st Century*, (R. S. Mishra, A. K. Mukherjee, and K. Linga Murty, eds.), p. 41, The Minerals, Metals and Materials Society, Warrendale, PA (1999)
223. De Hey, P., Sietsma, J., and Van den Beukel, A., *Acta Metall. Mater.*, 46:5873 (1998)
224. Kawamura, Y., Shibata, T., Inoue, A., and Masumoto, T., *Mater. Trans., JIM* 40:335 (1999)
225. Reger-Leonhard, A., Heilmaier, M., and Eckert, J., unpublished results.
226. Taub, A. I., *Acta Metall. Mater.*, 28:633 (1980)
227. Taub, A. I., and Luborsky, F. E., *Acta Metall.Mater.*, 29:1939 (1981)
228. Bakke, E., Busch, R., and Johnson, W. L., *Appl. Phys. Lett.*, 67:3260 (1995)
229. Einstein, A., *Ann. Physik (Leipzig)* 34:592 (1911)
230. Waniuk, T. A., Busch, R., Masuhr, A., and Johnson, W. L., *Acta Metall. Mater.*, 46:5229 (1998)
231. Taub, A. I., and Spaepen, F., *Acta Metall.*, 28:1781 (1980)
232. Grest, G. S., and Cohen, M. H., *Adv. Chem. Phys.*, 48:455 (1981)
233. Tsao, S. S., and Spaepen, F., *Acta Metall. Mater.* , 33:881 (1985)
234. Wilde, G., Görler, G. P., Jeropoulos, K., Willnecker, R., and Fecht, H. J., in: Mechanically Alloyed, Metastable and Nanocrystalline Materials, (M. D. Baró and S. Suriñach, eds.), 269–272:541, *Mater. Sci. Forum*, Trans. Tech. Publ., Zürich (1998)
235. Angell, C. A., *Science*, 267:1924 (1995)
236. Kübler, A., Eckert, J., Gebert, A., and Schultz, L., *J. Appl. Phys.*, 83:3438 (1998)
237. Weiss, M., Moske, M., and Samwer, K., *Appl. Phys. Lett.*, 69:3200 (1996)
238. Zappel, J., and Sommer, F., *J. Non-Cryst. Solids*, 205–207:494 (1996)
239. Busch, R., Bakke, E., and Johnson, W. L., in: Synthesis and Properties of Mechanically Alloyed and Nanocrystalline Materials, (D. Fiorani and M. Magini, eds.), 235–238:327, *Mater. Sci. Forum*, Trans. Tech. Publ., Zürich (1997)
240. Inoue, A., Zhang, T., and Kim, Y. H., *Mater. Trans. JIM*, 38:749 (1997)
241. Ke, M., Hackney, S. S., Milligan, W. W., and Aifantis, E. C., *Nanostruct. Mater.*, 5:689 (1995)

12

Nanostructured Electronics and Optoelectronic Materials

Raphael Tsu and Qi Zhang

1.0 INTRODUCTION

The availability of highly controlled deposition techniques and methods for nanometer-scale lithographic and device fabrications, stimulated the growth of interest in the study of quantum size effects in the electronic and optical properties of nanostructures, ranging from one-dimensionally confined superlattices to three-dimensionally confined quantum dots. In what follows, rather than covering a wide range of topics, we focus on semiconductors, more precisely, silicon. A fairly in-depth treatment of several topics is included, such as the dielectric constant, the capacitance, doping, and exciton binding energies of a nanoparticle, as well as several examples of applications such as porous silicon, nanoscale silicon particles embedded in an oxide matrix, superlattice of amorphous silicon sandwiched between thin oxides and superlattice of crystalline silicon sandwiched between adsorbed oxygen monolayers, as well as some quantum devices. The first part deals with the physics of nanostructured materials, and the second part deals with possible devices such as light emitting diodes, LED's, and quantum field effect transistors, QD-FET's.

2.0 PHYSICS OF NANOSTRUCTURED MATERIALS

2.1 Quantum Confinement: Superlattices and Quantum Wells

Superlattices and quantum wells were introduced as man-made quantum structures to engineer the quantum states for electrical and optical applications.[1][2] In twenty-five or more years, thousands of papers have been published. By 1997, more than 465 patents had been awarded on topics relating to the application of microelectronic and optoelectronic devices, and techniques for producing superlattice materials. In retrospect, the idea relies heavily on the availability of good heterojunctions, lattice matched systems,[3] and later, the strained layered systems.[4] It is appropriate to briefly discuss what is needed for the realization of a man-made quantum system. To realize quantum states in a given geometry, the size must be smaller or comparable to the coherence length of electrons, in order to exhibit quantum interference. This requirement eliminates doping as an effective means to achieve confinement, except at low temperatures,[5] because doping comes from charge separation which results in barriers generally far exceeding the coherence length of electrons at room temperatures. On the other hand, band-edge alignment of a heterojunction provides abrupt barrier height. This short range potential is the consequence of higher order multiples in the atomic potentials.

A new type of superlattice was proposed, the Epilayer Doping Superlattice (EDS), consisting of, for example, a couple of layers of Si in AlP.[6] The idea is fundamentally different from atomic plane-doped or δ-doped superlattices[7] where only a small fraction of a plane is occupied by doping or substitution. Another type of superlattice designed to incorporate extremely localized interaction, most promising for silicon, was introduced in 1993,[8] consisting of an effective barrier for silicon, formed by a suboxide with a couple of monolayers of oxygen atoms. This system, as a barrier for silicon, has been experimentally realized.[9] Localized interaction in a man-made quantum system is not new; for example, resonant tunneling involving localized defects was reported.[10] Recently, superlattices with extremely localized interaction in what we call the Hetero-Epilattice Superlattice (HES) has been successfully fabricated by sandwiching thin silicon epitaxial layers between monolayers of adsorbed oxygen.[11] What is most remarkable is the lack of stacking fault defects. A variation of this HES shows electroluminescence.[12] The HES can finally promote silicon to join the ranks of devices with quantum effects, thus far almost totally dominated by III-V semiconductors, except in the Si/Ge case where carrier confinement is in the germanium.[13]

2.2 Dielectric Constant of Nanoscale Silicon

As the physical size approaches several nanometers, reduction in the static dielectric constant, ε, becomes significant. Basically, ε measures screening which is not effective in a nanoparticle. Specifically, quantum confinement increases the separation of the energy states causing a lowering of the transitions, which sum to the value for ε. A modified Penn model, taking into account the quantum confinement induced discrete energy states, was applied to a sphere and to a wire. The calculated size-dependent ε is consistent with the wave vector dependent $\varepsilon(\mathbf{q})$.[14] This work was initially motivated by the need to understand the self-limiting mechanism in the electrochemically etched porous silicon. The argument goes as follows: A significant reduction of ε results in a large increase in the binding energy of dopants rendering the disappearance of extrinsic dopings.[15] Reference 15 was distributed during the Materials Research Society (MRS) meeting in 1992 which led to the more involved numerical calculations of Wang and Zunger[16] and Lannoo, et al.,[17] What followed was somewhat familiar. A more complete version of Ref. 15 was rejected by the reviewer who insisted that Rayleigh scattering was not taken into account! We had to rewrite our manuscript and resubmit the results in the present form of Ref. 14.

Strictly speaking, the dielectric function is only definable in an unbounded region of space. The wave-vector dependent, $\varepsilon(\mathbf{q})$, has been derived for semiconductors such as Si, Ge, and GaAs.[18] The use of this $\varepsilon(\mathbf{q})$ is essential for the calculation of screened shallow impurity potentials.

Replacing the free electron momentum by $k_{np} = \alpha_{np}/a$ with α_{np} being the nth roots of the spherical Bessel function, $j_p(ka) = 0$, we obtain the size-dependent dielectric constant $\varepsilon(a)$

Eq. (1) $$\varepsilon(a) = 1 + (\varepsilon_b - 1)/[1 + (\Delta E/E_g)^2]$$

in which ε_b is the bulk dielectric constant, E_g is the energy of the center of the imaginary part of the dielectric function, and ΔE is given by $\pi E_F/k_F a$. Taking the values for silicon, $\varepsilon_b = 12$, $E_g = 4$ eV, and filling the energy bands up to E_F of $4 \times 5 \times 10^{22}$ valence electrons per cm^3, or $E_F = 12.6$ eV, the computed $\varepsilon(a)$ is shown in Fig. 1. Note that our computed results when using $\varepsilon_b = 11.3$ agree almost perfectly with that of $\varepsilon(\mathbf{q})$ from Ref. 18 by putting $q = 2\pi/2a$ (from Fourier transform), and agree well with those of Refs. 16 and 17. This simple treatment gives a good physical insight, apart from the fact that the approach is far simpler to incorporate into other calculations, for example, the binding energy of a dopant in a nanoparticle, which we shall treat next.

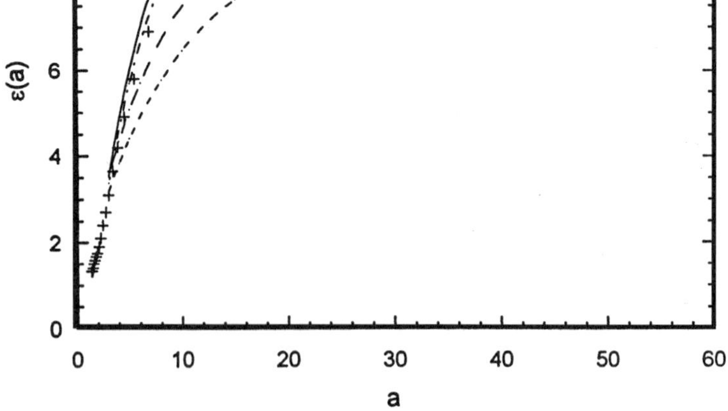

Figure 1. Plot of size-dependent static dielectric constant $\varepsilon(a)$ vs the radius a of silicon sphere in anstroms. The solid line is from the modified Penn model with $\varepsilon_b = 12$; the dashed-dot line is from the same model with $\varepsilon_b = 11.3$; the crosses are from Ref. 18; the longer dashes are from Ref. 17; and the shorter dashes are from Ref. 16.

Before we leave this subject, we point out that the size-dependent dielectric constant of a quantum wire using this model gives the same expression as long as we use an isotropic effective mass. This size-dependent dielectric constant should be important in other physical situations, such as electron-phonon interaction, polaritons, optical properties, etc.

2.3 Doping of a Nanoparticle

The doping of a quantum dot is an important issue. Since electrochemically etched porous silicon exhibits quantum confinement in photoluminescence,[19] the quantum size effect on doping, including interactions with induced charges at the dielectric discontinuity, requires investigation. Fundamentally, quantum confinement pushes up the allowed energies resulting in an increase in the binding energy, E_b, of shallow impurities such as the cases of quantum wells[20] and superlattices.[21] Theoretical treatments of the dielectric constant in quantum confined systems[22][23] show that a significant reduction takes place when the width of the quantum well is below 2 nm. Qualitatively, quantum confinement cuts down the motion of electrons, resulting in a reduction in screening. Using the Bohr model for shallow dopants, the binding energy is inversely proportional to the square of the dielectric constant, a reduction in the static dielectric constant greatly increases the binding energy to the extent that most nanoparticles show no extrinsic doping. In a quantum dot of radius a, the measured $\varepsilon(a)$ for porous silicon[15] was in fair agreement with the calculation given by Eq. (1) in Sec. 2.2. However, preliminary calculated binding energy[24] for dopants points out that this reduction in ε plays a small role in the final results because the larger increase is due to the induced polarization charges at the boundary of the dielectric discontinuity.[25] With ε_1 and ε_2 denoting the dielectric constants of the particle and the matrix, for $\varepsilon_1 > \varepsilon_2$, the induced charge of the donor is of the same sign resulting in an attractive interaction with the electron of the dot, pushing deeper the ground state energy of the donor resulting in an appreciable increase in E_b. For $\varepsilon_1 < \varepsilon_2$, the opposite is true; E_b is much reduced allowing possible extrinsic conductivity at room temperatures. Figure 2 shows the donor binding energy versus several values of the dielectric constant for the matrix, 1 for air or vacuum, 6 for water within the Helmholtz layer, etc. Note that at a dot radius of 2 nm, the former gives $E_b = 0.8$ eV, while the latter gives 0.2 eV, making it possible to show extrinsic doping at room temperatures.[26] The lack of extrinsic doping as the particle size is reduced by electrochemical etching serves as a limiting factor on the size reduction in etching in the dark. If etching is performed with the presence of light, electron-hole generation can lead to the continuous etching without limitation.[26] In electroluminescent (EL) diodes, it should be important to match the dielectric constants to facilitate extrinsic doping. In fact, this may be the reason why in porous silicon, PSi, and EL devices, the use of the SiC/PSi/Si pn junction[27] seems to work well; because the dielectric constant of SiC matches that of PSi, allowing the formation of a pn-junction.

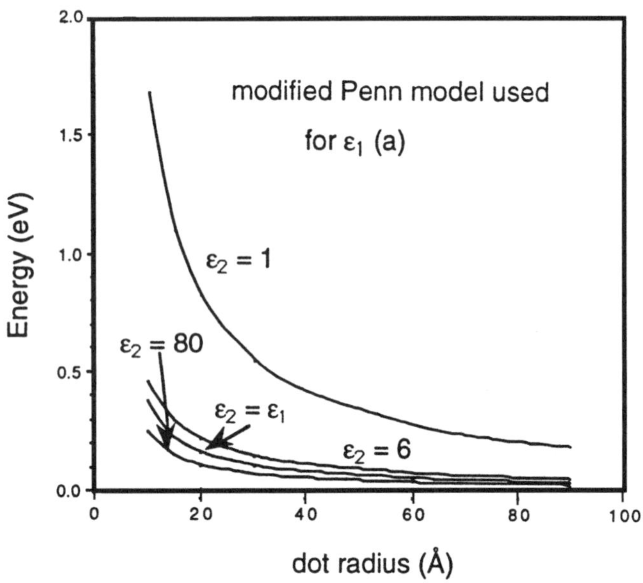

Figure 2. Donor binding energy vs dot radius in anstroms for several values of the dielectric constant of the matrix.

Extrinsic doping forms the backbone of all solid state devices with pn-junctions. We have shown that almost all shallow levels become deep in nanoscale particles, and induced charges at the dielectric interface between the quantum dot and its matrix affect, to a large degree, the binding energy of the dopants. Therefore, nanostructured materials in optoelectronic applications require more thought.

2.4 Excitonic Binding and Recombination Energies

Electrochemically etched porous silicon displays visible luminescence.[19] The role of quantum confinement in the porous silicon luminescence is established by the increase of the optical absorption gap,[28] and by the decrease of the Raman phonon frequency with the increase of the peak luminescence energy.[29] The quantum confinement effects in silicon nanocrystallites have been treated by tight-binding, effective mass, pseudopotential and first-principles local density approximations. However, in order to take into account the induced electrostatic polarization due to dielectric mismatch at the silicon crystallite boundary with the external medium, recombination and binding energies of excitons in silicon quantum dots may be calculated within essentially the same framework as the calculation of the binding energy of a quantum dot treated in Sec. 2.3.[30] The peak recombination energy in silicon quantum dots is quite insensitive to the nature of the external medium due to approximate cancellation of the polarization terms in the recombination energy of the excitons involved. No such cancellation is present for the binding energy of the excitons. Excitons in silicon crystallites surrounded by vacuum are electrostatically bound by about 1 eV. When immersed in water, the binding energy is dramatically decreased compared to vacuum as the doping case considered in the last section.

Recombination and binding energies of the excitons confined in silicon nanocrystallites are calculated within the effective mass approximation. This approximation has been already applied to evaluate the one- and two-electron ground state energies,[31] donor binding energy,[25] excitonic energy,[32] and absorption coefficient.[33] The envelop wave functions of both electron and hole are determined by the kinetic energy of each particle, which dominates the properties of excitons in quantum dots of the size in the range of several nanometers. The electrostatic terms, Coulomb interaction, polarization interaction, and electron-hole polarization self-energies are treated by perturbation, in a variational calculation. As before,[25] the binding energy is defined as the minimum energy required to break an exciton by removing both electron and hole from the same silicon quantum dot and placing them in separate silicon nanocrystals. The modified Penn model is used for the dielectric constant of the dot, see Sec. 2.1. Figure 3 shows the calculated values of the exciton binding energy, *solid;* and the recombination energy, *dashed;* vs dot radius for various ε values of the matrix. Note that all curves for the recombination energy are close to each other, due to cancellation effects discussed previously. Note that, for a radius of 1.5 nm, the exciton binding energy ranges from 82 meV ($\varepsilon_2 = 80$) to 1.03 eV ($\varepsilon_2 = 1$), a change of more than an order of magnitude. For the case of the same dielectric constants of the dot and the matrix, where only the Coulomb interaction is present, the exciton binding energy is 0.16 eV, over ten times higher than the bulk value of 14.7 meV. This major increase of the Coulomb interaction part of the exciton binding energy is caused by the increased overlap of the electron-hole wave functions. For all four values of the dielectric constant of the external matrix, the exciton binding energy is much greater than the characteristic thermal energy at room temperature, therefore, excitons confined to a quantum dot are well bound and stable irrespective of the surrounding matrix. Since the self-polarization and the polarization terms are very large, any theory not including the dielectric mismatch between the dot and the environment cannot be taken seriously. We need to have a strong appreciation in the magnitude of the exciton binding energy for a quantum dot with a radius below 15 nm. Excitons simply cannot be broken in the usual sense! The activation energy of the break-up of excitons, estimated from the slope of the luminescence decay in air with increasing temperature, is approximately 100–120 meV,[34] which is far less than the exciton electrostatic binding energy (1 eV in vacuum), and is thus related to the turning-on of some non-radiative recombination channels.

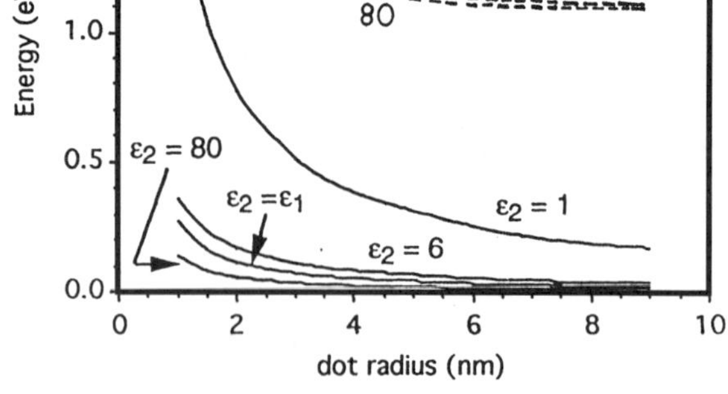

Figure 3. Exciton binding energy, *solid*, and recombination energy, *dashed*, vs dot radius in nm for several values of the dielectric constants of the matrix.

The photoluminescence in porous silicon at room temperature is due to recombination of excitons confined in silicon nanocrystals whose effective diameters are approximately 3 nm for nanowires and 3 nm for nanodots. The transition is still phonon assisted as in the bulk that involves an electron from the bottom of the conduction band and a hole from the top of the valence band, thus separated in the momentum space. In a well passivated system without surface trap states, a non-radiative channel may involve electron tunneling out of the quantum dot.[35]

It is instructive to compare exciton radiative recombination in silicon nanocrystallites and in bulk Si, both at room temperature. In the bulk, it is far likelier that the exciton will be broken by a phonon than to encounter the right phonon for phonon assistance. The exciton break up is facilitated by the quasi-continuum of available states in both valence and conduction bands. The electron and hole liberated from the exciton by a phonon, fly apart, thus disabling radiative recombination. The quasi-continuum of valence and conduction band states is modified into a discrete set of energy levels due to quantum confinement. Since the thermal phonons, without sufficient energy to break up these excitons, allow excitons enough time to wait for the right phonon with the necessary momentum, phonon-assisted radiative recombination occurs. In short, it was assumed that nanoscale particles allowed the relaxation of momentum conservation, or even suggested that the band structures may be made direct by nanostructuring. Now we understand the situation quite well: short of nanostructuring to a dimension of well under 1 nm, indirect band structure still dominates optical transitions. The apparent increase in the observed luminescent efficiency is due to the long-lived excitons due to quantum confinement. The direct gap in a silicon nanocrystallite is located at 2.9 eV.[36] Therefore, some of the observed weak and fast blue luminescence in non-oxidized Si may be from this component.

2.5 Capacitance in a Nanoparticle

The effects of charge accumulation in quantum confinement have been under intensive study involving quantum dots,[37] and energy states of a silicon nanoparticle.[38] It is possible to introduce as many electrons as one would like into a classical capacitor until breakdown occurs from the ever increasing voltage. The situation is quite different for a nanoscale capacitor where each electron introduced into a quantum capacitor must satisfy the Pauli exclusion principle. In particular, since it takes kinetic energy to localize an electron, as the dimension shrinks to nanometer regime, only a small number of electrons can be put into the system before its energy is higher than the barriers serving to confine them. The problem is quite complex even in a classical calculation since Green's function method must be used to calculate the potentials due to the presence of an electron, or electrons with all the interactions with induced charges on the boundaries of the dielectric mismatch. For a silicon sphere embedded in an oxide matrix, ε of silicon is 12 while that of the oxide is 4. We shall sketch the salient procedures used in the calculation.[31]

For a sphere, the use of the electrostatic Green's function allows the identification of the potential energy terms in both one- and two-electron cases. Even in the one-electron case, an electron interacts with its induced bound surface charge density, resulting in the self-polarization. Note that this term goes to zero without the dielectric discontinuity. In the two-electron case, there are four terms: self polarization terms for each electron, Coulomb interaction, and polarization interaction. The polarization term arises because one electron induces a bound surface charge density, which interacts with the second electron. There is one important consideration worthy of discussion. Evaluation of the electrostatic energy requires that one specifies electronic positions. For a single electron, the position in the center gives the minimum polarization energy. However, for two electrons, the classical ground-state energy is determined by the minimum of the competing repulsive components. Coulomb and polarization interactions push the electrons apart towards the well boundary, while the self-polarization terms push the electrons away from the boundary, closer to each other. The separation of the two electrons are obtained by a minimization of the total electrostatic energy. Table 1 lists the calculated values for one electron in a sphere of radius a, E^c_1, and for two electrons, E^c_2, and Δ^c, the difference.

Table 1. Classically Calculated One- and Two-Electron Electrostatic Energies

a (nm)	1	2	3	4
E^c_1 (eV)	0.12	0.06	0.04	0.03
E^c_2 (eV)	0.60	0.30	0.20	0.15
Δ^c (eV)	0.48	0.24	0.16	0.12

For the quantum-mechanical calculation, the zero order spherical Bessel function is used for both the one-and two-electron cases for the unperturbed ground state wave functions which apply to the Hamiltonian having only the kinetic energy term with infinite barrier boundary condition. All of the energies for the self-energy, the Coulomb energy, and the polarization energy are computed using perturbation similar to the treatment of the helium ground state. Below $a = 1$ nm, the validity of the effective mass formulation is questionable, and beyond $a = 4$ nm, the simple ground state wave function used is not a good approximation (higher order spherical Bessel functions need to be used for the perturbation calculation). The calculated results are tabulated in Table 2 for the total energies of the one- and two-electron cases, with the capacitance defined by $E_2 - E_1 = \Delta = e^2/2C$.

Table 2. One- and Two-Electron Ground State Energies From Quantum-Mechanics

a (nm)	1	2	3	4
E_1 (eV)	1.59	0.43	0.21	0.12
E_2 (eV)	3.64	1.09	0.57	0.37
Δ^{qm} (eV)	2.05	0.66	0.36	0.25
$C(qm)/C$(classical)	0.23	0.36	0.44	0.48

In Table 2, it is noted that the capacitance $C(qm)$ for a spherical particle of radius $a = 3$ nm is less than half the value of the capacitance for the classical particle. Physically, it is not meaningful to consider only electrostatically what happens when adding or subtracting an electron from the confining system because, quantum-mechanically, electrons have kinetic energies occupying their states. The kinetic energy scales with a^{-2}, while all the other terms scale with a^{-1}, therefore, our $C(qm)$ does not scale with a, unlike the calculated $C(qm)$ for larger systems where the kinetic term is negligibly small.[39][40] Apart from the larger size regime and the inclusion of more electrons, there is an important difference between the results presented here and those of Refs. 39 and 40. Although it is important to include the dielectric mismatch between the quantum dot and the matrix which can be accounted for in our approach, it is a formidable task to even extend this treatment to three- or four-electron cases. Nevertheless, our results should have important applications in nanoscale electronics, particularly in devices with few electrons. Qualitatively, at temperatures when the coherence length of the electrons is exceeded by the size, we argued[31] that the capacitance should approach the classical value. However, the problem is far more complex because statistical formulation needs to be considered.

2.6 Structure, Bonds, and Coordinations of Si Nanostructure: Porous Si and Si Clusters

Structure. Since the discovery of fairly efficient photoluminescence in the visible spectrum by Canham[19] in PSi, an electrochemically etched silicon, the debate is on as to the origin of the luminescence: quantum confinement,[28][41] a-SiH,[42] siloxene derivatives,[43] surface states,[44] etc. Our assessment is that, whenever careful researchers disagree, often several of these mechanisms are present. Which one dominates depends on special preparation resulting in differences in structures. More details and recent thoughts on this subject, as well as many other aspects of porous Si, may be found in the book cited,[45] and a review article.[46] Since structure, bonds, and coordination are the starting points of any physical models, we shall deal with it from our point of view, not necessarily from a vantage point. More inclusive discussions are found in Refs. 45 and 46.

From the correlation of the Raman shift with the upshifts of the photoluminescence (PL) peak, it was concluded that the red luminescence originates from nanostructures in the order of 2–3 nm.[29] Subsequently, using high resolution cross sectional transmission electron microscopy (TEM),[47] silicon nanocrystallites of 3–5 nm were identified in typical porous silicon.

Bonds and Coordinations. Freshly produced red, yellow, and green emitting porous Si specimens have been studied by near edge and extended x-ray absorption fine structure (NEXAFS and EXAFS).[48] The emission peaks are at 690, 580, and 520 nm, which almost cover the full visible range. The correlation between the coordination numbers of the first, second, and third Si neighbor shells, from Fourier transform fitting of EXAFS with both emission peak energies and optical bandgaps estimated by PLE (photoluminescence excitation), suggests that the nanostructures of the PSi are nanowires, rather than clusters of nanocrystallites, for porous silicon samples prepared with low resistive silicon wafer. Two types of quantum nanowire with one and one-plus a fractional dimensionality are proposed to interpret the correlation. Moreover, the order factors of the theoretical fit suggest that nanowires of the freshly produced PSi have crystalline cores.

To better understand the role of the structure of porous silicon in quantum confinement, many methods, including soft x-ray absorption,[49] XAS,[50] TEM,[47] and EXAFS,[51][52] have been used. It was found that electrochemical etching of silicon into porous silicon produces different structures depending on the resistivity of the silicon wafer used. This phenomenon may be understood using the self-limiting model of electrochemical etching in the dark.[26] In this model, etching requires the holes to come to the surface. Highly extrinsically doped silicon allows this to take place, while highly resistive materials allow the normally very slow etching in HF to compete with this low etching process resulting in the break-up of a thicker wire into clusters. In brief, highly resistive wafers result in more cluster-like PSi, while lower resistivity wafers produce more wire-like PSi. This is a good example how EXAFS can clearly distinguish the structures. The coordination number and distances of Si-Si shells from Ref. 48 are given in Tables 3 and 4. Figure 4 gives the dependence of dimensionality and size with PSi "color." See figure caption for details.

Table 3. The Coordination Numbers of Si-Si Shells

	1st $N_{Si\text{-}Si}$	2nd $N_{Si\text{-}Si}$	3rd $N_{Si\text{-}Si}$
red PSi	3.80±0.15	7.42±0.44	8.15±0.75
yellow PSi	3.65±0.13	5.74±0.38	6.47±0.67
green PSi	3.0±0.2	3.01±0.49	3.98±0.73
c-Si	4.0±0.1	12.00±0.35	12.00±0.58
a-Si	4.0±0.1	0	0

Table 4. The Distances of Si-Si Shells of PS

	1st R	2nd R	3rd R
red PSi	2.34	3.81	4.49
yellow PSi	2.34	3.81	4.50
green PSi	2.34	3.79	4.55

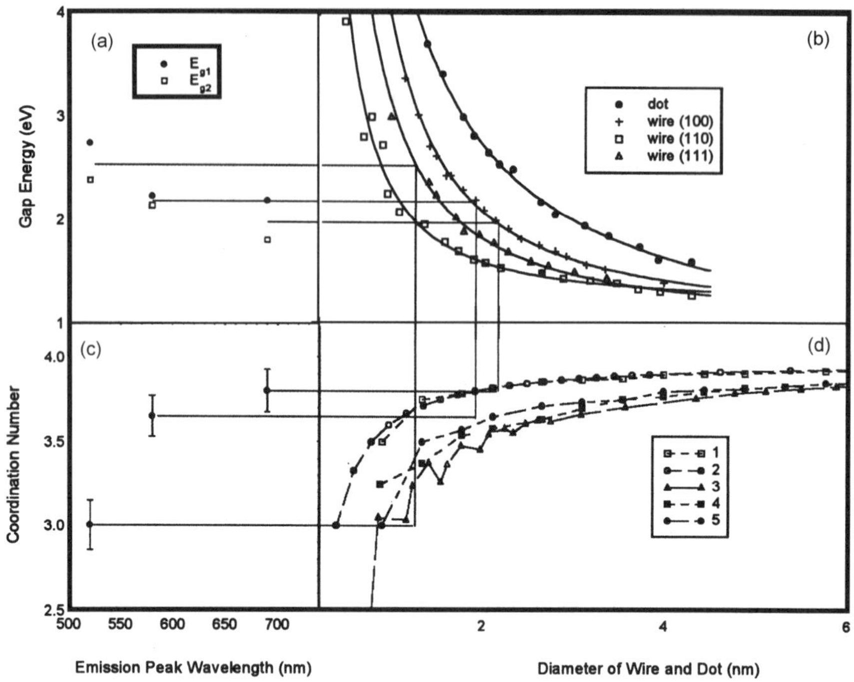

Figure 4. Dependence of dimensionality and size with PSi "color." (*a*) Bandgaps of the three types of emitting PSi: E_{g1} from PLE and E_{g2} from PL peaks. (*b*) Theoretical calculation of confined bandgap energy of Psi of wires and dots by LCAO. (*c*) First shell $N_{Si\text{-}Si}$ of red, yellow and green PSi from EXAFS fits. (*d*) Dimensionality and size dependence with the $N_{Si\text{-}Si}$ from *(i)* bare wire: curves 3 -cylinder (100), 4 and 5 - two types of hexagonal columns (111); *(ii)* wire network (one-plus-a-fraction dimension): curves 1 - (111) and 2 - (100), details see Ref. 48. As shown by the guiding lines, the PSi in this study favor the nanowire network and wire structures, with average diameters of 2.2, 1.9, and 1.3 nm, respectively for red, yellow, and green PSi.

Although more than one mechanism is likely to be involved in the luminescence of PSi, EXAFS and theoretical fits show a relatively simple picture for freshly produced PSi with PL peaks from 690 nm to 520 nm, which is nearly the full range we can obtain using only anodization processes. The nanocores of the PSi are crystalline and quantum confinement is the only mechanism widening the bandgap at this stage. A nanowire network (one-plus a fractional dimensionality) for red and yellow PSi, and nanowire (one dimensional or less) for green PSi, are suggested. The mean core sizes are 2.2, 1.9, and 1.3 nm respectively for red, yellow, and green PSi. It was pointed out in Refs. 46 and 48 that the quantum wire nanostructure is better suited for PSi application in EL quantum dots.

Strong and stable blue photoluminescence, visible to the naked eye under 0.4 μW of 300 nm and 2.7 μW of 370 nm excitation, has been observed for samples of Si clusters embedded in SiO_2 matrices, prepared by rf co-sputtering followed by N_2 annealing at 800°C. Si K-edge EXAFS and NEXAFS strongly suggest that Si nanoclusters have crystalline cores after annealing.[53]

The Si-Si bond length in the annealed sample is 2.35 ± 0.02 Å, as expected, whereas the Si-O distance is 1.58 ± 0.02 Å instead of the expected 1.62 Å. Our EXAFS data does not show direct evidence for crystallinity: the fitted Debye-Waller factors of the first shell Si-Si bonds at about 0.006 ± 0.003 $Å^2$ suggest a structure between amorphous and crystalline for which the factors are 0.004 and 0.007 $Å^2$, respectively. Alternatively, EXAFS suggests that the structure is composed of clusters which have crystalline cores surrounded by disordered bonds. Either way the absence of any features corresponding to second shell Si-Si-Si is not surprising, considering the small first shell Si-Si contribution and the rapid fall-off of EXAFS with distance. The formation of more Si-Si bonds in the Si-clustered samples is responsible for the increases in PL efficiency in the blue.

3.0 APPLICATIONS

3.1 Porous Silicon

Since Canham's discovery,[19] there have been continuous reports of porous silicon-based diodes (LEDs), as well as photodetectors, optical-logic gates, etc. The first PSi LEDs were those of Richter, et al.,[53] and Koshida and Koyama.[54] Figure 5 shows the device structure, which represents a typical early LED work. Its EL spectrum peaked at ~680 nm. This LED had a structure of semi-transparent metal–porous silicon layer–p-type silicon–Al electrode with an external quantum efficiency of only 10^{-5}% and with very limited device lifetime. Since then, the efficiency has been much improved to ~0.2%, close to a practically useful level of ~1% of industrial requirement.[55]–[57] These devices are represented by a structure of indium tin oxide (ITO)–p^+ PSi layer–n^- substrate, Al/poly-as-contact–n^+ PSi layer–p^- substrate, and a p^+nn^+ PSi structure, respectively.

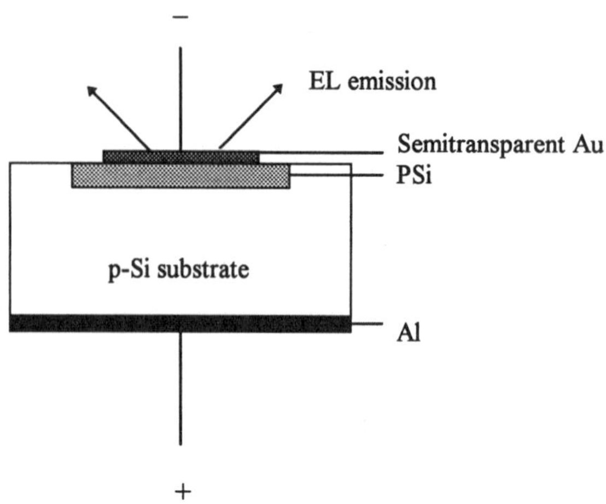

Figure 5. Schematic of one of the first PSi LED. *(Reprinted by permission of the publisher.)*

Both the PL and EL spectra of PSi are normally broad, similar to other types of nanostructures and nanoclusters. A dramatic reduction of the spectral width of EL and the radiative decay time is accomplished by combining micro-cavities to enhance LED performance,[58] leading to the possibility[59] of integration of PSi LED with standard bipolar circuitry. There are modest improvements in device lifetime.[60][61] The former involves encapsulating the porous silicon in aluminum and aluminum oxide, whereas the latter involves oxidizing the porous silicon surface to prevent further oxidization and to form stable passive layers with radiative centers. The longest reported device lifetime is several weeks either under DC operation or continuous pulse operation. Although oxide passivated structures show better stability, efficiencies are generally significantly lower. The fundamental shortcoming of porous silicon (they are extremely reactive and fragile) seems to be difficult to overcome, except recently it was shown that gently etched PSi has much improved morphology.[62]

A photodetector with a structure of Al/RTO (rapid thermal oxidized) PSi/p-Si/Al exhibited higher responsivity at 350 nm than an UV-enhanced Si photodiode with an external quantum efficiency of 75% at 740 nm.[63] A large optically induced absorption change in PSi has been demonstrated in an all-optical logic gates (invert- or NOR-gate function).[64]

3.2 Photoluminescence in nc-Si/SiO_2 Superlattices

To overcome the problem of structural robustness associated with the porous silicon, it was proposed that nanoparticles of silicon with sizes in the range of several nanometers sandwiched between thin oxide layers to form a superlattice may solve the problem of mechanical robustness while retaining the features of quantum confinement as in porous silicon.[65] In Ref. 55, the name IAG-superlattice was introduced, for Interface Adsorbed Gas-superlattice. This name originates from the scheme that oxygen is introduced via surface adsorption in order to prevent the formation of a very thick oxide. In this scheme, silicon up to 12 nm thick is deposited either in the amorphous phase or crystalline phase, followed by the in-situ growth of a thin oxide. This arrangement is repeated to acquire the desired volume of interaction. In practice, the maximum period of the superlattice structure tried was only nine periods. The reason to keep a relatively few periods is dictated by the need to further passivate interface defects, annealing in the presence of gas mixtures. In particular, annealing in H_2 or H_2+O_2 gives better results as shown in Fig. 6a with PL peaks at 1.7 eV and 2.34 eV. Figure 6b shows a cross-section TEM of the structure. It was found in surface Auger that the oxygen peaks coincide with the silicon dips, indicating that the structure indeed consists of silicon separated by regions with high oxygen content.[66] Moreover, the 2.34 eV peak is attributed to surface effects.[66] This brings up an important point in all nanostructured materials. In devices dictated by bulk, surface or interface regions are considered undesirable. As the particle size shrinks to nanometer regime, surface or interface regions become significant or even dominate over the "bulk," thus we need to reorient our views so that surface or interface regions are the focus of our considerations.

The grain size of the silicon nanoparticles was found to be approximately 3 nm using Raman scattering and checked by TEM.[65] The mechanism controlling the grain size is quite involved. We only touch the salient points here. Basically, unlike the amorphous-crystalline phase transition in bulk, in very thin structures, the phase transition is controlled by proximity effects rather than simple temperature.[67] These considerations prevented us from using extremely thin silicon layers as recently demonstrated in the work of Lockwood, et al.,[68] We postpone the discussion on what happens when we use a much-reduced thickness for the silicon layers until the treatment on the epitaxial Si/O superlattices.

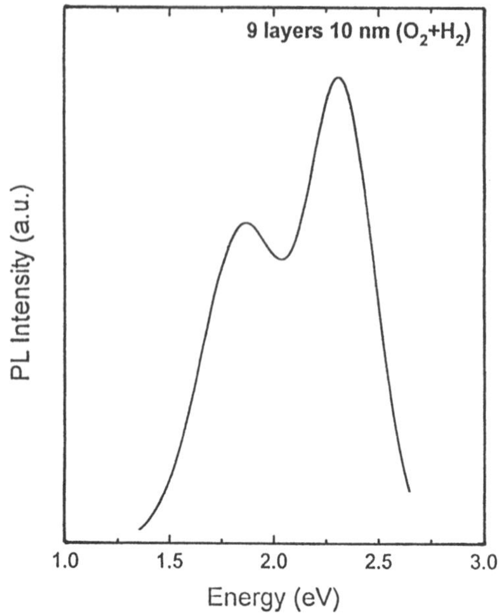

(a)

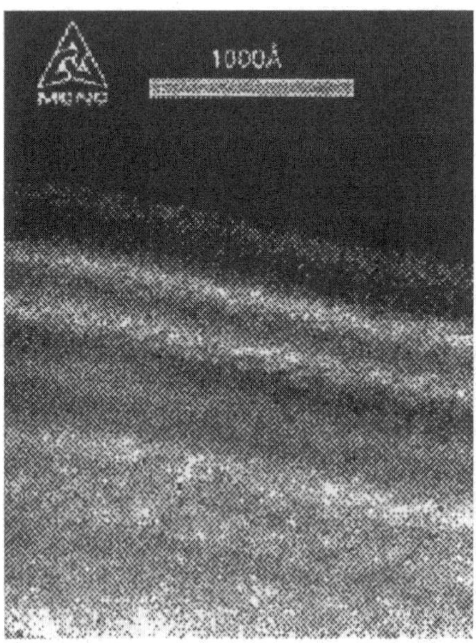

Figure 6. (*a*) PL intensity vs photon energy for a nine-period Si/IAG superlattice annealed in oxygen and hydrogen at 850°C; and (*b*) cross-section TEM of the sample.

3.3 Luminescence from Clusters

Elemental semiconductors Ge,[69] Si,[70] and C,[71] embedded in an SiO_2 matrix exhibited fairly strong and stable PL, with peaks ranging from IR to blue.[70][71] Samples for LED were fabricated consisting of 45 nm thick polycrystalline Si films deposited initially as amorphous films by e-beam evaporation onto 70 nm SiO_2 films thermally grown on n^+ Si substrates, followed by Ge implantation, to create a supersaturated solid solution of Ge in the SiO_2 film with approximately uniform Ge (~5 nm in diameter) concentration of 5%. The samples were subsequently annealed at 600°C, 1×10^{-6} torr, for 40 minutes to induce precipitation. The EL spectrum was broad and peaked at 1.2–1.4 eV.

Samples of silicon clusters can be prepared by sputtering SiO_2 onto silicon wafers without additional heating. After annealing at 800°C for 20–30 minutes in N_2, the typical PL spectra show typical quantum size effect in Fig. 7(a–c) with increasing Si to oxide ratio. A more detailed discussion on the EXAFS characterization and mechanisms of cluster size control may be found in Refs. 67 and 70.

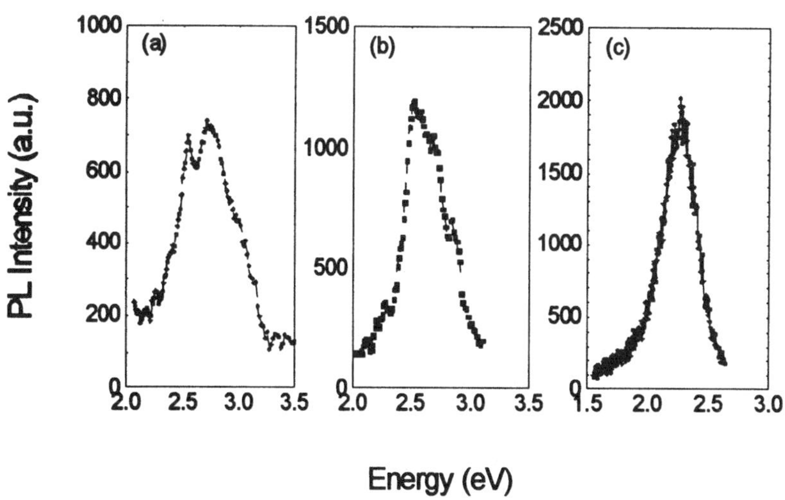

Figure 7. PL of silicon clusters in SiO_2 matrix. The intensites of PLs are related to the cluster densities found by TEM.

Stable blue and unstable UV PL from C clusters embedded in SiO_2 matrix has been observed.[71][72] The EXAFS analyses for O and C are more complex than Si. Applying the K-edge EXAFS to the system, it was found that the blue PL was related to C nanoclusters with local π-bonds. The unstable UV PL is thought to be related to silicon.

3.4 Hetero-Epilattice Si/O Superlattice

We introduce the term Hetero-Epilattice-Superlattice, HES, to describe a superlattice system consisting of epitaxial layers of silicon, for example, sandwiched between adsorbed disorder species. Several years ago, in search of a barrier system for silicon, where a lattice matched heterojunction is lacking except in the Si_xGe_{1-x} system,[73] it was proposed that perhaps the best and simplest way to build a barrier onto silicon is to utilize the concept of strain layer superlattice with sufficiently thin silicon layers.[8] Subsequently, it was realized that the best way to limit the thickness of the oxide which introduces disorder, is to limit the supply of oxygen by surface adsorption.[9] This is so because after a monolayer coverage of oxygen on a clean silicon surface, further oxygen adsorption is not possible without substantial heating to drive in the oxygen via diffusion. This method is, therefore, in the realm of self-organized crystal growth.[74]

HES is the outgrowth of the originally proposed barrier for silicon described in Ref. 8. Basically, the concept of a strain-layer superlattice is that, with a sufficiently thin epitaxial layer, the strain energy in each layer is below the energy needed for the growth of point defects or dislocations. It is important to note that dislocations have an activation energy for nucleation and a lower activation energy for growth. Therefore, in principle, it is possible to greatly exceed the energy requirement without actually generating defects. Figure 8 shows the scheme of a strain-layer epitaxial Si/O barrier system. Because the width $b \ll w$, E_{b1} and E_{b2} are much higher than E_1 and E_2, the quantum well states. This is the basis of a strain-layer barrier. The effective barrier height may be as high as 1 eV, so that it is possible to design the separation $E_2 - E_1$, depending on the width w, to be much greater than k_BT at room temperature.

The thickness at which SiO_2 can be tolerated for continuous epitaxy is the key for this kind of superlattice. What is under debate is whether the monolayer of SiO_2 on Si is ordered or disordered. Epitaxial growth of silicon may be continued after interruption with oxide growth.[75] The explanation involves the natural seeding provided by even the reconstructed Si(100) at low-pressure chemical vapor deposition (CVD) with low flow rate. Based on grazing angle x-ray diffraction, Rabedeau, et al.,[76] found evidence for low coverage 2 × 1 epitaxial structure at the SiO_2/Si interface for dry oxides grown on ordered Si surfaces at room temperature, however, the 2 × 1 structure does not survive thermal annealing.

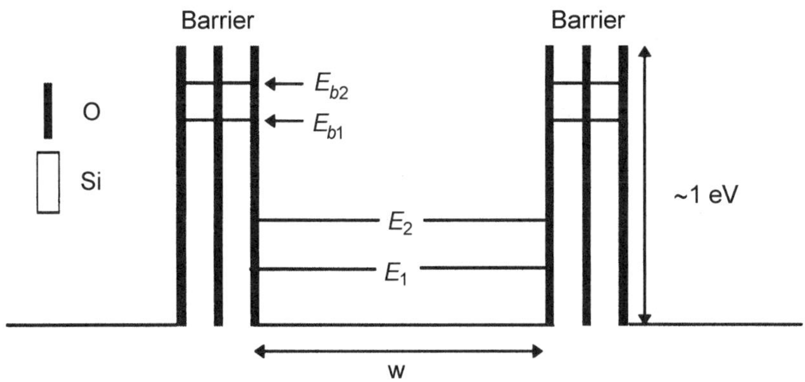

Figure 8. A scheme of a Si/O quantum well using a Si/O superlattice as the barrier. *(Reprinted by permission of the publisher.)*

Subsequently, the Si/O superlattice was formed and I-V measurements showed a barrier height of more than 1 eV.[9] However, more careful theoretical and experimental work showed that the barrier height is 0.5 eV,[77] which is sufficiently high for a variety of applications in electronic and optoelectronic applications. Basically, the method involves the adsorption of oxygen onto a clean Si surface followed by Si deposition of 1–2 nm thick, using the in-situ reflection high energy electron diffraction (RHEED) pattern for the monitoring of the 2 × 1 surface reconstruction as a measure of the restoration of a clean silicon surface before the next growth step. Recently, more detailed work about defects and degree of epitaxy, has been carried out on the hetero-epilattice with oxygen exposure.[11][77]

Figure 9 shows the typical in-situ RHEED (reflection high energy electron diffraction) pattern, used to monitor epitaxy. The 2 × 1 dimerization, the half order, is clearly shown in Fig. 9a from surface reconstruction of the Si(100). Exposure to 10 L of oxygen at 30°C does not eliminate the 2 × 1 structure (shown in Fig. 9b), until 1.1 nm of Si is deposited (shown in Fig. 9c), where the transformation from a 2-D to a 3-D pattern is quite evident. A second exposure at 10 L does not substantially alter the RHEED (shown in Fig. 9d), however, the 2 × 1 reappears after 8 nm of Si is deposited (shown in Fig. 9e). This series of in-situ RHEED demonstrates that epitaxial growth is continued beyond the adsorbed oxygen layer. A four-time repeated sequence with 10 L oxygen exposure followed by 1.1 nm of Si deposition at 550°C has been formed and the 2 × 1 reconstruction was brought back with the silicon layer of 12 nm, serving to restore the epitaxial structure as well as a capping for protection.

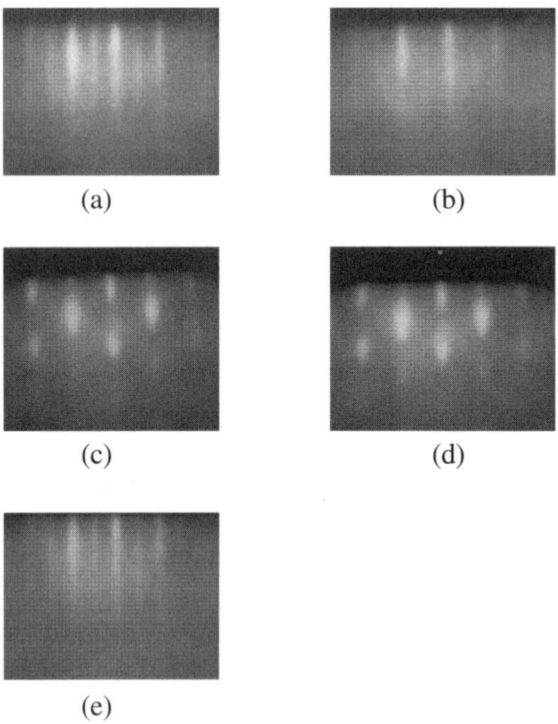

Figure 9. RHEED of Hetero-Epilattice Superlattice showing the re-establishment of epitaxy. The 2 × 1 reconstruction in (*a*) is not destroyed after oxygen exposure in (*b*), however few monolayers of Si deposition transforms the 2-D pattern to 3-D in (*c*), and (*d*). Full restoration is achieved after 8 nm of Si, shown in (*e*).

Similar structures were fabricated for a class of electroluminescent devices. Figure 10a shows the emission of visible light from the edge of an Al contact, yellowish-green to the naked eye, and Fig. 10b shows a spectrum of a peak at 2 eV with additional higher energy contributions. It is important to point out that this HES EL device is extremely stable and robust. With the application of 6–8 V, this EL diode has operated continuously for more than several months, representing, to our knowledge, the longest operation of a silicon nanostructured EL device to date. In fact, our life test is still continuing at this writing without detectable degradation![78]

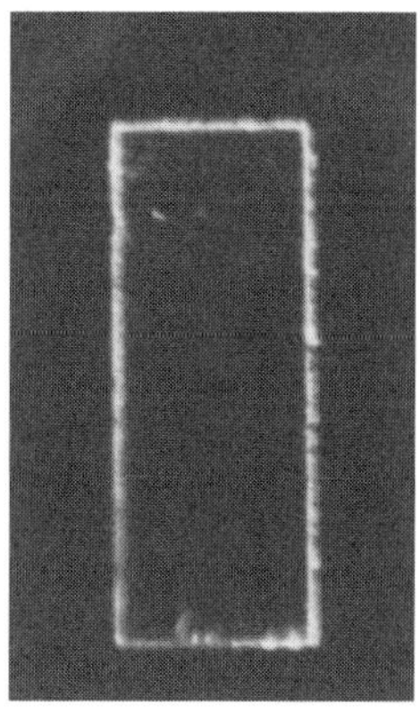

(a)

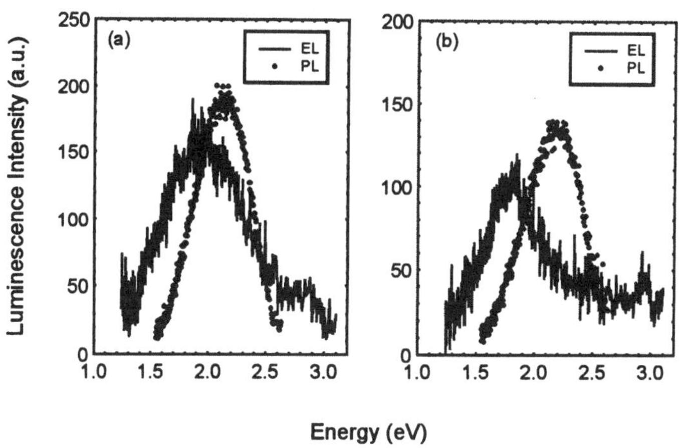

(b)

Figure 10. EL of LED using HES. (*a*) photograph of EL; (*b*) EL spectrum of a LED. The EL (*solid curve*) and PL (*dotted curve*, excited by 457.9 nm line of argon ion laser) spectra measured at room temperature of sample 1, *left*, at reverse bias $V = 14$ V and sample 2, *right*, at reverse bias $V = 20$ V.

3.5 Amorphous Silicon/Oxide Superlattice

Lu and Lockwood reported strong visible PL from amorphous-Si/SiO_2 superlattice.[68][78] The PL peak can be shifted from 1.7 eV to 2.3 eV when the thickness of amorphous silicon layer in the superlattice is 3 nm to 1 nm. The visible light emission was explained in terms of quantum confinement of electrons in the two-dimensional silicon layers. It is appropriate to discuss the difference of this a-Si/SiO_2 and the HES discussed in more detail in Sec. 3.4. As long as the amorphous silicon thickness is below that of the coherence length, from quantum phenomenon point of view, it is not so important whether the confined electrons can interfere in an amorphous layer or in a crystalline layer. However, amorphous bonding allows more flexibility resulting in a larger variety of bonding defects which ultimately manifest in increased recombination and scattering centers, in comparison to the crystalline case. Nevertheless, hydrogenated amorphous silicon photovoltaic materials have been steadily improved. Any premature judgment is somewhat risky at this stage.

3.6 nc-Si in an Oxide Matrix

Resonant tunneling involving discrete quantum states in nanocrystalline-Si (nc-Si) with a-SiO_2 barriers[38][79] gives rise to very rich features, which include Coulomb blockade[31] and avalanche multiplication[80] that results in slow oscillations in time due to the presence of negative resistance.[81] Basically, the idea is to create sufficiently small particles of silicon in the nanometer regime, tunneling via these quantized energy states should give rise to sharp features. The observed current peaks have line widths approximately given by k_BT.[81] There the mystery arises. In the structure for resonant tunneling, a layer of silicon consisting of nanoscale particles embedded in an SiO_2 matrix sandwiched between two oxide barriers, there should be a size distribution resulting in broad features, or even wiping out any conductance peaks. The answer to the mystery lies in the fact that nucleation and subsequent growth of the nanocrystallites are under some constraint conditions. The most important one is the thickness of the layer sandwiched between something else. The effects of a constraint condition have been discussed in relation to the control of particle size via crystallization from the amorphous phase presented in Sec. 3.2. We take this opportunity to reiterate that crystal growth in an extremely thin layer dominated by proximity effects is a new frontier of crystal growth.

Without going into too much detail in this section, it is important to point out various features normally one does not encounter. In forming many nanoscale particles in a layer for I–V measurements, these nanoparticles are all in parallel. Imagine, one tries to measure the transconductance of one hundred transistors having slightly different features connected in parallel. Fortunately, the situation is not entirely hopeless as presented in Ref. 80. Basically, the observed multitudes of current jumps may arise from very small difference in the particle size. Due to IR drop in the substrate, they appear at fairly well separated voltages. A simple theory, including the avalanche multiplication triggered by the resonant tunneling through each particle or a group of particles having approximately the same particle size, can indeed fit the measured conductance data as shown in Fig. 11, taken from Ref. 80. Results indicate another interesting point: particle size variation is in almost discrete steps. In retrospect, it seems to be obvious that in a small nanoscaled particle, size increase must be dictated by the size of the basic unit cell. Another interesting effect is the sharpening of the resonant conductance peak with light. Using phase sensitive I–V measurements, it was discovered that the drastic sharpening of the peak with light of photon energy above the bandgap may be due to the filling of charged traps.[82]

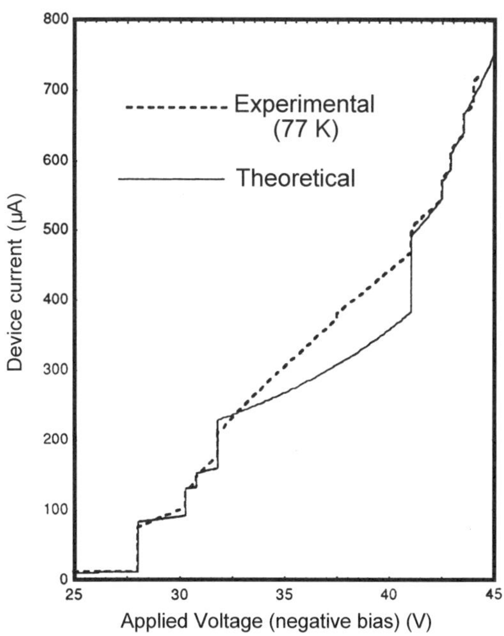

Figure 11. Theoretical and experimental current vs bias voltage. *(Taken from Ref. 80.)*

At this stage, it is natural to ask how such a complicated system may be applied to devices. As discussed, the formation of these nanocrystallites needs new studies in crystallization in a constraint situation, taking into account the proximity of hetero-materials on each side of the layer. In dealing with the presence of many particles having slightly different sizes in parallel; avalanche multiplication in the substrate where IR drops may lead to a multitude of structures belonging, seemingly, to the same basic energy state, etc. It should be realized that these complications arise because one is engaged in a general study, not targeted to an actual device. In an actual device, the fabrication technique has advanced to the point that only a handful of particles are in parallel, and with better basic understanding, the particle size distribution can be better controlled. Thus, it should be quite possible to utilize these features in the design of multiple logic, functional devices,[83] and even in the emerging field of quantum computing.[84]

3.7 Electronic Applications of HEL-Si/O Superlattices

In Sec. 3.4, we discussed the formation of the Hetero-Epilattice-Superlattice. We shall point out some of the possible incorporations of this technology into future electronic devices. A brief summary is outlined to give a better grip of what it is and how to make use of it.

a. It is a superlattice consisting of alternate layers of nanometer scale silicon layers sandwiched between adsorbed oxygen monolayers.

b. The epitaxy of silicon is continued beyond the adsorbed disordered layer of oxygen, forming a structurally perfect superlattice, as if no disordered layers were present.

c. Electrically, a barrier consisting of two adsorbed monolayers of oxygen with a thin layer of silicon, 1–2 nm thick, shows a barrier height of 0.5 eV in I–V measurement.

d. The silicon deposition with thickness in excess of 3 nm on an adsorbed oxygen monolayer is epitaxy and defect free. Therefore, such a system is ready to play an important role in the design of high speed and low power transistors.

Figure 12 shows a typical high-resolution cross-section TEM with 10 L of adsorbed oxygen.[77] The presence of clusters of oxygen with dimensions approximately 3 nm is quite evident. It is important to recognize the fact that the presence of oxygen clustering in the oxygen adsorbed layer does not invalidate a barrier; because electrons are De Broglie waves, they cannot leak out of the "hole" with dimensions smaller than a wavelength. Figure 13 shows a couple of examples of I–V. Note that there is a 1000:1 difference in the measured currents between the forward and reverse directions, acting as a good conductor and good insulator, respectively. This feature may be applied to 3-D integration of ICs. The HES is still in its research stage, nonetheless our results should open the door for a wide variety of future applications.

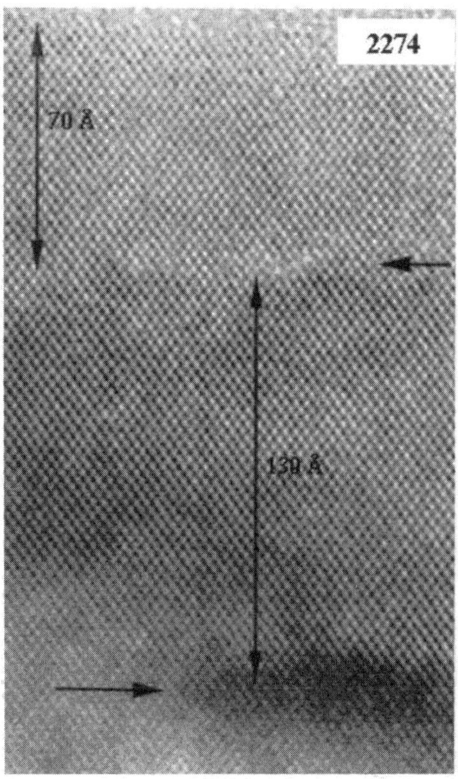

Figure 12. A typical high resolution cross-section TEM with 10 Langmuirs of adsorbed oxygen.

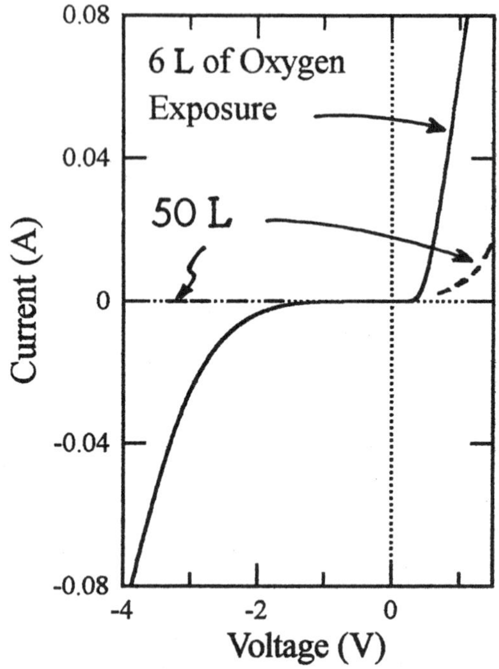

Figure 13. Measured I-V characteristic of a barrier formed by the superlattice. Dashed lines shows a case with 50 L of adsorbed oxygen.

3.8 Single Electron Transistor

There are a wide variety of quantum devices.[85] The single electron transistor, SET, is selected because the subject matter is somewhat close to the subjects we have presented thus far. The observation with current transport of discrete electronic states in a quantum dot set the stage of SET.[86] An excellent tutorial treatment may be found in an article in *Physics Today*.[87] Basically, SET promises extremely high density with ultrasmall power dissipation and the possibility of room temperature operation. The application of SET includes devices of nanoscaled memory, capacitance, and logic gate, etc.

Effects of single electron charging were observed in metal particles many years ago.[88] Single electron charging in a field effect transistor (FET) was observed more than ten years ago.[37] However, the first experimental study of a new nanometer FET with a single barrier in a one-dimensional channel was presented by Chou and Wang.[89] Not only is it not absolutely necessary to confine the electrons by two barriers, but a single barrier, under appropriate conditions, can result in single electron charging effects.

In 1996, Tiwari, et al., reported a memory device utilizing silicon nanocrystallites.[90] The device was fairly stable with no degradation in performance after more than 10^9 cycles of operation. It is impressive, though far short of the desired cyclability of 10^{13}. Figure 14 shows the schematic cross section and band profiles during injection (write cycle), storage, and removal (erasure) of an electron in the device. A thin tunneling oxide (1.1–1.8 nm thick) separates the inversion surface of an n-channel silicon FET from a distributed film of nanocrystals of silicon, 5 nm crystallites, at a density of 1×10^{12} cm^{-2}, covering the entire surface channel region. A thicker tunneling oxide (4.5 nm or higher) separates the nanocrystals from the control gate of the FET. Injection of an electron occurs from the inversion layer via direct tunneling when the control gate is forward biased relative to the source and drain. The resulting stored charge screens the gate charge and reduces the conduction in the inversion layer.

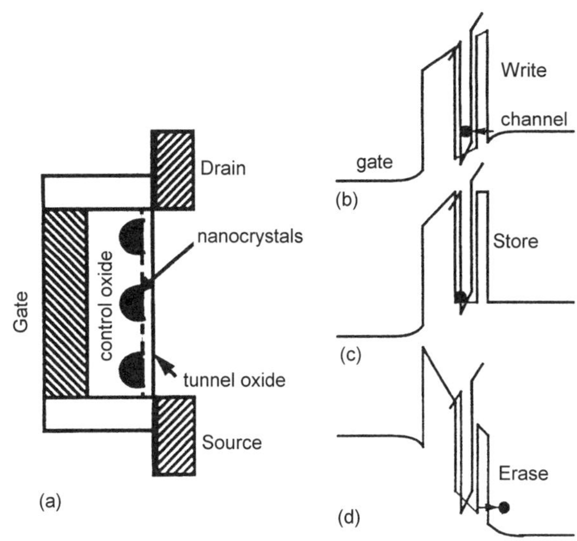

Figure 14. (*a*) A Schematic cross section and (*b*) band diagram during injection, (*c*) storage, and (*d*) removal of an electron from a nanocrystal. *(Reprinted by permission of the publisher.)*

The nanostructure made by the combination of electron beam lithography and reactive ion etching can also lead to the other approach to single-electron MOS memory,[91] operable at room temperature.[92] The structure consists of a narrow channel MOS FET with a width (~10 nm) smaller than the Debye screening length of a single electron, and a nanoscale polysilicon dot (7 × 7 nm) as the floating gate embedded between the channel and the control gate, as shown in Fig. 15. Even without adding a tunnel oxide between the channel and the polysilicon floating gate, the potential barrier still exists because of the thin native oxide at the grain boundary. As reported, this approach limited the fluctuation of the actual transistor channel width, the floating-gate size, and tunnel barrier thickness in Ref. 91, as well as the broad size distribution of silicon nanocrystals in Ref. 90. The storage of one electron leads to a discrete shift in the threshold voltage. A staircase relationship between the charging voltage and the threshold shift, exemplifies a self-limiting charging process.

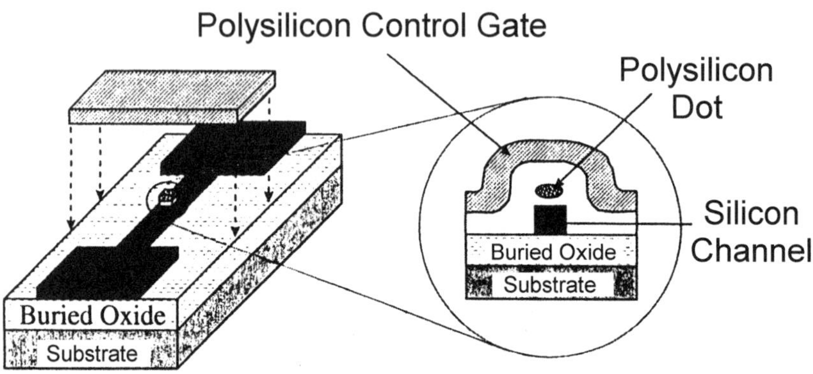

Figure 15. Schematic of a single-electron MOS memory that has a narrow silicon channel and a nanoscale polysilicon dot as the floating gate. The cross-section view illustrates the floating gate and the channel region. *(Reprinted by permission of the publisher.)*

Single-electron nano-capacitance devices utilize nanometer-size metal and semiconductor particles as the active device elements.[93][94] For sufficiently small (~ 2–4 nm in diameter) particles, the single-electron charging energies that characterize a particle are much greater that k_BT at room temperature, suggesting the possibility of operation of metal nanocrystal-based single-electron devices at ambient temperature. For example, Markovich, et al.,[95] reported a technique for fabricating single-electron capacitance devices from a two-dimensional close-packed organically functionalized Ag nanocrystal layer. Evidence of a Coulomb blockade, a step structure reminiscent of a Coulomb staircase, and memory charging effects were obtained. The step structures were reproducible from device to device over multiple scans and different frequencies. The authors argued that each of the steps reflected an increase in capacitance of the device due to collective single-electron (or hole) charging of the particles in the monolayer film. It is noted that the increase of capacitance as particle size is reduced below a couple of nanometers discussed in Sec. 2.5 is neglected here. Figure 16 shows the basic device structure, its equivalent circuit, and working function, in which PMMA (polymethylmethacrylate) serves as the dielectric spacer layer.

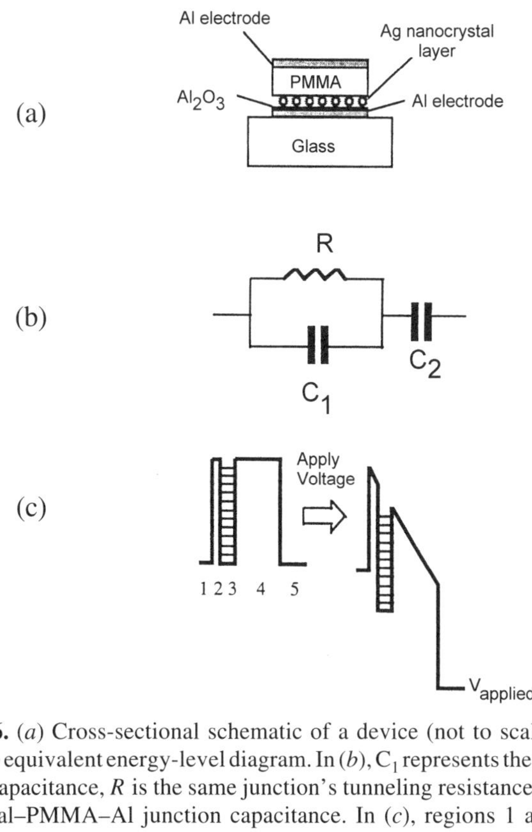

Figure 16. (*a*) Cross-sectional schematic of a device (not to scale); (*b*) equivalent RC circuit; (*c*) equivalent energy-level diagram. In (*b*), C_1 represents the Al/Al_2O_3–nanocrystal junction capacitance, *R* is the same junction's tunneling resistance, and C_2 represents the nanocrystal–PMMA–Al junction capacitance. In (*c*), regions 1 and 5 are Al electrode layers, region 2 is an Al_2O_3 layer (1 nm), region 3 is the metal nanocrystal monolayer (5 nm), and region 4 is the PMMA insulating spacer layer (30–40 nm). *(Reprinted by permission of the publisher.)*

Ohshima and Kiehi proposed a new approach to digital circuitry based on phase bi-stability in locked single-electron tunneling oscillations.[96]–[98] The scheme is based on binary logic states associated with the tunneling phase in single-electron tunnel junctions pumped at twice the tunneling frequency and activated by clocking the dc bias.

Figure 17 shows the proposed circuit in which tunneling junctions with capacitance, *C*, are pumped by a common ac source, V_p. The pump at ω_p supplies the drive for locking the oscillations and the circuit timing reference. The basic operation principle is the generation of single-electron tunneling oscillations at ω_{set} phase-locked to the pump, creating a multi-stability due to the indeterminate phase relationship between ω_{set} and $\omega_p = n\ \omega_{set}$, where *n* was 2. Phase information is clicked from gate to gate by sequentially clocking the dc bias in each gate. When the dc bias is raised, the gate is activated and sub-harmonic locking is established. Since the gates are capacitively coupled, the final state is influenced by the states of the previous gate. Therefore, Tunneling Phase Logic, TPL, the synchronization of single-electron tunneling events with the pump to produce a phase bi-stability, provides the properties needed for potential applications in digital logic circuitry.

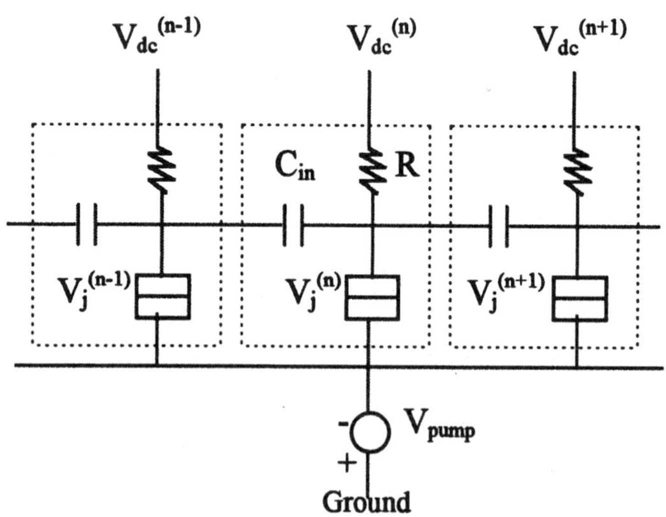

Figure 17. TPL of three coupled stages. Ultrasmall tunnel junctions (double-box symbols) with capacitance *C* are pumped by a common ac source and dc biased by clocked sources through a load resistance. *(Reprinted by permission of the publisher.)*

Self-assembled Si quantum dots have been tested for resonant tunneling devices operating at room temperature.[99] The size of the hemispherical Si quantum dots grown by low pressure CVD, 5–20 nm in diameters and 1–10 nm in height, leads to strong carrier confinement in the Si nanodot allowing the observation of negative differential conductance (NDC) at room temperature. The tunneling device consists of a double-barrier of n^+-Si(100)/SiO_2 (3-nm-thick)/Si nanodots/native SiO_2. This is an important development, therefore, this work, in our views, should be repeated at lower temperatures for a better NDC.

3.9 Quantum Dot Laser

The potential application of self-assembled quantum dots (QD) in optoelectronics is notable. This in-situ growth approach utilizes the island growth, Stranski-Krastanow (SK) mode, in highly strained hetero-epitaxial systems, such as InGaAs on GaAs. The defect-free self-assembled quantum dots can be simply fabricated by using conventional molecular beam epitaxy (MBE), metal-organic vapor phase epitaxy (MOVPE), or metal-organic chemical vapor deposition (MOCVD). The size distributions of these dots are uniform to within 10%, with densities as high as 10^{11} cm^{-2}. The PL efficiencies of the dots are greater than, or equal to, that of the reference quantum well.[100] By using these self-assembled quantum dots as active layer, low-threshold edge-emitting lasers have been fabricated. Soon afterwards, the QD vertical-cavity surface-emitting laser (VCSEL) appeared. VCSEL, using QDs in the active region, is more attractive, since it has the advantage of controlling both the electron and photon modes in a microcavity structure. The matching of the cavity mode with narrow bandwidth light emission from QDs results in a high-performance light source with very low threshold current. The bandwidth, which critically determines the temperature characteristics of the VCSEL, can be designed by controlling the dot size distribution. Therefore, the QD VCSEL may represent an optimum optical device utilizing QD.

Saito, et al.,[101] reported a InGaAs QD VCSEL when operated under continuous-wave current 32 μA at room temperature. The structure of VCSEL is shown in Fig. 18, grown by MBE. Bottom 18-period, n-doped, AlAs/GaAs distributed Bragg reflectors (DBRs) and a spacer layer of AlGaAs were first grown on a GaAs substrate. The thicknesses of AlAs and GaAs layers in the DBR were 64.9 and 51.5 nm, respectively. The interface layer between them was a graded composition AlGaAs layer (18 nm). Next, 10 periods of $In_{0.5}Ga_{0.5}As$ dot/$Al_{0.25}Ga_{0.75}As$ layer were grown at 520°C. The dots were obtained by alternate depositions of GaAs, 0.2 monolayer (ML) and InAs, 0.2 ML, including a 2 second pause. The tendency of the flattening of the AlGaAs cover layer on the dots results in the subsequent growth as good as the first layer. After the growth of the InGaAs dot active layer, a p-type AlGaAs spacer layer and top 14.5-period DBRs were grown at 600°C. A reflectivity spectrum of an as-grown wafer showed that the DBR mirror center and resonant cavity were positioned at a wavelength of around 960 nm. With a 3-nm height of the QD, effects of quantum confinement dominate over the thermal energy at room temperature.

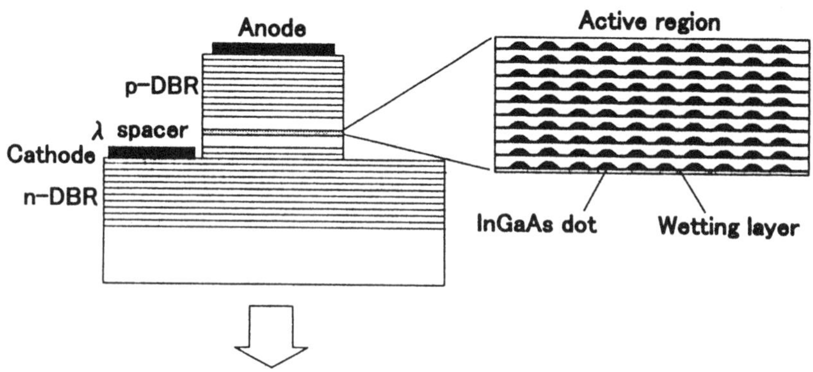

Figure 18. Quantum dot VCSEL with 10-period InGaAs dots in the active region. On the average, the dots were 28 nm in diameter and 3 nm in height with a density of $2 \times 10^{10} cm^{-2}$, covering 12% of the surface. *(Reprinted by permission of the publisher.)*

Huffaker, et al.,[102] recently reported a low threshold, oxide-confined (7 μm aperture) VCSEL based on a self-assembled QD active region. The operating condition was at room temperature under the pulsed threshold of 560 μA. The schematic cross section is shown in Fig. 19. The structure consists of a lower DBR of twenty-six n-type AlAs/GaAs quarter-wave pairs, a half-wave cavity spacer consisting of lower n-type and upper p-type $Al_{0.99}Ga_{0.01}As$ layers cladding the QD active region, and a single upper p-type GaAs quarter-wave layer followed by heavily p-type $Al_{0.75}Ga_{0.25}As$ and GaAs contact layers. The upper DBR is completed by six additional quarter-wave pairs of MgF/ZnSe deposited using electron beam evaporation. The last heavily p-doped layers are designed to be selectively etched from the VCSEL cavity prior to the MgF/ZnSe deposition. The QD active region is formed by depositing five monolayers of $In_{0.50}Ga_{0.35}Al_{0.15}As$ sandwiched between undoped GaAs and graded AlGaAs barriers. The substrate temperature is held at ~ 620°C for growth of the DBRs and cavity spacer, and reduced for the QD deposition and growth to ~ 500°C. A 20 s pause after the five monolayer deposition is used to allow the QD formation, monitored with RHEED. The 15% Al on the column III lattice sites is added to shift the room temperature QD emission wavelength to ~ 0.98 μm. Atomic force microscope (AFM) images showed the $In_{0.50}Ga_{0.35}Al_{0.15}As$ QDs formed with the six monolayer deposition to have lateral sizes of roughly 20 nm diameter with a high density of ~ 10^{11}cm^{-2}.

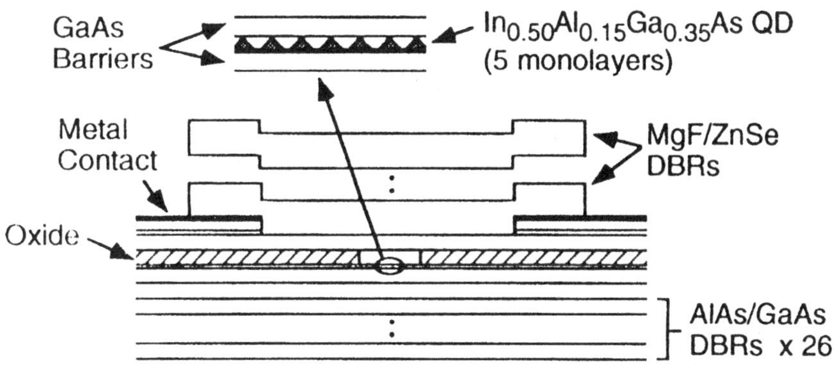

Figure19. Schematic illustration of the oxide-confined VCSEL containing the $In_{0.50}Ga_{0.35}Al_{0.15}As$ active region. *(Reprinted by permission of the publisher.)*

A purple (420 nm) laser diode, using self-assembled QDs in the active layer, has been investigated by Narukawa, et al.[103] This active layer consists of seven periods of undoped $In_{0.20}Ga_{0.80}N$ (3 nm)/$In_{0.05}Ga_{0.95}N$ (6 nm) MQW, sandwiched between GaN wave guiding layers (0.1 μm in each) and $Al_{0.15}Ga_{0.85}N$ cladding layers (0.4 μm in each). Using cross-section TEM and energy-dispersive x-ray (EDX), the authors found that the radiative recombination was from quantum self-assembled QDs. These QDs were in the In-rich region in the wells and had sizes from 3 to 5 nm with a peak at 3 nm in diameter.

Similar to the structure of QD laser, a simple approach to LED is the self-organized island formation,[104][105] with three monolayers of InAs islands sandwiched between 5.5 nm of GaAs layers. These LEDs emit light over a broad spectrum, with a typical linewidth of 120 nm, peaked between 1000 and 1100 nm.

4.0 EPILOGUE

As device size is reduced, several limits are beyond scaling. For example, barriers produced by doping cannot be scaled down due to the solid solubility limit of dopants. The abruptness of the heterojunctions allows scaling down to nanometer regime, however, quantum effects set in. Not all quantum effects are good for device performance. Lowering the carrier mobility resulting in degrading the carrier coherence length can eliminate quantum effects, but power dissipation and speed will be sacrificed. The game plan is, therefore, to utilize the quantum effects in seeking new designs of devices. Fundamentally, quantum effects are the manifestation of the wave nature of electrons, which controls constructive and destructive interferences, resulting in new flexibilities and possibilities. Quantized systems have discrete energy-momentum relations, which allow new applications. The reduction in the dielectric screening, the capacitance, the increase in dopant binding energies resulting in the difficulty in doping nanoparticles, may represent undesirable features. Usually lower capacitance translates into a reduction in the RC time constant, however, the characteristic time may not be dictated by RC, rather by tunneling time. The concept of charging a capacitor needs to be modified to include the recognition that it is not the storing of a charge, rather the storing of an electron, that is involved, and that the electron has kinetic energy in a quantum system. The widely separated energy states of electrons in a nanoparticle lead to steps or peaks in current. Therefore, with better understanding and designs, many undesirable features can be turned around to our advantage. The subject matter is huge. Our review is, thus, quite limited, mainly focused on a few systems that we have worked on during the past years.

ACKNOWLEDGMENTS

Part of the review describes work performed by the authors under the auspices of the US Army Research Office, Office of the Naval Research, and National Science Foundation. Qi Zhang also wishes to acknowledge the support of his work from CCLRC Darsbury Laboratories and BESSY.

REFERENCES

1. Esaki, L., and Tsu, R., *IBM J. Res. Devel.*, 14:61 (1970)
2. Tsu, R., and Esaki, L., *Appl. Phys. Letts.*, 22:562 (1973)
3. Cho, A. Y., *Appl. Phys. Lett.*, 19:467 (1971)
4. Matthews, J. W., and Blakeslee, A. E., *J. Cryst. Growth*, 32:265 (1976)
5. Dohler, G. H., Kunzel, H., Olego, D., Ploog, K., Ruden, P., Stoltz, H., and Abstreiter, G., *Phys. Rev. Lett.*, 47:864 (1981)
6. Tsu, R., *Mat. Res. Soc. Sypm. Proc.*, 102:219 (1988)
7. Zrenner, A., Reisinger, H., Koch, F., Ploog, K., *17th Int. Conf. Phys. Semi.* (D. Chadi and W. Harrison, eds.), p. 325, Springer, Berlin (1985)
8. Tsu, R., *Nature*, 364:19 (1993)
9. Tsu, R., Filios, A., Lofgren, C., Cahill, D., van Nostrand, J., and Wang, C. G., *Solid-State Electronics*, 40:221 (1996)
10. Dellow, M. W., Beton, P. H., Langerak, C. J. G. M., Foster, T. J., Main, P. C., Eaves, L., Henini, M., Beaumont, S. P., and Wilkinson, C. D. W., *Phys. Rev. Lett.*, 68:1754 (1992)
11. Tsu, R., Filios, A., Lofgren, C., Dovidenko, K., and Wang, C. G., to be published.
12. Zhang, Q., Lofgren, C., Filios, A., and Tsu, R., *Physica B,* 8:365 (2000)
13. Wegscheider, W., Eberl, K., Cerva, H., and Oppolzer, H., *Appl. Phys. Letts.*, 55:448 (1989)
14. Tsu, R., Basic, D., Ioriatti, L., *J. Appl. Phys.*, 82:1327 (1997)
15. Tsu, R., Ioriatti, L., Harvey, J. F., Shen, H., and Lux, R., *Mat. Res. Soc. Sypm. Proc.*, 283:2093 (1993)
16. Wang, L. W., and Zunger, A., *Phys. Rev. Lett.*, 73:1039 (1994)
17. Lannoo, M., Delerue, C., and Allan, G., *Phys. Rev. Lett.*, 74:3415 (1995)

18. Walter, J. P., and Cohen, M. L., *Phys. Rev. B*, 2:1821 (1970)
19. Canham, L. T., *Appl. Phys. Lett.*, 57:1046 (1990)
20. Bastard, G., *Phys. Rev. B*, 24:4714 (1981)
21. Ioriatti, L., and Tsu, R., *Surf. Sci.*, 174:420 (1986)
22. Tsu, R., and Ioriatti, L., *Superlatt. Microstruct.*, 1:295 (1985)
23. Kahen, K. B., Leburton, J. P., and Hess, K., *Superlatt. Microstruct.*, 1:289 (1985)
24. Tsu, R., and Babic, D., *Nato Workshop on Optical Properties of Low Dimensional Silicon Structures*, p. 203, *CNET*, France, Kluwer, Dordrect (March, 1–3, 1993)
25. Tsu, R., and Babic, D., *Appl. Phys. Lett.*, 64:1806 (1994)
26. Tsu, R., and Babic, D., *Porous Silicon Science and Technology, Les Houches* (J. C. Vial and J. Derrien, eds.), p. 111, Springer Verlag, Berlin (Feb. 8–12, 1994)
27. Matsumoto, et al., in: Proc. 1992, *Internat. Conf. Solid State Devices and Materials*, p. 478, Tsukuba, Jpn
28. Lehmann, V., and Gosele, U., *Appl. Phys. Lett.*, 58:856 (1991)
29. Tsu, R., Shen, H., and Dutta, M., *Appl. Phys. Lett.*, 60:112 (1992)
30. Babic, D., and Tsu, R., *Superlattices Microstruct.*, (Nov. 1997)
31. Babic, D., Tsu, R., and Greene, R. F., *Phys. Rev. B*, 45:14150 (1992)
32. Takagahara, T., and Takeda, K., *Phys. Rev. B*, 46:15578 (1992)
33. Hybertsen, M. S., *Phys. Rev. Lett.*, 72:1514 (1994)
34. Harvey, J. F., Shen, H., Lux, R. A., Dutta, M., Pannlapati, J., and Tsu, R., *Mat. Res. Soc. Symp. Proc.*, 256:175 (1992)
35. Vial, J. C., et al, *Phys. Rev. B*, 45:14171 (1992)
36. Chelikowsky, J. R., Wagener, T. J., Weaver, J. H., and Jin, A., *Phys. Rev. B*, 40:9644 (1989)
37. Meirav, U., Kastner, M. A., and Wind, S. J., *Phys. Rev. Lett.*, 65:771 (1991)
38. Ye, Q. Y., Tsu, R., and Nicollian, E. H., *Phys. Rev. B*, 44:1806 (1991)
39. Macucci, M., Hess, K., and Iafrate, G. J., *Phys. Rev. B*, 48:17354 (1993)
40. Iafrate, G. J., Hess, K., Krieger, J. B., and Macucci, M., *Phys. Rev. B*, 52:10737 (1995)
41. Cullis, A. G., and Canham, L. T., *Nature*, 353:335 (1991)
42. Vasques, R. P., Fathauer, R. W., George, T., and Ksendzov, A., *Appl. Phys. Lett.*, 60:1004 (1992)
43. Brandt, M. S., Fuchs, D. H., Stutzmann, M., Weber, J., and Cardona, M., *Solid State Comm.*, 81:307 (1992)

44. Ito, T., Ohta, T., Arakaki, O., and Hiraki, A., *Mat. Res. Soc. Symp.*, 256:127 (1992)

45. Feng, Z. C., and Tsu, R., (eds.), *Porous Silicon*, World Scientific, Singapore (1994)

46. Cullis, A. G., Canham, L. T., and Calcott, P. D. L., *J. Appl. Phys.*, 82:909 (1997)

47. Cole, M. W., Harvey, J. F., Lux, R. A., Eckart, D. W., and Tsu, R., *Appl. Phys. Lett.*, 60:2800 (1992)

48. Zhang, Q., and Bayliss, S. C., *J. Appl. Phys.*, 79:1351 (1996); Bayliss, S. C., Zhang, Q., and Hutt, D. A., *Phys. Stat. Sol. B*, 190:85 (1995)

49. Buren, T. V., Gao, Y., Tiedje, T., Dahn, J. R., and Way, B. M., *Appl. Phys. Lett.*, 60:3013 (1992)

50. Naudon, A., Goudeau, P., Halimaoui, A., Lambert, B., and Bomchil, G., *Appl. Phys. Lett.*, 75:780 (1994)

51. Berbezier, I., and Halimaoui, A., *J. Appl. Phys.*, 74:5421 (1993)

52. Sham, T. K., Jiang, D. T., Coulthard, I., Lorimer, J. W., Feng, X. H., Tan, K. H., Frigo, S. P., Rosenberg, R. A., Houghton, D. C., and Bryskiewicz, B., *Nature*, 367:331 (1993)

53. Pichter, A., Steiner, P., Kozlowski, F., and Lang, W., *IEEE Electron Device Lett.*, p. 691 (1991)

54. Koshida, N., and Koyama, H., *Appl. Phys. Lett.*, 60:347 (1992)

55. Loni, A., Simons, A. J., Cox, T. I., Calcott, P. J. D., and Canham, L. T., *Electronics Letters*, 31:1288 (1995)

56. Tsybeskov, L., Moore, K. L., Hall, D. G., and Fauchet, P. M., *Phys. Rev. B*, 54:R8361(1996)

57. Linros, J., and Lilic, N., *Appl. Phys. Lett.*, 66:3048 (1995)

58. Pavesi, L., Gardini, R., and Mazzoleni, C., *Solid State Com.*, 97:1 (1996)

59. Hirschman, K. D., Tsybeskov, L., Duttagupta, S. P., and Fauchet, P. M., *Nature*, 384:338 (1996)

60. Lazarouk, S., Jaguiro, P., Katsouba, S., Masini, G., La Monica, S., Maiello, G., and Ferrari, A., *Appl. Phys. Lett.*, 68:2108 (1996)

61. Tsybeskov, L., Duttagupta, S. P., Hirschman, K. D., and Pauchet, P. M., *Appl. Phys. Lett.*, 68:2058 (1996)

62. Filios, A., Hefner, S., and Tsu, R., *JVST B*, 14:3431 (1996)

63. Tsai, C., Li, K. H., Campbell, J. C., and Tasch, A., *Appl. Phys. Lett.*, 62:2818 (1993)

64. Matsumoto, T., Hasegawa, N., Tamaki, T., Ueda, K., Futagi, T., Mimura, H., and Kanemitsu,Y., *Jpn. J. Appl. Phys.*, 33:L35 (1994)

65. Tsu, R., Morais, J., and Bowhill, A., *Mat. Res. Soc. Symp. Proc.*, 358:825 (1995)
66. Morais, J., Doctoral Thesis, UNICAMP, Department of Physics, Campinas, Brazil, (1995); Also in Morais, J., Lander, R., and Tsu, R., to be published.
67. Tsu, R., Filios, A., Lofgren, C., Ding, J., Zhang, Q., Morais, J., and Wang, C. G., *Proc. ECS. 97–11, Quantum Confinement: Nanoscale Materials, Devices and System*, (M. Cahay, ed.) Montreal (May 4–7, 1997)
68. Lu, Z. H., Lockwood, D. J., and Barlbeau, J. M., *Nature*, 378:258 (1995)
69. Maeda, Y., *Phys. Rev. B*, 51:1658 (1995)
70. Zhang, Q., Bayliss, S. C., and Hutt, D. A., *Appl. Phys. Lett.*, 66:1977 (1995)
71. Zhang, Q., Bayliss, S. C., Frentrup, W., *Solid State Commun.*, 99:883 (1996); Zhang, Q., Bayliss, S. C., Al-Ajili, A., and Hutt, D. A., *Nucl. Instr. and Meth. in Phys. Res. B*, 97:329 (1994)
72. Hayashi, S., Kataoka, M., and Yamamoto, K., *Jpn. J. Appl. Phys.*, 32:L247 (1993)
73. Ismail, K., Meyerson, B. S., and Wang, P. J., *Appl. Phys. Lett.*, 55:973 (1991)
74. Fukuda, M., Nakagawa, K., Miyazaki, S., and Hirose, M., *Appl. Phys. Lett.*, 70:2291 (1997)
75. Meakin, D., Stobbs, M., Stoemenis, J., and Economou, N. A., *Appl. Phys. Lett.*, 52:1053 (1988)
76. Rabedeau, T. A., Tidswell, I. M., Pershan, P. S., Berk, J., and Freer, B. S., *Appl. Phys. Lett.*, 59:706 (1991)
77. Ding, J., and Tsu, R., *Appl. Phys. Lett.*, (Oct. 1997)
78. Baribeau, J. M., Lockwood, D. J., and Lu, Z. H., *Mater. Res. Soc. Symp. Proc.*, Vol. 382, Pittsburgh (1995)
79. Tsu, R., *SPIE J.*, 1361:313 (1990)
80. Boeringer, D. W., and Tsu, R., *Phys. Rev. B*, 51:13337 (1995)
81. Tsu, R., Li, X. L., and Nicollian, E. H., *Appl. Phys. Lett.*, 65:842 (1994)
82. Ding, C., and Tsu, R., *Mat. Res. Soc. Symp. Proc.*, 378:757 (1995)
83. Capasso, F., Mohammed, K., and Cho, A. Y., *IEEE Quantum Electronics*, QE 22:1853 (1986)
84. Lloyd, S., *Scientific Am.*, Vol. 140 (1995)
85. For a good recent review, the readers are referred to: Semiconductor Quantum Devices, Marc Cahay, in: *Adv. Electronics and Electron Phys.*, Acad. Press, 89:93 (1994)

86. Reed, M. A., Randall, J. N., Aggarwal, R. J., Matyi, R. J., Moore, T. M., and Wetsel, A. E., *Phys. Rev. Lett.*, 60:535 (1988)

87. Kastner, M., *Phys. Today*, Vol. 24 (Jan. 1993)

88. Zeller, T. R., and Giaever, I., *Phys. Rev.*, 20:1504 (1968)

89. Chou, S. Y., and Wang, Y., *Appl. Phys. Lett.*, 61:1591(1992)

90. Tiwari, S., Rana, F., Hanafi, H., Hartstein, A., Crabbe, E., and Chan, K., *Appl. Phys. Lett.*, 68:1377 (1996)

91. Yano, K., Ishii, T., Hashimoto, T., Kobayashi, T., Murai, F., and Seki, K., *IEEE Trans. Electron Devices*, 41:1628 (1994)

92. Guo, L., Leobandung, E., and Chou, S. Y., *Appl. Phys. Lett.*, 70:850 (1997)

93. Covin, V. L., Schlamp, M. C., and Alivisatos, A. P., *Nature*, Vol. 370 (1994)

94. Schmidt, H., Medeiros-Ribiero, G., Oestreich, M., Petroff, P. M., and Dohler, G. H., *Phys. Rev. B*, 54:11346 (1996)

95. Markovich, G., Leff, D. V., Chung, S. W., Soyez, H. M., Bunn, B., and Heath, J. R., *Appl. Phys. Lett.*, 70:3107 (1997)

96. Kiehi, R. A., and Ohshima, T., *Appl. Phys. Lett.*, 67:2494 (1995)

97. Ohshima, T., and Kiehl, R. A., *J. Appl. Phys.*, 80:912 (1996)

98. Ohshima, T., *Appl. Phys. Lett.*, 69:4059 (1996)

99. Fukuda, M., Nakagawa, K., Miyazaki, S., and Hirose, M., *Appl. Phys. Lett.*, 70:2291(1997)

100. Leonard, D., Pond, K., and Petroff, P. M., *Phys. Rev. B*, 50:11687 (1994)

101. Saito, H., Nishi, K., Ogura, I., Sugou, S., and Sugimoto, Y., *Appl. Phys. Lett.*, 69:3140 (1996)

102. Huffaker, D. L., Baklenov, O., Graham, L. A., Streetman, B. G., and Deppe, D. G., *Appl. Phys. Lett.*, 70:2356 (1997)

103. Narukawa, Y., Kawakami, Y., Funato, M., Fujita, S. Z., Fujita, S. G., and Nakamura, S., *Appl. Phys. Lett.*, 70:981 (1997)

104. Tabuchi, M., Noda, S., and Sasaki, A., in *Science and Technology of Mesoscopic Structures*, (S. Namba, C. Hamaguchi, and T. Ando, eds.), Springer, Tokyo (1992)

105. Solomon, G. S., Larson, M. C., and Harris, J. S., Jr., *Appl. Phys. Lett.*, 69:1897 (1996)

Index

A

B

D

J

K

L

Q

R

S

T

U

V

W